Chaouki Abdallah
Nov-22-1982

ANALYTIC SETS

Developed from lectures given at the *London Mathematical Society Instructional Conference* on Analytic Sets.
University College London, July 1978

ANALYTIC SETS

by

C. A. ROGERS
J. E. JAYNE
C. DELLACHERIE
F. TOPSØE
J. HOFFMAN-JØRGENSEN
D. A. MARTIN
A. S. KECHRIS
A. H. STONE

1980

ACADEMIC PRESS

A Subsidiary of Harcourt Brace Jovanovich, Publishers

London New York Toronto Sydney San Francisco

ACADEMIC PRESS INC. (LONDON) LTD.
24/28 Oval Road,
London NW1 7DX

United States Edition published by
ACADEMIC PRESS INC.
111 Fifth Avenue
New York, New York 10003

British Library Cataloguing in Publication Data
Instructional Conference on Analytic Sets,
University College, London, 1978
Analytic Sets.
1. Set theory—Congresses
2. Analytic spaces—Congresses
I. Title II. Rogers, Claude Ambrose
III. London Mathematical Society
511′.3 OA248 80-40647

ISBN 0–12–593150–6

Printed in Great Britain by
J. W. Arrowsmith Ltd., Bristol BS3 2NT

Contributors

C. Dellacherie *Department de Mathématique, Université de Rouen, Boîte Postale No. 67, 76130 Mont-Saint-Aignan, France*

J. Hoffmann-Jørgensen, *Matematiske Institute, Aarhus Universitets, Aarhus, Denmark*

J. E. Jayne, *Department of Mathematics, University College London, London WC1, England*

A. S. Kechris, *Department of Mathematics, California Institute of Technology, Pasadena, California 91125, U.S.A.*

D. A. Martin, *Department of Mathematics, University of California at Los Angeles, Los Angeles, California 90024, U.S.A.*

C. A. Rogers, *Department of Mathematics, University College London, London WC1, England*

A. H. Stone, *Department of Mathematics, University of Rochester, Rochester, New York 14627, U.S.A.*

F. Topsøe, *Matematiske Institute, Københavns Universitets, Universitetsparken 5, 2100 Copenhagen Ø, Denmark*

Preface

The London Mathematical Society's Instructional Conference on *Analytic Sets, their connections with Mathematical Logic and Applications in Functional Analysis* was held at University College London from the 16th to 29th July 1978. The Conference was organized by a small committee, with C. Dellacherie, J. Hoffmann-Jørgensen, J. E. Jayne, J. R. Ringrose and C. A. Rogers as members. From an early stage the committee decided that the speakers giving the instructional lectures would be asked to prepare, simultaneously with their lectures, contributions that would form parts of a modern book on Analytic Sets. The present book is the result of their labours before and after the conference. The first part on K-analytic sets is contributed by J. E. Jayne and C. A. Rogers, C. A. Rogers taking main responsibility for sections 1–4, and J. E. Jayne for sections 5, 6. The second part on 'Les ensembles analytiques' is by C. Dellacherie; note that this part has its own index, repeated, in part, in the main index. The third part on analytic spaces and their applications is by J. Hoffmann-Jørgensen and F. Topsøe, the first section being by F. Topsøe and J. Hoffmann-Jørgensen and the second section being by J. Hoffmann-Jørgensen. The fourth part on infinite games and effective descriptive set theory is by A. S. Kechris and D. A. Martin; D. A. Martin wrote sections 1–5 and A. S. Kechris wrote sections 6–9. This work by Kechris and Martin was supported by grants from the U.S. National Science Foundation. The fifth part on analytic sets in non-separable metric spaces is by A. H. Stone. The authors are named on the title page and on the first page of each part in the order in which their contributions appear in the book.

The conference was attended by about seventy mathematicians. The programme of instructional lectures was supported by a series of more specialized talks on research topics. The researches described are being published separately by their authors. A problem book was kept at the conference. The problems written in the book are published here together with some additional problems collected subsequently. Some 'stop press' information concerning the solution of some of the problems is included. While we believe that the remaining problems are unsolved at the time of going to press, some may well have been solved in the existing literature, and others may well be solved before the book is published.

The organizers are grateful to the London Mathematical Society for sponsoring this conference and for providing a substantial loan. They are also grateful to the European Research Office of the U.S. Army and to the Royal Society of London for generous grants.

September 1980 C. A. Rogers

Contents

Part 4. Infinite Games and Effective Descriptive Set Theory
by D. A. Martin and A. S. Kechris

Part 5. Analytic Sets in Non-separable Metric Spaces
by A. H. Stone

Part 1

K-analytic sets

C. A. ROGERS AND J. E. JAYNE

1 History of the theory of analytic sets

1.1 Pre-history: Lebesgue

As Lebesgue's memoir "Sur les fonctions représentables analytiquement" (Lebesgue, 1905) contains many ideas that were to prove of great importance in the subsequent development of the theory of analytic sets, it seems to be a convenient starting point for our discussion. One of Lebesgue's objects was to establish the existence of functions of each Baire class and of functions that belong to no Baire class. Let us confine our attention to functions defined on subsets of an interval I of points $x = (x_1, x_2, \dots, x_n)$ in $\mathbb{R}^n$ satisfying

$$a_1 \leqslant x_1 \leqslant b_1,\ a_2 \leqslant x_2 \leqslant b_2, \dots, a_n \leqslant x_n \leqslant b_n,$$

with $a_1 < b_1$, $a_2 < b_2, \dots, a_n < b_n$. The Baire functions of class 0 are the continuous functions defined on closed subsets of I and left undefined on the rest of I (if any). For each countable ordinal $\alpha < 0$, a function f is a Baire function of class α if (a) there is a sequence $\{f_i\}$ of functions, f_i being of Baire class $\beta_i < \alpha$, for $1 \leqslant i$, and f is defined as

$$\lim_{i\to\infty} f_i,$$

for just those points x of I at which the sequence $\{f_i\}$ is defined and convergent, and (b) f is not of Baire class $\beta < \alpha$. It follows, by transfinite induction, that a Baire function of class α is defined on a Borel subset of I

that is both of multiplicative class at most $\alpha+1$ and of additive class at most $\alpha+1$.

Lebesgue remarks (*loc. cit.*, pp. 207–208) that, if α is a countable ordinal, it is possible to define an injective map of the functions of class less than α into the interval I. A Cantor-type diagonal argument then leads to the construction on I of a function f_α that is not of Baire class less than α. If this f_α is a function in the Baire classification, then f_α, or some other function g_α, must be of class α; but perhaps f_α is not a Baire function at all.

Lebesgue overcomes this difficulty. He constructs a function $f(x, t)$ of class α or less, that is defined on a Borel subset of $[0, 1]\times[0, 1]$ of multiplicative class $\alpha+1$ or less, and that is universal for the Baire functions of class α on $I=[0, 1]$, in that, given any function g of Baire class α or less, defined on a subset of $[0, 1]$, there is a t_g with $0\leqslant t_g\leqslant 1$, such that g and the restriction of $f(x, t)$ to the line $t=t_g$ coincide, both in their domains of definitions and in their values. Now take $\varphi(x)=1$ when $f(x, x)$ is defined and has the value 0, and take $\varphi(x)=0$ otherwise. This definition yields a function φ defined on $[0, 1]$ that is of class at most $\alpha+1$. If φ were of class α or less, $f(x, t_\varphi)$ would coincide with $\varphi(x)$, for $0\leqslant t\leqslant 1$, and would, in particular, be defined for all t. This is impossible as $\varphi(t_\varphi)$ is not equal to $f(t_\varphi, t_\varphi)$. Hence φ is precisely of class $\alpha+1$. In this way Lebesgue constructs functions of class α, whenever α is a successor ordinal; he constructs such functions, whenever α is a limit ordinal, by piecing together functions corresponding to the previous successor ordinals. He emphasizes that this process is only effective if the ordinal α is effectively enumerable.

Baire (1900, p. 71) had already remarked that the whole family of Baire functions has the power c of the continuum, so that, by Cantor's work, the family of all functions on I, having power exceeding c, must contain a function that does not belong to the Baire classification. As this does not lead to a construction or specification of a function that is not a Baire function, Lebesgue is dissatisfied. He takes the view that an object is only defined when it is possible to single it out by using a finite number of words: "Un objet est défini ou donné quand on a prononcé un nombre fini de mots s'appliquent à cet objet et à celui-là seulement; c'est-à-dire quand on a nommé une propriété caractéristique de l'objet." (*loc. cit.*, p. 205).

Using the functions of Baire class α that he has already constructed, Lebesgue proceeds to define, in a unique way, a function $\varphi(x, t)$ that is universal for the whole family of Baire functions on $[0, 1]$, in that corresponding to each Baire function g defined on the whole of $[0, 1]$ there is a t_g, with $0\leqslant t_g\leqslant 1$, so that the restriction of $\varphi(x, t)$ to the line $t=t_g$ coincides with g. Clearly this function φ cannot belong to the family of Baire functions.

It seems paradoxical that, while the construction of the function of Baire class α depends on α being effectively enumerable, in the construction of the function φ, that is not a Baire function, each ordinal α arises, together with a natural enumeration of its predecessors, so that no choice process is needed.

1.2 Pre-history: Alexandrov and Hausdorff

In 1916 P. S. Alexandrov and F. Hausdorff independently verified the truth of the Continuum Hypothesis for the class of Borel sets of the real line. Both authors used a rather complicated process for constructing each Borel set from closed (or open) sets by a countable sequence of operations without any explicit use of transfinite numbers. This process was modelled on a method that Lebesgue gives for the construction of Baire functions (*loc. cit.*, pp. 208–209), that we now describe. Let T be a "tree" consisting of a "root" φ and system of finite sequences

$$t = t_1, t_2, \ldots, t_n$$

of positive integers, n being a variable positive integer, satisfying the conditions

(a) if $t_1, t_2, \ldots, t_n \in T$, then $t_1, t_2, \ldots, t_{n-1} \in T$;
(b) if $t_1, t_2, \ldots, t_n \in T$, then $t_1, t_2, \ldots, t_{n-1}, q \in T$, for $q \geqslant 1$;
(c) for no infinite sequence $t_1, t_2, \ldots$ do we have $t_1, t_2, \ldots, t_n \in T$ for all $n \geqslant 1$.

We call a sequence $t_1, t_2, \ldots, t_n$ a terminal sequence of T, if for no $q \geqslant 1$ do we have $t_1, t_2, \ldots, t_n, q \in T$.

Lebesgue shows that each Baire function f can be obtained by choosing a tree T, by choosing a continuous function $f_{t_1, t_2, \ldots, t_n}$ corresponding to each terminal element of T, and then defining functions for the other elements of T by the scheme

$$f_{t_1, t_2, \ldots, t_{n-1}} = \lim_{q \to \infty} f_{t_1, t_2, \ldots, t_{n-1}, q},$$

$$f = f_\varphi = \lim_{q \to \infty} f_q,$$

of definition by infinite descent. When this scheme is translated to yield a scheme for constructing Borel sets it becomes rather complicated, as each limit operation needs to be replaced by a double operation consisting of a countable union followed by a countable intersection (or vice versa). Using this representation for the Borel set and some ingenuity, Alexandrov and Hausdorff obtain their results. We do not go into detail as we shall describe a simpler scheme and proof in Sections 1.3 and 3.5.

Had Alexandrov and Hausdorff been less gifted, this result might have been the first triumph of the theory of analytic sets, rather than part of the pre-history of the subject.

1.3 Souslin and Lusin

W. Sierpiński (1950, pp. 28–30) gives an account of the curious birth of the theory of analytic sets. In 1916, Nicolas Lusin, then a young teacher in the University of Moscow, had asked his pupil, Mikhail Souslin, to study the Memoir of Lebesgue "Sur les fonctions représentables analytiquement". Souslin noticed that at one point (*loc. cit.*, p. 192) Lebesgue had made use of a remark, without giving any justification, and that the remark could not be justified. As it happened, Sierpiński was present when Souslin told Lusin of his observation and gave him the manuscript of his first work. Lusin took seriously the young student who claimed that he had found a mistake in the work of an eminent mathematician. Sierpiński remarks that he was one of the first, immediately after Lusin, to read the manuscripts of Souslin, and that he knew well how Lusin aided his pupil and guided his researches; it would be more just to call analytic sets "sets of Souslin and Lusin" than to call them "Souslin sets".

Souslin was not content to show that Lebesgue's proof was invalid; he set himself to rebuild a satisfactory theory. In a very short time he made amazing progress. He only published one short note in *Comptes Rendue* (Souslin, 1917) announcing some of his results and giving only the briefest indications of their proofs. We gather a little further information from a note by Lusin that appeared with Souslin's note in *Comptes Rendue* (Lusin, 1917) and from a paper by Lusin and Sierpiński (1918).

Souslin introduced the sets, that we nowadays call Souslin sets, as the sets of the form

$$\bigcup_{t_1,\, t_2, \ldots} \bigcap_{n=1}^{\infty} I_{t_1 t_2 \ldots t_n},$$

with $I_{t_1 t_2 \ldots t_n}$ a real interval, for each sequence $t_1, t_2, \ldots, t_n$ of positive integers, the union being taken over the uncountable system of sequences of positive integers. He shows that all Borel sets are Souslin sets. He found an example of a Souslin set that is not a Borel set. He established that, if a set and its complement are both Souslin sets, then the set is a Borel set. Each Souslin set on the line is the projection of a $\mathcal{G}_\delta$-set in the plane. The projection of a Souslin set in $\mathbb{R}^n$ onto $\mathbb{R}^{n-1}$ is a Souslin set. A Souslin set is either countable or contains a perfect subset (and so is of the cardinal of the continuum).

Lusin (1917) drew various conclusions for the theory of functions from Souslin's work, in particular, a function f that is defined implicitly by taking $f(x)$ to be the unique y, satisfying

$$\varphi(x, y) = 0,$$

where φ is a Baire function for which the solution *is* unique, is necessarily a Baire function. Furthermore, every Souslin set is Lebesgue-measurable.

Lusin and Sierpiński (1918) give the first detailed proofs of some of the more difficult results mentioned above. They make it plain that Souslin and Lusin first obtained the results by use of a geometrical theory of indices introduced by Souslin and simplified by Lusin. They prove that the complement of a Souslin set is the union of $\aleph_1$ Borel sets. They give a construction (using the axiom of choice) of a set of cardinal $\aleph_1$ in $[0, 1]$ that is transformed into a set of Lebesgue measure zero by any homeomorphism of $[0, 1]$. They also show that the Lebesgue-measurable sets are stable under the Souslin operation. They announce Lusin's result that the application of the Souslin operation to Souslin sets leads only to Souslin sets.

1.4 Development, the first twenty years

Initially, the examples of analytic sets, that are not Borel sets, seemed rather strange. But it was soon realized that analytic sets, that are not Borel sets, do arise in a natural way in most branches of analysis. They arise typically as projections onto one component, X say, of a Borel set, that is not a $\mathscr{K}_\sigma$-set in a Cartesian product space $X \times Y$; in special circumstances such a projection will be a Borel set; in general, it will not be a Borel set, but it will be an analytic set, if X and Y are themselves analytic. The theory was developed rapidly by Lusin, Sierpiński and many others. The emphasis was largely on the questions: "When is a set that is an analytic set, constructed in a particular way, necessarily a Borel set?" and "When can an analytic set be replaced by a Borel set that is in some way equivalent or nearly equivalent to it?".

The two main tools of the theory were the first separation theorem (essentially due to Souslin (1917)) and the second separation theorem (due to Lusin (1930), see also Kuratowski (1936)). Let us state these theorems in their simplest forms; later we will be concerned with many generalizations and refinements.

The analytic sets in a complete separable metric space, a Polish space, for short, are the sets obtained by application of the Souslin operation to the closed sets of the space; they are equally the sets in the space that can be obtained as a continuous image of the space $\mathbb{N}^{\mathbb{N}}$ (of sequences of positive integers, with the topology of pointwise convergence).

THEOREM 1.4.1 *Let A and B be disjoint analytic sets in a Polish space. Then there are disjoint Borel sets C and D with*

$$A \subset C, \qquad B \subset D.$$

THEOREM 1.4.2 *Let A and B be analytic sets in a Polish space X. Then there are sets C and D, that are complements of analytic sets, with*

$$A\backslash B \subset C, \qquad B\backslash A \subset D,$$
$$C \cap D = \varnothing, \qquad C \cup D = X\backslash(A \cap B).$$

In the first separation theorem the Borel sets separate the disjoint analytic sets. In the second separation theorem the complements of analytic sets separate the remainders of the analytic sets after the suppression of their common part.

These results, and a refinement of the second separation theorem due to Novikov (1934), were much used in the study of the projection on X of a set B in a Cartesian product $X \times Y$ of two Polish spaces. What conditions on B, often expressed in terms of its sections parallel to Y, ensure that $\mathrm{proj}_X B$ is a Borel set? If B is a Borel set and P is a given set-theoretic property, what can one say about the set-theoretical character of the set of x in X, for which the section

$$B^{(x)} = \{y: (x, y) \in B\},$$

has property P? As an example, let us quote the Kunugui–Arsenin theorem (see Arsenin, 1940).

THEOREM 1.4.3 *Let B be a Borel set in the product $X \times Y$ of two Polish spaces. The set of x in X, for which $B^{(x)}$ is σ-compact in Y, is the complement of an analytic set in X. If $B^{(x)}$ is σ-compact for each x in X, the projection of B on X is a Borel set.*

Considerable effort was spent on the uniformization problem. If B is a set of a given set-theoretical character in a Cartesian product $X \times Y$, can one find a subset U of B, preferably of the same set-theoretical character, that "uniformizes" B, in that, for each x in X with $B^{(x)} \neq \varnothing$, we find that $U^{(x)}$ consists of a single point? After a line of attack had been opened up by Lusin and Novikov (1935), the following result was obtained by Kondô (1939).

THEOREM 1.4.4 *Let C be the complement of an analytic set in the product $X \times Y$ of two Polish spaces. Then C contains a set U, that is the complement of an analytic set, and that uniformizes C.*

The workers at this time, notably Sierpiński and Kunugui, noticed that some of the results of the theory could be cast in purely set-theoretic or combinatorial forms, whilst others seemed to be topological in nature. Most of the work was done in metric spaces; the theory worked most satisfactorily in Polish spaces.

Some notable contributions to the study of Borel sets in non-separable metric spaces were made by Montgomery (1934) and Kuratowski (1934); later this was developed further by A. H. Stone (see this volume Part 5, below).

The work of Kondô and Novikov became more and more involved with mathematical logic, and we cannot attempt to summarize it here. But it should just be mentioned that in his announcement Gödel (1938) says that it is consistent with the usual axioms for set theory, for there to be a Polish space X with an analytic, non-Borel, subset A, such that $X\backslash A$ has no perfect subset. A proof, based on Gödel's book (1940), was given by Novikov (1951). This result to some extent confirms Lusin's view (Lusin, 1930) that the complements of analytic sets are much less accessible for mathematical study than are the analytic sets themselves.

1.5 Development, topological spaces

The theory of analytic sets was well developed by 1938, and many sets arising naturally in analysis (and a few arising unnaturally in number theory) had been found to be analytic or coanalytic without being Borel. V. E. Šneĭder (1945*a*, *b*, 1948) studied the Souslin-$\mathscr{F}$ sets in compact Hausdorff spaces, showing that some of the classical results, known to hold in Polish spaces, hold also in compact Hausdorff spaces where each closed set is a $\mathscr{G}_\delta$-set.

At this stage, the theory had had very few applications in analysis and functional analysis. This changed abruptly. J. von Neumann published and used his measurable selection theorem in 1949 (he had obtained it some years earlier). Roy O. Davies used the theory in 1952 to help him solve an outstanding problem in the theory of Hausdorff measures. G. Choquet in 1953, having developed an extension of the theory to topological spaces in 1951, was able to prove his capacitability theorem. Each of these investigations gave rise to many developments and applications. In some remarkable papers G. W. Mackey (1957*a*, *b*) applied the theory to the study of locally compact groups.

M. Sion (1960*a*, *b*, *c*, 1961, 1963) took part in the development of Choquet's theory using a definition that appeared to be slightly more general. Z. Frolík (1961*a*, *b*, *c*, 1967, 1970*a*, *b*) gave a detailed account of the theory using a definition apparently more general than those of Choquet and Sion. They proved that many of the results of the classical theory have appropriate generalizations; other results were and remain resistant to such generalization.

The theories were brought together in 1976, when J. E. Jayne showed the equivalence of the definitions of Choquet, Sion and Frolík, in Hausdorff spaces (see Section 2.8 below).

1.6 Development, recursive function theory

In the 1930s logicians, especially Kleene, developed the theory of recursive functions, obtaining many striking results. In the 1950s J. W. Addison and A. Mostowski first recognized that there was a strong analogy between Kleene's results and those of the classical theory of analytic sets. Later Addison brought the two theories together and laid the foundations of effective descriptive theory of sets that is described in Part 4 of this volume.

2 Stability

2.1 Summary of notation and definitions

We use $\mathbb{N}$ to denote the set of natural numbers $1, 2, 3, \ldots$; we regard $\mathbb{N}$ as having the discrete topology. We use $\mathbb{N}^{\mathbb{N}}$ for the space of sequences

$$\sigma = \{\sigma_1, \sigma_2, \sigma_3, \ldots\}$$

of natural numbers, with the Cartesian product topology, or, equivalently, with the topology of coordinate-wise convergence. We identify $\mathbb{N}^{\mathbb{N}}$ with the space I of irrational numbers between 0 and 1 by means of the continued fraction map $c : \mathbb{N}^{\mathbb{N}} \to I$, given by

$$c(\sigma) = \frac{1}{\sigma_1 +} \; \frac{1}{\sigma_2 +} \; \frac{1}{\sigma_3 +} \; \cdots .$$

The topology of $\mathbb{N}^{\mathbb{N}}$ coincides with the topology that I inherits from the real line.

We use $\mathbb{N}^{(\mathbb{N})}$ to denote the space of finite sequences s of positive sequences, including the empty sequence $\emptyset$ of zero length. We use $l(s)$ to denote the length of the sequence s. This space has the discrete topology.

If $\sigma \in \mathbb{N}^{\mathbb{N}}$ and $n \geq 0$ we use $\sigma|n$ (read "sigma restricted to n" as an abbreviation of "sigma restricted to its first n terms") to denote the element

$$\emptyset, \quad \text{if } n = 0,$$

or

$$\sigma_1, \sigma_2, \ldots, \sigma_n, \quad \text{if } n \geq 1\,,$$

of $\mathbb{N}^{(\mathbb{N})}$ of length n.

For each s in $\mathbb{N}^{(\mathbb{N})}$ we will use $I(s)$ to denote the "Baire interval" of points σ of $\mathbb{N}^{\mathbb{N}}$ with $\sigma|l(s) = s$. The Baire intervals $\{I(s)\}_{s\in\mathbb{N}^{(\mathbb{N})}}$ are a family of clopen sets forming a base for the topology of $\mathbb{N}^{\mathbb{N}}$.

If S is any family of sets in a space X, we use $\mathcal{S}_\sigma$, $\mathcal{S}_+$, $\mathcal{S}_\delta$ and $\mathcal{S}_c$ to denote respectively the families of countable unions, countable disjoint unions, countable intersections and complements of sets of the family $\mathcal{S}$. If X is a topological space, we use $\mathcal{F}(X)$, $\mathcal{G}(X)$, $\mathcal{K}(X)$ and $\mathcal{Z}(X)$, or just $\mathcal{F}$, $\mathcal{G}$, $\mathcal{K}$ and $\mathcal{Z}$, to denote, respectively, the families of closed sets, open sets, compact sets and zero sets of continuous real-valued functions of X.

If $\mathcal{S}$ is any family of sets, we say that a set is a Souslin-$\mathcal{S}$ set if it has a representation in the form

$$\bigcup_{\sigma\in\mathbb{N}^{\mathbb{N}}} \bigcap_{n=1}^{\infty} S(\sigma|n)\,, \tag{1}$$

with $S(\sigma|n) \in \mathcal{S}$ for all finite sequences $\sigma|n$ of positive integers. For simplicity we shall use the formula

$$\bigcup_{\sigma} \bigcap_{n} S(\sigma|n) \tag{2}$$

as an abbreviation for (1). We say that the set (1) is obtained from the family $\mathcal{S}$ by the Souslin operation; we use $\mathbf{S}(\mathcal{S})$ to denote the family of Souslin-$\mathcal{S}$ sets obtained in this way.

If it so happens that for all σ, τ in $\mathbb{N}^{\mathbb{N}}$, with $\sigma \neq \tau$, we have

$$\left(\bigcap_{n} S(\sigma|n)\right) \cap \left(\bigcap_{n} S(\tau|n)\right) = \varnothing\,,$$

we say that the set (2) has a disjoint Souslin-$\mathcal{S}$ representation; and we use $\mathbf{S}_d(\mathcal{S})$ to denote the family of these sets.

We say that a map F from a topological space X to the power set of a topological space Y is upper semi-continuous, if

> for each x in X and each open set G of Y containing $F(x)$, there is a neighbourhood U of x with $F(U) \subset G$

If we define the outer inverse image function F^{-1} by taking $F^{-1}(V)$ to be the set of all x in X for which

$$F(x) \cap V \neq \varnothing\,,$$

this definition is equivalent to the requirement

$F^{-1}(V)$ is closed in X, whenever V is closed in Y.

We remark that it would be more descriptive to call this sort of continuity "outer semi-continuity" rather than "upper semi-continuity"; but we shall bow to tradition.

A set in a Hausdorff space will be called an analytic set if it is empty or is the continuous image of $\mathbb{N}^{\mathbb{N}}$, or, equivalently, if it is the continuous image of some Polish space.

A set in a Hausdorff space X will be called a K-analytic set if it is of the form

$$\bigcup_{\sigma} K(\sigma),$$

where K is an upper semi-continuous map from $\mathbb{N}^{\mathbb{N}}$ to the compact sets of X.

A set in a Hausdorff space will be called a Lusin set if it is an injective continuous image of a closed subset of $\mathbb{N}^{\mathbb{N}}$, or, equivalently, of a Polish space.

A set in a Hausdorff space X will be called a K-Lusin set if it is of the form

$$\bigcup_{\sigma} K(\sigma),$$

where K is an upper semi-continuous map from $\mathbb{N}^{\mathbb{N}}$ to the compact sets of X, with the property

$$K(\sigma) \cap K(\tau) = \emptyset,$$

whenever $\sigma, \tau \in \mathbb{N}^{\mathbb{N}}$ and $\sigma \neq \tau$.

A map f of one Hausdorff space X into another Y is called a proper map if

(a) f is continuous;
(b) $f^{-1}(y)$ is compact for each y in Y; and
(c) $f(E)$ is closed for each closed set F in X.

We remark thab(and (c) taken together ensure that f^{-1} is an upper semi-continuous map from Y back to the compact sets of X. To verify this, we need merely note that f is the outer inverse image function to f^{-1}.

2.2 Discussion of the definitions

In a compact metric space the Souslin-$\mathscr{K}$ sets, analytic sets, and the Souslin-$\mathscr{F}$ sets all coincide. In a Polish space that is not σ-compact, the Souslin-$\mathscr{K}$ sets are a proper subset of the analytic sets, but the analytic sets and the Souslin-$\mathscr{F}$ sets coincide. In certain separable metric spaces (e.g. if the space is a non-analytic subset of $\mathbb{R}$), the Souslin-$\mathscr{K}$ sets are a proper subset of the analytic sets and the analytic sets are a proper subset of the Souslin-$\mathscr{F}$ sets. For many investigations, the class of Souslin-$\mathscr{K}$ sets seems to be unnecessarily restrictive, the class of Souslin-$\mathscr{F}$ sets does not seem to be restrictive enough, but the class of analytic sets seems to be just right.

In Hausdorff spaces we have a similar situation. The Souslin-$\mathscr{K}$ sets are, in general, a proper subset of the K-analytic sets, and, in general, the K-analytic sets are a proper subset of the Souslin-$\mathscr{F}$ sets. The K-analytic sets seem to be the most useful class to study.

Lusin clearly regarded the analytic sets as "positively constructable", a status he was not prepared to grant to complements of analytic sets or to more general projective sets. If one is prepared to regard a compact Hausdorff space as an object that is sufficiently well-known to use as a basis for further construction, we can regard the K-analytic sets as "positively constructed" from the compact Hausdorff spaces. For many purposes this seems to be a legitimate point of view; however, it does not seem to be useful if one seeks to select representative points from a class of subsets of the space.

2.3 Stability of the Souslin operation

If $\mathscr{S}$ is a family of sets in a space X, which does not need a topology, we use $\mathbf{S}(\mathscr{S})$ and $\mathbf{S}_d(\mathscr{S})$ to denote the classes of sets obtained by applying the Souslin operation and the disjoint Souslin operation to the class $\mathscr{S}$. We prove

THEOREM 2.3.1 *Let $\mathscr{S}$ be a family of sets in a space X. Then*

$$\mathbf{S}(\mathbf{S}(\mathscr{S})) = \mathbf{S}(\mathscr{S})$$

and

$$\mathbf{S}_d(\mathbf{S}_d(\mathscr{S})) = \mathbf{S}_d(\mathscr{S}) .$$

While some new proofs have been given for this result (see, for example P. A. Meyer, 1966) we give an old proof, which is essentially that given by Sierpiński (1933).

LEMMA 2.3.2 *There are bijections*

$$\nu : \mathbb{E} \times \mathbb{N} \to \mathbb{N}$$

and

$$\chi : \mathbb{N}^{\mathbb{N}} \times (\mathbb{N}^{\mathbb{N}})^{\mathbb{N}} \to \mathbb{N}^{\mathbb{N}} ,$$

with the property that, for each $m, n \geqslant 1$, a knowledge of

$$\chi(\sigma; \tau(1), \tau(2), \ldots)|\nu(m, n)$$

determines $\sigma|m$ and $\tau(m)|n$ uniquely.

Proof For each positive integer l, the Diophantine equation

$$l = 2^{m-1}(2n-1)$$

has a unique solution for the pair (m, n) of positive integers. Let $(\varphi(l), \psi(l))$ denote this solution. Now

$$\nu(m, n) = 2^{m-1}(2n-1)$$

is a bijection from $\mathbb{N}\times\mathbb{N}$ to $\mathbb{N}$ with inverse

$$\nu^{-1}(l) = (\varphi(l), \psi(l)) .$$

When $\sigma \in \mathbb{N}^{\mathbb{N}}$ and

$$\tau(m) = \tau_1^{(m)}, \tau_2^{(m)}, \tau_3^{(m)}, \ldots, \qquad m = 1, 2, \ldots,$$

is a sequence in $(\mathbb{N}^{\mathbb{N}})^{\mathbb{N}}$, we write

$$\chi(\sigma; \tau(1), \tau(2), \ldots) = \nu(\sigma_1, \tau_{\psi}^{(\varphi(1))}), \nu(\sigma_2, \tau_{\psi(2)}^{(\varphi(2))}), \ldots, \nu(\sigma_l, \tau_{\psi(l)}^{(\varphi(l))}), \ldots .$$

Clearly χ maps $\mathbb{N}^{\mathbb{N}}\times(\mathbb{N}^{\mathbb{N}})^{\mathbb{N}}$ into $\mathbb{N}^{\mathbb{N}}$. Further, for any ρ in $\mathbb{N}^{\mathbb{N}}$, the equation

$$\chi(\sigma; \tau(1), \tau(2), \ldots) = \rho$$

has the unique solution

$$\sigma_m = \varphi(\rho_m) ,$$
$$\tau_n^{(m)} = \psi(\rho_{\nu(m, n)}) .$$

As we always have $m \leqslant \nu(m, n)$ and $\nu(m, n') \leqslant \nu(m, n)$ when $n' < n$, we see that a knowledge of

$$\chi(\sigma; \tau(1), \tau(2), \ldots)|\nu(m, n)$$

determines $\sigma|m$ and $\tau(m)|n$ uniquely.

Proof of Theorem 2.3.1 Let T be a set of $\mathbf{S}(\mathbf{S}(\mathscr{S}))$. Then T has a representation

$$T = \bigcup_{\sigma} \bigcap_{m} T(\sigma|m) , \tag{3}$$

with each set $T(\sigma|m)$, having a representation

$$T(\sigma|m) = \bigcup_{\tau} \bigcap_{n} S(\sigma|m; \tau|n) , \tag{4}$$

with each set $S(\sigma|m; \tau|n)$ in $\mathscr{S}$.

Given any sequence r of l positive integers, $l \geq 1$, and using the notation of the lemma, we can find

$$\sigma;\ \tau(1), \tau(2), \ldots$$

in $\mathbb{N}^{\mathbb{N}} \times (\mathbb{N}^{\mathbb{N}})^{\mathbb{N}}$ with

$$r = \chi(\sigma;\ \tau(1), \tau(2), \ldots)|l\,.$$

Although $\sigma;\ \tau(1), \tau(2), \ldots$ are not uniquely determined by r, the partial sequences

$$\sigma|\varphi(l) \quad \text{and} \quad \tau(\varphi(l))|\psi(l)$$

are determined by r, as

$$\nu(\varphi(l), \psi(l)) = l\,.$$

So we can define the set $R(r)$ of $\mathcal{S}$ by the formula

$$R(r) = S(\sigma|\varphi(l);\ \tau(\varphi(l))|\psi(l))\,. \tag{5}$$

Now $\mathbf{S}(\mathcal{S})$ contains the set

$$\begin{aligned}
\bigcup_{\rho}\bigcap_{l} R(\rho|l) &= \bigcup_{\sigma;\tau(1),\tau(2),\ldots}\ \bigcap_{l} R(\chi(\sigma;\ \tau(1), \tau(2), \ldots)|l) \\
&= \bigcup_{\sigma;\tau(1),\tau(2),\ldots}\ \bigcap_{l} S(\sigma|\varphi(l);\ \tau(\varphi(l))|\psi(l)) \\
&= \bigcup_{\sigma;\tau(1),\tau(2),\ldots}\ \bigcap_{m,n} S(\sigma|m;\ \tau(m)|n) \\
&= \bigcup_{\sigma}\ \bigcup_{\tau(1),\tau(2),\ldots}\ \bigcap_{m}\bigcap_{n} S(\sigma|m;\ \tau(m)|n) \\
&= \bigcup_{\sigma}\bigcap_{m}\ \bigcup_{\tau(m)}\bigcap_{n} S(\sigma|m;\ \tau(m)|n) \\
&= \bigcup_{\sigma}\bigcap_{m} T(\sigma|m) \\
&= T\,.
\end{aligned}$$

Thus $\mathbf{S}(\mathbf{S}(\mathcal{S})) \subset \mathbf{S}(\mathcal{S})$. By taking all the sets $T(\sigma|m)$ in the representation (3) to be a fixed set of $\mathcal{S}$, it follows that $\mathbf{S}(\mathcal{S}) \subset \mathbf{S}(\mathbf{S}(\mathcal{S}))$. So $\mathbf{S}(\mathbf{S}(\mathcal{S})) = \mathbf{S}(\mathcal{S})$ as required.

Now suppose that T belongs to $\mathbf{S}_d(\mathbf{S}_d(\mathcal{S}))$. Then T has a representation in the form (3) with each set $T(\sigma|m)$ having a representation of the form (4). Further, we may suppose that the representations have the properties that, if $\sigma \neq \hat{\sigma}$, then

$$\Big(\bigcap_{m} T(\sigma|m)\Big) \cap \Big(\bigcap_{m} T(\hat{\sigma}|m)\Big) = \varnothing\,,$$

and, for all σ, m, if $\tau \neq \hat{\tau}$, then

$$\left(\bigcap_n S(\sigma|m;\tau|n)\right) \cap \left(\bigcap_n S(\sigma|m;\hat{\tau}|n)\right) = \varnothing .$$

As before, we define $R(r)$ by (5) and obtain the representation

$$T = \bigcup_\rho \bigcap_l R(\rho|l) . \tag{6}$$

Suppose that $\rho \neq \hat{\rho}$. Write

$$\chi^{-1}(\rho) = \sigma;\, \tau(1),\, \tau(2),\, \ldots,$$
$$\chi^{-1}(\hat{\rho}) = \hat{\sigma};\, \hat{\tau}(1),\, \hat{\tau}(2),\, \ldots .$$

If $\sigma \neq \hat{\sigma}$, then

$$\begin{aligned}\left(\bigcap_l R(\rho|l)\right) \cap \left(\bigcap_l R(\hat{\rho}|l)\right) &\subset \left(\bigcap_m T(\sigma|m)\right) \cap \left(\bigcap_m T(\hat{\sigma}|m)\right)\\ &= \varnothing .\end{aligned}$$

If $\sigma = \hat{\sigma}$, then, for some $m^* \geqslant 1$, $\tau(m^*) \neq \hat{\tau}(m^*)$, and

$$\begin{aligned}&\left(\bigcap_l R(\rho|l)\right) \cap \left(\bigcap_l R(\hat{\rho}|l)\right)\\ &\qquad \subset \left(\bigcap_n S(\sigma|m^*;\tau(m^*)|n)\right) \cap \left(\bigcap_n S(\sigma|m^*;\hat{\tau}(m^*)|n)\right)\\ &\qquad = \varnothing .\end{aligned}$$

Thus the representation (6) is a disjoint Souslin representation and

$$\mathbf{S}_d(\mathbf{S}_d(\mathcal{S})) \subset \mathbf{S}_d(\mathcal{S}) .$$

To prove that in fact $\mathbf{S}_d(\mathbf{S}_d(\mathcal{S})) = \mathbf{S}_d(\mathcal{S})$ we have to consider a rather trivial difficulty. It may happen that whenever we choose a countable sequence $\{S_i\}$ from $\mathcal{S}$ we have

$$\bigcap_i S_i \neq \varnothing .$$

In this case $\mathbf{S}_d(\mathcal{S})$ and $\mathbf{S}_d(\mathbf{S}_d(\mathcal{S}))$ are empty. So suppose that we can choose a sequence $\{N_i\}$ from S with

$$\bigcap_i N_i = \varnothing .$$

Taking $S(\sigma|n)=N_n$, for all σ and n, we see that $\emptyset\in\mathbf{S}_d(\mathscr{S})$). Now, if $S\in\mathbf{S}_d(\mathscr{S})$, we can take

$$T(\sigma|m)=\begin{cases}S, & \text{if } \sigma_1=\sigma_2=\cdots=\sigma_m=1\,,\\ \emptyset\,, & \text{otherwise}\,,\end{cases}$$

and see that $S\in\mathbf{S}_d(\mathbf{S}_d(\mathscr{S}))$. Thus, in all cases, $\mathbf{S}_d(\mathbf{S}_d(\mathscr{S}))=\mathbf{S}_d(\mathscr{S})$.

We may now consider the following corollary to Theorem 2.3.1

COROLLARY 2.3.3 *The family* $\mathbf{S}(\mathscr{S})$ *is closed under countable intersection and countable union. The family* $\mathbf{S}_d(\mathscr{S})$ *is closed under countable intersection and countable disjoint union.*

Proof Let $\{T_i\}$ be any sequence of sets of $\mathbf{S}(\mathscr{S})$. Write

$$T(\sigma|m)=T_m\,,\qquad \text{all } \sigma, m\,,$$
$$T'(\sigma|m)=T_{\sigma_1}\,,\qquad \text{all } \sigma, m\,.$$

Then

$$\bigcap_m T_m=\bigcup_\sigma\bigcap_m T(\sigma|m)\,,$$
$$\bigcup_s T_s=\bigcup_\sigma\bigcap_m T'(\sigma|m)\,,$$

belong to $\mathbf{S}(\mathbf{S}(\mathscr{S}))$ and so to $\mathbf{S}(\mathscr{S})$.

Now let $\{T_i\}$ be any sequence of sets of $\mathbf{S}_d(\mathscr{S})$. Then $\mathbf{S}_d(\mathscr{S})$ is not empty, and so includes $\emptyset$. In this case we write

$$T''(\sigma|m)=\begin{cases}T_m\,, & \text{if } \sigma_1=\sigma_2=\cdots=\sigma_m=1\,,\\ \emptyset\,, & \text{otherwise}\,,\end{cases}$$

$$T'''(\sigma|m)=\begin{cases}T_{\sigma_1}\,, & \text{if } \sigma_2=\sigma_3=\cdots=\sigma_m=1\,,\\ \emptyset\,, & \text{otherwise}\,.\end{cases}$$

Now

$$\bigcap_m T_m=\bigcup_\sigma\bigcap_m T''(\sigma|m)\,,$$
$$\bigcup_s T_s=\bigcup_\sigma\bigcap_m T'''(\sigma|m)\,;$$

the intersection belongs to $\mathbf{S}_d(\mathbf{S}_d(\mathscr{S}))$ and so to $\mathbf{S}_d(\mathscr{S})$, and, when the union is disjoint, it belongs to $\mathbf{S}_d(\mathbf{S}_d(\mathscr{S}))$ and so to $\mathbf{S}_d(\mathscr{S})$.

2.4 Polish spaces

In this section we show that each Polish space is an analytic set. We obtain this result as a corollary of a theorem due to Kuratowski (1934).

THEOREM 2.4.1 *Let X be a non-empty Polish space that is dense-in-itself. Then X is the continuous injective image of $\mathbb{N}^{\mathbb{N}}$ under a map that maps open sets of $\mathbb{N}^{\mathbb{N}}$ to $\mathcal{F}_\sigma$-subsets of X.*

COROLLARY 2.4.2 *Each Polish space X is a Lusin set, being a continuous injective image of a closed subset of $\mathbb{N}^{\mathbb{N}}$, under a map that maps open sets of $\mathbb{N}^{\mathbb{N}}$ to $\mathcal{F}_\sigma$-subsets of X.*

COROLLARY 2.4.3 *Each Polish space X is analytic, being empty or a continuous image of $\mathbb{N}^{\mathbb{N}}$.*

The main idea of the proof of Theorem 2.4.1 is first to split the space up into little bits, and then to use a construction reminiscent of the construction of the Peano curve; one of the main difficulties is to ensure that the "little bits" remain non-empty.

Proof of Theorem 2.4.1 We first prove that, *if F_1 and F_2 are closed sets in X, with $F_2 \backslash F_1$ non-empty and dense-in-itself, and $\delta > 0$, then there is a sequence $\{H_s\}$ of closed sets with*

$$\operatorname{diam} H_s < \delta, \qquad s \geqslant 1 ;$$

$$\bigcup_{s=1}^{\infty} H_s = F_2 \backslash F_1 ;$$

and

$$H_s \backslash \bigcup_{r<s} H_r$$

non-empty and dense-in-itself, for each $s \geqslant 1$.

As X is metric, each point of $F_2 \backslash F_1$ belongs to an open set, of diameter less than δ, whose closure does not meet F_1. As X is separable, we can choose a sequence $\{G_r\}$ of open sets, of diameter less than δ, covering $F_2 \backslash F_1$, and with

$$(\operatorname{cl} G_r) \cap F_1 = \varnothing , \qquad r \geqslant 1 .$$

Consider the sequence of sets

$$J_r = \operatorname{cl}(F_2 \cap Gr) , \qquad r \geqslant 1 .$$

Each set J_r is closed, and

$$\bigcup_{r=1}^{\infty} J_r = F_2 \backslash F_1 .$$

Suppose that $s \geqslant 1$, and that

$$x \in J_s \Big\backslash \bigcup_{r<s} J_r .$$

Let U be any open set containing x. As

$$x \notin \bigcup_{r<s} J_r \quad \text{and} \quad x \notin F_1 ,$$

the set U contains an open set V, containing x, and meeting neither

$$\bigcup_{r<s} J_r \quad \text{nor} \quad F_1 .$$

As

$$x \in J_s = \text{cl}\,(F_2 \cap G_s) ,$$

V meets $F_2 \cap G_s$ and so meets $(F_2 \backslash F_1) \cap G_s$. As $F_2 \backslash F_1$ is dense-in-itself and meets the open set $V \cap G_s$, there is a point $y \neq x$ in $V \cap G_2 \cap (F_2 \backslash F_1)$. Now $y \neq x$ and

$$y \in \Big(J_s \Big\backslash \bigcup_{r<s} J_r\Big) \cap U .$$

Hence

$$\Big(J_s \Big\backslash \bigcup_{r<s} J_r\Big)$$

is dense-in-itself.

We may suppress from the sequence $J_1, J_2, \ldots$ all sets with

$$J_s \Big\backslash \bigcup_{r<s} J_r = \emptyset .$$

When we do this, we obtain a finite or infinite sequence $J_1, J_2, \ldots$ of closed sets, with union $F_2 \backslash F_1$, such that each set

$$J_s \Big\backslash \bigcup_{r<s} J_r ,$$

is non-empty and dense-in-itself. In the case when the sequence $J_1, J_2, \ldots$ is infinite, it suffices to take $H_s = J_s$ for $s \geqslant 1$. In the case when the sequence $J_1, J_2, \ldots$ is finite, we need to use an extra idea.

Suppose that $F_2 \backslash F_1$ is covered by the finite sequence $J_1, J_2, \ldots, J_n$, with $n \geq 1$. Let x_0 be any point of

$$J_n \Big\backslash \bigcup_{r<n} J_r.$$

As this set is dense-in-itself, we can choose a sequence of distinct points $x_1, x_2, \ldots$ of the set, all distinct from x_0, converging to x_0. Then we can choose a sequence $G_{n+1}, G_{n+2}, \ldots$ of open sets, with disjoint closures, that meet neither F_1 nor $\bigcup_{r<n} J_r$, and with

$$x_r \in G_{n+r}, \qquad r \geq 1,$$

$$\operatorname{diam} G_{n+r} < \delta 2^{-r}, \qquad r \geq 1.$$

Write

$$H_s = J_s, \qquad 1 \leq s < n,$$

$$H_n = \operatorname{cl}\left(J_n \Big\backslash \bigcup_{r=1}^{\infty} \operatorname{cl} G_{n+2r}\right),$$

$$H_{n+s} = \operatorname{cl}(J_n \cap G_{n+2s}), \qquad s \geq 1.$$

It is easy to verify that the sequence $\{H_s\}$ of closed sets has union $F_2 \backslash F_1$, and that each set

$$H_s \Big\backslash \bigcup_{r<s} H_r, \qquad s \geq 1,$$

is non-empty and dense-in-itself. Note that we have used only every other set of the sequence $G_{n+1}, G_{n+2}, \ldots$ in order to ensure that the set J_n is dense at the point x_0. Thus we have established the italicized result stated in the first paragraph of this proof.

Using this result, in the case $F_2 = X$, $F_1 = \emptyset$, we may write

$$X = \bigcup_{s_1=1}^{\infty} F(s_1),$$

with $F(s_1)$ a closed set of diameter less than $1/2$, for each $s_1 \geq 1$, and with each set

$$F(\sigma_1) \Big\backslash \bigcup_{s_1<\sigma_1} F(s_1), \qquad \sigma_1 \geq 1,$$

non-empty and dense-in-itself.

Now suppose that, for some $n \geq 1$, we have chosen a family

$$F(\sigma|n) \qquad \sigma \in \mathbb{N}^{\mathbb{N}},$$

of closed sets, each of diameter less than 2^{-n}, with

$$\bigcup_{\sigma_h=1}^{\infty} F(\sigma_1, \ldots, \sigma_h) = F(\sigma_1, \ldots, \sigma_{h-1}) \Big\backslash \bigcup_{s<\sigma_{h-1}} F(\sigma_1, \ldots, \sigma_{h-2}, s), \tag{7}$$

for $1 \leqslant h \leqslant n$, with the convention $F(\varnothing) = X$, and with each set

$$F(\sigma_1, \ldots, \sigma_h) \Big\backslash \bigcup_{s<\sigma_h} F(\sigma_1, \ldots, s), \qquad 1 \leqslant h \leqslant n, \tag{8}$$

non-empty and dense-in-itself. Then, applying the italicized result to each pair of closed sets,

$$F(\sigma_1, \ldots, \sigma_n), \qquad \bigcup_{s<\sigma_n} F(\sigma_1, \ldots, s),$$

we can choose a family

$$F(\sigma|n+1), \qquad \sigma \in \mathbb{N}^{\mathbb{N}},$$

of closed sets, each of diameter less than 2^{-n-1}, with

$$\bigcup_{\sigma_{n+1}=1}^{\infty} F(\sigma_1, \ldots, \sigma_{n+1}) = F(\sigma_1, \ldots, \sigma_n) \Big\backslash \bigcup_{s<\sigma_n} F(\sigma_1, \ldots, s)$$

and with each set

$$F(\sigma_1, \ldots, \sigma_{n+1}) \Big\backslash \bigcup_{s<\sigma_{n+1}} F(\sigma_1, \ldots, s)$$

non-empty and dense-in-itself.

Hence we can use induction to choose closed sets

$$F(\sigma|n), \qquad \sigma \in \mathbb{N}^{\mathbb{N}}, \qquad n \geqslant 1,$$

satisfying (7), the sets $F(\sigma|n)$ having diameter less than 2^{-n}, and with all the sets (8) non-empty and dense-in-themselves.

Now, for each σ in $\mathbb{N}^{\mathbb{N}}$, the sequence $F(\sigma|n)$, $n = 1, 2, \ldots$ is a decreasing sequence of non-empty closed sets with diameters tending to zero. As X is complete the set

$$F(\sigma) = \bigcap_{n=1}^{\infty} F(\sigma|n)$$

is non-empty and contains a single point. We define the function f from $\mathbb{N}^{\mathbb{N}}$ to X by the condition that

$$\{f(\sigma)\} = F(\sigma),$$

for all σ in $\mathbb{N}^{\mathbb{N}}$. We show that f maps $\mathbb{N}^{\mathbb{N}}$ bijectively and continuously onto X.

Let σ, τ be any two distinct elements of $\mathbb{N}^{\mathbb{N}}$. Let $m = m(\alpha, \tau)$ be the largest integer for which

$$\sigma|m = \tau|m .$$

Then

$$f(\sigma) \in \bigcap_{n=1}^{\infty} F(\sigma|n) \subset F(\sigma|m) ,$$

$$f(\tau) \in \bigcap_{n=1}^{\infty} F(\tau|n) \subset F(\tau|m) = F(\sigma|m) .$$

As $F(\sigma|m)$ has diameter less than 2^{-m}, the distance between $f(\sigma)$ and $f(\tau)$ is less than 2^{-m}. Hence f is continuous.

To prove that f is injective, consider any two distinct elements σ, τ of $\mathbb{N}^{\mathbb{N}}$. As before, let m be the largest integer for which $\sigma|m = \tau|m$. Then $\sigma_{m+1} \neq \tau_{m+1}$. We may suppose that $\sigma_{m+1} < \tau_{m+1}$. Then

$$f(\sigma) \in F(\sigma) = \bigcap_{n=1}^{\infty} F(\sigma|n) \subset F(\sigma|m+1) ,$$

and

$$\begin{aligned} f(\tau) \in F(\tau) &= \bigcap_{n=1}^{\infty} F(\tau|n) \\ &\subset F(\tau|m+2) \\ &\subset \bigcup_{s=1}^{\infty} F(\sigma_1, \sigma_2, \ldots, \sigma_m, \tau_{m+1}, s) \\ &\subset F(\sigma_1, \sigma_2, \ldots, \sigma_m, \tau_{m+1}) \Big\backslash \bigcup_{s<\tau_{m+1}} F(\sigma_1, \sigma_2, \ldots, \sigma_m, s) . \end{aligned}$$

As $\sigma_{m+1} < \tau_{m+1}$ this ensures that $f(\sigma) \neq f(\tau)$. So f is injective.

To prove that f is surjective, consider any point x of X. As

$$X = \bigcup_{s_1=1}^{\infty} F(s_1) ,$$

and

$$F(\sigma_1, \ldots, \sigma_n) \Big\backslash \bigcup_{s<\sigma_n} F(\sigma_1, \ldots, s) = \bigcup_{\sigma_{n+1}=1}^{\infty} F(\sigma_1, \ldots, \sigma_{n+1}) ,$$

we can choose $\tau_1, \tau_2, \ldots$ inductively so that, for each $n \geq 1$, τ_n is the least integer $t \geq 1$ for which

$$x \in F(\tau_1, \tau_2, \ldots, \tau_{n-1}, t) .$$

Taking $\tau = \tau_1, \tau_2, \ldots$, we have

$$x \in F(\tau)\,,$$

so that $x = f(\tau)$. Thus f maps $\mathbb{N}^{\mathbb{N}}$ surjectively onto X.

It remains to prove that f maps open subsets of $\mathbb{N}^{\mathbb{N}}$ to $\mathscr{F}_\sigma$-sets in X. Each open set of $\mathbb{N}^{\mathbb{N}}$ is the union of a countable sequence of Baire intervals, i.e. of sets of the form

$$I(\sigma|n) = \{\tau | \tau \in \mathbb{N}^{\mathbb{N}} \text{ and } \tau|n = \sigma|n\}.$$

But, as in the proof that f is surjective,

$$f(I(\sigma|n)) = F(\sigma_1, \ldots, \sigma_n) \setminus \bigcup_{s<\sigma_n} F(\sigma_1, \ldots, s)\,,$$

which is an $\mathscr{F}_\sigma$-set, that is dense-in-itself. Hence f maps each open set in $\mathbb{N}^{\mathbb{N}}$ to an $\mathscr{F}_\sigma$-set that is dense-in-itself.

Proof of Corollary 2.4.2 As $\mathbb{N}^{\mathbb{N}}$ contains a countably infinite discrete closed set, the result is trivial, if X is a countable Polish space. So suppose that X is an uncountable Polish space. Let C be the union of all the countable open sets of X. As X is separable, C is itself countable. Further, the perfect kernel $D = X \setminus C$ is closed, non-empty and dense-in-itself. By the theorem, we can choose a continuous injective map f from the Baire interval

$$I(1) = \{\tau \colon \tau \in \mathbb{N}^{\mathbb{N}} \text{ and } \tau_1 = 1\}\,,$$

onto D. Choose a discrete closed subset J of $\mathbb{N}^{\mathbb{N}} \setminus I(1)$ with cardinal equal to that of C. Then f extends to yield a continuous bijection of the closed subset

$$I(1) \cup J$$

of $\mathbb{N}^{\mathbb{N}}$ onto $X = D \cup C$.

Proof of Corollary 2.4.3 We use the method for proving Corollary 2.4.2, but, in this case, we split the set $\mathbb{N}^{\mathbb{N}} \setminus I(1)$ into a union of non-empty clopen sets, one for each point of C, and extend f by mapping the whole clopen set onto the corresponding point of C.

Remark Corollary 2.4.2 can be proved by a rather different method. By a standard result of general topology (see, for example, Engelking, 1968, pp. 189–190) every Polish space X is homeomorphic to a subset Y of $[0, 1)^{\mathbb{N}}$. Define a map φ from $\mathbb{N}^{\mathbb{N}}$ to $[0, 1)$ by taking

$$\varphi(\sigma) = 1 - 2^{-\sigma_1} - 2^{-\sigma_1 - \sigma_2} - 2^{-\sigma_1 - \sigma_2 - \sigma_3} - \cdots$$

for all σ in $\mathbb{N}^{\mathbb{N}}$. Then φ is a continuous bijection from $\mathbb{N}^{\mathbb{N}}$ to $[0, 1)$ and φ maps each Baire interval of $\mathbb{N}^{\mathbb{N}}$ to a half-open interval of $[0, 1)$. Let Φ be the map of $(\mathbb{N}^{\mathbb{N}})^{\mathbb{N}}$ to $[0, 1)^{\mathbb{N}}$ obtained by application of φ to each coordinate separately. Let θ be a homeomorphism mapping $\mathbb{N}^{\mathbb{N}}$ onto $(\mathbb{N}^{\mathbb{N}})^{\mathbb{N}}$ and let ψ be a homeomorphism mapping Y to X. It is easy to verify that

$$Z = \theta^{-1}(\Phi^{-1}(Y))$$

is closed in $\mathbb{N}^{\mathbb{N}}$ and that $\psi \circ \Phi \circ \theta$, when restricted to Z provides the required bijective map from Z to X.

2.5 Stability of *K*-analytic sets

In this section we establish the main stability properties of the K-analytic sets. We shall find it easy to prove

THEOREM 2.5.1 *An image, in a Hausdorff space, under a compact-valued upper semi-continuous map, of a K-analytic set is a K-analytic set.*

In order to prove stability under the Souslin operation it is convenient to prove two preliminary results.

THEOREM 2.5.2 *A K-analytic set in a Hausdorff space is a Souslin-$\mathcal{F}$ set in that space.*

THEOREM 2.5.3 *The intersection of a K-analytic set and a Souslin-$\mathcal{F}$ set, in a Hausdorff space, is a K-analytic set.*

The stability result takes the form

THEOREM 2.5.4 *Application of the Souslin operation to the K-analytic sets of a Hausdorff space yields only K-analytic sets.*

It follows from these results that, in a Hausdorff space, the Souslin-$\mathcal{K}$ sets are K-analytic and the K-analytic sets are Souslin-$\mathcal{F}$. Although all three families are invariant under the Souslin operation, only the family of K-analytic sets is invariant under the operation of continuous mapping into a Hausdorff space.

We also prove

THEOREM 2.5.5 *If X_i, $i = 1, 2, \ldots$, are K-analytic spaces, then*

$$\prod_{i=1}^{\infty} X_i$$

is also a K-analytic space.

Proof of Theorem 2.5.1 Let F be a compact-valued upper semi-continuous map from a set A in a Hausdorff space X to the compact sets of a Hausdorff space Y. Let K be any compact subset of A. We first show that

$$F(K) = \bigcup_{x \in K} F(x)$$

is compact in Y. Let $\{G_\beta\}_{\beta \in B}$ be any open cover of $F(K)$. For each x in K, the compact set $F(x)$ is covered by $\{G_\beta\}_{\beta \in B}$, and we can choose a finite set $B(x)$ contained in B with

$$F(x) \subset \bigcup_{\beta \in B(x)} G_\beta .$$

Let F^{II} denote the "inner inverse image" function of F defined by

$$F^{\text{II}}(Z) = \{x : F(x) \subset Z\} .$$

As F is upper semi-continuous, the set

$$U(x) = F^{\text{II}}\Big(\bigcup_{\beta \in B(x)} G_\beta\Big) \tag{9}$$

is an open set containing x. As K is compact and is covered by the open family $\{U(x)\}_{x \in K}$, we can choose a finite subset H of K with

$$K \subset \bigcup_{x \in H} U(x) .$$

Now

$$\begin{aligned} F(K) &\subset \bigcup_{x \in H} F\Big(F^{\text{II}}\Big(\bigcup_{\beta \in B(x)} G_\beta\Big)\Big) \\ &\subset \bigcup_{x \in H} \bigcup_{\beta \in B(x)} G_\beta . \end{aligned}$$

Hence $F(K)$ is compact.

Now suppose that A is K-analytic and has the representation

$$A = \bigcup_{\sigma} K(\sigma)$$

with K an upper semi-continuous map from $\mathbb{N}^{\mathbb{N}}$ to the compact sets of X. Now

$$F(A) = \bigcup_{\sigma} F(K(\sigma))$$

and $F \circ K$ maps $\mathbb{N}^{\mathbb{N}}$ to the compact sets of Y. We just need to prove that $F \circ K$ is upper semi-continuous. Let $\sigma \in \mathbb{N}^{\mathbb{N}}$ and let G be an open set in Y with

$$F(K(\sigma)) \subset G .$$

As F is upper semi-continuous the set

$$F^{\text{II}}(G)$$

is an open set in X containing $K(\sigma)$. As K is semi-continuous, we can choose an open neighbourhood V of σ with

$$K(V) \subset F^{\text{II}}(G) .$$

This ensures that

$$F(K(V)) \subset F(F^{\text{II}}(G)) \subset G .$$

Thus $F \circ K$ is upper semi-continuous, and $F(A)$ is K-analytic.

Proof of Theorem 2.5.2 Let A be a K-analytic set in a Hausdorff space X. Then $A = K(\mathbb{N}^{\mathbb{N}})$ with K an upper semi-continuous map from $\mathbb{N}^{\mathbb{N}}$ to the compact sets of X. We show that A is obtained by applying the Souslin operation to the family

$$F(s) = \text{cl}\, K(I(s)), \qquad s \in \mathbb{N}^{(\mathbb{N})} ,$$

of closures of images of the Baire intervals. Clearly, for each σ in $\mathbb{N}^{\mathbb{N}}$,

$$K(\sigma) \subset \bigcap_n F(\sigma|n) .$$

Suppose that f is in

$$\bigcap_n F(\sigma|n) ,$$

but is not in $K(\sigma)$. As $K(\sigma)$ is compact and X is Hausdorff, we can choose an open set G and a closed set H with

$$K(\sigma) \subset G \subset H \quad \text{and} \quad f \notin H . \tag{10}$$

Provided n^* is sufficiently large

$$K(I(\sigma|n^*)) \subset G \subset H ,$$

so that

$$f \in \bigcap_n F(\sigma|n) \subset F(\sigma|n^*) = \text{cl}\, K(I(\sigma|n^*)) \subset H ,$$

contrary to (10). Hence

$$K(\sigma) = \bigcap_n F(\sigma|n) \tag{11}$$

and

$$A = K(\mathbb{N}^{\mathbb{N}}) = \bigcup_\sigma K(\sigma) = \bigcup_\sigma \bigcap_n F(\sigma|n) ,$$

is a Souslin-$\mathscr{F}$ set.

We may now state a corollary to Theorem 2.5.2.

COROLLARY 2.5.6 *The K-analytic set has the representations*

$$A = K(\mathbb{N}^{\mathbb{N}}) = \bigcup_{\sigma} \bigcap_{n} F(\sigma|n),$$

with

$$F(\sigma|n) = \operatorname{cl} K(I(\sigma|n)),$$

$$K(\sigma) = \bigcap_{n=1}^{\infty} F(\sigma|n) = \bigcap_{n=1}^{\infty} \operatorname{cl} K(I(\sigma|n)), \tag{12}$$

for all $\sigma \in \mathbb{N}^{\mathbb{N}}$, $n \in \mathbb{N}$.

Proof of Theorem 2.5.3 Let $A = K(\mathbb{N}^{\mathbb{N}})$ be a K-analytic set, with K an upper semi-continuous map from $\mathbb{N}^{\mathbb{N}}$ to the compact sets of a Hausdorff space X. Let

$$S = \bigcup_{\sigma} \bigcap_{n} F(\sigma|n),$$

with

$$F(\sigma|n) \in F, \qquad \sigma \in \mathbb{N}^{\mathbb{N}}, \qquad n \geqslant 1,$$

be a Souslin-$\mathscr{F}$ set in X. Then

$$\begin{aligned} A \cap S &= \Big(\bigcup_{\sigma} K(\sigma)\Big) \cap \Big(\bigcup_{\tau} \bigcap_{n} F(\tau|n)\Big) \\ &= \bigcup_{\sigma,\tau} \Big[K(\sigma) \cap \bigcap_{n} F(\tau|n)\Big] \\ &= \bigcup_{\sigma,\tau} H(\sigma, \tau), \end{aligned}$$

with

$$H(\sigma, \tau) = K(\sigma) \cap \Big(\bigcap_{n} F(\tau|n)\Big).$$

Now H maps $\mathbb{N}^{\mathbb{N}} \times \mathbb{N}^{\mathbb{N}}$ to the compact sets of X. As $\mathbb{N}^{\mathbb{N}} \times \mathbb{N}^{\mathbb{N}}$ is homeomorphic to $\mathbb{N}^{\mathbb{N}}$, it will suffice to show that H is upper semi-continuous. Suppose that for some (σ, τ) in $\mathbb{N}^{\mathbb{N}} \times \mathbb{N}^{\mathbb{N}}$ and for some open set G we have

$$H(\sigma, \tau) \subset G.$$

Now

$$K(\sigma) \cap \Big(\bigcap_{r=1}^{n} F(\tau|r)\Big), \qquad n = 1, 2, \ldots,$$

is a decreasing sequence of compact sets, with intersection $H(\sigma, \tau)$ contained in the open set G. So, for some $n^* \geqslant 1$,

$$K(\sigma) \cap \left(\bigcap_{r=1}^{n^*} F(\tau|r) \right) \subset G$$

and

$$K(\sigma) \subset G \cup \left(X \Big\backslash \bigcap_{r=1}^{n^*} F(\tau|r) \right).$$

As K is upper semi-continuous, there is an $m^* \geqslant 1$, with

$$K(I(\sigma|m^*)) \subset G \cup \left(X \Big\backslash \bigcap_{r=1}^{n^*} F(\tau|r) \right),$$

so that

$$K(I(\sigma|m^*)) \cap \left(\bigcap_{r=1}^{n^*} F(\tau|r) \right) \subset G$$

and

$$H(I(\sigma|m^*)) \times I(\tau|n^*) \subset G.$$

Thus H is upper semi-continuous, and $A \cap S$ is K-analytic.

Proof of Theorem 2.5.4 First consider a sequence $\{A_s\}$ of K-analytic sets. Let $A_s = K_s(\mathbb{N}^{\mathbb{N}})$, $s \geqslant 1$, with K_s an upper semi-continuous map from $\mathbb{N}^{\mathbb{N}}$ to the compact sets of the Hausdorff space X. Define the map $H : \mathbb{N}^{\mathbb{N}} \to \mathscr{K}(X)$ by taking

$$H(\sigma) = K_{\sigma_1}(\sigma_2, \sigma_3, \ldots), \qquad \sigma \in \mathbb{N}^{\mathbb{N}}.$$

It is easy to see that H is upper semi-continuous and that

$$\bigcup_{s=1}^{\infty} A_s = \bigcup_{\sigma} H(\sigma) = H(\mathbb{N}^{\mathbb{N}}).$$

Thus the union of any sequence of K-analytic sets is a K-analytic set.

Now suppose that for each $\sigma \in \mathbb{N}^{\mathbb{N}}$ and $n \geqslant 1$, the set

$$A(\sigma|n)$$

is K-analytic in the Hausdorff space X. We show that

$$S = \bigcup_{\sigma} \bigcap_{n} A(\sigma|n)$$

is a K-analytic set. By Theorem 2.5.2 each set $A(\sigma|n)$ is a Souslin-$\mathcal{F}$ set. By Theorem 2.3.1, the set S is a Souslin-$\mathcal{F}$ set. But

$$S = S \cap \bigcup_{s=1}^{\infty} A(s)$$

and so is the intersection of a Souslin-$\mathcal{F}$ set with a K-analytic set. It follows, by Theorem 2.5.3, that S is a K-analytic set.

Proof of Theorem 2.5.5 Let X_i be a Hausdorff space with $X_i = K_i(\mathbb{N}^{\mathbb{N}})$, and K_i an upper semi-continuous map from $\mathbb{N}^{\mathbb{N}}$ to the compact sets of X_i, for $i \geq 1$. Then

$$\begin{aligned}\prod_{i=1}^{\infty} X_i &= \prod_{i=1}^{\infty} \bigcup_{\sigma(i)} K_i(\sigma(i)) \\ &= \bigcup_{\sigma(1),\sigma(2),\ldots} \prod_{i=1}^{\infty} K_i(\sigma(i)) \\ &= \bigcup_{\sigma(1),\sigma(2),\ldots} K(\sigma(1), \sigma(2), \ldots),\end{aligned}$$

with

$$K(\sigma(1), \sigma(2), \ldots) = \prod_{i=1}^{\infty} K_i(\sigma(i)).$$

Here K is a map from $(\mathbb{N}^{\mathbb{N}})^{\mathbb{N}}$ to the compact sets of

$$X = \prod_{i=1}^{\infty} X_i .$$

As $(\mathbb{N}^{\mathbb{N}})^{\mathbb{N}}$ is homeomorphic to $\mathbb{N}^{\mathbb{N}}$, it suffices to prove that K is upper semi-continuous.

Consider any $\sigma(1), \sigma(2), \ldots$ in $(\mathbb{N}^{\mathbb{N}})^{\mathbb{N}}$ and any open set G in X with

$$K(\sigma(1), \sigma(2), \ldots) \subset G .$$

We first consider the possibility that

$$K(\sigma(1), \sigma(2), \ldots) = \prod_{i=1}^{\infty} K_i(\sigma(i)) = \emptyset .$$

Then, for some $i^* \geq 1$,

$$K_{i^*}(\sigma(i^*)) = \emptyset .$$

By the upper semi-continuity, there is an $m^* \geq 1$, with

$$K_{i^*}(I(\sigma(i^*)|m^*)) = \emptyset .$$

So

$$K(\tau(1), \tau(2), \ldots) = \varnothing \subset G,$$

for all $\tau(1), \tau(2), \ldots$ in $(\mathbb{N}^{\mathbb{N}})^{\mathbb{N}}$ sufficiently close to $\sigma(1), \sigma(2), \ldots$.

Now we may suppose that

$$\varnothing \neq K(\sigma(1), \sigma(2), \ldots) \subset G.$$

Now G is a union of basic open sets in the Cartesian product topology on X. As $K(\sigma(1), \sigma(2), \ldots)$ is compact it is contained in a finite union of basic open sets, all contained in G. Hence, for some $n^* \geqslant 1$, and some open set G^* in

$$X^* = \prod_{i=1}^{n^*} X_i,$$

we have

$$\varnothing \neq K(\sigma(1), \sigma(2), \ldots) \subset G^* \times \prod_{i=n^*+1}^{\infty} X_i \subset G.$$

Hence

$$\varnothing \neq \prod_{i=1}^{n^*} K_i(\sigma(i)) \subset G^*.$$

As the sets $K_i(\sigma(i))$, $1 \leqslant i \leqslant n^*$, are compact, we can now choose open sets G_i^*, $1 \leqslant i \leqslant n^*$, in X_i, $1 \leqslant i \leqslant n^*$, with

$$K_i(\sigma(i)) \subset G_i^*, \qquad 1 \leqslant i \leqslant n^*,$$

$$\prod_{i=1}^{n^*} G_i^* \subset G^*.$$

By the upper semi-continuity of K_i, we can choose m_i^* with $1 \leqslant i \leqslant n^*$ so that

$$K_i(I(\sigma(i)|m_i^*)) \subset G_i^*.$$

Now for all $\tau(1), \tau(2), \ldots$ in $(\mathbb{N}^{\mathbb{N}})^{\mathbb{N}}$ sufficiently close to $\sigma(1), \sigma(2), \ldots$ we have

$$\tau(i)|m_i^* = \sigma(i)|m_i^*, \qquad 1 \leqslant i \leqslant n^*,$$

and

$$K(\tau(1), \tau(2), \ldots) \subset \left(\prod_{i=1}^{n^*} G_i^*\right) \times \left(\prod_{i=n^*+1}^{\infty} X_i\right) \subset G.$$

Hence K is upper semi-continuous and X, being Hausdorff, is a K-analytic space.

2.6 Projections and representations as projections

In this section we give a number of results concerning the representation of K-analytic sets and Souslin sets as projections, and corresponding results on stability under the projection operation.

If $\mathscr{S}$ and $\mathscr{T}$ are classes of sets in spaces X and Y, we use $\mathscr{S} \times \mathscr{T}$ to denote the class of "rectangles" of the form $S \times T$ in $X \times Y$ with $S \in \mathscr{S}$ and $T \in \mathscr{T}$.

THEOREM 2.6.1 *Each K-analytic set in a Hausdorff space X can be represented as the projection on X of a closed K-analytic set in* $X \times \mathbb{N}^{\mathbb{N}}$.

THEOREM 2.6.2 *Let $\mathscr{S}$ be any class of sets in a space X. Let $\mathscr{I}$ denote the class of Baire intervals in* $\mathbb{N}^{\mathbb{N}}$. *Then any Souslin-$\mathscr{S}$ set in X can be presented as the projection on X of a* $(\mathscr{S} \times \mathscr{I})_{\sigma\delta}$*-set in* $X \times \mathbb{N}^{\mathbb{N}}$.

THEOREM 2.6.3 *Any Souslin-$\mathscr{F}$ set in X can be represented as the projection on X of an $\mathscr{F}_{\sigma\delta}$-set in* $X \times [0, 1]$.

THEOREM 2.6.4 *Let X and Y be Hausdorff spaces. Then the projection on X of a K-analytic set in $X \times Y$ is K-analytic.*

THEOREM 2.6.5 *Let $\mathscr{S}$ be any class of sets, containing $\varnothing$, in a space X. Let Y be a K-analytic space. Then the projection on X of a Souslin-$(\mathscr{S} \times \mathscr{F})$ set in $X \times Y$ is a Souslin-$\mathscr{S}$ set.*

THEOREM 2.6.6 *Let X be a topological space and let Y be a K-analytic space. Then the projection on X of any Souslin-$\mathscr{F}$ set in $X \times Y$ is a Souslin-$\mathscr{F}$ set in X.*

THEOREM 2.6.7 *Let X and Y be completely regular (Hausdorff) spaces and suppose that Y is K-analytic. Let $\mathscr{Z}^*$ be the class of Lindelöf zero sets in $X \times Y$. Then the projection on X of a Souslin-$\mathscr{Z}^*$ set in $X \times Y$ is a Souslin-$\mathscr{Z}$ set in X.*

Theorems 2.6.1 and 2.6.2 are very easy consequences of results that we have already proved. Theorem 2.6.3 is an easy consequence of Theorem 2.6.2. Theorem 2.6.5 and Theorem 2.6.6 depend on familiar ideas. (Theorem 2.6.6 seems to be first proved explicitly by Rogers and Willmott (1966).) Theorem 2.6.7 is due to Jayne; we omit its proof.

Proof of Theorem 2.6.1 Let A be a K-analytic set in the Hausdorff space X. We may suppose that

$$A = K(\mathbb{N}^{\mathbb{N}}),$$

with K an upper semi-continuous map from $\mathbb{N}^{\mathbb{N}}$ to the compact sets of X. Then A is the projection on X of the generalized graph

$$H=\bigcup_{\sigma}\{K(\sigma)\times\{\sigma\}\} \tag{13}$$

in $X\times\mathbb{N}^{\mathbb{N}}$. If $x\times\sigma$ is any point of $X\times\mathbb{N}^{\mathbb{N}}$ not in H, then $x\notin K(\sigma)$. As X is Hausdorff and $K(\sigma)$ is closed, we can choose open sets U and G with $x\in U$, $K(\sigma)\subset G$, $U\cap G=\emptyset$. By the upper semi-continuity of K, we can choose an open set I in $\mathbb{N}^{\mathbb{N}}$ with

$$\sigma\in I,\qquad K(I)\subset G.$$

If $y\times\tau$ were common to H and to $U\times I$ we would have

$$y\in K(\tau)\subset K(I)\subset G$$

as well as $y\in U$. Hence $U\times I$ does not meet H. Thus $(X\times\mathbb{N}^{\mathbb{N}})\backslash H$ is open and H is closed. It follows immediately from (13)—well, almost immediately from (13)—that H is a K-analytic set in $X\times\mathbb{N}^{\mathbb{N}}$.

Proof of Theorem 2.6.2 Let S in X have the Souslin-$\mathcal{S}$ representation

$$S=\bigcup_{\sigma}\bigcap_{n}S(\sigma|n).$$

The set

$$H=\bigcap_{n}\bigcup_{\sigma}[S(\sigma|n)\times I(\sigma|n)]$$

is clearly a $(\mathcal{S}\times\mathcal{I})_{\sigma\delta}$-set in $X\times\mathbb{N}^{\mathbb{N}}$. Now $y\in\mathrm{proj}_X H$, if, and only if, there is a τ in $\mathbb{N}^{\mathbb{N}}$ with the property

(a) for each $n\geqslant 1$ there is a σ_n in $\mathbb{N}^{\mathbb{N}}$ with both $y\in S(\sigma_n|n)$ and $\tau|n=\sigma_n|n$.

But (a) is satisfied, if, and only if

(b) for each $n\geqslant 1$, $y\in S(\tau|n)$.

Thus $y\in\mathrm{proj}_X H$, if, and only if,

$$y\in\bigcup_{\tau}\bigcap_{n}S(\tau|n)=S.$$

Hence $S=\mathrm{proj}_X H$ and S has the required representation.

Proof of Theorem 2.6.3 Let S have the Souslin-$\mathcal{F}$ representation

$$S=\bigcup_{\sigma}\bigcap_{n}F(\sigma|n),$$

with all sets $F(\sigma|n)$ closed in X. As in the proof of Theorem 2.6.2, $S=\operatorname{proj}_X H$, with

$$H=\bigcup_n\bigcap_\sigma[F(\sigma|n)\times I(\sigma|n)].$$

We can regard $\mathbb{N}^{\mathbb{N}}$ as being identified with the set I of irrational numbers in the interval $[0, 1]$ of $\mathbb{R}$. Each Baire interval $I(\sigma|n)$ is the intersection of I with the closed sub-interval $J(\sigma|n)$ of $[0, 1]$ with end-points

$$\frac{1}{\sigma_1+}\ \frac{1}{\sigma_2+}\ \cdots\ \frac{1}{\sigma_n}\quad\text{and}\quad\frac{1}{\sigma_1+}\ \frac{1}{\sigma_2+}\ \cdots\ \frac{1}{\sigma_n+1}.$$

Now we can regard H as a subset of $X\times[0, 1]$ and we have

$$\begin{aligned}H&=\bigcap_n\bigcup_\sigma[F(\sigma|n)\times I(\sigma|n)]\\&=\Big(\bigcap_n\bigcup_\sigma[F(\sigma|n)\times J(\sigma|n)]\Big)\cap(X\times I).\end{aligned}$$

As I is an $\mathscr{F}_{\sigma\delta}$-set in $[0, 1]$, H is the intersection of two $\mathscr{F}_{\sigma\delta}$-sets in $X\times[0, 1]$ and so is an $\mathscr{F}_{\sigma\delta}$-set in $X\times[0, 1]$. Hence $S=\operatorname{proj}_X H$ has a representation of the required form.

Proof of Theorem 2.6.4 As the projection from $X\times Y$ to X is a continuous map from $X\times Y$ to X, the result follows from Theorem 2.5.1.

Proof of Theorem 2.6.5 Let H be a Souslin-$(\mathscr{S}\times\mathscr{F})$ set in $X\times Y$ with the representation

$$H=\bigcup_\tau\bigcap_n[S(\tau|n)\times F(\tau|n)],$$

with $S(\tau|n)\in\mathscr{S}$ and $F(\tau|n)\in\mathscr{F}$, for all $\tau\in\mathbb{N}^{\mathbb{N}}$, $n\in\mathbb{N}$. Now as Y is K-analytic we have

$$Y=K(\mathbb{N}^{\mathbb{N}}),$$

with K an upper semi-continuous map from $\mathbb{N}^{\mathbb{N}}$ to the compact sets of Y. As in the proof of Theorem 2.5.3, the map H from $\mathbb{N}^{\mathbb{N}}\times\mathbb{N}^{\mathbb{N}}$ to the compact sets of Y, defined by

$$H(\sigma, \tau)=K(\sigma)\cap\Big(\bigcap_n F(\tau|n)\Big),$$

is upper semi-continuous.

Now

$$H = \bigcup_{\tau} \bigcap_{n} [S(\tau|n) \times F(\tau|n)]$$

$$= \bigcup_{\tau} \Big(\bigcap_{n} S(\tau|n) \Big) \times \Big(\bigcap_{n} F(\tau|n) \Big)$$

$$= \bigcup_{\sigma,\tau} \Big(\bigcap_{n} S(\tau|n) \Big) \times \Big(K(\sigma) \cap \bigcap_{n} F(\tau|n) \Big)$$

$$= \bigcup_{\sigma,\tau} \Big(\bigcap_{n} S(\tau|n|) \times H(\sigma, \tau) \Big) .$$

Hence

$$\mathrm{proj}_X H = \mathrm{proj}_X \bigcup_{\sigma,\tau} \Big(\bigcap_{n} S(\tau|n) \Big) \times H(\sigma, \tau)$$

$$= \bigcup_{\sigma,\tau} \mathrm{proj}_X \Big(\bigcap_{n} S(\tau|n) \Big) \times H(\sigma, \tau)$$

$$= \bigcup_{\sigma,\tau} \Big(\bigcap_{n} S(\tau|n) \Big) \cap E(\sigma, \tau) , \tag{14}$$

where

$$E(\sigma, \tau) = \begin{cases} X, & \text{if } H(\sigma, \tau) \neq \emptyset , \\ \emptyset , & \text{if } H(\sigma, \tau) = \emptyset . \end{cases}$$

Now write

$$E(\sigma|n, \tau|n) = \begin{cases} X, & \text{if } H(I(\sigma|n) , I(\tau|n)) \neq \emptyset , \\ \emptyset , & \text{if } H(I(\sigma|n) , I(\tau|n)) = \emptyset . \end{cases}$$

If

$$\bigcap_{n=1}^{\infty} E(\sigma|n, \tau|n) = \emptyset ,$$

then one of the sets $E(\sigma|n, \tau|n)$ is empty, so that $H(\sigma, \tau) = \emptyset$ and $E(\sigma, \tau) = \emptyset$. On the other hand, if $E(\sigma, \tau) = \emptyset$, then $H(\sigma, \tau) = \emptyset$, and, by the upper semi-continuity, $H(I(\sigma|n), I(\tau|n)) = \emptyset$ for some $n \geq 1$, and consequently

$$\bigcap_{n=1}^{\infty} E(\sigma|n, \tau|n) = \emptyset .$$

Hence, for all σ, τ,

$$E(\sigma, \tau) = \bigcap_{n=1}^{\infty} E(\sigma|n, \tau|n) .$$

Substituting this in (14) we obtain

$$\text{proj}_X H = \bigcup_{\sigma,\tau} \bigcap_n (S(\tau|n) \cap E(\sigma|n, \tau|n)) .$$

As

$$S(\tau|n) \cap E(\sigma|n, \tau|n) \in \mathscr{S} ,$$

this formula provides a Souslin-$\mathscr{S}$ representation for $\text{proj}_X H$.

Proof of Theorem 2.6.6 Let H be a Souslin-$\mathscr{F}$ set in $X \times Y$ with the representation

$$H = \bigcup_{\tau} \bigcap_n F(\tau|n) ,$$

with $F(\tau|n) \in \mathscr{F}(X \times Y)$, for all $\tau \in \mathbb{N}^{\mathbb{N}}$, $n \in \mathbb{N}$. As Y is K-analytic we have

$$Y = K(\mathbb{N}^{\mathbb{N}}) ,$$

with K an upper semi-continuous map from $\mathbb{N}^{\mathbb{N}}$ to the compact sets of Y. Then the projection of H onto X takes the form

$$\begin{aligned} \text{proj}_X H &= \text{proj}_X \bigcup_{\tau} \bigcap_n F(\tau|n) \\ &= \text{proj}_X \bigcup_{\sigma} \bigcup_{\tau} \bigcap_n \{F(\tau|n) \cap (X \times K(\sigma))\} \\ &= \bigcup_{\sigma} \bigcup_{\tau} \text{proj}_X \bigcap_n \{F(\tau|n) \cap (X \times K(\sigma))\} \\ &= \bigcup_{\sigma} \bigcup_{\tau} E(\sigma, \tau) , \end{aligned}$$

with

$$E(\sigma, \tau) = \text{proj}_X \bigcap_n \{F(\tau|n) \cap A(\sigma|n)\} .$$

Write

$$J(\sigma^*, \tau^*, \sigma) = \text{cl} \bigcup_{\sigma|n=\sigma^*|n} \ \bigcup_{\tau|n=\tau^*|n} E(\sigma, \tau) .$$

Then, for each σ^*, τ^* in $\mathbb{N}^{\mathbb{N}}$, we have

$$E(\sigma^*, \tau^*) \subset \bigcap_n J(\sigma^*, \tau^*, n) .$$

We prove that, in fact,

$$E(\sigma^*, \tau^*) = \bigcap_n J(\sigma^*, \tau^*, n). \tag{15}$$

Consider any point j that is not in the projection

$$E(\sigma^*, \tau^*) = \text{proj}_X \bigcap_n \{F(\tau^*|n) \cap (X \times K(\sigma^*))\}.$$

We have

$$(\{j\} \times Y) \cap \bigcap_n \{F(\tau^*|n) \cap (X \times K(\sigma^*))\} = \varnothing.$$

Hence

$$\bigcap_n \{F(\tau^*|n) \cap (\{j\} \times K(\sigma^*))\} = \varnothing.$$

As $\{j\} \times K(\sigma^*)$ is compact, we can choose $m \geqslant 1$ so that

$$\left(\bigcap_{n=1}^m F(\tau^*|n)\right) \cap (\{j\} \times K(\sigma^*)) = \varnothing.$$

As

$$\bigcap_{n=1}^m F(\tau^*|n)$$

is closed, and $K(\sigma^*)$ is compact, we can choose a rectangle $U \times V$, with U, V open and

$$j \in U, \qquad K(\sigma^*) \subset V,$$

$$(U \times V) \cap \bigcap_{n=1}^m F(\tau^*|n) = \varnothing.$$

By the upper semi-continuity of K, we can choose $l \geqslant m$, so that

$$K(\sigma) \subset V \quad \text{if} \quad \sigma|l = \sigma^*|l.$$

Then, if $\sigma|l = \sigma^*|l$ and $\tau|l = \tau^*|l$, we have

$$K(\sigma) \subset V,$$

$$(U \times V) \cap \bigcap_{n=1}^\infty F(\tau|n) = \varnothing,$$

so that

$$U \cap E(\sigma, \tau) = U \cap \text{proj}_X \bigcap_{n=1}^\infty \{F(\tau|n) \cap (X \times K(\sigma))\} = \varnothing,$$

and

$$j \notin J(\sigma^*, \tau^*, l) = \text{cl} \bigcup_{\sigma|l=\sigma^*|l} \bigcup_{\tau|l=\tau^*|l} E(\sigma, \tau) .$$

Hence (15) holds, and $\text{proj}_X H$ has the representation

$$\text{proj}_X H = \bigcup_{\sigma} \bigcup_{\tau} \bigcap_{n} J(\sigma, \tau, n) .$$

As each set $J(\sigma, \tau, n)$ is closed and depends only on the values of $\sigma|n$, $\tau|n$, and as there is a homeomorphism of $\mathbb{N}^{\mathbb{N}} \times \mathbb{N}^{\mathbb{N}}$ to $\mathbb{N}^{\mathbb{N}}$ that maps the Baire rectangles

$$I(\sigma|n) \times I(\tau|n)$$

to Baire intervals of order n, it follows that $\text{proj}_X H$ is a Souslin-$\mathscr{F}$ set.

2.7 The Lindelöf property of K-analytic sets

In this section we prove a very simple and very useful result about K-analytic sets. Sion (1960*b*) obtains this result for his definition of K-analytic. Using the present definition the result is a direct consequence of the upper semi-continuity.

THEOREM 2.7.1 *A K-analytic set in a Hausdorff space is Lindelöf.*

Proof Let $A = K(\mathbb{N}^{\mathbb{N}})$ with K an upper semi-continuous map from $\mathbb{N}^{\mathbb{N}}$ to the compact sets of a Hausdorff space X. Let $\{G_\beta\}_{\beta \in B}$ be any open cover of A. For each σ in $\mathbb{N}^{\mathbb{N}}$ the compact set $K(\sigma)$ is covered by the family $\{G_\beta\}_{\beta \in B}$. So we can choose a finite subset $B(\sigma)$ of B so that

$$K(\sigma) \subset \bigcup_{\beta \in B(\sigma)} G_\beta .$$

By the upper semi-continuity, for each σ in $\mathbb{N}^{\mathbb{N}}$ we can choose an open set $U(\sigma)$ with

$$K(U(\sigma)) \subset \bigcup_{\beta \in B(\sigma)} G_\beta .$$

As $\mathbb{N}^{\mathbb{N}}$ is Lindelöf, we can choose a countable sequence $\sigma(1), \sigma(2), \ldots$ so that

$$\mathbb{N}^{\mathbb{N}} \subset \bigcup_{i=1}^{\infty} U(\sigma(i)) .$$

Then

$$A = K(\mathbb{N}^{\mathbb{N}}) \subset \bigcup_{i=1}^{\infty} K(U(\sigma(i)))$$

$$\subset \bigcup_{i=1}^{\infty} \bigcup_{\beta \in B(\sigma(i))} G_\beta .$$

Hence A is Lindelöf.

Remark Of course we have just proved the general result that the property of being Lindelöf is preserved under compact-valued upper semi-continuous maps.

The same proof yields

THEOREM 2.7.2 *The image in a Hausdorff space of a compact set under a compact-valued upper semi-continuous map is compact.*

2.8 Equivalence of the definitions of K-analytic sets

We have used the Frolík definition of a K-analytic set: a set that is the image in a Hausdorff space X, of $\mathbb{N}^{\mathbb{N}}$ under an upper semi-continuous map to the compact sets of X. Consider the apparently much larger class $\mathscr{A}^*$ of all sets that are the image, in a Hausdorff space X, of a Souslin-$\mathscr{K}$ set S, in a Hausdorff space Y, under an upper semi-continuous map from S to the compact sets of X. Any compact set in Y is trivially K-analytic. By Theorem 2.5.4, each Souslin-$\mathscr{K}$ set in Y is K-analytic. By Theorem 2.5.1, each set of the class $\mathscr{A}^*$ is K-analytic. Thus the extra generality is only apparent.

Choquet (1951) defined the analytic sets to be the class $\mathscr{A}_c$ of all sets that are continuous images in Hausdorff spaces of $\mathscr{F}_{\sigma\delta}$-sets in compact Hausdorff spaces. Clearly $\mathscr{A}^*$ includes $\mathscr{A}_c$. In this section we give Jayne's proof (Jayne, 1976) that all K-analytic sets, in fact, belong to the class $\mathscr{A}_c$. We first obtain Jayne's results on the graph of a K-analytic representation.

If X is a K-analytic space, with a representation $X = K(\mathbb{N}^{\mathbb{N}})$, with K an upper semi-continuous map from $\mathbb{N}^{\mathbb{N}}$ to the compact sets of X, the set

$$G = \bigcup_{\sigma} \{K(\sigma) \times \{\sigma\}\}, \tag{16}$$

regarded as a subspace of $X \times \mathbb{N}^{\mathbb{N}}$, is called the graph of the representation. We first prove some results of Jayne (*loc. cit.*) concerning this graph.

THEOREM 2.8.1 *Let G be the graph of a representation $X = K(\mathbb{N}^{\mathbb{N}})$ of a K-analytic space X. The projection π from G to $\mathbb{N}^{\mathbb{N}}$ maps closed sets to closed* *sets. Further, G is a completely regular K-analytic space and is a $\mathcal{G}_\delta$-subset and an $\mathcal{F}_{\sigma\delta}$-subset of its Stone–Čech compactification.*

COROLLARY 2.8.2 *Since the space G is Čech-complete, it is of the second category in itself.*

Proof of Theorem 2.8.1 We first show that the map, mapping σ in $\mathbb{N}^\mathbb{N}$ to the compact set

$$K(\sigma)\times\{\sigma\}$$

in $X\times\mathbb{N}^\mathbb{N}$, is upper semi-continuous. Suppose that $\sigma\in\mathbb{N}^\mathbb{N}$ and that U is an open set in $X\times\mathbb{N}^\mathbb{N}$ containing $K(\sigma)\times\{\sigma\}$. Now $K(\sigma)\times\{\sigma\}$ is compact and U is a union of open rectangles. So we can choose a finite system

$$V_r\times W_r, \qquad r=1,2,\ldots,n,$$

of open rectangles with

$$K(\sigma)\times\{\sigma\}\subset\bigcup_{r=1}^{n} V_r\times W_r\subset U, \qquad \sigma\in W_r, \qquad 1\leqslant r\leqslant n.$$

Then

$$K(\sigma)\times\{\sigma\}\subset V\times W\subset U$$

with

$$V=\bigcup_{r=1}^{n} V_r, \qquad W=\bigcap_{r=1}^{n} W_r.$$

As K is upper semi-continuous, we have

$$K(\tau)\subset V, \qquad \tau\in W,$$

and

$$K(\tau)\times\{\tau\}\subset V\times W\subset U,$$

for all τ sufficiently close to σ. Thus the map is upper semi-continuous. Hence G is K-analytic.

Now, if F is closed in $X\times\mathbb{N}^\mathbb{N}$, the set of σ in $\mathbb{N}^\mathbb{N}$, for which $\pi^{-1}(\sigma)=K(\sigma)\times\{\sigma\}$ lies in the open complement of F, is itself open. Hence $\pi(F\cap G)$ is closed, and π maps closed sets to closed sets.

We next show that G is a regular space. Let g be any point of G and let F be a closed set of $X\times\mathbb{N}^\mathbb{N}$ with $g\notin F$. Write $g=(h,\gamma)$ with $h\in X$, $\gamma\in\mathbb{N}^\mathbb{N}$, and let F_γ be the set of points x of X with $(x,\gamma)\in F$. Then F_γ is closed in X and h does not belong to the compact set

$$K(\gamma)\cap F_\gamma.$$

As X is Hausdorff we can choose open sets U and V of X with

$$h \in U, \qquad K(\gamma) \cap F_\gamma \subset V.$$

Now

$$K(\gamma) \times \{\gamma\} \subset (V \times \mathbb{N}^{\mathbb{N}}) \cup [(X \times \mathbb{N}^{\mathbb{N}}) \backslash F].$$

As the map $\sigma \to K(\sigma) \times \{\sigma\}$ is upper semi-continuous, we can choose a Baire interval Γ in $\mathbb{N}^{\mathbb{N}}$, with

$$\gamma \in \Gamma$$

and

$$\bigcup_{\sigma \in \Gamma} K(\sigma) \times \{\sigma\} \subset (V \times \mathbb{N}^{\mathbb{N}}) \cup [(X \times \mathbb{N}^{\mathbb{N}}) \backslash F].$$

As U and V are disjoint, we have

$$g = (h, \gamma) \in U \times \Gamma,$$
$$G \cap F \subset (V \times \Gamma) \cup (X \times (\mathbb{N}^{\mathbb{N}} \backslash \Gamma)).$$

As Γ is both open and closed in $\mathbb{N}^{\mathbb{N}}$, the sets

$$U \times \Gamma \quad \text{and} \quad (V \times \Gamma) \cup (X \times (\mathbb{N}^{\mathbb{N}} \backslash \Gamma))$$

are disjoint open sets in $X \times \mathbb{N}^{\mathbb{N}}$. Hence the graph G is a regular space, when regarded as a subspace of $X \times \mathbb{N}^{\mathbb{N}}$.

Since G is K-analytic, it is Lindelöf as well as regular, and so is normal and completely regular (see, for example, Kelley, 1955, pp. 112–115).

We now identify $\mathbb{N}^{\mathbb{N}}$ with the space I of irrational numbers between 0 and 1, regarded as a subspace of the real unit interval [0, 1]. Let π now denote the projection mapping each point (h, γ) of G to the point γ, regarded as a real number in [0, 1]. As G is completely regular, the bounded continuous real-valued function $\hat{\pi}$ has a continuous extension π to the Stone–Čech compactification $\beta(G)$ of G.

Consider any point e of $\beta(G) \backslash G$ and any point σ of I. Then

$$e \notin G \cap \hat{\pi}^{-1}(\sigma) = \pi^{-1}(\sigma) = K(\sigma) \times \{\sigma\}.$$

As $K(\sigma) \times \{\sigma\}$ is a compact set of G, we can choose a closed neighbourhood F of e in $\beta(G)$ that does not meet $G \cap \hat{\pi}^{-1}(\sigma)$. Then $\sigma \notin \pi(F \cap G)$. As F is a closed neighbourhood of e in $\beta(G)$, the point e belongs to the closure in $\beta(G)$ of $F \cap G$, and $\hat{\pi}(e) \in \hat{\pi}(\mathrm{cl}\, F \cap G)$. So $\hat{\pi}(e)$ belongs to the closure in [0, 1] of

$$\hat{\pi}(F \cap G) = \pi(F \cap G).$$

As π maps closed sets of G to closed sets of I, the point $\hat{\pi}(e)$ belongs to $\pi(F \cap G)$ or to $[0, 1]\backslash I$. As σ belongs to I but not to $\pi(F \cap G)$ we conclude that $\hat{\pi}(e) \neq \sigma$. Hence $\hat{\pi}$ maps no point of $\beta(G)\backslash G$ to any point of I. Thus

$$G = \pi^{-1}(I) = \hat{\pi}^{-1}(I) .$$

As I is a $\mathscr{G}_\delta$-set in $[0, 1]$, it follows that G is a $\mathscr{G}_\delta$-set in $\beta(G)$. As I is an $\mathscr{F}_{\sigma\delta}$-set in $[0, 1]$, it follows that G is an $\mathscr{F}_{\sigma\delta}$-set in $\beta(G)$. This completes the proof.

Proof of Corollary 2.8.2 As π is a continuous closed map and the inverse image of each point of $\mathbb{N}^{\mathbb{N}}$ is compact, this map π is a proper map. As G is a completely regular space, that is a $\mathscr{G}_\delta$-set in its Stone–Čech compactification, G is Čech-complete. The standard proof of the Baire category theorem applies and shows that G is of the second category in itself (see, for example, Engelking, 1977, pp. 251–253).

THEOREM 2.8.3 *A K-analytic set in a Hausdorff space is the continuous image of some $\mathscr{F}_{\sigma\delta}$-set in some compact Hausdorff space.*

Proof By Theorem 2.8.1, the K-analytic set X is the continuous image of the graph G of one of its representations, and G is an $\mathscr{F}_{\sigma\delta}$-set in the compact Hausdorff space βG.

We use $\mathscr{S}_+$ to denote the family of all sets that are countable disjoint unions of sets of the family $\mathscr{S}$. Sierpiński (1925) shows that the irrationals in $[0, 1]$ form an $\mathscr{F}_{+\delta+\delta}$-set in $[0, 1]$, see Theorem 5.7.1. Jayne's method now yields

THEOREM 2.8.4 *Any K-Lusin set in a Hausdorff space is a continuous injective image of an $\mathscr{F}_{+\delta+\delta}$-set in a compact Hausdorff space.*

Proof Such a set X has a representation

$$X = K(\mathbb{N}^{\mathbb{N}}) ,$$

with K an upper semi-continuous map from $\mathbb{N}^{\mathbb{N}}$ to the compact sets of the Hausdorff space X, with

$$K(\sigma) \cap K(\tau) = \varnothing , \tag{17}$$

whenever σ, τ are distinct points of $\mathbb{N}^{\mathbb{N}}$. By the proof of Theorem 2.8.1, the graph G is a K-analytic completely regular space and

$$G = \hat{\pi}^{-1}(I) .$$

As I is an $\mathscr{F}_{+\delta+\delta}$-set in $[0, 1]$, the set G is an $\mathscr{F}_{+\delta+\delta}$-set in $\beta(G)$. The condition (17) ensures that the natural projection from G to X is injective. Hence X is the continuous injective image of the $\mathscr{F}_{+\delta+\delta}$-set G in the compact Hausdorff space $\beta(G)$.

2.9 Measurable sets and sets with the Baire property

Consider an outer measure μ defined, with non-negative values, which may be $+\infty$, on all sets of a space X. The sets of μ-measure zero form what we shall call a "null class"; they are a class $\mathcal{N}$ of sets with the properties

(a) if N belongs to $\mathcal{N}$ and S is any subset of N, then S belongs to $\mathcal{N}$;
(b) if $N_1, N_2, \ldots$ is a sequence of sets of $\mathcal{N}$, then

$$\bigcup_{i=1}^{\infty} N_i$$

belongs to $\mathcal{N}$.

The class $\mathcal{M}$ of μ-measurable sets, defined by use of Carathéodory's criterion, is closed under the operations of countable union and countable intersection, and contains the complements of all sets of measure zero. Of course $\mathcal{M}$ is also closed under the taking of complements; but we shall not need to use this fact. If the measure μ is a regular σ-finite measure, then given any set S in X there is a set M in $\mathcal{M}$ with

$$S \subset M,$$

and such that $M \backslash M'$ is of measure zero whenever M' in $\mathcal{M}$ satisfies

$$S \subset M' \subset M.$$

The reader familiar with measure theory, treated from the outer measure point of view, will be familiar with these properties, but will feel that they constitute a perverse summary of some of the elementary results. We have stated the properties in this way so that we can discuss simultaneously the class of μ-measurable sets, and the class of sets with the Baire property.

A set in a topological space is said to be of the first category if it can be expressed as a countable union of nowhere dense sets. Clearly the class $\mathcal{N}$ of sets of the first category satisfies the requirements for a null class. A set M is said to have the Baire property, if there is an open set G that approximates M, in that, the symmetric difference

$$M \triangle G = (M \backslash G) \cup (G \backslash M)$$

is of the first category. It is clear that the countable union of a sequence of sets with the Baire property has the Baire property. For any open set G the set

$$\mathrm{cl}\, G \backslash G$$

is automatically nowhere dense. Hence, if M has the Baire property there is an open G with $M \triangle G$ of the first category, and the formula

$$\begin{aligned}(X\backslash M)\triangle \operatorname{int}(X\backslash G) &= [(X\backslash M)\backslash \operatorname{int}(X\backslash G)] \cup [\operatorname{int}(X\backslash G)\backslash(X\backslash M)] \\ &= [(X\backslash M) \cap \mathrm{cl}\, G] \cup [(\operatorname{int}(X\backslash G)) \cap M] \\ &\subset (G\backslash M) \cup [(\mathrm{cl}\, G)\backslash G] \cup (M\backslash G),\end{aligned}$$

shows that $X\backslash M$ has the Baire property. Thus the class $\mathcal{M}$ of sets with the Baire property is closed under countable intersections as well as under countable unions. Further $\mathcal{M}$ contains the complements of all first-category sets.

At the end of this section we shall prove

LEMMA 2.9.1 *Let A be any set in a topological space X. Then there is a set M with the Baire property, with $A \subset M$, and such that: if M' has the Baire property and $A \subset M' \subset M$, then $M\backslash M'$ is of the first category.*

We see that the first-category sets form a null class, and the sets with the Baire property form a class with just the properties that we have discussed in connection with regular σ-finite measures. We can now state our main theorem, due to Szpilrajn-Marczewski (1929, 1933), that will show that both the class of measurable sets, for such a measure, and the class of sets with the Baire property, are closed under the Souslin operation.

THEOREM 2.9.2 *Let $\mathcal{M}$ be a class of sets that is closed under countable unions and countable intersections, and that contains the complements of the sets of a null class $\mathcal{N}$. Suppose that, for each set S, there is a set M in $\mathcal{M}$ with $S \subset M$ and such that, if M' in $\mathcal{M}$ satisfies $S \subset M' \subset M$, then $M\backslash M' \in \mathcal{N}$. Then $\mathcal{M}$ is closed under the Souslin operation.*

COROLLARY 2.9.3 *Let μ be a σ-finite regular measure. Then the class of μ-measurable sets is closed under the Souslin operation.*

COROLLARY 2.9.4 *The class of subsets of a topological space that have the Baire property is closed under the Souslin operation.*

The assumptions that the measure μ is σ-finite and regular are not necessary. For proofs of more general results, see S. Saks (1937), E. J. Mickle (1958) or C. A. Rogers (1970*b*). The result for the Baire property is due to O. Nikodym (1925).

Proof of Theorem 2.9.2 Consider any set of the form

$$A = \bigcup_{\sigma} \bigcap_{n} B(\sigma|n),$$

with each set

$$B(\sigma|n), \qquad \sigma \in \mathbb{N}^{\mathbb{N}}, \qquad n \geq 1,$$

a member of $\mathcal{M}$. For each finite sequence s of $\mathbb{N}^{(\mathbb{N})}$ of length $l(s) \geq 0$, write

$$A(s) = \bigcup_{\substack{\tau \\ \tau|l(s)=s}} \bigcap_{n=1}^{\infty} B(\tau|n),$$

and choose a set $C(s)$ of $\mathcal{M}$ approximating $A(s)$, in that,

$$A(s) \subset C(s),$$

and, if M' in $\mathcal{M}$ satisfies

$$A(s) \subset M' \subset C(s),$$

then $M \backslash M' \in \mathcal{N}$. Write

$$D(s) = \bigcap_{r=0}^{l(s)} [B(s|r) \cap C(s|r)], \tag{18}$$

with the convention

$$B(\emptyset) = X.$$

Then, for each s in $\mathbb{N}^{(\mathbb{N})}$, the set $D(s)$ belongs to $\mathcal{M}$, satisfies

$$A(s) \subset D(s) \subset C(s),$$

and approximates $A(s)$ in the above sense.

For each s in $\mathbb{N}^{(\mathbb{N})}$, the set

$$\bigcup_{t=1}^{\infty} D(s, t)$$

belongs to $\mathcal{M}$, and satisfies

$$A(s) = \bigcup_{t=1}^{\infty} A(s, t) \subset \bigcup_{t=1}^{\infty} D(s, t) \subset D(s).$$

As $D(s)$ approximates $A(s)$, this ensures that, for each s in $\mathbb{N}^{(\mathbb{N})}$,

$$N(s) = D(s) \setminus \bigcup_{t=1}^{\infty} D(s, t) \in \mathcal{N}.$$

Hence

$$L = \bigcup_{s \in \mathbb{N}^{(\mathbb{N})}} N(s) \in \mathcal{N}.$$

Write

$$D = D(\varnothing) = C(\varnothing).$$

We prove that $D \setminus L \subset A$. Let x be any point of D that is not in L. As

$$x \in D(\varnothing), \quad \text{but} \quad x \notin D(\varnothing) \setminus \bigcup_{t=1}^{\infty} D(t),$$

we can choose t_1 so that

$$x \in D(t_1),$$

but then

$$x \notin D(t_1) \setminus \bigcup_{t=1}^{\infty} D(t_1, t).$$

Proceeding inductively, we can choose $t_1, t_2, t_3, \ldots$, so that

$$x \in D(t_1, t_2, \ldots, t_n), \qquad n \geq 1.$$

Then, taking $\tau = t_1, t_2, t_3, \ldots$, we have

$$x \in D(\tau|n) \subset B(\tau|n),$$

for all $n \geq 1$. Hence $x \in A$. So $D \setminus L \subset A$ and

$$D \setminus A \subset L \in \mathcal{N}.$$

As $\mathcal{N}$ is a null class, $D \setminus A \in \mathcal{N}$ and

$$X \setminus (D \setminus A) \in \mathcal{M}.$$

Now $A = A(\varnothing) \subset C(\varnothing) = D$, so that

$$A = D \cap (X \setminus (D \setminus A)) \in \mathcal{M}.$$

Thus $\mathcal{M}$ is closed under the Souslin operation.

Proof of Lemma 2.9.1 Let $\mathcal{P}$ be the class of all open sets P for which $A \cap P$ is the first category. Write

$$E(A) = \bigcup_{P \in \mathcal{P}} P, \qquad D(A) = X \setminus E(A).$$

Then $E(A)$ is open and $D(A)$ is closed. We prove that $A \backslash D(A)$ is a set of the first category. By Zorn's lemma we can choose a disjoint sub-family $\{P_\beta\}_{\beta \in B}$ of $\mathscr{P}$ that is maximal amongst such disjoint sub-families. Consider the set

$$Q = E(A) \Big\backslash \left(\bigcup_{\beta \in B} P_\beta \right) .$$

If Q were dense in a non-empty open set U, then the set

$$U \cap E(A) = U \cap \left(\bigcup_{P \in \mathscr{P}} P \right)$$

would be non-empty, and Q would be dense in this open subset of U. So, for some P in $\mathscr{P}$, the set $U \cap P$ would be non-empty, and Q would be dense in $U \cap P$. As

$$U \cap P \subset E(A) ,$$

and Q is relatively closed in $E(A)$, Q must contain $U \cap P$. Since $A \cap P$ is of the first category, $A \cap U \cap P$ is of the first category and $U \cap P$ belongs to P. But $U \cap P$, being a subset of Q, is disjoint from the family $\{P_\beta\}_{\beta \in B}$. This contradicts the maximality of this family. We conclude that Q is nowhere dense.

We now show that the set

$$A \cap \left(\bigcup_{\beta \in B} P_\beta \right)$$

is of the first category. For each β in B, the set $A \cap P_\beta$ is of the first category, and so

$$A \cap P_\beta = \bigcup_{i=1}^{\infty} N_\beta^{(i)} ,$$

for $N_\beta^{(1)}, N_\beta^{(2)}, \ldots$ suitable nowhere dense sets. Write

$$N^{(i)} = \bigcup_{\beta \in B} N_\beta^{(i)} , \qquad i \geqslant 1 .$$

Then

$$A \cap \left(\bigcup_{\beta \in B} P_\beta \right) = \bigcup_{i=1}^{\infty} N^{(i)} .$$

Further, as

$$N_\beta^{(i)} \subset P_\beta ,$$

and the open sets $\{P_\beta\}_{\beta\in B}$ are disjoint, it follows easily that each set $N^{(i)}$, $i\geqslant 1$, is nowhere dense. Thus

$$A\cap\Big(\bigcup_{\beta\in B} P_\beta\Big)$$

is a first-category set.

Since

$$A\backslash D(A)=A\cap E(A)\subset Q\cup\Big[A\cap\Big(\bigcup_{\beta\in B} P_\beta\Big)\Big],$$

the set $A\backslash D(A)$ is of the first category. So $A\backslash D(A)$ is a countable union of nowhere dense sets. Passing to the closures of these nowhere dense sets, $A\backslash D(A)$ is contained in an $\mathscr{F}_\sigma$-set, F_0 say, of the first category.

Now the set

$$M=F_0\cup D(A)$$

is an $\mathscr{F}_\sigma$-set. So M has the Baire property. Further,

$$A\subset(A\backslash D(A))\cup D(A)\subset M.$$

We show that M is the required approximation to A.

Consider any set M', with the Baire property, satisfying $A\subset M'\subset M$. Then defining $D(M')$ from M', just as $D(A)$ is obtained from A, we easily see that $A\subset M'$ implies $D(A)\subset D(M')$. Hence

$$M\backslash M'=(F_0\backslash M')\cup(D(A)\backslash M')\subset F_0\cup(D(M')\backslash M').$$

Now F_0 is of the first category. So $M\backslash M'$ will be of the first category, provided we can show that $D(M')\backslash M'$ is of the first category. As M' has the Baire property, there is an open set G' with

$$(M'\backslash G')\cup(G'\backslash M')$$

of the first category. Writing

$$F'=\operatorname{cl} G'$$

we have

$$M'\cap(X\backslash F')=M'\backslash\operatorname{cl} G'\subset M'\backslash G',$$

and $M'\cup(X\backslash F')$ is of the first category. So $X\backslash F'$ is one of the open sets whose union is $E(M')$, and

$$X\backslash F'\subset E(M')\quad\text{and}\quad D(M')\subset F'.$$

Hence

$$\begin{aligned} D(M')\backslash M' &\subset F'\backslash M' \\ &= \text{cl } G'\backslash M' \\ &\subset (\text{cl } G'\backslash G') \cup (G'\backslash M'), \end{aligned}$$

and $D(M')\backslash M'$ is of the first category. Thus $M\backslash M'$ is of the first category, as required.

This proof is essentially the standard one; it will be found, for example in Kuratowski (1966), pp. 80–90, embedded in an account of the general theory.

COROLLARY 2.9.5 *If A is any set in the topological space X, the set*

$$A\backslash \text{int } D(A)$$

is of the first category in X.

Proof We have shown that the sets $A\backslash D(A)$ and $D(A)\backslash \text{int } D(A)$ are of the first category in X. The result follows as $A\backslash \text{int } D(A)$ is contained in the union of these two sets.

2.10 The Souslin graph theorem

In this section we prove a simple version of a recent generalization of the closed graph theorem. The main generalization is due to L. Schwartz (1966). For more detailed accounts see M. De Wilde (1978) and F. Treves (1967). Our treatment is close to that of Frolík (1968). The result is one of the simplest "automatic continuity" results of the type discussed by Hoffmann–Jørgensen in Part 3, Section 2 of the present volume.

THEOREM 2.10.1 *Let X and Y be Hausdorff topological vector spaces, and let φ be a linear map of X into Y. Suppose that*

(a) *X is of the second category;*
(b) *Y is K-analytic; and*
(c) *the graph of φ is a Souslin-$\mathscr{F}$ set in $X \times Y$.*

Then φ is continuous.

We first prove two lemmas.

LEMMA 2.10.2 *Let U be a neighbourhood of the origin o of a topological vector space X. Then there is a closed neighbourhood V of o with*

$$V-V\subset U.$$

Proof As X is a topological vector space, the function $(x, y)\mapsto x-y$ is continuous from $X\times X$ to X. So there is a neighbourhood N of (o, o) in $X\times X$ with $x-y\in U$ whenever $(x, y)\in N$. So there are open sets P, Q of X containing o, with

$$(o, o)\in P\times Q\subset N.$$

Write $R=P\cap Q$. Then R is open and

$$o\in R, \qquad R-R\subset U.$$

Repeating the argument we can choose an open set S with

$$o\in S, \qquad S-S\subset R.$$

Consider the set

$$V=\text{cl}\, S.$$

Clearly V is a closed neighbourhood of o. If v is any point of V, as S is open and contains o, the set $v+S$ meets S. So we can choose s_1, s_2 in S with

$$s_1=v+s_2.$$

Now

$$v=s_1-s_2\in S-s.$$

Hence

$$V\subset S-S\subset R,$$

and

$$V-V\subset R-R\subset U.$$

Thus V satisfies our requirements.

LEMMA 2.10.3 *Let Z be a second category set in a Hausdorff topological vector space X. If Z has the Baire property, then $Z-Z$ contains o in its interior.*

Proof As Z has the Baire property, we can choose an open set G with

$$Z\triangle G=(Z\backslash G)\cup(G\backslash Z)$$

a set of the first category. As Z is of the second category and

$$Z \subset G \cup (Z \triangle G),$$

it follows that G is of the second category and so is not empty. Now

$$H = G - G$$

is open and contains o.

Consider any point h of H. We have

$$h = g_1 - g_2$$

with $g_1 \in G$, $g_2 \in G$. So

$$(-g_1 + G) \cap (-g_2 + G)$$

is an open set containing o, and

$$Z \subset X = \bigcup_{n=1}^{\infty} n[(-g_1 + G) \cap (-g_2 + G)].$$

As Z is of the second category, the set

$$(-g_1 + G) \cap (-g_2 + G)$$

must also be of the second category. But this set differs from the set

$$(-g_1 + Z) \cap (-g_2 + Z)$$

by a set of the first category. Hence $(-g_1 + Z) \cap (-g_2 + Z)$ is of the second category and is non-empty. So we can find z_1 and z_2 in Z with

$$-g_1 + z_1 = -g_2 + z_2.$$

Hence

$$h = g_1 - g_2 = z_1 - z_2.$$

As h may be any point of H,

$$H \subset Z - Z,$$

and o is an interior point of $Z - Z$.

Proof of Theorem 2.10.1 As φ is linear, the continuity will follow, provided we can prove that, if U is any neighbourhood of o in Y, then $\varphi^{-1}(U)$ is a neighbourhood of o in X. By Lemma 2.10.2 we can choose a closed neighbourhood V of o in Y with

$$V - V \subset U.$$

Let G be the graph of φ in $X\times Y$, and suppose that G is a Souslin-$\mathscr{F}$ set in $X\times Y$. As V is closed in Y,

$$G\cap(X\times V)$$

is a Souslin-$\mathscr{F}$ set in $X\times Y$. It follows, by Theorem 2.6.6, that the projection of $G\cap(X\times V)$ onto X is a Souslin-$\mathscr{F}$ set in X. This projection is just $\varphi^{-1}(V)$. By Theorem 2.9.2, this set $\varphi^{-1}(V)$ has the Baire property. As o is an interior point of V in Y,

$$\bigcup_{n=1}^{\infty} nV=Y.$$

Hence

$$\bigcup_{n=1}^{\infty} n\varphi^{-1}(V)=X,$$

and $\varphi^{-1}(V)$ is necessarily of the second category. Taking $Z=\varphi^{-1}(V)$ in Lemma 2.10.3, shows that

$$Z-Z=\varphi^{-1}(V)-\varphi^{-1}(V)=\varphi^{-1}(V-V)\subset\varphi^{-1}(U)$$

is a neighbourhood of o in X. Hence $\varphi^{-1}(U)$ is a neighbourhood of o, as required.

Essentially the same proof yields

THEOREM 2.10.4 *Let X and Y be Hausdorff topological groups, and let φ be a homomorphism of X into Y. Suppose that*

(a) *X is of the second category;*
(b) *Y is K-analytic; and*
(c) *the graph of φ is a Souslin-$\mathscr{F}$ set in $X\times Y$.*

Then φ is continuous.

2.11 Banach spaces that are K-analytic in their weak topology

In this section we give brief indications of the way that the theory of K-analytic sets is being introduced into the theory of Banach spaces. The theory of weakly compactly generated Banach spaces has been studied for a number of years (see, for example, J. Diestel, 1975). A Banach space X is said to be weakly compactly generated, if X is the norm-closure of the linear span of some weakly compact set. Note that for convex sets in a Banach space, the norm-closure coincides with the weak-closure. If X is a separable

Banach space, there is a norm-dense countable sequence r_n, $n=1, 2, \ldots$, of non-zero points. It is easy to verify that the set

$$C=\{0\}\cup\bigcup_{n=1}^{\infty}\left\{\frac{1}{n\|r_n\|}r_n\right\}$$

is norm-compact and weak-compact in X. Further, the norm-closure of the linear span of C is X. Thus X is weakly compactly generated. If X is a reflexive Banach space, then X, in its natural embedding in its second dual X^{**}, coincides with X^{**}. The unit ball B^{**} in X^{**} is automatically compact in the weak-star topology on X^{**}. As $X=X^{**}$ the weak-star topology of X^{**} coincides with the weak topology in X. Hence B^{**} is compact in the weak topology of X. So

$$X=\bigcup_{n=1}^{\infty} nB^{**}$$

is a $\mathcal{K}_\sigma$-set in the weak topology of X, and X is weakly compactly generated.

Not only are the separable Banach spaces and the reflexive Banach spaces weakly compactly generated, a number of other classes of Banach spaces are weakly compactly generated.

Very recently Vašak (1976) and Talagrand (1975, 1979*a*, 1979*b*, 198?) have investigated the Banach spaces that are K-analytic in their weak topology. They find that many of the results of the theory of weakly compactly generated Banach spaces can be retained and that, in some cases, the proofs are simplified. We give Talagrand's proof that weakly compactly generated Banach spaces are K-analytic in their weak topology (modifying it slightly to avoid the need to appeal to a difficult result of Krein).

THEOREM 2.11.1 *Let X be a weakly compactly generated Banach space. Then X, in its natural embedding in X^{**}, is a $\mathcal{K}_{\sigma\delta}$-set of X^{**} in its weak-star topology. Hence X is a K-analytic set in its own weak topology.*

Proof Let C be a weakly compact set in X whose linear span is norm-dense in X. We regard X as naturally embedded in its second dual X^{**}. Then C, being weakly compact in X, is weak-star compact in X^{**}.

Let B and B^{**} be the unit balls of X and X^{**}. Then $B=X\cap B^{**}$. Further, B^{**} is weak-star compact in X^{**}.

For each finite sequence $r_1, r_2, \ldots, r_k$, the set

$$C(r_1, r_2, \ldots, r_k)=r_1C+r_2C+\cdots+r_kC$$

is weakly compact in X and weak-star compact in X^{**}. Let $C^{(n)}$, $n=1, 2, \ldots$, be an enumeration of the sets $C(r_1, r_2, \ldots, r_k)$ with $r_1, r_2, \ldots, r_k$ all rational. As the linear span of C is norm-dense in X we have

$$X \subset \left(\bigcup_{n=1}^{\infty} C^{(n)}\right) + \frac{1}{p} B$$
$$\subset \left(\bigcup_{n=1}^{\infty} C^{(n)}\right) + \frac{1}{p} B^{**}.$$

Thus

$$X \subset \bigcap_{p=1}^{\infty} \bigcup_{n=1}^{\infty} \left(C^{(n)} + \frac{1}{p} B^{**}\right).$$

Consider any point y of

$$\bigcap_{p=1}^{\infty} \bigcup_{n=1}^{\infty} \left(C^{(n)} + \frac{1}{p} B^{**}\right).$$

Then, for each $p \geqslant 1$, there is a point x_p of

$$\bigcup_{n=1}^{\infty} C^{(n)} \subset X,$$

with

$$\|x_p - y\| \leqslant \frac{1}{p}.$$

Hence $\{x_p\}$ is a Cauchy sequence for the norm on X^{**} and so for the norm on X. So the sequence converges to a point of X, and $y \in X$. Thus

$$X = \bigcap_{p=1}^{\infty} \bigcup_{n=1}^{\infty} \left(C^{(n)} + \frac{1}{p} B^{**}\right).$$

This shows that X is a $\mathcal{K}_{\sigma\delta}$-set in the weak-star topology of X^{**}. Hence X is K-analytic in the weak-star topology on X^{**}. Now $X = K(\mathbb{N}^{\mathbb{N}})$ with K a weak-star upper semi-continuous map to the weak-star compact sets of X^{**}. But the weak topology on X coincides with the relativization of the weak-star topology on X^{**} to X. Hence X is K-analytic for the weak topology on X.

The same arguments prove:

THEOREM 2.11.2 *A Banach space X is K-analytic in its weak topology, if, and only if, it is, in its natural embedding in its second dual X^{**}, a Souslin-$\mathcal{K}$ set in X^{**} in the weak-star topology on X^{**}.*

When a Banach space X is separable it is clearly Lusin in its weak topology, and so K-Lusin in its weak topology. It is somewhat surprising that the converse holds. Namely

THEOREM 2.11.3 *If a Banach space X is K-Lusin in its weak topology, then X is separable.*

Proof Let X_w denote X with its weak topology and let

$$K:\mathbb{N}^{\mathbb{N}}\to\mathscr{K}(X_w)\,,\qquad K(\mathbb{N}^{\mathbb{N}})=X_w$$

be a disjoint upper semi-continuous map. For some $\tau\in\mathbb{N}^{\mathbb{N}}$ we must have the origin $o\in K(\tau)$, and that

$$X_w\backslash K(\tau)=K(\mathbb{N}^{\mathbb{N}}\backslash\{\tau\})$$

is K-Lusin. Thus, as a Lindelöf open subset of the completely regular space X_w, $X_w\backslash K(\tau)$ is the countable union of closed sets. Hence $K(\tau)$ is a compact $\mathscr{G}_\delta$-subset of X_w.

Since weakly compact sets are norm bounded, this implies that the origin o is a $\mathscr{G}_\delta$-set in X_w. By the definition of the weak topology, this implies that a countable set of continuous linear functionals on X_w separate the points of X_w. By Theorem 5.1.7, below, we get that X_w is then the continuous image of $\mathbb{N}^{\mathbb{N}}$. Thus X_w has a countable dense set D. The countable set of all convex sums of elements in D with rational coefficients will be a dense subset of X; i.e. X is separable.

3 The first separation theorem and approximation from within

3.1 Introduction

The K-analytic spaces are a common generalization of the Polish spaces and of the compact Hausdorff spaces. In a compact *Hausdorff space every decreasing sequence of non-empty closed sets has a non-empty intersection*; in a Polish space we have the same conclusion, *provided the diameters of the sets tend to zero.* In a K-analytic space we have a rather more complicated result.

LEMMA 3.1.1 *Let X be a K-analytic space of the form $X=K(\mathbb{N}^{\mathbb{N}})$, with K an upper semi-continuous map from $\mathbb{N}^{\mathbb{N}}$ to the compact sets of X. Let $\{F_r\}$ be a decreasing sequence of closed sets in X. Suppose that for some σ in $\mathbb{N}^{\mathbb{N}}$,*

$$F_r\cap K(I(\sigma|r))\neq\varnothing\,, \tag{19}$$

for each $r \geqslant 1$. *Then*

$$\left(\bigcap_{r=1}^{\infty} F_r\right) \cap K(\sigma) \neq \varnothing .$$

Proof Suppose that

$$\left(\bigcap_{r=1}^{\infty} F_r\right) \cap K(\sigma) = \varnothing .$$

Then the decreasing sequence $\{F_r \cap K(\sigma)\}$ of compact sets has empty intersection. So, for some $r \geqslant 1$,

$$F_r \cap K(\sigma) = \varnothing ,$$

and $K(\sigma)$ is contained in the open set $X \backslash F_r$. By the upper semi-continuity of K, there is an $s \geqslant 1$, with

$$K(I(\sigma|s)) \subset X \backslash F_r .$$

Taking $t = \max(r, s)$,

$$F_t \cap K(I(\sigma|t)) = \varnothing ,$$

contrary to (19). This proves the lemma.

We shall use this idea to prove the first separation theorem and to show that K-analytic sets can be approximated from within by compact sets. We discuss also variants of the first separation theorem, including a result of Novikov, and give some applications.

3.2 Notation and definitions

If $\mathscr{S}$ is any family of sets in a space X, the smallest family of sets of X, containing $\mathscr{S}$ and closed under the operations of countable union and complementation, will be called the family of Borel-$\mathscr{S}$ sets and denoted by $\mathscr{S}_{(\sigma, c)}$. The smallest family of sets of X, containing $\mathscr{S}$ and closed under the operations of countable union and countable intersection, will be called the family of Borelian-$\mathscr{S}$ sets and denoted by $\mathscr{S}_{(\sigma, \delta)}$. When X is a topological space,

$$\mathscr{G}_{(\sigma, c)} = \mathscr{F}_{(\sigma, c)}$$

and this family is just the family of Borel sets in X. In a Polish space the families

$$\mathscr{G}_{(\sigma, c)} = \mathscr{F}_{(\sigma, c)} , \qquad \mathscr{F}_{(\sigma, \delta)} , \qquad \mathscr{G}_{(\sigma, \delta)}$$

all coincide; in more general spaces these three families are different, we have to work with the family most suited to our particular objectives.

If A and B are sets and $\mathscr{C}$ is a family of sets, we say that A can be separated from B by $\mathscr{C}$, when there is a set C in $\mathscr{C}$ with

$$A \subset C, \qquad C \cap B = \emptyset .$$

If A and B are sets and $\mathscr{C}$ and $\mathscr{D}$ are families of sets, we say that A and B can be separated by $\mathscr{C}$ and $\mathscr{D}$, when there are sets C in $\mathscr{C}$ and D in $\mathscr{D}$ with

$$A \subset C, \qquad C \cap D = \emptyset , \qquad B \subset D .$$

3.3 The first separation theorem

In this section we first state and prove a variant of the first separation theorem due to Frolík (1967).

THEOREM 3.3.1 *Let A be a K-analytic set and let B be a Souslin-$\mathscr{F}$ set in a Hausdorff space X. If $A \cap B = \emptyset$, then A can be separated from B by a Borelian-$\mathscr{G}$ set, i.e. there is a Borelian-$\mathscr{G}$ set, C say, with*

$$A \subset C, \qquad C \cap B = \emptyset .$$

Before giving the proof it will be convenient to have a simple lemma.

LEMMA 3.3.2 *Suppose that for each $n \geqslant 1$, the sets A_n and B can be separated by the families of sets $\mathscr{C}$ and $\mathscr{D}$. Then the sets*

$$\bigcup_{n=1}^{\infty} A_n \quad \textit{and} \quad B$$

can be separated by the families $\mathscr{C}_\sigma$ and $\mathscr{D}_\delta$.

Proof For each $n \geqslant 1$, there are sets C_n in $\mathscr{C}$ and D_n in $\mathscr{D}$ with

$$A_n \subset C_n , \qquad C_n \cap D_n = \emptyset , \qquad B \subset D_n .$$

Then, writing

$$C = \bigcup_{n=1}^{\infty} C_n , \qquad D = \bigcap_{n=1}^{\infty} D_n ,$$

we have

$$\bigcup_{n=1}^{\infty} A_n \subset C, \qquad C \cap D = \emptyset , \qquad B \subset D .$$

This is the required separation.

Proof of Theorem 3.3.1 Let the analytic set A have the representation

$$A = K(\mathbb{N}^{\mathbb{N}}),$$

with K an upper semi-continuous map from $\mathbb{N}^{\mathbb{N}}$ to the compact sets of X. Let B have the Souslin-$\mathscr{F}$ representation

$$B = \bigcup_{\sigma} \bigcap_{n} F(\sigma|n),$$

with each set $F(\sigma|n)$ closed.

For each finite sequence s of positive integers, i.e. for each s in $\mathbb{N}^{(\mathbb{N})}$, let

$$A(s) = K(I(s)),$$

$$B(s) = \bigcup_{\substack{\sigma \\ \sigma|l(s)=s}} \bigcap_{n} F(\sigma|n),$$

with the natural conventions

$$A(\varnothing) = A, \qquad B(\varnothing) = B.$$

Let us call a pair of finite sequences s, t "good", if $A(s)$ can be separated from $B(t)$ by a Borelian-$\mathscr{G}$ set. In these terms we need to show that the pair $\varnothing, \varnothing$ is good.

We suppose that the pair $\varnothing, \varnothing$ is "bad", i.e. not "good", and we seek a contradiction. Suppose that we have some bad pair s, t. If the pair s, σ and t were good, for each integer $\sigma \geqslant 1$, then, for each $\sigma \geqslant 1$, the sets $A(s, \sigma)$ and $B(t)$ could be separated by the families of Borelian-$\mathscr{G}$ sets and of Borelian-$\mathscr{F}$ sets. By Lemma 3.3.2, the sets

$$A(s) = \bigcup_{\sigma=1}^{\infty} A(s, \sigma) \quad \text{and} \quad B(t)$$

could also be separated by the families of Borelian-$\mathscr{G}$ sets and Borelian-$\mathscr{F}$ sets; and the pair s, t would be good. Thus, given a bad pair s, t in $\mathbb{N}^{(\mathbb{N})}$, we can choose $\sigma \geqslant 1$ so that s, σ and t are a bad pair. Similarly we can choose $\tau \geqslant 1$, so that s and t, τ is a bad pair.

Now, starting from the assumption that the pair $\varnothing, \varnothing$ is bad, we can choose $\sigma_1, \tau_1, \sigma_2, \tau_2, \ldots$ in turn so that the pairs

$$\sigma|n, \tau|(n-1), \qquad n \geqslant 1,$$

and

$$\sigma|n, \tau|n, \qquad n \geqslant 1,$$

are all bad.

As $A \cap B = \varnothing$, we have

$$K(\sigma) \cap \bigcap_n F(\tau|n) = \varnothing .$$

Applying Lemma 3.3.2 to the decreasing sequence

$$F_n = \bigcap_{r=1}^{n} F(\tau|r)$$

of closed sets, it follows that for some $r \geqslant 1$,

$$K(I(\sigma|r)) \cap \bigcap_{m=1}^{r} F(\tau|m) = \varnothing .$$

As

$$A(\sigma|r) = K(I(\sigma|r))$$

and

$$B(\tau|r) \subset \bigcap_{m=1}^{r} F(\tau|m) ,$$

we have

$$A(\sigma|r) \subset G, \qquad G \cap B(\tau|r) = \varnothing ,$$

with

$$G = X \setminus \bigcap_{m=1}^{\infty} F(\tau|m)$$

and so open. This is impossible as the pair $\sigma|r$, $\tau|r$ is bad. The required result follows.

A minor modification of this proof yields a multiple separation theorem.

THEOREM 3.3.3 *Let $A_1, A_2, \ldots$ be a finite or infinite sequence of disjoint K-analytic sets in a Hausdorff space X. Then there is a sequence $B_1, B_2, \ldots$ of disjoint Borelian-$\mathscr{G}$ sets with $A_i \subset B_i$ for all possible i.*

Proof First consider the case of two disjoint K-analytic sets A_1 and A_2. Let A_i have the representation

$$A_i = K_i(\mathbb{N}^{\mathbb{N}}) ,$$

with K_i an upper semi-continuous map from $\mathbb{N}^{\mathbb{N}}$ to the compact sets of X, $i = 1, 2$. For each s in $\mathbb{N}^{(\mathbb{N})}$ write

$$A_i(s) = K_i(I(s)) , \qquad i = 1, 2 .$$

Let us call a pair s, t of elements of $\mathbb{N}^{(\mathbb{N})}$ good if $A_1(s)$ and $A_2(t)$ can be separated by a pair of sets that are both Borelian-$\mathscr{G}$ sets. As in the proof of Theorem 3.3.1, we can suppose that the pair $\emptyset, \emptyset$ is bad, and choose σ, τ in $\mathbb{N}^{\mathbb{N}}$ so that the pair

$$\sigma|n \quad \text{and} \quad \tau|n$$

is bad for each $n \geq 1$. As $A_1 \cap A_2 = \emptyset$, we have

$$K_1(\sigma) \cap K_1(\tau) = \emptyset .$$

As X is Hausdorff, we can choose open sets G_1, G_2 with

$$K_1(\sigma) \subset G_1 , \qquad G_1 \subset G_2 = \emptyset , \qquad K_2(\tau) \subset G_2 .$$

By the upper semi-continuity of the maps K_1, K_2 we can choose n so large that

$$A_1(\sigma|n) \subset G_1 , \qquad G_1 \cap G_2 = \emptyset , \qquad A_2(\tau|n) \subset G_2 .$$

This is impossible as the pair $\sigma|n$ and $\tau|n$ is bad. Hence the pair $\emptyset, \emptyset$ is good rather than bad.

Applying this result to each pair A_r, A_s from the sequence, we can choose, for each $r \neq s$, Borelian-$\mathscr{F}$ sets $B_{r,s}, B_{s,r}$ with

$$A_r \subset B_{r,s} , \qquad B_{r,s} \cap B_{s,r} = \emptyset , \qquad A_s \subset B_{s,r} .$$

Now the sets

$$B_r = \bigcap_s B_{r,s}$$

form a disjoint sequence of Borelian-$\mathscr{G}$ sets, with

$$A_r \subset B_r$$

for all possible values of r.

3.4 Novikov's approximation theorem

In this section we prove a variant of a theorem of Novikov (1934). Until recently, proofs of this theorem depended on the theory of sieves, or on the use of transfinite indices; but these have been avoided by Mokobodzki (1976) and Saint-Raymond (1976*a*); our variant is due to Roy. O. Davies, J. E. Jayne, A. J. Ostaszewski and C. A. Rogers (1977).

THEOREM 3.4.1 *Let $\{A_n\}$ be a sequence of K-analytic sets in a Hausdorff space X, and suppose that*

$$A = \bigcap_{n=1}^{\infty} A_n$$

is a Borelian-$\mathscr{G}$ set. Then there is a sequence $\{B_n\}$ of Borelian-$\mathscr{G}$ sets with

$$A_n \subset B_n, \qquad n \geq 1,$$

$$\bigcap_{n=1}^{\infty} A_n = \bigcap_{n=1}^{\infty} B_n.$$

Although this theorem is often called a separation theorem, it seems more natural to regard it as an approximation of a sequence of K-analytic sets, from without, by a sequence of Borelian-$\mathscr{G}$ sets.

Proof of Theorem 3.4.1 We first reduce the general case to the case when the sequence $\{A_n\}$ has an empty intersection. The set

$$X \backslash A$$

is a Borelian-$\mathscr{F}$ set and so is a Souslin-$\mathscr{F}$ set (by Theorem 2.3.1). Hence each set

$$A_n^* = A_n \backslash A$$

is a K-analytic set (by Theorem 2.5.3). Further,

$$\bigcap_{n=1}^{\infty} A_n^* = \varnothing.$$

Suppose that we can choose a sequence $\{B_n^*\}$ of Borelian-$\mathscr{G}$ sets with

$$A_n^* \subset B_n^*,$$

$$\bigcap_{n=1}^{\infty} B_n^* = \varnothing;$$

then the Borelian-$\mathscr{G}$ sets

$$B_n = B_n^* \cup A, \qquad n \geq 1,$$

satisfy

$$A_n \subset B_n, \qquad n \geq 1,$$

$$\bigcap_{n=1}^{\infty} A_n = A = \bigcap_{n=1}^{\infty} B_n.$$

Thus the general case will follow, once we have established the special case when

$$\bigcap_{n=1}^{\infty} A_n = \varnothing.$$

We now confine our attention to this case.

Let A_n have the representation $A_n = K_n(\mathbb{N}^{\mathbb{N}})$, with K_n an upper semi-continuous map from $\mathbb{N}^{\mathbb{N}}$ to the compact sets of X. For each s in $\mathbb{N}^{(\mathbb{N})}$ and each $n \geq 1$, write

$$A_n(s) = K_n(I(s)) .$$

Call a sequence $s_1, s_2, \ldots$ in $\mathbb{N}^{(\mathbb{N})}$ good if there are Borelian-$\mathscr{G}$ sets $\{C_n\}$ with

$$A_n(s_n) \subset C_n , \qquad \bigcap_{n=1}^{\infty} C_n = \emptyset .$$

In this terminology, the theorem follows if we can prove that the sequence $\emptyset, \emptyset, \emptyset, \ldots$ is good.

We suppose that the sequence $\emptyset, \emptyset, \emptyset, \ldots$ is bad and we seek a contradiction. Consider the more general situation when we know that a sequence $s_1, s_2, s_3, \ldots$ in $\mathbb{N}^{(\mathbb{N})}$ is bad. Suppose $j \geq 1$ is given. If, for each $\sigma \geq 1$, the sequence

$$s_1, s_2, \ldots, s_{j-1}, (s_j, \sigma), s_{j+1}, \ldots \tag{20}$$

were good, we would be able to find Borelian-$\mathscr{G}$ sets $C_{i,\sigma}$, $i \geq 1$, $\sigma \geq 1$ with

$$A(s_i) \subset C_{i,\sigma} , \qquad i \neq j , \ \sigma \geq 1 ,$$

$$A(s_j, \sigma) \subset C_{j,\sigma} , \qquad \sigma \geq 1 ,$$

$$\bigcap_{i=1}^{\infty} C_{i,\sigma} = \emptyset , \qquad \sigma \geq 1 .$$

Then the sets

$$C_i = \bigcap_{\sigma=1}^{\infty} C_{i,\sigma} , \qquad i \neq j ,$$

$$C_j = \bigcup_{\sigma=1}^{\infty} C_{j,\sigma}$$

would be Borelian-$\mathscr{G}$ sets with

$$A(s_i) \subset C_i , \qquad i \geq 1 ,$$

$$\bigcap_{i=1}^{\infty} C_i = \emptyset ,$$

and the sequence $s_1, s_2, s_3, \ldots$ would be good rather than bad. Thus, given the bad sequence $s_1, s_2, s_3, \ldots$ and the integer $j \geq 1$, we can always choose $\sigma \geq 1$ to ensure that the sequence (20) is bad.

Now, starting from the bad sequence $\emptyset, \emptyset, \emptyset, \ldots$, and using a diagonal process, we can choose the elements of a sequence $\sigma^{(1)}, \sigma^{(2)}, \sigma^{(3)}, \ldots$ in $\mathbb{N}^{\mathbb{N}}$, so that, for each $n \geqslant 1$, the sequence

$$\sigma^{(1)}|n, \sigma^{(2)}|(n-1), \ldots, \sigma^{(n)}|1, \emptyset, \emptyset, \ldots \tag{21}$$

is bad. As

$$\bigcap_{n=1}^{\infty} A_n = \emptyset ,$$

we have

$$\bigcap_{n=1}^{\infty} K_n(\sigma^{(n)}) = \emptyset .$$

As the sets $K_n(\sigma^{(n)})$ are compact, we can choose $m \geqslant 1$, so that

$$\bigcap_{n=1}^{m} K_n(\sigma^{(n)}) = \emptyset . \tag{22}$$

Now, if $K_1, K_2, \ldots, K_m$ are compact sets in the Hausdorff space X with

$$\bigcap_{n=1}^{m} K_n = \emptyset ,$$

it is always possible to choose open sets $G_1, G_2, \ldots, G_m$ with

$$K_n \subset G_n , \qquad 1 \leqslant n \leqslant m ,$$

$$\bigcap_{n=1}^{m} G_n = \emptyset .$$

This is immediate if $m = 1$ or 2. Suppose that it holds for m equal to some $k \geqslant 2$, and consider the case $m = k+1$. By the case $m = k$, it is possible to choose open sets $G_1, G_2, \ldots, G_{k-1}, G'_k$ with

$$K_1 \subset G_1, K_2 \subset G_2, \ldots, K_{k-1} \subset G_{k-1}, K_k \cap K_{k+1} \subset G'_k ,$$

$$\left(\bigcap_{r=1}^{k-1} G_r\right) \cap G'_k = \emptyset .$$

As

$$(K_k|G'_k) \cap (K_{k+1}|G'_k) = \emptyset ,$$

we can choose open sets U_k, U_{k+1} with

$$K_k|G'_k \subset U_k , \qquad U_k \cap U_{k+1} = \emptyset , \qquad K_{k+1}|G'_k \subset U_{k+1} .$$

Writing $G_k = G'_k \cup U_k$ and $G_{k+1} = G'_k \cup U_{k+1}$, we verify that $K_r \subset G_r$ for $1 \leq r \leq k+1$ and that

$$\bigcap_{r=1}^{k+1} G_r = \varnothing .$$

The required result follows by induction.

Returning to the formula (22), and using the result of the last paragraph, we can choose open sets $G_1, G_2, \ldots, G_m$ with

$$K_n(\sigma^{(n)}) \subset G_n, \qquad 1 \leq n \leq m,$$

$$\bigcap_{n=1}^{m} G_n = \varnothing .$$

By the upper semi-continuity of the maps K_n, $1 \leq n \leq m$, we can choose $M \geq 1$ so that

$$A_n(\sigma^{(n)}|M) \subset G_n, \qquad 1 \leq n \leq m.$$

Write

$$G_n = X, \qquad n > m.$$

Now, provided $l \geq m + M$, we have

$$\begin{aligned}
&A_1(\sigma^{(1)}|l) \subset G_1, \\
&A_2(\sigma^{(2)}|(l-1)) \subset G_2, \\
&\vdots \\
&A_m(\sigma^{(m)}|(l-m+1)) \subset G_m, \\
&\vdots \\
&A_l(\sigma^{(l)}|l) \subset G_l, \\
&A_n(\varnothing) \subset G_n, \qquad n > l.
\end{aligned}$$

Thus the sequence (21) with $n = l$ is good rather than bad. This contradiction yields the required conclusion.

For a survey of variants and extensions of this result see the paper by Davies, Jayne, Ostaszewski and Rogers (1977); note that some of the results in this paper were found independently by J. Saint-Pierre (1977).

3.5 Approximation from within

There are many ways in which an analytic set or a K-analytic set can be approximated from within by compact sets. Some of the results are discussed by Dellacherie in Part 2, Chapitre I, of the present volume. In this section we first prove, under certain assumptions, that an uncountable K-analytic space has cardinal at least that of the continuum. Some further approximation results are stated at the end of the section.

THEOREM 3.5.1 *Let X be an uncountable K-analytic space. Suppose that each point of X can be represented as the intersection of a decreasing sequence of sets that are alternately open and closed. Then X contains a compact subset with cardinal at least (and at most) that of the continuum. Further the compact subset will admit a continuous map onto $2^{\mathbb{N}}$.*

Remarks The example of the compact Hausdorff space consisting of the ordinals up to and including ω_1, with the order topology, shows that some extra assumption, like the representation condition for the points of the space, is necessary. The condition that we have imposed on the points of X ensures that the resulting compact set is first countable; by a result of Arhangel'skiĭ (1969) the compact set will have the cardinal of the continuum. For further results see section 5.3.

Proof of Theorem 3.5.1 We first prove

(α) *if A is an uncountable K-analytic subset of X, it is possible to choose disjoint closed subsets H_0 and H_1 with $A \cap H_0$ and $A \cap H_1$ both uncountable.*

Suppose that for each point a of A there is an open set $G(a)$ with

$$a \in G(a),$$

and

$$G(a) \cap A \quad \text{countable}.$$

Then, as A is Lindelöf, A would be covered by a countable family chosen from $\{G(a)\}_{a \in A}$ and so would itself be countable. Hence we can choose a point a in A with the property that

$$A \cap G \text{ is uncountable}$$

whenever G is an open set containing a. Now write

$$\{a\} = G_1 \cap F_1 \cap G_2 \cap F_2 \cap \cdots$$

with

$$G_1 \supset F_1 \supset G_2 \supset F_2 \supset \cdots,$$

$$G_n \text{ open}, \qquad n \geq 1,$$

$$F_n \text{ closed}, \qquad n \geq 1.$$

Then

$$A = \{a\} \cup \left(\bigcup_{n=1}^{\infty} (A|G_n) \right).$$

As A is uncountable, we can choose $n \geq 1$ so that

$$A \backslash G_n$$

is uncountable. As

$$a \in A \cap G_{n+1} \subset A \cap F_{n+1},$$

the set $A \cap F_{n+1}$ is uncountable. Now the closed sets

$$H_0 = X \backslash G_n, \qquad H_1 = F_{n+1}$$

are disjoint with $A \cap H_0$ and $A \cap H_1$ both uncountable. This proves (α).

As X is K-analytic, it has a representation $X = K(\mathbb{N}^{\mathbb{N}})$ with K an upper semi-continuous map from $\mathbb{N}^{\mathbb{N}}$ to the compact sets of X. For each s in $\mathbb{N}^{(\mathbb{N})}$, write

$$A(s) = K(I(s)).$$

We note that, for each s in $\mathbb{N}^{(\mathbb{N})}$, we have

$$A(s) = \bigcup_{\sigma=1}^{\infty} A(s, \sigma).$$

Hence

(β) if $s \in \mathbb{N}^{(\mathbb{N})}$, and H is a closed set, with

$$A(s) \cap H$$

uncountable, then there is a $\sigma \geq 1$, with

$$A(s, \sigma) \cap H$$

uncountable.

We now make an inductive choice of closed sets

$$H(\eta_1), H(\eta_1, \eta_2), H(\eta_1, \eta_2, \eta_3), \ldots, \qquad \eta_1, \eta_2, \eta_3, \ldots = 0 \text{ or } 1,$$

and of positive integers

$$s(\eta_1), s(\eta_1, \eta_2), s(\eta_1, \eta_2, \eta_3), \ldots, \qquad \eta_1, \eta_2, \eta_3, \ldots = 0 \text{ or } 1.$$

We arrange that, for each infinite sequence

$$\eta = \eta_1, \eta_2, \eta_3, \ldots,$$

made up of "0"s and "1"s, the sets

$$A(\sigma(\eta)|n) \cap H(\eta|n), \qquad n \geqslant 1,$$

are all uncountable, with

$$\sigma(\eta) = s(\eta_1), s(\eta_1, \eta_2), s(\eta_1, \eta_2, \eta_3), \ldots .$$

Further,

$$H(\eta|1) \supset H(\eta|2) \supset H(\eta|3) \supset \cdots,$$

and, for fixed n, the sets

$$H(\eta_1, \eta_2, \ldots, \eta_n), \qquad \eta_1, \eta_2, \ldots, \eta_n = 0 \text{ or } 1, \tag{23}$$

are all disjoint.

The construction is initiated by using (α) with $A = X$ and choosing $H(0)$, $H(1)$ to be disjoint uncountable closed subsets of X. Then $s(0)$, $s(1)$ are chosen by use of (β) so that

$$A(s(0)) \cap H(0), \qquad A(s(1)) \cap H(1)$$

are uncountable. When the construction has proceeded as far as the sets

$$H(\eta_1, \eta_2, \ldots, \eta_n), \qquad \eta_1, \eta_2, \ldots, \eta_n = 0 \text{ or } 1,$$

and the integers

$$s(\eta_1, \eta_2, \ldots, \eta_n), \qquad \eta_1, \eta_2, \ldots, \eta_n = 0 \text{ or } 1,$$

we apply (α) to the uncountable analytic sets

$$A(s(\eta_1), \ldots, s(\eta_1, \eta_2, \ldots, \eta_n)) \cap H(\eta_1, \eta_s, \ldots, \eta_n)$$

and choose disjoint closed sets

$$H(\eta_1, \eta_2, \ldots, \eta_n, 0), \qquad H(\eta_1, \eta_2, \ldots, \eta_n, 1),$$

which we may suppose subsets of $H(\eta_1, \eta_2, \ldots, \eta_n)$, so that the sets

$$A(s(\eta_1), \ldots, s(\eta_1, \eta_2, \ldots, \eta_n)) \cap H(\eta_1, \eta_2, \ldots, \eta_n, \eta_{n+1}), \eta_{n+1} = 0 \text{ or } 1,$$

are uncountable. Then, using (β), we can choose the required positive integers

$$s(\eta_1, \eta_2, \ldots, \eta_n, \eta_{n+1}), \qquad \eta_{n+1} = 0 \text{ or } 1.$$

The construction can be effected inductively in this way.

Now, for each η in $\{0, 1\}^{\mathbb{N}}$, we have a decreasing sequence

$$H(\eta|1), H(\eta|2), H(\eta|3), \ldots, \tag{24}$$

and, for each $n \geqslant 1$,

$$H(\eta|n) \cap A(\sigma(\eta)|n) = H(\eta|n) \cap K(I(\sigma(\eta)|n))$$

is not empty. Thus, by Lemma 3.1.1, we conclude that the set

$$L(\eta) = \left(\bigcap_{n=1}^{\infty} H(\eta|n)\right) \cap K(\sigma(\eta)) \tag{25}$$

is not empty.

We study the set

$$H = \bigcup_{\eta} \left(\bigcap_{n=1}^{\infty} H(\eta|n)\right) \cap K(\sigma(\eta)) = L(\{0, 1\}^{\mathbb{N}}).$$

In the first place, as the sets (23) are disjoint, the non-empty sets (25), corresponding to different sequences η in $\{0, 1\}^{\mathbb{N}}$, are disjoint. Hence H has cardinal at least as large as that of the continuum.

We show that the map L, from $\{0, 1\}^{\mathbb{N}}$ to the non-empty compact sets of X, is upper semi-continuous. Suppose that $\eta \in \{0, 1\}^{\mathbb{N}}$ and that G is open in X with

$$L(\eta) \subset G.$$

Then

$$\left(\bigcap_{n=1}^{\infty} H(\eta|n)\right) \cap K(\sigma(\eta)) \subset G.$$

As $K(\sigma(\eta))$ is compact and (24) is a decreasing sequence of closed sets, there is an $m \geqslant 1$, with

$$H(\eta|m) \cap K(\sigma(\eta)) \subset G.$$

As $H(\eta|m)$ is closed, and K is upper semi-continuous, there is an $n \geqslant m$ with

$$H(\eta|m) \cap K(I(\sigma(\eta)|n)) \subset G.$$

As $\sigma(\eta)|n$ is determined by knowledge of the first n members of the sequence η, we have

$$L(I(\eta|n)) \subset H(\eta|n) \cap K(I(\sigma(\eta)|n)) \subset G.$$

Thus L is upper semi-continuous.

As $\{0, 1\}^{\mathbb{N}}$ is compact and L is upper semi-continuous, $H = L(\{0, 1\}^{\mathbb{N}})$ is compact, and L maps closed sets of $\{0, 1\}^{\mathbb{N}}$ to compact sets in H. As L maps distinct points of $\{0, 1\}^{\mathbb{N}}$ to disjoint subsets of H, there is a unique map p from H onto $\{0, 1\}^{\mathbb{N}}$ with

$$h \in L(p(h)),$$

for all h in H. As L maps closed sets of $\{0, 1\}^{\mathbb{N}}$ to closed sets of H, the map p is continuous. As L is upper semi-continuous, p maps closed sets of H to closed sets of $\{0, 1\}^{\mathbb{N}}$ and the inverse images of points of $\{0, 1\}^{\mathbb{N}}$ under p are compact. Hence p is a proper map from H onto $\{0, 1\}^{\mathbb{N}}$ which is, of course, just $2^{\mathbb{N}}$.

COROLLARY 3.5.2 *When X is an uncountable analytic space, it contains a closed subset homeomorphic to $2^{\mathbb{N}}$.*

Proof In this case the upper semi-continuous map K can be chosen to map into the subsets of X with at most one point. In this case L maps $\{0, 1\}^{\mathbb{N}}$ to the one point sets of X, and the proper map p becomes a homeomorphism.

We remark that J. E. Jayne and C. A. Rogers (1979), extending an early result of Hurewicz (1928) (see also section 6.3), have recently proved

THEOREM 3.5.3 *Let X be a perfectly normal K-analytic space that is not σ-compact. Then X contains a closed subset that admits a proper map onto $\mathbb{N}^{\mathbb{N}}$.*

Further generalizing a recent result of Saint-Raymond (1975), they prove

THEOREM 3.5.4 *Let A be a K-analytic set in a perfectly normal Lindelöf space X. If A is not a subset of any σ-compact set in X, then A contains a subset that is closed in X and admits a proper map onto $\mathbb{N}^{\mathbb{N}}$.*

Perhaps a more typical example of approximation from within is the following result proved by Roy O. Davies (1952).

THEOREM 3.5.5 *Let A be an analytic set in $\mathbb{R}^n$, with positive Hausdorff μ^h-measure, with h a continuous strictly increasing function, with $h(0)=0$. Then A contains compact subsets K of finite μ^h-measure with*

$$\sup_{K \subset A} \mu^h(K) = \mu^h(A) .$$

D. G. Larman (1967*a*) extended Davies' result from $\mathbb{R}^n$ to a wider class of spaces, that he calls β-spaces, and that share with $\mathbb{R}^n$ certain metrical covering properties. The results are rather tricky and technical to prove. One should not expect the proofs to be too simple as Roy O. Davies and C. A. Rogers (1969) construct a compact metric space X of infinite μ^h-measure, for a suitable continuous increasing function h, with $h(0)=0$, in a way that ensures that no subset of X has finite positive μ^h-measure. For further work on the problem, see Roy O. Davies (1970).

There are similar results when one starts with an analytic set A of non-σ-finite μ^h-measure and seeks within A either a compact subset of non-σ-finite μ^h-measure, or a family of $2^{\aleph_0}$ disjoint closed subsets each of positive μ^h-measure. See Roy O. Davies (1956, 1968) and D. G. Larman (1967*b*).

One of the best known results on approximation from within is, of course, Choquet's capacitability theorem (1953), which is treated by Dellacherie in Part 2 of this volume. A Choquet capacity, which is slightly different from the capacities considered by Dellacherie, on a Hausdorff space X is a map

$$I:\mathcal{P}(X)\to\mathbb{R}\cup\{\pm\infty\},$$

where $\mathcal{P}(X)$ denotes the set of all subsets of X, such that

(i) if $A\subset B$, then $I(A)\leqslant I(B)$;
(ii) if $X_1\subset X_2\subset\cdots$, then

$$\lim_{n\to\infty} I(X_n)=I\left(\bigcup_{n=1}^{\infty} X_n\right);$$

(iii) if $K_1\supset K_2\supset\cdots$ are compact sets, then

$$\lim_{n\to\infty} I(K_n)=I\left(\bigcap_n K_n\right).$$

A subset E of X is called I-capacitable if

$$I(E)=\sup\{I(K): K\subset E, K \text{ compact}\}.$$

Choquet's capacitability theorem asserts that *if* $A\in\mathbf{S}(\mathcal{K}(X))$, *then* A *is* I*-capacitable.* We will make use of this theorem in Section 5.4.

3.6 *K*-Lusin sets

In this section we use the first separation theorem to show that K-Lusin sets are Borel sets; we obtain a partial converse in Section 5.3 below, see Theorem 5.8.10.

THEOREM 3.6.1 *A K-Lusin set in a Hausdorff space is a Borel set of the space.*

Note that the question of whether or not a set is a K-Lusin set can be decided by looking at its internal topological properties, without reference to the space within which it lies; so the theorem shows that a K-Lusin set is an "absolute Borel set relative to Hausdorff spaces" in that it remains a Borel set whenever it is topologically embedded or re-embedded in a Hausdorff space.

Proof of Theorem 3.6.1 Let A be a K-Lusin set in a Hausdorff space X. Then

$$A = K(\mathbb{N}^{\mathbb{N}})$$

with K an upper semi-continuous map from $\mathbb{N}^{\mathbb{N}}$ to the compact sets of X, and with

$$K(\sigma) \cap K(\tau) = \varnothing , \qquad \sigma \neq \tau ; \quad \sigma, \tau \in \mathbb{N}^{\mathbb{N}} .$$

For each s in $\mathbb{N}^{(\mathbb{N})}$ write

$$A(s) = \bigcup_{\substack{\sigma \\ \sigma|l(s)=s}} K(\sigma) = K(I(s)) .$$

For each fixed $l \geq 1$, the family

$$A(s) , \qquad s \in \mathbb{N}^{(\mathbb{N})} , \qquad l(s) = l ,$$

is a countable family of disjoint K-analytic sets. By Theorem 3.3.3, for each $l \geq 1$, we can choose a corresponding disjoint family

$$B(s) , \qquad s \in \mathbb{N}^{(\mathbb{N})} , \qquad l(s) = l , \tag{26}$$

of Borelian-$\mathscr{G}$ sets, with

$$A(s) \subset B(s) , \qquad s \in \mathbb{N}^{(\mathbb{N})} , \qquad l(s) = l .$$

For each s in $\mathbb{N}^{(\mathbb{N})}$, write

$$C(s) = \operatorname{cl} A(s) ,$$

and

$$D(s) = C(s) \cap \Big(\bigcap_{0 \leq r \leq l(s)} B(s|r) \Big) .$$

Consider the set

$$D = \bigcap_n \bigcup_\sigma D(\sigma|n) .$$

As the union is effectively taken over the countable system of finite sequences $\sigma|n$, this set D is a Borel set in X. Since

$$A \subset \bigcup_\sigma A(\sigma|n) \subset \bigcup_\sigma D(\sigma|n) ,$$

for each $n \geq 1$, we have $A \subset D$. Now the required conclusion will follow, if we can show that $D \subset A$.

Consider any point d^* in D. For each $n \geqslant 1$, there will be an s_n^* in $\mathbb{N}^{(\mathbb{N})}$ of length n, with

$$d^* \subset D(s_n^*) . \tag{27}$$

As the family (26), with $l = n$, is disjoint, s_n^* is uniquely determined by (27). But (27) ensures that

$$d^* \subset B(s_n^*|r) , \qquad 0 \leqslant r \leqslant n .$$

Hence, using the fact that the family (26) is disjoint,

$$s_n^*|r = s_r^* , \qquad 0 \leqslant r \leqslant n .$$

These consistency relations ensure the existence of a σ^* in $\mathbb{N}^{\mathbb{N}}$ with

$$s_n^* = \sigma^*|n , \qquad n \geqslant 1 .$$

Now

$$d^* \in D(\sigma^*|n) \subset \text{cl } A(\sigma^*|n) ,$$

for all $n \geqslant 1$. By Corollary 2.5.6 we have

$$d^* \in K(\sigma^*) \subset A .$$

Hence $D \subset A$ as required.

3.7 Convex Borel sets

Let X be a topological vector space over the real field. It is clear that the union of any increasing sequence of convex sets in X is also convex. Further, the intersection of any decreasing sequence of convex sets in X is also convex. So it is natural to study the system $\mathscr{C}$ of subsets of X characterized by the conditions

(i) $\mathscr{C}$ contains the closed convex sets of X;
(ii) $\mathscr{C}$ is closed under the operation of countable increasing union;
(iii) $\mathscr{C}$ is closed under the operation of countable decreasing intersection;
(iv) $\mathscr{C}$ is minimal subject to (i), (ii) and (iii) above.

It is easy to verify that such a system $\mathscr{C}$ exists and is unique. Further, $\mathscr{C}$ can be identified as the system of all sets $\mathscr{C}$ that have a representation as the terminal element of a transfinite sequence

$$C_1, C_2, \ldots, C_\omega, C_{\omega+1}, \ldots, C_\tau = C ,$$

with τ a countable ordinal and with each C_α, $1 \leqslant \alpha \leqslant \tau$, being

(i) a closed convex set; or
(ii) the union of an increasing sequence $C_{\beta(1)}, C_{\beta(2)}, \ldots$ with $1 \leq \beta(1) < \beta(2) < \cdots < \tau$; or
(iii) the intersection of a decreasing sequence $C_{\beta(1)}, C_{\beta(2)}, \ldots$ with $1 \leq \beta(1) < \beta(2) < \cdots < \tau$.

Thus the sets of $\mathscr{C}$ are the sets that can be obtained, in this way, from the closed convex sets by a countable transfinite sequence of monotonic Borel operations, without leaving the class of convex sets at any stage. So it is natural to say that the sets of $\mathscr{C}$ are the convexly generated convex Borel sets.

V. Klee (1951) asked whether a convex Borel set is necessarily convexly generated. This was proved in $\mathbb{R}^2$ by Klee (1953), in $\mathbb{R}^3$ by D. G. Larman (1971), in $\mathbb{R}^n$, for $n \geq 1$, by D. Preiss (1973), and was disproved in vector spaces of infinite dimension by P. Holický (1974). In this section we give Preiss's proof of the result in $\mathbb{R}^n$.

THEOREM 3.7.1 *A convex Borel set in $\mathbb{R}^n$ is convexly generated.*

We note that if B is a convex Borel set in $\mathbb{R}^n$ then both B and $E = \mathbb{R}^n \backslash B$ are analytic sets. Hence Theorem 3.7.1 is an immediate consequence of

THEOREM 3.7.2 *Let B be a convex analytic set in $\mathbb{R}^n$ and let E be an analytic set in $\mathbb{R}^n$ that does not meet B. Then there is a convexly generated set C with*

$$B \subset C, \qquad C \cap E = \varnothing .$$

Proof We first show that the family $\mathscr{C}$ of convexly generated sets is closed under countable intersections. Let $\mathscr{C}^*$ denote the family of all sets C of $\mathscr{C}$ with

$$C \cap D \in \mathscr{C}$$

for all closed convex sets D. It is easy to verify that $\mathscr{C}^*$ contains all closed convex sets and is closed under countable increasing unions and under countable decreasing intersections. By the minimality of $\mathscr{C}$, we must have $\mathscr{C}^* = \mathscr{C}$. Now take $\mathscr{C}^\dagger$ to be the family of all sets C of $\mathscr{C}$ with

$$C \cap D \in \mathscr{C}$$

for all sets D of $\mathscr{C}$. Using the result just proved, $\mathscr{C}^\dagger$ contains all the closed convex sets. Again it is easy to verify that $\mathscr{C}^\dagger$ is closed under countable increasing unions and countable decreasing intersections. By the minimality of $\mathscr{C}$, we must have $\mathscr{C}^\dagger = \mathscr{C}$. This shows that the intersection of two sets of $\mathscr{C}$

belongs to $\mathscr{C}$. So $\mathscr{C}$ is closed under finite intersections. If $C_1, C_2, \ldots$ is any sequence of sets of $\mathscr{C}$, then

$$\bigcap_{m=1}^{\infty} C_m = \bigcap_{m=1}^{\infty} \left(\bigcap_{r=1}^{m} C_r \right),$$

and we have, on the right, a decreasing intersection of sets of $\mathscr{C}$. Hence $\mathscr{C}$ is closed under the operation of countable intersection.

We need a variant of Lemma 3.3.1.

Let $\{A_m\}$ be an increasing sequence of sets and suppose that, for each $m \geqslant 1$, the set A_m can be separated from a fixed set B by the family $\mathscr{C}$ of convexly generated sets. Then

$$\bigcup_{m=1}^{\infty} A_m \tag{28}$$

can be separated from B by $\mathscr{C}$.

To prove this, choose a sequence of sets $\{C_m\}$ in C with

$$A_m \subset C_m, \qquad C_m \cap B = \varnothing,$$

for all $m \geqslant 1$. Then

$$A_m \subset A_r \subset C_r, \qquad r \geqslant m,$$

and

$$\bigcup_{m=1}^{\infty} A_m \subset \bigcup_{m=1}^{\infty} \bigcap_{r=m}^{\infty} C_r, \qquad \left(\bigcup_{m=1}^{\infty} \bigcap_{r=m}^{\infty} \right) C_r \cap B = \varnothing.$$

As the sequence

$$\bigcap_{r=m}^{\infty} C_r, \qquad m \geqslant 1,$$

is an increasing sequence of convexly generated sets, the set

$$\bigcup_{m=1}^{\infty} \bigcap_{r=m}^{\infty} C_r$$

is convexly generated and so the union (28) is separated from B by $\mathscr{C}$.

Now consider a convex analytic set B in $\mathbb{R}^n$ and an analytic set E in $\mathbb{R}^n$ that does not meet B. Choose maps b and e mapping $\mathbb{N}^{\mathbb{N}}$ continuously onto B and E. For each s in $\mathbb{N}^{(\mathbb{N})}$ write

$$B(s) = b(J(s)),$$

$$E(s) = e(I(s)),$$

where $I(s)$, as usual, denotes the Baire interval corresponding to s, and where $J(s)$ denotes the set of σ in $\mathbb{N}^{\mathbb{N}}$ satisfying

$$1 \leq \sigma_i \leq s_i, \qquad 1 \leq i \leq l(s),$$

with the convention

$$J(\varnothing) = \mathbb{N}^{\mathbb{N}}.$$

We say that a pair of finite sequences s, t of $\mathbb{N}^{(\mathbb{N})}$ is "good" if $B(s)$ can be separated from $E(t)$ *by* $\mathscr{C}$. As

$$B(\varnothing) = b(J(\varnothing)) = B,$$

$$E(\varnothing) = e(I(\varnothing)) = E,$$

we need to prove that the pair $\varnothing$, $\varnothing$ is good. Suppose that the pair $\varnothing$, $\varnothing$ is bad and seek a contradiction. When we know that a pair s, t is bad, we note that the sequence

$$B(s, \sigma), \qquad \sigma = 1, 2, 3, \ldots,$$

is increasing. It follows from the variant form of Lemma 3.3.2, that we can choose $\sigma \geq 1$, so that the pair s, σ and t is bad. Similarly, Lemma 3.3.2 itself (with the $\mathscr{C}$ of that lemma taken to be the complements of the convexly generated sets, and the $\mathscr{D}$ of that lemma taken to be the convexly generated sets) shows that, given the bad pair s, t, we can choose $\tau \geq 1$, so that the pair s and t, τ is bad.

Starting from the bad pair $\varnothing, \varnothing$, we can choose positive integers $\sigma_1, \tau_1, \sigma_2, \tau_2, \ldots$ in turn so that the pairs

$$\sigma|m, \qquad \tau|m, \qquad m \geq 1,$$

are all bad.

Let $J(\sigma)$ denote the set of all ν in $\mathbb{N}^{\mathbb{N}}$ with

$$1 \leq \nu_i \leq \sigma_i, \qquad i \geq 1.$$

Then $J(\sigma)$ is a compact subset of $\mathbb{N}^{\mathbb{N}}$. So

$$B(\sigma) = b(J(\sigma))$$

is a compact subset of the convex set B. Hence the convex hull, conv $B(\sigma)$ of $B(\sigma)$ (i.e. the smallest convex set containing $B(\sigma)$) is contained in B. As $B(\sigma)$ is a compact subset of $\mathbb{R}^n$, its convex hull is compact (this result depends on Carathéodory's theorem, that each point of the convex hull can be expressed as a convex combination of no more than $n+1$ of the points of the original set; it does not extend to infinite dimensional spaces). As E does not meet B, the point $e(\tau)$ does not lie in the compact convex set

$$\operatorname{conv} B(\sigma)$$

contained in B. So we can choose a compact convex set K in $\mathbb{R}^n$ with

$$B(\sigma) \subset \operatorname{int} K, \qquad e(\tau) \notin K.$$

It follows, from the uniform continuity of b on $J(\sigma)$, and from the continuity of e at τ, that

$$B(\sigma|m) = B(J(\sigma|m)) \subset K, \qquad K \cap E(\tau|m) = \emptyset,$$

for all sufficiently large m. This ensures that the pair $\sigma|m$, $\tau|m$ is good as well as bad. The contradiction completes the proof of Theorem 3.7.2

Dellacherie (see Part 2 of this volume) observes that the operation taking a set in $\mathbb{R}^n$ to its convex hull is a capacity operation. This enables him to obtain Preiss's theorem within the general framework of the theory of capacity operations and of multi-capacities.

4 Universal sets

4.1 Introduction

Let $\mathcal{S}$ be a class of sets in a space X. We say that a set U, in a Cartesian product $X \times Y$ of X with a second space Y, is universal for the class $\mathcal{S}$ if

(a) for each y in Y, the set $U^{(y)}$ belongs to $\mathcal{S}$; and
(b) for each set S in $\mathcal{S}$ there is a y in Y with $U^{(y)} = S$.

Here we use $U^{(y)}$ to denote the set of all points x of X with

$$(x, y) \in U.$$

We shall only be concerned with the case when $\mathcal{S}$ is a class of sets that are in some way topologically respectable, and we shall always require that U be topologically respectable in $X \times Y$.

We will first take X to be any second countable space and Y to be $2^{\mathbb{N}}$ (or equivalently to be the Cantor discontinuum C). We construct universal sets U of class $\mathcal{S}$ in $X \times 2^{\mathbb{N}}$ that are universal for the sets of X of class $\mathcal{S}$, in the cases when $\mathcal{S}$ is any one of the Borel classes, or when $\mathcal{S}$ is the class of Souslin-$\mathcal{F}$ sets.

When X is a Hausdorff space that contains a homeomorphic image of $2^{\mathbb{N}}$, these universal sets enable us to find Borel sets in X of all classes as well as an analytic non-Borel set. Similar ideas lead to the construction of an example showing that there are disjoint complements of analytic sets that cannot be separated by Borel sets.

A set U in a space X will be said to be universal, in the generalized sense, for a class $\mathscr{S}$ of sets in a space Y, if given any set S of $\mathscr{S}$, there is a relatively closed subset of U that is homeomorphic to S. Clearly in the case of a set U in $X \times Y$ that is universal for a class $\mathscr{S}$ of X, the set U is universal, in the generalized sense, for the sets of $\mathscr{S}$ in X. It will be easy to show that a set U in a Polish space X that is universal, in the generalized sense, for the analytic sets of $2^{\mathbb{N}}$, cannot be a Borel set. This will provide us with a general method of proving that some well-known examples of analytic sets are not Borel sets, by showing that they are universal, in the generalized sense, for analytic sets of $2^{\mathbb{N}}$.

Recently J. Steel (cf. the account in Martin and Kechris in Section 5.3 of Part 4 of this volume) has shown that the axiom of analytic determinacy implies that every analytic non-Borel set in a Polish space is universal, in the generalized sense, for the analytic subsets of $2^{\mathbb{N}}$. It follows from a result of Gödel (1938) that the axiom of constructibility implies the existence of an analytic non-Borel set in $2^{\mathbb{N}}$ which does not have this generalized universal property.

4.2 The construction of universal sets

As we have mentioned, the theory of universal sets has its roots in Lebesgue's discussion of universal functions. For some of the original accounts see Lusin (1925, 1930) and Sierpiński (1929, 1952). Our starting point will be

THEOREM 4.2.1 *Let X be a second countable space. Then there is an open set in $X \times 2^{\mathbb{N}}$ that is universal for the open sets of X.*

This will enable us to discuss the Borel classes

$$\mathscr{B}_0 = \mathscr{G}, \mathscr{B}_1 = \mathscr{F}, \mathscr{B}_2 = \mathscr{F}_\sigma, \mathscr{B}_3 = \mathscr{G}_\delta, \mathscr{B}_4 = \mathscr{G}_{\delta\sigma}, \mathscr{B}_5 = \mathscr{F}_{\sigma\delta}, \ldots,$$
$$\mathscr{B}_\omega, \mathscr{B}_{\omega+1}, \ldots, \mathscr{B}_\alpha, \ldots, \qquad \alpha < \omega_1 .$$

We name the Borel classes so that

(a) $\mathscr{B}_0 = \mathscr{G}$;
(b) $\mathscr{B}_{\alpha+1}$ is the family of complements of sets in $\mathscr{B}_\alpha$, if α is even (limit ordinals count as even);
(c) $\mathscr{B}_{\alpha+1}$ is the family of countable unions of sets of $\mathscr{B}_\alpha$, if α is odd; and
(d) $\mathscr{B}_\alpha$ is the family of countable unions of sets chosen from the families $\mathscr{B}_\beta$ with $\beta < \alpha$, when α is a limit ordinal.

We prove

THEOREM 4.2.2 *Let X be a second countable space. Let $\mathscr{S}$ be one of the classes $\mathscr{B}_\alpha$, $0 \leqslant \alpha < \omega_1$, or the class Souslin-$\mathscr{F}$. Then there is a set in $X \times 2^{\mathbb{N}}$ of class $\mathscr{S}$ that is universal for the sets of class $\mathscr{S}$ in X.*

Proof of Theorem 4.2.1 As X is second countable, there is a sequence $G_1, G_2, \ldots$ of open sets in X forming a base for the open sets of X. Each point y of $2^{\mathbb{N}}$ is a sequence $y_1, y_2, \ldots$, with $y_i = 0$ or 1, $i \geqslant 1$. Consider the set U in $x \times 2^{\mathbb{N}}$ of all points (x, y) with the property

$$\text{for some } i \geqslant 1, \qquad y_i = 1 \quad \text{and} \quad x \in G_i .$$

If (x, y) is such a point of U, we can choose $i \geqslant 1$ with $y_i = 1$ and $x \in G_i$. Let Y denote the set of y' in $2^{\mathbb{N}}$ with

$$y'_j = y_j, \qquad 1 \leqslant j \leqslant i .$$

Now, if

$$(x', y') \in G_i \times Y ,$$

we have

$$y'_1 = y_i = 1 \quad \text{and} \quad x' \in G_i ,$$

so that $(x', y') \in U$. Thus U is open in $X \times 2^{\mathbb{N}}$.

We show that U is universal for the open sets of X. As U is open in $X \times 2^{\mathbb{N}}$, each section $U^{(y)}$, $y \in 2^{\mathbb{N}}$, is open in X. Conversely, if G is open in X, it is the union of those sets of the base $G_1, G_2, G_3, \ldots$ that it contains. Define $y_1, y_2, \ldots$ by the conditions

$$y_i = \begin{cases} 1, & \text{if } G_i \subset G ; \\ 0, & \text{if } G_i \not\subset G . \end{cases}$$

Then $y = y_1, y_2, \ldots$ belongs to $2^{\mathbb{N}}$, and $x \in U^{(y)}$, if, and only if, for some $i \geqslant 1$, we have

$$y_i = 1 \quad \text{and} \quad x \in G_i .$$

Thus $x \in U^{(y)}$, if, and only if, $x \in G$, and $U^{(y)} = G$. So U is an open set in $X \times 2^{\mathbb{N}}$, and U is universal for the open sets of X.

Proof of Theorem 4.2.2 (a) We first note that if U is a set of class $\mathscr{S}$ in $X \times 2^{\mathbb{N}}$ and U is universal for the sets of class $\mathscr{S}$ in X, then

$$U' = (X \times 2^{\mathbb{N}}) \backslash U$$

is of the class $\mathscr{S}_c$ of complements of the class $\mathscr{S}$, and U' is universal for the sets of the class $\mathscr{S}_c$ of complements in X.

(b) Suppose that U is a set of class $\mathcal{S}$ in $\mathbb{N}\times X\times 2^{\mathbb{N}}$ and that U is universal for the class $\mathcal{S}$ in $\mathbb{N}\times X$. Consider the projection

$$V = \mathrm{proj}_{X\times 2^{\mathbb{N}}}\, U$$

of U onto $X\times 2^{\mathbb{N}}$. Each section $V^{(y)}$ is clearly an $\mathcal{S}_\sigma$-set in X. But given any $\mathcal{S}_\sigma$-set

$$S = \bigcup_{i=1}^{\infty} S_i\,, \qquad S_i \in \mathcal{S}, \quad i \geq 1\,,$$

the set

$$\bigcup_{i=1}^{\infty} (\{i\}\times S_i)$$

is a set of class $\mathcal{S}$ in $\mathbb{N}\times X$. As U is universal, there is a y in $2^{\mathbb{N}}$ with

$$U^{(y)} = \bigcup_{i=1}^{\infty} (\{i\}\times S_i)\,.$$

Hence

$$S = \bigcup_{i=1}^{\infty} S_i = \mathrm{proj}_X \bigcup_{i=1}^{\infty} (\{i\}\times S_i)$$
$$= \mathrm{proj}_X\, U^{(y)} = V^{(y)}\,.$$

Thus V is an $\mathcal{S}_\sigma$-set in $X\times 2^{\mathbb{N}}$ that is universal for the $\mathcal{S}_\sigma$-sets of X.

(c) Let $\mathcal{S}_1, \mathcal{S}_2, \ldots$ be a sequence of classes of sets, and let $\mathcal{T}$ denote the class of all sets of the form

$$T = \bigcup_{i=1}^{\infty} T_i\,, \qquad T_i \in \mathcal{S}_i, \quad i \geq 1\,.$$

Suppose that for each $i \geq 1$, there is a set U_i of class $\mathcal{S}_i$ in $X\times 2^{\mathbb{N}}$, with U_i universal for the sets of class $\mathcal{S}_i$ in X. Let $Y_1, Y_2, \ldots$ denote a sequence of copies of $2^{\mathbb{N}}$. Let V_i, $i \geq 1$, be the set of all points $(x, y_1, y_2, \ldots)$ in

$$X \times \prod_{j=1}^{\infty} Y_j\,,$$

with

$$(x, y_i) \in U_i\,.$$

Write

$$V = \bigcup_{i=1}^{\infty} V_i\,.$$

Then, for each $(y_1, y_2, \ldots)$ in $(2^{\mathbb{N}})^{\mathbb{N}}$, we have

$$V^{(y_1, y_2, \ldots)} = \bigcup_{i=1}^{\infty} U_i^{(y_i)},$$

and $V(y_1, y_2, \ldots)$ is a set of class $\mathcal{T}$ in X. Conversely, given any set T of class $\mathcal{T}$ we have

$$T = \bigcup_{i=1}^{\infty} T_i, \qquad T_i \in \mathcal{S}_i, \qquad i \geq 1,$$

and we can choose $y_1, y_2, \ldots$ in $2^{\mathbb{N}}$ so that

$$T_i = U_i^{(y_i)}, \qquad i \geq 1.$$

This ensures that

$$T = \bigcup_{i=1}^{\infty} T_i = \bigcup_{i=1}^{\infty} U_i^{(y_i)} = V(y_1, y_2, \ldots).$$

Hence V in $X \times (2^{\mathbb{N}})^{\mathbb{N}}$ is a set of class $\mathcal{T}$ that is universal for the sets of class $\mathcal{T}$ in X. As $(2^{\mathbb{N}})^{\mathbb{N}}$ is homeomorphic to $2^{\mathbb{N}}$, we can obtain a set W in $X \times 2^{\mathbb{N}}$ that is of class $\mathcal{T}$ and that is universal for the sets of class $\mathcal{T}$ in X.

(d) Let $\mathcal{S}$ be a class of sets. Let $\mathcal{P}$ denote the class of projections of sets of class $\mathcal{S}$ in the Cartesian product of the space with $2^{\mathbb{N}}$. Let U be a set of class $\mathcal{S}$ in $2^{\mathbb{N}} \times X \times 2^{\mathbb{N}}$ and suppose that U is universal for the sets of class $\mathcal{S}$ in $2^{\mathbb{N}} \times X$. Consider the projection

$$V = \mathrm{proj}_{X \times 2^{\mathbb{N}}} U$$

of U onto $X \times 2^{\mathbb{N}}$. Clearly V is a set of class $\mathcal{P}$ that is universal for the sets of class $\mathcal{P}$ in X.

(e) By Theorem 4.2.1, given any second countable space X, there is a set U in $X \times 2^{\mathbb{N}}$, with U open in $X \times 2^{\mathbb{N}}$ and U universal for the open sets of X. By (a) above, we get the same result for the closed sets of X. If X is second countable, so is $\mathbb{N} \times X$, and there is a closed set in $\mathbb{N} \times X \times 2^{\mathbb{N}}$ that is universal for the closed sets of $\mathbb{N} \times X$. Hence, by (b) above, the result holds for the class of $\mathcal{F}_\sigma$-sets. It follows, in this way, by transfinite induction, using (c) above, at the limit ordinals, that the result holds for all Borel classes $\mathcal{B}_\alpha$, $0 \leq \alpha < \omega_1$.

(f) If X is second countable, so is $2^{\mathbb{N}} \times X$, and (e) above yields an $\mathcal{F}_{\sigma\delta}$-set U in $2^{\mathbb{N}} \times X \times 2^{\mathbb{N}}$ that is universal for the $F_{\sigma\delta}$-sets of $2^{\mathbb{N}} \times X$. Now $2^{\mathbb{N}}$ contains an $\mathcal{F}_{\sigma\delta}$-set homeomorphic to $\mathbb{N}^{\mathbb{N}}$. Hence any Souslin-$\mathcal{F}$ set in X can first be represented as the projection on X of a closed set in $\mathbb{N}^{\mathbb{N}} \times X$ and can then be represented as the projection on X of an $\mathcal{F}_{\sigma\delta}$-set in $2^{\mathbb{N}} \times X$. It follows by (d) above that the projection of U onto $X \times 2^{\mathbb{N}}$ is a Souslin-$\mathcal{F}$ set in $X \times 2^{\mathbb{N}}$ that is universal for the Souslin-$\mathcal{F}$ sets of X.

Remark The methods used to prove Theorem 4.2.2 suffice to prove the corresponding results for the classes of projective sets.

4.3 The existence of Borel sets in each class and of analytic sets that are not Borel sets

In this section we first study the space $2^{\mathbb{N}}$ and prove

THEOREM 4.3.1 *The space $2^{\mathbb{N}}$ contains an analytic set that is not a Borel set. For each α with $1 \leqslant \alpha < \omega_1$, the space $2^{\mathbb{N}}$ contains a set of class $\mathscr{B}_\alpha$ that belongs to no $\mathscr{B}_\beta$ with $0 \leqslant \beta < \alpha$.*

This result has an immediate corollary:

THEOREM 4.3.2 *Let X be a Hausdorff space that contains a set homeomorphic to $2^{\mathbb{N}}$. Then X contains an analytic set that is not a Borel set. Further, for each α with $1 \leqslant \alpha < \omega_1$, the space X contains a set of class $\mathscr{B}_\alpha$ that is not of any class $\mathscr{B}_\beta$ with $0 \leqslant \beta < \alpha$.*

Proof of the first assertion of Theorem 4.3.1 By Theorem 4.2.2 there is a Souslin-$\mathscr{F}$ set U in $2^{\mathbb{N}} \times 2^{\mathbb{N}}$ that is universal for the Souslin-$\mathscr{F}$ sets of $2^{\mathbb{N}}$. Let Δ be the diagonal of $2^{\mathbb{N}} \times 2^{\mathbb{N}}$ and consider the set

$$A = \Delta \cap U.$$

Then A is a Souslin-$\mathscr{F}$ set in $2^{\mathbb{N}} \times 2^{\mathbb{N}}$ and so is an analytic set. Suppose, for sake of argument, that A is a Borel set. Then

$$B = \Delta \backslash A$$

is also a Borel set, and so is a Souslin-$\mathscr{F}$ set. But B is homeomorphic to its projection B' onto the first coordinate space of the product $2^{\mathbb{N}} \times 2^{\mathbb{N}}$. So B' is a Souslin-$\mathscr{F}$ set in $2^{\mathbb{N}}$ and we can find a y in $2^{\mathbb{N}}$ with

$$U^{(y)} = B'.$$

If $y \in U^{(y)}$ then $(y, y) \in A$ and $(y, y) \notin B$ and $y \notin B' = U^{(y)}$. If $y \notin U^{(y)}$ then $(y, y) \notin A$ and $(y, y) \in B$ and $y \in B' = U^{(y)}$. This contradiction shows that A is an analytic set in $2^{\mathbb{N}} \times 2^{\mathbb{N}}$ that is not a Borel set. As $2^{\mathbb{N}}$ is homeomorphic to $2^{\mathbb{N}} \times 2^{\mathbb{N}}$, there is an analytic set in $2^{\mathbb{N}}$ that is not a Borel set.

Before proving the second part of Theorem 4.3.1, it will be convenient to prove

LEMMA 4.3.3 *Suppose that the open sets of a space X are all $\mathscr{F}_\sigma$-sets. Then, if $0 \leq \alpha < \omega_1$, we have*

$$\mathscr{B}_\alpha \subset \mathscr{B}_{\alpha+2}.$$

Further, if $0 \leq \alpha < \gamma < \omega_1$ and γ is a limit ordinal

$$\mathscr{B}_\alpha \subset \mathscr{B}_{\gamma+1}.$$

Proof If $\alpha = 0$, then

$$\mathscr{B}_0 = \mathscr{G} \subset \mathscr{F}_\sigma = \mathscr{B}_2.$$

Suppose that $0 \leq \alpha < \omega_1$ and that we know that

$$\mathscr{B}_\alpha \subset \mathscr{B}_{\alpha+2}.$$

If α is even,

$$\mathscr{B}_{\alpha+1} = (\mathscr{B}_\alpha)_c \subset (\mathscr{B}_{\alpha+2})_c = \mathscr{B}_{\alpha+3}.$$

If α is odd,

$$\mathscr{B}_{\alpha+1} = \mathscr{B}_{\alpha\sigma} \subset (\mathscr{B}_{\alpha+2})_\sigma = \mathscr{B}_{\alpha+3}.$$

Now suppose that $0 < \gamma < \omega_1$, and that γ is a limit ordinal. Suppose further that

$$\mathscr{B}_\alpha \subset \mathscr{B}_{\alpha+2}$$

for all α with $0 \leq \alpha < \gamma$. Then $\mathscr{B}_\gamma$ consists of all countable unions of sets chosen from the classes $\mathscr{B}_\alpha$, $0 \leq \alpha < \gamma$. As each class contains $\varnothing$, it follows, in particular, that $\mathscr{B}_\gamma$ contains all of the sets of all $\mathscr{B}_\alpha$ with $0 \leq \alpha < \gamma$. Now all complements of sets of $\mathscr{B}_\alpha$ belong to $\mathscr{B}_{\alpha+1}$ or to $\mathscr{B}_{\alpha+2}$ (according to α being even or odd). So all sets of $\mathscr{B}_\alpha$ belong to

$$\mathscr{B}_{\gamma+1} = (\mathscr{B}_\gamma)_c.$$

Thus

$$\mathscr{B}_\alpha \subset \mathscr{B}_{\gamma+1}, \qquad 0 \leq \alpha < \gamma.$$

Now each set of $\mathscr{B}_\gamma$ is a countable union of sets in $\mathscr{B}_\alpha$, $0 \leq \alpha < \gamma$, and so is a countable union of sets of $\mathscr{B}_{\gamma+1}$. Hence

$$\mathscr{B}_\gamma \subset (\mathscr{B}_{\gamma+1})_\sigma = \mathscr{B}_{\gamma+2}.$$

The results follow by transfinite induction.

Proof of the second part of Theorem 4.3.1 First note that, if x is a point of $2^{\mathbb{N}}$, then $\{x\}$ is a closed set that is not open. Suppose that $2 \leq \alpha < \omega_1$ and that α is even. By Theorem 4.2.2 there is a set U of class $\mathscr{B}_\alpha$ in $2^{\mathbb{N}} \times 2^{\mathbb{N}}$ that is universal for the $\mathscr{B}_\alpha$-sets of $2^{\mathbb{N}}$. Consider the set

$$A = \Delta \cap U .$$

Then A is a $\mathscr{B}_\alpha$-set in $2^{\mathbb{N}} \times 2^{\mathbb{N}}$. First suppose that

$$A \in \mathscr{B}_{\alpha+1} .$$

Then

$$B = \Delta \backslash A \in \mathscr{B}_\alpha .$$

So the projection B' of B onto the first coordinate space is also a $\mathscr{B}_\alpha$-set, this time in $2^{\mathbb{N}}$. Just as in the first part of this proof we obtain a contradiction. Hence

$$A \notin \mathscr{B}_{\alpha+1} .$$

By Lemma 4.3.3, we conclude that A can belong to no $\mathscr{B}_\beta$ with $0 \leq \beta < \alpha$.

As

$$A \in \mathscr{B}_\alpha , \qquad A \notin \mathscr{B}_{\alpha+1} ,$$

the set $B = \Delta \backslash A$ has

$$B \in \mathscr{B}_{\alpha+1} , \qquad B \notin \mathscr{B}_\alpha .$$

If α is a limit ordinal, we have

$$\mathscr{B}_\beta \subset \mathscr{B}_\alpha , \qquad 0 \leq \beta < \alpha ,$$

so that

$$B \notin \mathscr{B}_\beta , \qquad 0 \leq \beta < \alpha + 1 .$$

If $\alpha = \gamma + 1$, with γ necessarily odd,

$$\mathscr{B}_\beta \subset (\mathscr{B}_\gamma)_\sigma = \mathscr{B}_\alpha , \qquad 0 \leq \beta \leq \gamma ,$$

and

$$B \notin \mathscr{B}_\beta , \qquad 0 \leq \beta < \alpha + 1 .$$

Thus, in each case, we have

$$B \in \mathscr{B}_{\alpha+1} , \qquad B \notin \mathscr{B}_\beta , \qquad 0 \leq \beta < \alpha + 1 .$$

So we have the required type of set in $2^{\mathbb{N}} \times 2^{\mathbb{N}}$ for odd ordinals as well as for even ordinals.

As $2^{\mathbb{N}} \times 2^{\mathbb{N}}$ is homeomorphic to $2^{\mathbb{N}}$, the sets of the required types exist in $2^{\mathbb{N}}$.

4.4 Some counter-examples

In this section we give some examples to show that some results that might be plausibly conjectured are not true. They add point to theorems proved elsewhere, showing that they cannot be improved in certain directions.

THEOREM 4.4.1 *There are analytic sets* A_0 *and* A_1 *in* $2^{\mathbb{N}}$ *such that the sets* $A_0 \backslash A_1$ *and* $A_1 \backslash A_0$ *cannot be separated by Borel sets.*

THEOREM 4.4.2 *There are disjoint complementary analytic sets in* $2^{\mathbb{N}}$ *that cannot be separated by Borel sets.*

These results show that Lusin's second separation theorem cannot be improved.

THEOREM 4.4.3 *There is a* $\mathscr{G}_\delta$*-set in* $X \times Y$, *with* $X = Y = 2^{\mathbb{N}}$, *that cannot be uniformized, with respect to* Y, *by any analytic set in* $X \times Y$.

Proof of Theorem 4.4.1 By Theorem 4.2.2 there is a Souslin-$\mathscr{F}$ set S in $\{0, 1\} \times 2^{\mathbb{N}} \times 2^{\mathbb{N}}$ that is universal for the Souslin-$\mathscr{F}$ sets in $\{0, 1\} \times 2^{\mathbb{N}}$. Now S is the union of two "sections"

$$\{0\} \times S_0 \quad \text{and} \quad \{1\} \times S_1 ,$$

with S_0 and S_1 Souslin-$\mathscr{F}$ sets in $2^{\mathbb{N}} \times 2^{\mathbb{N}}$. We suppose that $S_0 \backslash S_1$ and $S_1 \backslash S_0$ can be separated by Borel sets in $2^{\mathbb{N}} \times 2^{\mathbb{N}}$, and we seek a contradiction. Suppose then that B_0 and B_1 are Borel sets with

$$S_0 \backslash S_1 \subset B_0 , \qquad B_0 \cap B_1 = \varnothing , \qquad S_1 \backslash S_0 \subset B_1 .$$

Replacing B_1 by $2^{\mathbb{N}} \times 2^{\mathbb{N}} \backslash B_0$, if necessary, we may suppose that

$$B_0 \cup B_1 = 2^{\mathbb{N}} \times 2^{\mathbb{N}} .$$

As B_0 is a Borel set,

$$B_0 \in \mathscr{B}_\tau$$

for some ordinal τ with $0 \leq \tau < \omega_1$. By Theorem 4.3.1 there is a Borel set C_0 in $2^{\mathbb{N}}$ that is not of type $\mathscr{B}_\tau$. Write $C_1 = C \backslash C_0$. Then C_0 and C_1 are Souslin-$\mathscr{F}$ sets in $2^{\mathbb{N}}$ and

$$[\{0\} \times C_0] \cup [\{1\} \times C_1]$$

is a Souslin-$\mathcal{F}$ set in $\{0, 1\} \times 2^{\mathbb{N}}$. As S is universal, there is a y in $2^{\mathbb{N}}$, with

$$[\{0\} \times C_0] \cup [\{1\} \times C_1] = S^{(y)}.$$

This implies that

$$C_0 = S_0^{(y)}, \qquad C_1 = S_1^{(y)}.$$

As C_0, C_1 are disjoint,

$$C_0 = S_0^{(y)} \backslash S_1^{(y)} \subset B_0^{(y)},$$
$$C_1 = S_1^{(y)} \backslash S_0^{(y)} \subset B_1^{(y)}.$$

But $C_0 \cup C_1 = 2^{\mathbb{N}}$ and $B_0^{(y)} \cap B_1^{(y)} = \emptyset$. Hence the inclusions cannot be strict and we must have

$$C_0 = B_0^{(y)}, \qquad C_1 = B_1^{(y)}.$$

As B_0 is of Class $\mathcal{B}_\tau$ in $2^{\mathbb{N}} \times 2^{\mathbb{N}}$, so $B_0^{(y)}$ is of class $\mathcal{B}_\tau$ in $2^{\mathbb{N}}$. This is impossible by the choice of C_0.

Thus the sets S_0, S_1 are Souslin-$\mathcal{F}$ sets in $2^{\mathbb{N}} \times 2^{\mathbb{N}}$ and the sets $S_0 \backslash S_1$ and $S_1 \backslash S_0$ cannot be separated by Borel sets in $2^{\mathbb{N}} \times 2^{\mathbb{N}}$. As $2^{\mathbb{N}} \times 2^{\mathbb{N}}$ is homeomorphic to $2^{\mathbb{N}}$, the same result holds in $2^{\mathbb{N}}$. Of course, the Souslin-$\mathcal{F}$ sets of $2^{\mathbb{N}}$ coincide with the analytic sets of $2^{\mathbb{N}}$.

Proof of Theorem 4.4.2 By Theorem 4.4.1, there are analytic sets S_0, S_1 in $2^{\mathbb{N}}$ with the property that the sets $S_0 \backslash S_1$ and $S_1 \backslash S_0$ cannot be separated by Borel sets. By Lusin's second separation theorem (see Theorem 1.4.2) there are complementary analytic sets C_0 and C_1 with

$$S_0 \backslash S_1 \subset C_0, \qquad S_1 \backslash S_0 \subset C_1,$$
$$C_0 \cap C_1 = \emptyset, \qquad C_0 \cup C_1 = 2^{\mathbb{N}} \backslash (S_0 \cap S_1).$$

As $S_0 \backslash S_1$ and $S_1 \backslash S_0$ cannot be separated by Borel sets, the disjoint complementary analytic sets C_0 and C_1 cannot be separated by Borel sets.

Proof of Theorem 4.4.3 By Theorem 4.4.2, there are disjoint complementary analytic sets C_0, C_1 in $2^{\mathbb{N}}$ that cannot be separated by Borel sets. Write

$$A_0 = 2^{\mathbb{N}} \backslash C_0, \qquad A_1 = 2^{\mathbb{N}} \backslash C_1.$$

By Theorem 2.6.1 there are $\mathcal{G}_\delta$-sets G_0 and G_1 in $2^{\mathbb{N}} \times 2^{\mathbb{N}}$ with

$$\operatorname{proj}_X G_0 = A_0, \qquad \operatorname{proj}_X G_1 = A_1,$$

and with $2^{\mathbb{N}} \times 2^{\mathbb{N}}$ regarded as $X \times Y$ with $X \times Y = 2^{\mathbb{N}}$. We may suppose that

$$G_0 \subset 2^{\mathbb{N}} \times Y_0, \qquad G_1 \subset 2^{\mathbb{N}} \times Y_1,$$

where Y_i, $i=0, 1$, is the set of points $y=y_1, y_2, \dots$ in $2^{\mathbb{N}}$ with $y_1=i$. Then $G_0 \cup G_1$ is a $\mathcal{G}_\delta$-set in $X \times Y$.

We suppose that U is an analytic set that uniformizes $G_0 \cup G_1$, that is to say, a subset U of $G_0 \cup G_1$ with

$$\text{proj}_X U = \text{proj}_X G_0 \cup G_1$$

and

$$U^{(x)} \text{ a single point for each } x \text{ in } \text{proj}_X G_0 \cup G_1 .$$

Then

$$U \cap \{X \times Y_0\} \subset G_0 ,$$
$$U \cap \{X \times Y_1\} \subset G_1 .$$

So

$$S_0 = \text{proj}_X (U \cap \{X \times Y_0\})$$

and

$$S_1 = \text{proj}_X (U \cap \{X \times Y_1\})$$

are analytic sets with

$$\left.\begin{aligned} S_0 &\subset A_0 = 2^{\mathbb{N}} \backslash C_0 ,\\ S_1 &\subset A_1 = 2^{\mathbb{N}} \backslash C_1 ,\\ S_0 &\cap S_1 = \varnothing , \end{aligned}\right\} \tag{29}$$

but

$$\begin{aligned} S_0 \cup S_1 &= \text{proj}_X (U \cup \{X \times (Y_0 \cup Y_1)\}) \\ &= \text{proj}_X U = \text{proj}_X (G_0 \cup G_1) \\ &= A_0 \cup A_1 \\ &= X \backslash (C_0 \cap C_1) \\ &= X . \end{aligned}$$

As S_0, S_1 are disjoint analytic sets with union $2^{\mathbb{N}}$, these sets S_0, S_1 are necessarily Borel sets (Theorem 3.3.1). Further, (29) implies that

$$C_1 \subset S_0 , \qquad C_0 \subset S_1 ,$$

so that C_0, C_1 are separated by Borel sets, contrary to their choice. Thus the uniformization is impossible.

4.5 Analytic sets that are not Borel and the generalized universal property

Recall that a set U is said to be universal, in the generalized sense, for the analytic subsets of $2^{\mathbb{N}}$, if, given any analytic subset A of $2^{\mathbb{N}}$, there is a relatively closed subset B of U that is homeomorphic to A. Using Theorem 4.3.1, it will be easy to prove

THEOREM 4.5.1 *Let A be an analytic set in a Polish space X, and suppose that A is universal, in the generalized sense, for the analytic subsets of $2^{\mathbb{N}}$. Then A is not a Borel set in X.*

Of course, the set constructed in the proof of Theorem 4.2.2 to be universal for the analytic subsets of $2^{\mathbb{N}}$ is also universal, in the generalized sense, for the analytic subsets of $2^{\mathbb{N}}$.

One of the simplest examples of an analytic set that is not a Borel set is the set discussed by C. Dellacherie and P. A. Meyer (1975). They introduce the space of "stopping times" and show that the set of stopping times that assume the value $+\infty$ is an analytic subset that is not a Borel set. We prefer to work with the equivalent space of trees, and with the set of non-path finite trees.

We regard the space $\mathbb{N}^{(\mathbb{N})}$, of all finite sequences of positive integers (including the empty sequence $\varnothing$) as a tree with a root. Each finite sequence in $\mathbb{N}^{(\mathbb{N})}$ is a vertex of the tree; $\varnothing$ is its root. Two finite sequences are to be joined by an edge, just when one sequence can be formed from the other by deletion of its last element. Let $\mathbb{T} = \mathbb{T}(\mathbb{N}^{(\mathbb{N})})$ denote the space of all sub-trees of $\mathbb{N}^{(\mathbb{N})}$ that contain the root $\varnothing$. We construct a metric on $\mathbb{T}$. Let

$$\varphi_0 = \varnothing;\ \varphi_1 = 1;\ \varphi_2 = 1, 1;\ \varphi_3 = 2;\ \varphi_4 = 1, 1, 1;$$

$$\varphi_5 = 1, 2;\ \varphi_6 = 2, 1;\ \varphi_7 = 3;\ \varphi_8 = 1, 1, 1. 1; \ldots,$$

be the enumeration of the finite sequences $s_1, s_2, \ldots, s_l$ of $\mathbb{N}^{(\mathbb{N})}$ obtained by taking in turn the groups with a common value for

$$s_1 + s_2 + \cdots + s_l,$$

each group being ordered lexicographically. The distance between two distinct trees T_1, T_2 of $\mathbb{T}$ is then taken to be

$$d(T_1, T_2) = 2^{-n},$$

where n is the least integer for which φ_n belongs to

$$T_1 \triangle T_2 = (T_1 \backslash T_2) \cup (T_2 \backslash T_1).$$

There is a natural map ν mapping the tree T of $\mathbb{T}$ to the sequence

$$\nu(T) = \nu_1(T), \nu_2(T), \nu_3(T), \ldots$$

in $2^{\mathbb{N}}$ defined by

$$\nu_n(T) = 1, \quad \text{if} \quad \varphi_n \in T\,;$$

$$\nu_n(T) = 0, \quad \text{if} \quad \varphi_n \notin T\,.$$

It is easy to verify that ν is a continuous injective map from $\mathbb{T}$ to a closed subset $\mathbb{T}^*$ of $2^{\mathbb{N}}$. As $2^{\mathbb{N}}$ is compact, it follows that ν is a homeomorphism and that $\mathbb{T}$ is compact.

A tree is said to be path-finite if it contains no infinite succession of incident edges and vertices. Let $\mathbb{P} = \mathbb{P}(\mathbb{N}^{(\mathbb{N})})$ denote the set of path-finite trees of $\mathbb{T}$ and let $\mathbb{L} = \mathbb{L}(\mathbb{N}^{(\mathbb{N})})$ denote the set $\mathbb{T}\backslash\mathbb{P}$ of non-path-finite trees in $\mathbb{T}$. The set $\mathbb{L}$ of non-path-finite trees corresponds exactly to the set of stopping times that assume the value $+\infty$, studied by Dellacherie and Meyer (1975). We prove

THEOREM 4.5.2 *The set $\mathbb{L}$ in $\mathbb{T}$ is universal, in the generalized sense, for the analytic sets of $2^{\mathbb{N}}$.*

From these results we have the immediate corollary

THEOREM 4.5.3 *Let A be an analytic set in a Polish space X. If A contains a relatively closed subset that is homeomorphic to $\mathbb{L}$, then A is universal, in the generalized sense, for the analytic sets of $2^{\mathbb{N}}$, and A is an analytic set that is not a Borel set.*

It may at first sight appear that this theorem would seldom enable one to determine whether or not a given analytic set is a Borel set. In practice the method seems to work very smoothly. Rogers (1977) discusses two examples in some detail. Lusin (1927) introduced the set D of all real numbers having continued fraction expansions of the form

$$\frac{1}{a_1+}\ \frac{1}{a_2+}\ \cdots\ \frac{1}{a_r+}\ \cdots,$$

with $a_1, a_2, \ldots$ a sequence of positive integers, restricted by the condition that, for some strictly increasing sequence

$$r(1) < r(2) < \cdots$$

of positive integers, $a_{r(i)}$ divides $a_{r(i+1)}$ for $i = 1, 2, \ldots$. He proved, in fact, that a slightly different set is analytic but not Borel; his methods suffice to prove that D itself is analytic but not Borel.

Hurewicz (1930) studied the space $\mathscr{K}(X)$ of compact subsets of an uncountable complete separable metric space X. In the space $\mathscr{K}(X)$, Hurewicz studies the set of $\mathscr{N}(X)$ of uncountable compact subsets of X and shows that $\mathscr{N}(X)$ is an analytic set that is not a Borel set in $\mathscr{K}(X)$.

Rogers shows that both D and $\mathscr{N}(X)$ contain relatively closed subsets homeomorphic to $\mathbb{L}$, so that they are universal, in the generalized sense, for the analytic subsets of $2^{\mathbb{N}}$. An example due to Mazurkiewicz (1936) is discussed in Section 4.6, below.

Jayne and Rogers (1979) show, in particular, that, if two sets in zero-dimensional Polish spaces are each universal, in the generalized sense, for the analytic sets of $2^{\mathbb{N}}$, then the two sets correspond under a Borel-isomorphism, under which $\mathscr{F}_\sigma$-sets correspond to $\mathscr{F}_\sigma$-sets. Now Gödel (1938) announced, and Novikov (1951) proved (using the results of Gödel (1940)) that the axiom of constructibility implies the existence in $2^{\mathbb{N}}$ of an analytic set A whose complement contains no perfect subset. Maitra and Ryll-Nardzewski (1970) prove that this analytic non-Borel set A cannot be Borel isomorphic to the standard example of an analytic set in $2^{\mathbb{N}} \times 2^{\mathbb{N}}$ that is universal for the analytic sets of $2^{\mathbb{N}}$. We give a proof due to J. E. Jayne (and published here for the first time) that is more elementary than that of Maitra and Ryll-Nardzewski.

THEOREM 4.5.4 *Assume that A is an analytic set in $2^{\mathbb{N}}$ whose complement contains no perfect subset. Then any relatively closed subset of A that is a Borel set in $2^{\mathbb{N}}$ is necessarily a $\mathscr{G}_\delta$-set in $2^{\mathbb{N}}$. Further, A can be Borel isomorphic to no analytic set in $2^{\mathbb{N}}$ that is universal, in the generalized sense, for the analytic sets of $2^{\mathbb{N}}$.*

These results contrast sharply with recent results of Steel and Harrington, who show that the axiom of analytic determinacy is equivalent both to the proposition that all analytic non-Borel sets in $2^{\mathbb{N}}$ are Borel isomorphic and also to the proposition that every analytic non-Borel set in $2^{\mathbb{N}}$ is universal, in the generalized sense, for the analytic sets of $2^{\mathbb{N}}$.

Proof of Theorem 4.5.1 Let A be an analytic set in a Polish space X, and suppose that A is universal, in the generalized sense, for the analytic sets of $2^{\mathbb{N}}$. To prove that A is not a Borel set, we suppose that A is a Borel set and seek a contradiction. Suppose that A is a Borel set of class $\mathscr{B}_\alpha$. Consider any Borel set B in $2^{\mathbb{N}}$. By the generalized universal property, there is a relatively closed set C in A that is homeomorphic to B. Let $C = A \cap F$, with F closed in $2^{\mathbb{N}}$. By Lavrentiev's theorem the homeomorphism from B in $2^{\mathbb{N}}$ to C in F extends to a homeomorphism from a $\mathscr{G}_\delta$-set $\hat{B}$ in $2^{\mathbb{N}}$ to a $\mathscr{G}_\delta$-set $\hat{C}$ in F. Now $C = A \cap F$ is of class $\mathscr{B}_\alpha$ in F and so is also of class $\mathscr{B}_\alpha$ in $\hat{C}$. Hence B is of class $\mathscr{B}_\alpha$ in the $\mathscr{G}_\delta$-set $\hat{B}$ and so is of class at most $\mathscr{B}_{\alpha+2}$ in $2^{\mathbb{N}}$. As B was any Borel set in $2^{\mathbb{N}}$, this contradicts Theorem 4.3.1. We are forced to conclude that A is not a Borel set in X.

Proof of Theorem 4.5.2 We study the set $\mathbb{L}$ of non-path-finite trees in the space $\mathbb{T}$ of all trees. Let A be any analytic set in $2^{\mathbb{N}}$. We use the metric defined on $2^{\mathbb{N}}$ by

$$d(h, k) = 2^{-n},$$

where n is the least positive integer with

$$h_n \neq k_n,$$

when $h \neq k$ and

$$h = h_1, h_2, \ldots, \qquad h_i = 0 \text{ or } 1,$$

$$k = k_1, k_2, \ldots, \qquad k_i = 0 \text{ or } 1.$$

Now A has a Souslin representation

$$A = \bigcup_{\sigma} \bigcap_{n} F(\sigma|n),$$

with

$$F(\sigma|1) \supset F(\sigma|2) \supset F(\sigma|3) \supset \cdots, \qquad \sigma \in \mathbb{N}^{\mathbb{N}},$$

and with each set $F(\sigma|n)$ open, closed and of diameter less than 2^{-n}. Write $F(\varnothing) = 2^{\mathbb{N}}$.

Let T be the map from $2^{\mathbb{N}}$ to $\mathbb{T}$ defined by taking $T(h)$ to be the set of all s in $\mathbb{N}^{(\mathbb{N})}$ with

$$h \in F(s).$$

The relation

$$2^{\mathbb{N}} = F(\varnothing) \supset F(\sigma|1) \supset F(\sigma|2) \supset \cdots$$

for each σ in $\mathbb{N}^{\mathbb{N}}$ ensures that $T(h)$ is always a tree of $\mathbb{T}$.

If $s \in \mathbb{N}^{(\mathbb{N})}$ and $F(s)$ is neither $\varnothing$ nor $2^{\mathbb{N}}$, then the sets

$$F(s) \quad \text{and} \quad 2^{\mathbb{N}} \backslash F(s)$$

are non-empty, disjoint and compact and so there is a positive distance $\delta(s)$ between the two sets. When $F(s)$ is empty or the whole of $2^{\mathbb{N}}$, we write $\delta(s) = 1$. If h, k are two points at distance less than $\delta(s)$ apart, then h and k are both in $F(s)$ or neither is in $F(s)$. If h, k are at distance less than

$$\min\{\delta(\varphi_1), \delta(\varphi_2), \ldots, \delta(\varphi_n)\}$$

the points h, k lie in the same sub-family of sets

$$F(\varphi_1), F(\varphi_2), \ldots, F(\varphi_n),$$

and the trees $T(h)$, $T(k)$ contain the same selection from the possible vertices

$$\varphi_1, \varphi_2, \ldots, \varphi_n$$

and

$$d(T(h), T(k)) < 2^{-n}.$$

Hence T is continuous. As $2^{\mathbb{N}}$ is compact, the image $\mathscr{X} = T(2^{\mathbb{N}})$ is compact.

Now consider the set $\mathscr{A} = T(A)$. For each a in A, there is a σ in $\mathbb{N}^{\mathbb{N}}$ with

$$a \in F(\sigma|n), \qquad n \geq 1,$$

so that

$$\sigma|n \in T(a), \qquad n \geq 1,$$

and $T(a)$ is not path-finite. Conversely, if $T(x)$ is non-path-finite, there is a σ in $\mathbb{N}^{\mathbb{N}}$ with

$$\sigma|n \in T(x), \qquad n \geq 1,$$

so that

$$x \in F(\sigma|n), \qquad n \geq 1,$$

and $x \in A$. Thus we have

$$\mathscr{A} = \mathbb{L} \cap \mathscr{X}$$

and $\mathscr{A}$ is a relatively closed subset of $\mathbb{L}$.

Now consider any fixed tree T^* in $\mathscr{A}$. We have $T^* = T(a^*)$ for some a^* in A. As T^* is not path-finite, we choose σ^* in $\mathbb{N}^{\mathbb{N}}$ with

$$\sigma^*|n \in T(a^*), \qquad n \geq 1.$$

For fixed n, provided $T = T(a)$ in $\mathscr{A}$ is sufficiently close to T^*, we have

$$\sigma^*|n \in T,$$

so that $T = T(a)$ with

$$a^* \in F(\sigma^*|n), \qquad a \in F(\sigma^*|n)$$

and

$$d(a^*, a) < 2^{-n}.$$

Hence the map T, when restricted to A is one-to-one and has a continuous inverse. Thus $T|_A$ is a homeomorphism mapping the analytic set A of $2^{\mathbb{N}}$ to the relatively closed subset $\mathscr{A}$ of $\mathbb{L}$. Now $\mathbb{L}$ is universal, in the generalized sense, for the analytic sets of $2^{\mathbb{N}}$.

Proof of Theorem 4.5.3 This follows immediately from Theorems 4.5.1 and 4.5.2.

Proof of Theorem 4.5.4 Let A be an analytic set in $2^{\mathbb{N}}$ whose complement contains no perfect subset. Let F be a relatively closed subset of A that is a Borel set in $2^{\mathbb{N}}$. Then $F = A \cap \hat{F}$ with A a closed set in $2^{\mathbb{N}}$. Then $\hat{F} \backslash F$ is a Borel set contained in the complement of A. Hence $\hat{F} \backslash F$ contains no perfect subset, and so (by Theorem 3.5.1) is countable. So $\hat{F} \backslash F$ is an $\mathscr{F}_\sigma$-set and $F = \hat{F} \backslash (\hat{F} \backslash F)$ is a $\mathscr{G}_\delta$-set.

Now suppose that f is a Borel isomorphism of A to a set B in $2^{\mathbb{N}}$ that is universal, in the generalized sense, for the analytic sets of $2^{\mathbb{N}}$. By a generalization of Lavrentiev's theorem, due to Kuratowski (1934), f extends to a Borel isomorphism $\hat{f}$ of a Borel set $\hat{A}$ containing A to a Borel set $\hat{B}$ containing B. Let $\hat{B}$ belong to the class $\mathscr{B}_\alpha$. By the universal property, there will be a relatively closed subset F of B that is homeomorphic to a Borel set in $2^{\mathbb{N}}$ that does not belong to $\mathscr{B}_{\alpha+6}$. So F is not a relative $\mathscr{B}_{\alpha+6}$-set in B. Hence F is not a $\mathscr{B}_{\alpha+6}$-set in $2^{\mathbb{N}}$. As F is relatively closed in B,

$$F = B \cap \hat{F}$$

with $\hat{F}$ closed in $2^{\mathbb{N}}$. As

$$F = B \cap \hat{F} \subset \hat{B} \cap \hat{F},$$

we have

$$F = (\hat{B} \cap \hat{F}) \backslash ((\hat{B} \cap \hat{F}) \backslash F).$$

Hence

$$(\hat{B} \cap \hat{F}) \backslash F$$

is not of class $\mathscr{B}_{\alpha+4}$; it follows that $(\hat{B} \cap \hat{F}) \backslash F$ is uncountable. As $\hat{f}$ is a Borel isomorphism,

$$\hat{f}^{-1}((\hat{B} \cap \hat{F}) \backslash F)$$

is an uncountable Borel set contained in

$$\hat{f}^{-1}(\hat{B} \backslash B) = \hat{A} \backslash A \subset 2^{\mathbb{N}} \backslash A.$$

Thus $2^{\mathbb{N}} \backslash A$ contains a perfect subset. This contradiction completes the proof.

4.6 Continuous functions that are not differentiable

Let $C([0, 1])$ denote the Banach space of all continuous real-valued functions f on the unit interval $[0, 1]$ of the real line, with the supremum norm

$$\|f\| = \sup_{0 \leq x \leq 1} |f(x)| .$$

Mazurkiewicz (1936) shows that the set of functions of $C([0, 1])$ that are not differentiable is an analytic subset of $C([0, 1])$ that is not a Borel subset of $C([0, 1])$. In this section we prove

THEOREM 4.6.1 *Both the set of functions of $C([0, 1])$ that are not differentiable, and the set of monotonic increasing functions of $C([0, 1])$ that are not differentiable, are analytic subsets of $C([0, 1])$ that are not Borel subsets of $C([0, 1])$.*

To obtain this result, it will be convenient to discuss certain subsets of the space $\mathbb{T} = \mathbb{T}(2^{(\mathbb{N})})$ of all trees on $2^{(\mathbb{N})}$. Let $\mathbb{M}$ denote the set of all trees T on $2^{(\mathbb{N})}$ defined by the formula

$$\mathbb{M} = \{T: (\forall s)((s \in 2^{(\mathbb{N})} \,\&\, (s, 0) \in T) \;\Rightarrow\; (s, 0, 0) \in T \vee (s, 0, 1) \in T\} \setminus \bigcup_{s \in 2^{(\mathbb{N})}} \{T: (\forall n)((s, [0]_n) \in T\} .$$

Here we use $(s, [0]_n)$ to denote the finite sequence of zeros and ones obtained by adjunction of a sequence of n zeros to the sequence s. This formula expresses formally the requirements that each branch of the tree, that terminates, should terminate with the digit one, and that no infinite branch through the tree should be ultimately made up solely with the digit zero. The formula makes it clear that $\mathbb{M}$ is a $\mathscr{G}_\delta$-set in $\mathbb{T}(2^{(\mathbb{N})})$. Let $\mathbb{L} = \mathbb{L}(2^{(\mathbb{N})})$ denote the set of non-path-finite trees in $\mathbb{T}(2^{(\mathbb{N})})$. We prove

LEMMA 4.6.2 *There is a homeomorphism mapping $\mathbb{T}(\mathbb{N}^{(\mathbb{N})})$ onto $\mathbb{M}$ that maps $\mathbb{L}(\mathbb{N}^{(\mathbb{N})})$ onto $\mathbb{M} \cap \mathbb{L}(2^{(\mathbb{N})})$. The set $\mathbb{M} \cap \mathbb{L}(2^{(\mathbb{N})})$ is an analytic non-Borel subset of $\mathbb{T}(2^{(\mathbb{N})})$ and is universal, in the generalized sense, for the analytic subsets of $2^{\mathbb{N}}$.*

Proof of Lemma 4.6.2 The second statement will follow immediately from Theorem 4.5.3, once we have proved the first statement. Let ψ be the map from $\mathbb{N}^{(\mathbb{N})}$ to $2^{(\mathbb{N})}$ defined by

$$\psi(\varnothing) = \varnothing ,$$

$$\psi(s_1, s_2, \ldots, s_n) = [0]_{s_1 - 1}, 1, [0]_{s_2 - 1}, 1, \ldots, 1, [0]_{s_n - 1}, 1, \qquad n \geq 1 ,$$

with $[0]_t$ denoting a sequence of t zeros, when $t \geqslant 0$. (Of course, when $t=0$, we need only one comma, rather than two, to separate the two ones on either side of $[0]_0$.) Given any finite sequence $h_1, h_2, \ldots, h_n$ in $2^{(\mathbb{N})}$ with $h_n = 1$, there is a unique way of collecting together consecutive sequences of zeros and of inserting dummy sequences of zeros, of zero length, between any two consecutive ones; and so there is a unique way of expressing $h = h_1, h_2, \ldots, h_n$ in the form

$$[0]_{t_1-1}, 1, [0]_{t_2-1}, 1, \ldots, 1, [0]_{t_r-1}, 1\,,$$

with $t_i \geqslant 1$, $0 \leqslant i \leqslant r$. Thus $h = \psi(t_1, t_2, \ldots, t_r)$ for a unique $t_1, t_2, \ldots, t_r$ in $\mathbb{N}^{(\mathbb{N})}$. Hence ψ maps $\mathbb{N}^{(\mathbb{N})}$ homeomorphically into a subset $\mathbb{I}$ of $2^{(\mathbb{N})}$; the subset $\mathbb{I}$ consisting of $\varnothing$ together with those sequences that terminate with the digit 1.

Given any tree T in $\mathbb{T}(\mathbb{N}^{(\mathbb{N})})$ we obtain a corresponding tree $\psi(T)$ in $\mathbb{T}(2^{(\mathbb{N})})$ by taking $\psi(T)$ to be the least tree in $\mathbb{T}(2^{(\mathbb{N})})$ containing the set

$$\psi(s)\,, \qquad s \in T\,.$$

It is easy to verify that ψ is the required homeomorphism mapping $\mathbb{T}(\mathbb{N}^{(\mathbb{N})})$ onto $\mathbb{M}$ and mapping $\mathbb{L}(\mathbb{N}^{(\mathbb{N})})$ onto $\mathbb{M} \cap \mathbb{L}(2^{(\mathbb{N})})$.

Proof of Theorem 4.6.1 Let φ denote the real-valued function, defined on $\mathbb{R}$ by

$$\varphi(x) = 0\,, \qquad x \leqslant 0\,,$$
$$\varphi(x) = 3x^2 - 2x^3\,, \qquad 0 < x \leqslant 1\,,$$
$$\varphi(x) = 1\,, \qquad 1 < x\,.$$

Then φ is monotonic increasing and differentiable for all x,

$$0 \leqslant \varphi(x) \leqslant 1\,, \qquad \text{for all } x\,,$$
$$0 \leqslant \varphi'(x) \leqslant 1\tfrac{1}{2}\,, \qquad \text{for all } x\,,$$

and

$$1\tfrac{1}{3} \leqslant \varphi'(x)\,, \qquad \tfrac{1}{3} \leqslant x \leqslant \tfrac{2}{3}\,.$$

For each h in the set $\mathbb{I}$ of sequences of $2^{(\mathbb{N})}$ that are either the empty sequence $\varnothing$ or terminate with the digit 1, write

$$t(h) = h_1 + 2h_2 + 4h_3 + \cdots + 2^{n-1}h_n\,,$$

with the natural convention

$$t(\varnothing) = 0\,.$$

Then t is an injective map of $\mathbb{I}$ into the set of non-negative integers. For each h in $\mathbb{I}$ write

$$\varphi_h(x) = 2\times 3^{-t(h)} + 9^{-l(h)}\varphi(9^{l(h)}(x-\xi(h)))\,,$$

with

$$\xi(h) = \sum_{i=1}^{l(h)} (2h_i + 3)9^{-i}\,,$$

and with

$$\varphi_\varnothing(x) = 0\,, \qquad \text{for all } x\,.$$

Note that, for all h, the function φ_h is continuous and monotonic increasing, and is constant on either side of the interval

$$J(h) = [\xi(h), \xi(h) + 9^{-l(h)}]\,.$$

Further, the differential coefficient of φ_h is at least $1\frac{1}{3}$ in the interval

$$K(h) = [\xi(h) + \tfrac{1}{3}\times 9^{-l(h)}, \xi(h) + \tfrac{2}{3}\times 9^{-l(h)}]\,.$$

Note that the definition of ξ ensures that

$$J(k) \subset K(h)$$

when k, h belong to $\mathbb{I}$ with $l(k) > l(h)$ and $k|l(h) = h$.

For each tree T in the subset $\mathbb{M}$ of trees on $2^{(\mathbb{N})}$, write

$$\chi(T) = \sum_{h\in Y\cap\mathbb{I}} \varphi_h\,.$$

We show that χ maps $\mathbb{M}$ homeomorphically to a $\mathscr{G}_\delta$-subset of $C([0, 1])$. For each h in $\mathbb{I}$,

$$\|\varphi_h\| \leq 2\times 3^{-t(h)} + 9^{-l(h)}\,.$$

Hence

$$\sum_{h\in\mathbb{I}} \|\varphi_h\| \leq \left(\sum_{n=1}^{\infty} 2\times 3^{-n}\right) + \left(\sum_{l=1}^{\infty} 2^l \times 9^{-l}\right) < 2\,.$$

This ensures that the map χ from $\mathbb{M}$ to $C([0, 1])$ is continuous. Now the value

$$\chi(T)(0) = \sum_{\varnothing\neq h\in T\cap\mathbb{I}} 2\times 3^{-t(h)}\,,$$

of $\chi(T)$ at the origin determines the precise set of integers assumed by $t(h)$ on $T\cap\mathbb{I}$, and so determines T in $\mathbb{M}$ precisely. Thus χ is injective. But a knowledge of $\chi(T)(0)$ to within $3^{-2^{n+1}}$, determines the values, less than 2^{n+1}, assumed by $t(h)$ on $T\cap\mathbb{I}$, and so determines all elements of T of length not exceeding n. Hence χ^{-1} is continuous and χ is a homeomorphism. As $\mathbb{M}$ is a $\mathscr{G}_\delta$-set in the complete metric space $\mathbb{T}(2^{(\mathbb{N})})$, the homeomorphic image $\chi(\mathbb{M})$ is a $\mathscr{G}_\delta$-set in $C([0, 1])$.

Consider the function $\chi(T)$ for any T in $\mathbb{M}$. First suppose that for some x in $[0, 1]$ the point x belongs to infinitely many of the intervals $K(h)$ with $h \in T \cap \mathbb{I}$. Then the intervals must be nested and there is an

$$\eta = \eta_1, \eta_2, \eta_3, \ldots$$

in $2^{\mathbb{N}}$, infinitely many of the components $\eta_1, \eta_2, \eta_3, \ldots$ having the value 1, with

$$x \in K(\eta|n), \quad \text{whenever} \quad \eta_n = 1 .$$

In the first place, we have

$$\chi(T)'(x) \geqslant 1\tfrac{1}{3} \sum_{i=1}^{n} \eta_i ,$$

for each $n \geqslant 1$, so that $\chi(T)$ has no differential coefficient at x. Second, the tree T is non-path-finite, and so belongs to $\mathbb{M} \cap \mathbb{L}(2^{(\mathbb{N})})$.

Now, suppose that for each x in $[0, 1]$, the point x belongs to only a finite number of the intervals $J(h)$ with $h \in T$. In this case, in the neighbourhood of each x, the sum

$$\sum_{h \in T \cap \mathbb{I}} \varphi_n ,$$

reduces to a finite sum of differentiable functions, and so is itself differentiable. In this case T is a path-finite tree and so is not in $\mathbb{M} \cap \mathbb{L}(2^{(\mathbb{N})})$.

It follows that χ maps $\mathbb{M} \cap \mathbb{L}(2^{(\mathbb{N})})$ homeomorphically to $\chi(\mathbb{M}) \cap \mathbb{L}^*$, where $\mathbb{L}^*$ is the set of those functions of $C([0, 1])$ that are not differentiable (i.e. that are not differentiable at at least one point of $[0, 1]$).

Let $r_1, r_2, \ldots$ be an enumeration of the rational numbers in $[0, 1]$. The condition that the function f of $C([0, 1])$ should not be differentiable at the point x is

$$(\exists n)(\forall m)(\exists h)(\exists k)\Big(0 < |x - r_h| < 2^{-m} \;\&\; 0 < |x - r_k| < 2^{-m}$$
$$\& \left|\frac{f(x) - f(r_h)}{x - r_h} - \frac{f(x) - f(r_k)}{x - r_k}\right| > 2^{-n}\Big) .$$

This condition defines a $\mathscr{G}_{\delta\sigma}$-set in $C([0, 1])$. Hence the set $\mathbb{L}^*$ of non-differentiable functions is an analytic subset of $C([0, 1])$. But $\chi(\mathbb{M}) \cap \mathbb{L}^*$ is homeomorphic to $\mathbb{M} \cap \mathbb{L}(2^{(\mathbb{N})})$ and so to $\mathbb{L}(\mathbb{N}^{(\mathbb{N})})$ and is universal, in the generalized sense, for the analytic subsets of $2^{\mathbb{N}}$. Thus $\chi(\mathbb{M}) \cap \mathbb{L}^*$ is not Borel and so $\mathbb{L}^*$ is not Borel. Similarly the set $\mathbb{L}^{**}$ of monotonic increasing non-differentiable functions of $C([0,1])$ meets $\chi(\mathbb{M})$ in the same set $\chi(\mathbb{M}\cap\mathbb{L}(2^{(\mathbb{N})})$ and so is analytic but not Borel.

Remark The sets $\mathbb{L}^*$ and $\mathbb{L}^{**}$ must surely be universal, in the generalized sense, for the analytic subsets of $2^{\mathbb{N}}$, but the above proof does not seem to yield this; it only shows that both sets have relative $\mathscr{G}_\delta$-subsets with the generalized universal property.

5 *K*-Lusin spaces (and *K*-analytic spaces continued)

In Sections 1–4 we have intentionally limited the number of notions introduced and theorems presented in order to emphasize more clearly some of the most basic results of the theory. This will no longer be the case in the remaining two sections. Distinguishing the wood from the trees will be increasingly difficult. These last sections are intended to give a rather detailed discussion of several recent topics of research, which, in addition to being instructive, will, we hope, be a useful source of reference.

This section gives the basic properties of *K*-Lusin spaces which follow from the proofs of the analogous theorems for *K*-analytic spaces given in the preceding sections, along with some additional elementary properties of both *K*-Lusin and *K*-analytic spaces. The role of proper maps is considered. The fact that the property of being *K*-analytic or *K*-Lusin is completely determined by compact subsets is brought out. The questions as to when a *K*-analytic space contains a non-empty compact perfect subset, when a *K*-analytic (*K*-Lusin) space is analytic (Lusin), and when a *K*-analytic space is metrizable are examined in considerable detail. Following this are two subsections focusing on disjoint countable unions and the disjoint Souslin operation. Finally, some consequences for the sets of extreme points of compact convex sets are pointed out.

5.1 Basic structure

Recall from Section 2.1 that a Hausdorff space X is called a *K*-Lusin space if there is an upper semi-continuous map

$$F:\mathbb{N}^{\mathbb{N}}\to\mathscr{K}(X)\,,\qquad X=\bigcup\{F(\sigma):\sigma\in\mathbb{N}^{\mathbb{N}}\}$$

such that for each $\sigma\neq\tau$ in $\mathbb{N}^{\mathbb{N}}$

$$F(\sigma)\cap F(\tau)=\varnothing\,.$$

A Hausdorff space is called a Lusin space if it is the injective continuous image of a closed subset of $\mathbb{N}^{\mathbb{N}}$.

We shall find it useful to consider a larger class of spaces than the K-analytic ones in order to see more clearly to what extent we actually need analyticity in some of the following theorems. We will call a Hausdorff space K-countably determined if it is the image of some subset of $\mathbb{N}^{\mathbb{N}}$ under an upper semi-continuous compact-valued map. Thus all K-analytic and all separable metric spaces are K-countably determined, and these spaces are all Lindelöf. The terminology comes from the fact that for a subset E of a compact Hausdorff space X the following are equivalent:

(a) E is K-countably determined;
(b) there is a sequence $\{K_n : n \in \mathbb{N}\}$ of compact subsets of X such that for each $x \in E$ and $y \in X \backslash E$ there is an $m \in \mathbb{N}$ such that $x \in K_m$ and $y \notin K_m$;
(c) there is a subset S of $\mathbb{N}^{\mathbb{N}}$ and compact subsets $\{K(\sigma|n) : \sigma|n \in \mathbb{N}^{\mathbb{N}}\}$ of X such that

$$E = \bigcup_{\sigma \in S} \bigcap_{n} K(\sigma|n).$$

Proof (a) implies (c). If $F : S \to \mathcal{K}(X)$ is the map given by (a), then

$$E = \bigcup_{\sigma \in S} \bigcap_{n} \operatorname{cl} F(\{\tau \in S : \tau|n = \sigma|n\}),$$

where the closure is taken in X.

(c) implies (a). If

$$E = \bigcup_{\sigma \in S} \bigcap_{n} K(\sigma|n),$$

then the map $F : S \to \mathcal{K}(X)$, defined by

$$F(\sigma) = \bigcap_{n} K(\sigma|n),$$

has the required properties.

(c) implies (b). The countable family $\{K(\sigma|n) : \sigma|n \in \mathbb{N}^{(\mathbb{N})}\}$ has the required property.

(b) implies (c). For $(\sigma_1, \sigma_2, \ldots, \sigma_n) = \sigma|n$, define

$$K(\sigma|n) = \bigcap_{i=1}^{n} K_{\sigma_i}.$$

The first fundamental characterizations of K-Lusin spaces are analogous to, and follow from the proof of, Theorem 2.8.1 for K-analytic spaces.

THEOREM 5.1.1 (Jayne, 1976) *For a Hausdorff space X the following are equivalent*:

(a) *X is a K-Lusin space*;
(b) *X is the injective continuous image of a Lindelöf $\mathscr{G}_\delta$-subset of some compact Hausdorff space*;
(c) *X is the injective continuous image of a (cozero)$_\delta$ subset of some compact Hausdorff space.*

The equivalence of (b) and (c) in this theorem is due to the elementary fact that a subset of a compact Hausdorff space is the countable intersection of cozero sets, if, and only if, it is a Lindelöf $\mathscr{G}_\delta$-subset. To see this, first suppose that B is a Lindelöf $\mathscr{G}_\delta$-subset of a compact space K, and write

$$B = \bigcap_{n=1}^{\infty} G_n,$$

with G_n open in K. Since K is completely regular, each G_n is the union of cozero sets. Since $B \subset G_n$ is Lindelöf, a countable number of these cozero sets cover B. Since the countable union of cozero sets is also a cozero set, this implies that there is a cozero set C_n with $B \subset C_n \subset G_n$. Hence

$$B = \bigcap_{n=1}^{\infty} C_n.$$

Conversely, since each cozero set is an $\mathscr{F}_\sigma$-set, it is K-analytic; and so, a countable intersection of cozero sets is K-analytic and consequently Lindelöf.

Note that every (cozero)$_\delta$ set is a $\mathscr{Z}_{\sigma\delta}$-set, and, as we will see in Proposition 5.2.2, a $\mathscr{Z}_{+\delta+\delta}$-set.

The proof of Theorem 2.5.2 gives

THEOREM 5.1.2 *If B is a K-Lusin subset of a Hausdorff space X, then $B \in S_d(\mathscr{F}(X))$.*

Analogous to Theorem 2.5.1 we have

THEOREM 5.1.3 *The image of a K-Lusin space under a disjoint upper semi-continuous map is K-Lusin.*

From the proof of Theorem 2.5.4 we get

THEOREM 5.1.4 *If $\mathscr{L}$ denotes the family of K-Lusin subsets of a Hausdorff space X, then $S_d(\mathscr{L}) = \mathscr{L}$. Consequently $\mathscr{L}_{(+,\delta)} = \mathscr{L}$. Similarly, if $\mathscr{L}^*$ denotes the family of Lusin subsets of X, then $S_d(\mathscr{L}^*) = \mathscr{L}^*$, and so $\mathscr{L}^*_{(+,\delta)} = \mathscr{L}^*$.*

The proof of Theorem 2.5.5 for K-analytic spaces gives

THEOREM 5.1.5 *If $\{X_n : n \in \mathbb{N}\}$ are K-Lusin (resp. K-countably determined, analytic, Lusin) spaces, then so is their product $\prod_{n=1}^{\infty} X_n$.*

Part (a) of the next theorem is Theorem 3.3.1, the proof of which also gives parts (b) and (c). This theorem clearly exhibits the relation between the separation properties of points and closed sets in a space X and the separation properties of K-analytic sets and sets in $\mathbf{S}(\mathscr{F}(X))$.

THEOREM 5.1.6 (First separation principle) *Let X be a Hausdorff space, A a K-analytic subset of X, and $C \in S(\mathscr{F}(x))$ with $A \cap C = \varnothing$.*

(a) (Theorem 3.3.1) *There is a set $B \in \mathscr{G}_{(\sigma,\delta)}(X)$ such that*

$$A \subset B \subset X \backslash C.$$

If C is K-analytic, then there are sets $B_1, B_2 \in \mathscr{G}_{(\sigma,\delta)}(X)$ such that

$$A \subset B_1 . \qquad C \subset B_2 , \qquad B_1 \cap B_2 = \varnothing .$$

(b) *If X is regular, then there are sets $B_1, B_2 \in \mathscr{F}_{(\sigma,\delta)}(X)$ such that*

$$A \subset B_1 , \qquad C \subset B_2 , \qquad B_1 \cap B_2 = \varnothing .$$

(c) (Frolík, 1967) *If X is completely regular, then there are sets $B_1, B_2 \in \mathscr{Z}_{(\sigma,\delta)}$ such that*

$$A \subset B_1 , \qquad C \subset B_2 , \qquad B_1 \cap B_2 = \varnothing ;$$

equivalently, there is a set $B \in \mathscr{Z}_{(\sigma,\delta)}(X)$ such that

$$A \subset B \subset X \backslash C.$$

As we will see in Proposition 5.2.2, we always have

$$\mathscr{Z}_{(\sigma,\delta)}(X) = \mathscr{Z}_{(+,\delta)}(X) .$$

$\mathscr{Z}_{(\sigma,\delta)}(X)$ is a σ-algebra, called "the σ-algebra of Baire subsets of X", and one easily obtains from the last theorem that, if a K-analytic space X is completely regular, then

$$\mathscr{Z}_{(\sigma,\delta)}(X) = \{B \subset X : B, X \backslash B \in \mathscr{F}_{(\sigma,\delta)}(X)\} .$$

A bijection f of a space X onto a space Y is called a Baire isomorphism if $f[\mathscr{Z}_{(\sigma,\delta)}(X)] = \mathscr{Z}_{(\sigma,\delta)}(Y)$. These isomorphisms are discussed in some detail in Section 6. It is known that every metrizable K-Lusin space is Baire isomorphic to a compact metric space; in fact, every uncountable such space is Baire isomorphic to [0, 1]. It appears likely that every completely regular K-Lusin space is Baire isomorphic to a compact space, but this problem remains open. The next theorem shows that it would suffice to prove that every space which is the countable intersection of cozero sets in its Stone–Čech compactification is Baire isomorphic to a compact space.

THEOREM 5.1.7 *Let X be a K-Lusin space and let*

$$\mathscr{B} = \{B \subset X : B, X\backslash B \in \mathscr{F}_{(\sigma,\delta)}(X)\}.$$

There exists a proper K-Lusin space Y, which is the countable intersection of cozero sets in its Stone–Čech compactification, and a continuous bijection f of Y onto X such that

$$f[\mathscr{Z}_{(\sigma,\delta)}(Y)] = \mathscr{B}.$$

If X is completely regular, then $\mathscr{B} = \mathscr{Z}_{(\sigma,\delta)}(Y)$ *and f is a Baire isomorphism.*

Proof Let Y and $f; Y \to X$ be the space and continuous bijection given by Theorem 5.1.1. If $E \in \mathscr{Z}_{(\sigma,\delta)}(Y)$, then $f(E)$ and $f(Y\backslash E)$ are disjoint K-analytic sets. Hence, by Theorem 5.1.6(a) there are sets $B_1, B_2 \in \mathscr{G}_{(\sigma,\delta)}(X)$ such that

$$f(E) \subset B_1 \subset X\backslash f(X\backslash E) = f(E),$$

$$f(X\backslash E) \subset B_2 \subset X\backslash f(E) = f(X\backslash E).$$

Thus $f(E) = B_1 = X\backslash B_2$ and $f(X\backslash E) = B_2 = X\backslash B_1$, and so $f(E)$ and $f(X\backslash E)$ are in $\mathscr{F}_{(\sigma,\delta)}(X)$.

If $H \in \mathscr{B}$, then, since f is continuous,

$$f^{-1}(H) \in \{B \in Y : B, Y\backslash B \in \mathscr{F}_{(\sigma,\delta)}(Y)\}.$$

Thus, by Theorem 5.1.6(c), $f^{-1}(H) \in \mathscr{Z}_{(\sigma,\delta)}(Y)$.

Finally, if X is completely regular, then it follows from Theorem 5.1.6(c) that $\mathscr{B} = \mathscr{Z}_{(\sigma,\delta)}(X)$.

The following general form of Novikov's approximation theorem comes from the proof of Theorem 3.4.1 and is due to Roy O. Davies, J. E. Jayne, A. J. Ostaszewski and C. A. Rogers (1977).

THEOREM 5.1.8 *Let X be a Hausdorff space and* (A_n) *a sequence of K-analytic sets in X such that*

$$\bigcap_{n=1}^{\infty} A_n = \varnothing.$$

(a) (Theorem 3.4.1) *There is a sequence* $(B_n) \subset \mathscr{G}_{(\sigma,\delta)}(X)$ *such that for all* $n \in \mathbb{N}$

$$A_n \subset B_n \quad \textit{and} \quad \bigcap_{n=1}^{\infty} B_n = \varnothing.$$

(b) *If* X *is regular, then there is a sequence* $(B_n)\subset\mathscr{F}_{(\sigma,\delta)}(X)$ *such that for all* $n\in\mathbb{N}$

$$A_n\subset B_n \quad \textit{and} \quad \bigcap_{n=1}^{\infty} B_n=\varnothing\,.$$

(c) *If* X *is completely regular, then there is a sequence* $(B_n)\subset\mathscr{Z}_{(+,\delta)}(X)$ *such that for all* $n\in\mathbb{N}$

$$A_n\subset B_n \quad \textit{and} \quad \bigcap_{n=1}^{\infty} B_n=\varnothing\,.$$

5.2 Properness

Recall that a proper map is a continuous map which takes closed sets to closed sets and has compact point inverses. Thus the inverse of a proper map is an upper semi-continuous compact valued map. More precisely we have (see the proof of Theorem 2.8.1 for the necessity)

PROPOSITION 5.2.1 *If* X *and* Y *are Hausdorff spaces and* $F:X\to\mathscr{K}(Y)$ *is any map, then* F *is upper semi-continuous, if, and only if,*

$$\pi:\mathrm{gr}\,F\to X$$

is a proper map, where $\mathrm{gr}\,F\subset X\times Y$ *denotes the graph of* F *and* π *denotes the projection on* X.

Another basic property, for our purposes, the sufficiency of which is contained in the proof of Theorem 2.8.1, is

PROPOSITION 5.2.2 *Let* X *and* Y *be Hausdorff spaces and* $f:X\to Y$ *a surjective proper map. Then* X *is regular, if, and only if,* Y *is regular.*

It now follows that if a Hausdorff space X admits a proper map onto an analytic (resp. Lusin, arbitrary) subset of $\mathbb{R}^{\mathbb{N}}$, then X is a regular K-analytic (resp. K-Lusin, K-countably determined) space. We will call such spaces proper K-analytic (resp. proper K-Lusin, proper K-countably determined). These spaces have especially good properties, which more closely resemble the separable metric situation. Numerous characterizations of proper K-analytic and proper K-Lusin spaces are given in Theorems 5.8.7 and 5.8.9. For example, the class of proper K-analytic spaces is the smallest class of Hausdorff spaces containing all compact Hausdorff spaces, all analytic subsets of $\mathbb{R}^{\mathbb{N}}$, all finite or countable products, and all closed subspaces.

Another property of proper maps which we will make use of is

PROPOSITION 5.2.3 *If f is a continuous map of a compact Hausdorff space X onto a Hausdorff space Y and the restriction of f to a dense subspace E of X is proper, then* $f^{-1}\circ f(E)=E$.

Proof Let $y\in f(E)$ and $x\in X\backslash E$. Since $f^{-1}(y)\cap E$ is compact, there is a closed neighbourhood F of x in X such that

$$F\cap f^{-1}(y)\cap E=\varnothing .$$

Since $F\cap E$ is closed in E, $f(E\cap E)$ is closed in $f(E)$. Thus $(\mathrm{cl}\, f(F\cap E))\cap f(E)=f(F\cap E)$, where $\mathrm{cl}\, f(F\cap E)$ is the closure of $f(F\cap E)$ in Y. Hence $y\notin \mathrm{cl}\, f(F\cap E)$. Now $x\in \mathrm{cl}\,(F\cap E)$, and so $f(x)\in \mathrm{cl}\, f(F\cap E)$. Consequently, $f(x)\neq y$; i.e. $f^{-1}\circ f(E)=E$.

If X is a proper K-analytic space and $f:X\to Y\subset[0,1]^{\mathbb{N}}$ a proper map onto Y, then

$$Y=\bigcup_{\sigma}\bigcap_{n} Z(\sigma|n)\,,\qquad Z(\sigma|n)\in\mathscr{Z}([0,1]^{\mathbb{N}})\,,$$

and the continuous extension $\hat{f}$ of f to the Stone–Čech compactification βX of X onto the closure $\bar{Y}$ of Y in $[0,1]^{\mathbb{N}}$ has the property (by Proposition 5.2.3)

$$X=\bigcup_{\sigma}\bigcap_{n}\hat{f}^{-1}(Z(\sigma|n))\,,\qquad \hat{f}^{-1}(Z(\sigma|n))\in\mathscr{Z}(\beta X)\,.$$

Thus $X\in\mathbf{S}(\mathscr{Z}(\beta X))$.

Conversely, if $X\in\mathbf{S}(\mathscr{Z}(\beta X))$, then we can write

$$X=\bigcup_{\sigma}\bigcap_{n} Z(f_{\sigma|n})\,,$$

where $Z(f_{\sigma|n})$ is the zero set of a continuous function

$$f_{\sigma|n}:\beta X\to\mathbb{R}\,.$$

Re-index the $\{f_{\sigma|n}:\sigma|n\in\mathbb{N}^{(\mathbb{N})}\}$ as $\{f_n: n\in\mathbb{N}\}$ and define $f:\beta X\to\mathbb{R}^{\mathbb{N}}$ by

$$f(x)=(f_1(x),f_2(x),\ldots)\,.$$

Then $f^{-1}\circ f(X)=X$ and consequently $f|_X$ is a proper map onto $f(X)$.

Similarly we have that X is a proper K-Lusin space, if, and only if, $X\in\mathscr{Z}_{(\sigma,\delta)}(\beta X)$; and that the following are equivalent:

(a) X is proper K-countably determined;
(b) there is a sequence $\{Z_n: n\in\mathbb{N}\}\subset\mathscr{Z}(\beta X)$, such that for each $x\in X$ and $y\in\beta X\backslash X$ there is an $m\in\mathbb{N}$ such that

$$z\in Z_m\quad\text{and}\quad y\notin Z_m\,;$$

(c) there is a subset S of $\mathbb{N}^{\mathbb{N}}$ and sets

$$\{Z(\sigma|n) : \sigma|n \in \mathbb{N}^{(\mathbb{N})}\} \subset \mathscr{Z}(\beta X)$$

such that

$$X = \bigcup_{\sigma \in S} \bigcap_{n} Z(\sigma|n) .$$

The following three progressively pathological, though simple, examples will be used to illustrate most of the abnormalities of non-metrizable regular K-analytic spaces.

Example 5.2.1 The simplest example of a *Lusin space which is not proper K-Lusin* (or even proper K-countably determined) is the countable discrete space $\mathbb{N}$ plus one point $x \in \beta\mathbb{N} \backslash \mathbb{N}$, with the relative topology induced by $\beta\mathbb{N}$. $\mathbb{N} \cup \{x\}$ is not proper K-countably determined, since if it were, there would be a zero set $Z \in \mathscr{Z}(\beta(\mathbb{N} \cup \{x\}))$ with

$$x \in Z \subset \mathbb{N} \cup \{x\} .$$

Therefore, since

$$\{x\} = Z \cap \bigcap_{n=1}^{\infty} (\beta(\mathbb{N} \cup \{x\}) \backslash \{n\})$$

we would have that x is a $\mathscr{G}_\delta$-subset of $\beta(\mathbb{N} \cup \{x\})$ (which is equal to $\beta\mathbb{N}$). But this is not the case. Hence $\mathbb{N} \cup \{x\}$ is not K-countably determined.

Example 5.2.2 A completely regular σ-compact space which is neither K-Lusin nor proper K-countably determined. Let Y be the discrete union of countably many copies of the one-point compactification of an uncountable discrete set. Let X be the set Y with the countable set of compactifying points identified. Give X the quotient topology. Then X is σ-compact, but neither K-Lusin nor K-countably determined, since, if there were a subset S of $\mathbb{N}^{\mathbb{N}}$ and an upper semi-continuous map

$$F : S \to \mathscr{K}(X) , \qquad X = \bigcup_{\sigma \in \mathbb{N}^{\mathbb{N}}} F(\sigma) ,$$

then we would have that

$$F(S \backslash \{\sigma_0\}) = X \backslash F(\sigma_0) ,$$

where σ_0 is the unique point of S such that $F(\sigma_0)$ contains the identified set of compactifying points, is a Lindelöf space. But this is not the case, since $X \backslash F(\sigma_0)$ is discrete and uncountable.

The next example is due to M. Talagrand (1979*a*).

Example 5.2.3 A non-σ-compact space X which is a $\mathcal{K}_{\sigma\delta}$-set in βX, which is neither K-Lusin nor K-countably determined. Let p denote a point not in $\mathbb{N}^{\mathbb{N}}$. Give the set $X = \{p\} \cup \mathbb{N}^{\mathbb{N}}$ the topology in which each point of $\mathbb{N}^{\mathbb{N}}$ is open and p has, as an open neighbourhood base, the complements of sets which are closed and discrete in $\mathbb{N}^{\mathbb{N}}$ in its usual topology. With this topology X is a completely regular Hausdorff space. The map $F: \mathbb{N}^{\mathbb{N}} \to \mathcal{K}(X)$, defined by

$$F(\sigma) = \{p, \sigma\},$$

is clearly upper semi-continuous. Thus X is a K-analytic space. Consequently X is a regular Lindelöf space, and so a normal space.

The infinite compact subsets of X are precisely the sets of the form $\{p\} \cup E$, where E is a relatively compact subset of $\mathbb{N}^{\mathbb{N}}$ in its usual topology. Each set of this form is compact since the complement of an open set in X containing p is closed and discrete in $\mathbb{N}^{\mathbb{N}}$ and so must intersect E in a finite set. Conversely, if K is an infinite compact set in X, then clearly $p \in K$ and each sequence in $K \backslash \{p\}$ is not closed and discrete in $\mathbb{N}^{\mathbb{N}}$ (i.e. has a convergent subsequence in the usual topology on $\mathbb{N}^{\mathbb{N}}$). Consequently $K \backslash \{p\}$ is relatively compact. It follows that X is not σ-compact (in fact, contains no dense σ-compact subsets) and contains no non-empty compact perfect subsets.

We claim that X is a $\mathcal{K}_{\sigma\delta}$-subset of βX; specifically that

$$X = \bigcup_{\sigma} \bigcap_{n} \mathrm{cl}\, I(\sigma|n) = \bigcap_{n} \bigcup_{\sigma} \mathrm{cl}\, I(\sigma|n), \tag{$*$}$$

where $I(\sigma|n) = \{\tau \in \mathbb{N}^{\mathbb{N}}: \tau|n = \sigma|n\}$ and "cl" denotes closure in βX. This equation follows from three observations: (i) for each $\sigma \in \mathbb{N}^{\mathbb{N}}$,

$$\{p\} \cup \bigcap_{n} I(\sigma|n) = \bigcap_{n} \mathrm{cl}\, I(\sigma|n);$$

(ii) for each $\sigma|n \neq \tau|n$,

$$\mathrm{cl}\, I(\sigma|n) \cap \mathrm{cl}\, I(\tau|n) = \{p\};$$

(iii) the union and intersection in $(*)$ can be interchanged. The first two follow from the normality of X. The third is contained in Proposition 5.7.3. contained in Proposition 5.2.3.

This space is neither K-Lusin nor proper K-countably determined for the same reason that Example 5.1.2 is not.

5.3 *k*-spaces

A Hausdorff space X is called a k-space if each subset, with the property that its intersection with every compact subset of X is compact, is a closed set. Locally compact spaces and first countable spaces are k-spaces. In fact,

k-spaces are precisely those Hausdorff spaces which are quotients of locally compact Hausdorff spaces.

For each Hausdorff space X we denote by X_k the set X with the topology whose closed sets are precisely those which intersect each compact subset of X in a compact set. The space X_k, called the k-modification of X, is a k-space, has the same compact sets as X, and the identity map of X_k onto X is a continuous map whose restriction to each compact subset is a homeomorphism.

PROPOSITION 5.3.1 *Let X and Y be Hausdorff spaces and $f: X \to Y$ a surjective proper map. Then X is a k-space, if, and only if, Y is a k-space. Consequently, every proper K-countably determined (in particular, proper K-analytic) space is a k-space.*

Since the closed subspace of a k-space is clearly a k-space, Proposition 5.3.1, along with Proposition 5.2.1, implies

PROPOSITION 5.3.2 *If X is a k-space, Y a Hausdorff space, and $f: X \to \mathcal{K}(Y)$, then the graph of f in $X \times Y$ is a k-space.*

Suppose X is a k-space and $f: X \to Y$ is a map whose restriction to each compact subset of X is continuous. Let F be a closed set in Y, and for each compact set K in X let $f|_K$ denote the restriction of f to K. Then

$$f^{-1}(F) \cap K = (f|_K)^{-1}(F)$$

for each compact set K in X. Thus $f^{-1}(F)$ is closed in X, i.e. f is continuous. We have proved

PROPOSITION 5.3.3 *If X and Y are Hausdorff spaces and $f: X \to Y$ is a map whose restriction to each compact subset of X is continuous, then f is a continuous map from X_k to Y_k.*

The next theorem shows that the property of being K-analytic, K-Lusin, etc., is completely determined by the compact subsets of the space. For K-analytic spaces this fact was first noted by Frolík (1970*c*, p. 79).

THEOREM 5.3.4 *A Hausdorff space X is K-analytic (resp. analytic, K-Lusin, Lusin, K-countably determined), if, and only if, its k-modification X_k is K-analytic (resp. analytic, K-Lusin, Lusin, K-countably determined).*

Proof If X_k is K-analytic, then X is K-analytic, since the identity map of X_k onto X is continuous. Conversely, suppose that X is K-analytic and $F: \mathbb{N}^{\mathbb{N}} \to$

$\mathcal{K}(X)$, $F(\mathbb{N}^{\mathbb{N}}) = X$ is an upper semi-continuous map. By Proposition 5.3.2, the graph of F is a k-space. By Proposition 5.3.3, the projection of the graph of F onto X is a continuous map onto X_k. Hence X_k is K-analytic. This proof applies equally well to the other types of spaces.

Of the three examples in Section 5.2, Examples 5.2.2 and 5.2.3 are k-spaces, but Example 5.2.1, namely

$$X = \mathbb{N} \cup \{x\} \subset \beta\mathbb{N},$$

is not a k-space. Its k-modification X_k is discrete, since all compact subsets of X are finite.

5.4 Compact perfect subsets

This section is a continuation of Section 3.5. We call a subset of a Hausdorff space perfect if it is non-empty, closed and dense-in-itself. According to Theorem 3.5.1, if X is an uncountable K-analytic space, with the property that each point is the intersection of a decreasing sequence of sets that are alternately open and closed, then X contains a compact subset which admits a continuous map onto the Cantor set. If X is regular, then this condition on the points is simply that each point is a $\mathcal{G}_\delta$-set. In view of the next proposition, the conclusion of Theorem 3.5.1 is equivalent to X containing a compact perfect subset.

PROPOSITION 5.4.1 *Let X be a compact Hausdorff space.*

(a) *If X contains a perfect subset, then X admits a continuous map onto the unit interval* $[0, 1]$.
(b) *If X contains no perfect subsets and* $f: X \to Y$ *is a continuous map onto a Hausdorff space Y, then Y contains no perfect subsets.*

Proof (a) Let P be a perfect subset of X. If $x, y \in P$, then there exist disjoint closed neighbourhoods of x and y in X. The closures of the interiors of these neighbourhoods intersected with P are disjoint perfect subsets of P, denoted by P_1^1 and P_1^2. Now, continuing in this way, suppose the perfect sets $P_n^1, P_n^2, \ldots, P_n^{2^n}$ have been defined. For each k, $1 \leq k \leq 2^n$, let P_{n+1}^{2k-1} and P_{n+1}^{2k} be disjoint perfect subsets of P_n^k. Let

$$Q_n = \bigcup_{k=1}^{2^n} P_n^k \quad \text{and} \quad Q = \bigcap_{n=1}^{\infty} Q_n.$$

Then Q is a non-empty compact subset, since Q_n is closed in X for each n and $Q_{n+1} \subset Q_n$. Define f_n on Q to be the characteristic function of the set

$$\bigcup_{k=1}^{2^{n-1}} P_n^{2k} \cap Q .$$

Then f_n is continuous on Q, since the above set is clopen in Q. Now define $f:Q \to [0, 1]$ by

$$f(x) = \sum_{n=1}^{\infty} \frac{1}{2^n} f_n(x) .$$

Then f is a continuous function of Q onto $[0, 1]$. By the Tietze extension theorem we can extend f to a continuous function of X onto $[0, 1]$.

(b) Suppose Y contains a perfect subset P. Let K be the intersection of all closed subsets F of X such that $f(F) = P$. By hypothesis K contains an isolated point x. Write $K' = K \setminus \{x\}$. Then K' is compact and $f(K')$ is a proper compact subset of P. Thus $f(x)$ is an isolated point of P, which is impossible. Thus Y contains no perfect subsets.

Examples of compact spaces without perfect subsets are one-point compactifications of discrete spaces, and spaces of ordinals less than or equal to a fixed ordinal with the order topology. Clearly the countable union of compact spaces without perfect subsets contains no compact perfect subset. Example 5.2.3 is a non-σ-compact regular K-analytic space which contains no compact perfect subsets. Before characterizing regular K-analytic spaces which contain compact perfect subsets, we extend Proposition 5.4.1(b) to regular K-analytic spaces.

THEOREM 5.4.2 *If X is a regular K-analytic space which contains no compact perfect subsets and f is a continuous map of X onto a regular Hausdorff space Y, then Y contains no compact perfect subsets.*

Proof Let $\hat{f}: \beta X \to \beta Y$ be the continuous extension of f to the Stone–Čech compactification βX of X. Suppose P is a compact perfect subset of Y. According to Proposition 5.4.1(a), there is a continuous surjection $g: P \to [0, 1]$.

Let μ denote Lebesgue measure on $[0, 1]$ and μ^* its associated outer measure. Then

$$I: \mathcal{P}(X) \to \mathbb{R} ,$$

where $\mathcal{P}(X)$ is the class of all subsets of X, defined by

$$I(E) = \mu^*(g(f(E) \cap P)),$$

is a Choquet capacity on E (see Section 3.5). Since $X \in \mathbf{S}(\mathscr{K}(\beta X))$ (Theorem 2.5.2), it follows that X is I-capacitable; in particular, there must be a compact subset K of X with

$$I(K) = \mu^*(g(f(K) \cap P)) > 0 .$$

Thus $g(f(K) \cap P)$ is an uncountable compact subset of $[0, 1]$, and so, by Corollary 3.5.2, contains a copy of the Cantor set, which is perfect. It now follows from Proposition 5.4.1(b) that K contains a compact perfect subset.

THEOREM 5.4.3 *For a regular K-analytic space X the following are equivalent:*

(a) *X contains a compact perfect subset;*
(b) *X admits a continuous real-valued function with uncountable range;*
(c) *X admits a continuous function onto an uncountable regular Hausdorff space Y in which every point is a $\mathscr{G}_\delta$-set;*
(d) $\mathscr{Z}_{(\sigma,\delta)}(X) \neq \mathscr{Z}_{\sigma\delta}(X)$.

If X is a regular K-Lusin space (or more generally, the countable-to-one continuous image of a space which is the disjoint upper semi-continuous compact valued image of an analytic subset of $\mathbb{N}^{\mathbb{N}}$), then (a)–(d) *are equivalent to*:

(e) *X is not the countable union of compact subspaces which contain no perfect subsets. In particular, every non-σ-compact K-Lusin space contains a compact perfect subset.*

The statements (a) *and* (e) *are also equivalent without regularity, when X is a K-Lusin space.*

Proof (a) implies (b). This follows from Proposition 5.4.1(a) and the Tietze extension theorem.

(b) implies (c). Immediate.

(c) implies (d). By Theorem 3.5.1, Y contains a compact perfect subset. Thus, by Theorem 5.4.2, X must contain a compact perfect subset K. It will suffice to show that

$$\mathscr{Z}_{(\sigma,\delta)}(K) \neq \mathscr{Z}_{\sigma\delta}(X) .$$

By Proposition 5.4.1, there is a continuous surjection $f: K \to [0, 1]$. By Theorem 4.3.2, there is a set $B \subset [0, 1]$ such that

$$B \notin \mathscr{Z}_{\sigma\delta\sigma\delta}([0, 1]) .$$

We will show that $f^{-1}(B) \notin \mathscr{Z}_{\sigma\delta}(K)$.

Suppose to the contrary that $f^{-1}(B)\in\mathscr{Z}_{\sigma\delta}(K)$. Let $\{Z(f_{nm})\colon n, m\in\mathbb{N}\}$ be zero sets of continuous real-valued functions on K such that

$$f^{-1}(B)=\bigcap_{n=1}^{\infty}\bigcup_{m=1}^{\infty}Z(f_{nm}).$$

Re-index $\{f_{nm}\colon n, m\in\mathbb{N}\}$ by $\{f_n : n\in\mathbb{N}\}$ and define a continuous function

$$\Phi : K\to[0, 1]\times\mathbb{R}^{\mathbb{N}}$$

by

$$\Phi(x)=(f(x), f_1(x), f_2(x), \ldots).$$

Then

$$\Phi(f^{-1}(B))\in\mathscr{Z}_{\sigma\delta}(\Phi(K)).$$

Let π denote the projection of $[0, 1]\times\mathbb{R}^{\mathbb{N}}$ onto $[0, 1]$. Then

$$\pi(\Phi(K))=[0, 1],$$

and

$$\pi^{-1}(B)\cap\Phi(K)=\Phi(f^{-1}(B)).$$

By a theorem of Kuratowski and Ryll-Nardzewski (1965) (see Dellacherie, Part 2, Chapitre II, Section 2 in this volume) there is a function $g:[0, 1]\to\Phi(K)$ such that $g([0, 1])$ is a $\mathscr{G}_\delta$-set in $\Phi(K)$, such that $g^{-1}(G)$ is a $\mathscr{Z}_\sigma$-set in $[0, 1]$ for each open subset G of $\Phi(K)$, and such that, for each $x\in[0, 1]$,

$$g(x)\in\pi^{-1}(x)\cap\Phi(K).$$

Thus

$$g(B)\notin\mathscr{Z}_{\sigma\delta}(\Phi(K)),$$

since $B\notin\mathscr{Z}_{\sigma\delta\sigma\delta}([0, 1])$. But

$$g(B)=g([0, 1])\cap\Phi(f^{-1}(B))\in\mathscr{Z}_{\sigma\delta}(\Phi(K)).$$

This contradiction implies that $f^{-1}(B)\notin\mathscr{Z}_{\sigma\delta}(K)$. Clearly $f^{-1}(B)\in\mathscr{Z}_{(\sigma,\delta)}(K)$.

(d) implies (a). Let $B\in\mathscr{Z}_{(\sigma,\delta)}(X)$ and let $\{f_n\colon n\in\mathbb{N}\}$ be continuous real-valued functions on X such that

$$B\in\{Z(f_n)\colon n\in\mathbb{N}\}_{(\sigma,\delta)}.$$

Define $f:X\to\mathbb{R}^{\mathbb{N}}$ by $f(x)=(f_1(x), f_2(x), \ldots)$. Then

$$f^{-1}\circ f(B)=B.$$

If X does not contain a compact perfect subset, then Theorem 5.4.2 together with Corollary 3.5.2 imply that $f(X)$ is countable. This, in turn, implies that $f(B)$ and $f(X \backslash B)$ are countable and so

$$B = f^{-1} \circ f(B), \qquad X \backslash B = f^{-1} \circ f(X \backslash B)$$

are in $\mathscr{Z}_\sigma(X)$. So, in fact, $\mathscr{Z}_{(\sigma,\delta)}(X) = \mathscr{Z}_\sigma(X)$.

(e) implies (a). Suppose X is a K-Lusin space which is not the countable union of compact subspaces which contain no perfect subsets. Thus we can suppose that X is not σ-compact. Let $F : \mathbb{N}^{\mathbb{N}} \to \mathscr{K}(X)$, $F(\mathbb{N}^{\mathbb{N}}) = X$, be a disjoint upper semi-continuous map. The projection

$$\pi_X : \text{gr } F \to X$$

of the graph of F, gr F, is a continuous bijection. Since X is not σ-compact, gr F is not σ-compact. The projection

$$\pi : \text{gr } F \to E = \{\sigma \in \mathbb{N}^{\mathbb{N}} : F(\sigma) \neq \varnothing\}$$

is, according to Proposition 5.2.1, a proper map onto the closed subset E of $\mathbb{N}^{\mathbb{N}}$, and so E uncountable. Hence, by Corollary 3.5.2, E contains a compact perfect subset K. As π is a proper map, $\pi^{-1}(K)$ is a compact subset of gr F, which, by Proposition 5.4.1(b), must contain a compact perfect subset. Finally, since the restriction of π_X to $\pi^{-1}(K)$ is a homeomorphism, it follows that X contains a compact perfect subset.

The proof for the other type of K-analytic space follows from the above argument and the fact that the countable-to-one continuous image of a compact perfect space is perfect.

(a) implies (e). Let K be a compact perfect subset of X. If X is the countable union of compact subsets which contain no perfect subsets, then so is K. Thus, by Proposition 5.4.1(b) and Corollary 3.5.2, the range of every continuous real-valued function on K is countable, contrary to Proposition 5.4.1(a).

5.5 Analytic and Lusin spaces

It is often useful to know when a K-analytic or K-Lusin space is in fact analytic or Lusin. We have already used Theorem 5.5.1 in Section 2.11 to prove that every Banach space which is K-Lusin in its weak topology is in fact Lusin in that topology. This section is devoted to characterizing analytic and Lusin spaces among K-analytic and K-Lusin spaces.

An immediate consequence of the next theorem is that metrizable K-analytic (K-Lusin) spaces are analytic (Lusin).

THEOREM 5.5.1 *Let X be a K-analytic (resp. K-Lusin) space. The following statements are equivalent:*

(a) *X is an analytic (resp. Lusin) space;*
(b) *X is the continuous image of a separable metric space;*
(c) *$X \times X$ is hereditarily Lindelöf;*
(d) *$\Delta = \{(x, x) : x \in X\}$ is in $\mathscr{G}_{(\sigma,\delta)}(X \times X)$, where $\mathscr{G}$ denotes the family of open sets in $X \times X$;*
(e) *$(X \times X) \backslash \Delta$ is in $\mathscr{F}_{(\sigma,\delta)}(X \times X)$;*
(f) *$(X \times X) \backslash \Delta$ is in $\mathbf{S}(\mathscr{F}(X \times X))$;*
(f′) *$(X \times X) \backslash \Delta$ is K-analytic (resp. K-Lusin);*
(g) *$(X \times X) \backslash \Delta$ is Lindelöf;*
(h) *there is a sequence $\{(A_n, A'_n) : n \in \mathbb{N}\}$ of pairs of K-analytic (resp. K-Lusin) subsets of X such that*
 (1) *$A_n \cup A'_n = X$ for each $n \in \mathbb{N}$, and*
 (2) *for each $x \neq x'$ in X, there is an $n \in \mathbb{N}$ such that $x \notin A_n$ and $x' \notin A'_n$.*

If, furthermore, X is regular, then (a)–(h) *are equivalent to each of the following:*

(i) *there is a sequence $\{G_n : n \in \mathbb{N}\}$ of open sets in X such that if $x \neq y$ in X, then there is an $n \in \mathbb{N}$ such that $x \in G_n$ and $y \notin G_n$;*
(j) *there is a sequence $\{f_n : n \in \mathbb{N}\}$ of continuous real-valued functions on X which separate the points (i.e. if $x \neq x'$ in X, then there is an $n \in \mathbb{N}$ such that $f_n(x) \neq f_n(x')$);*
(k) *there is a continuous map $f : X \times X \to \mathbb{R}^{\mathbb{N}}$ such that $f^{-1} \circ f(\Delta) = \Delta$;*
(l) *$(X \times X) \backslash \Delta$ is in $\mathbf{S}(\mathscr{Z}(X \times X))$;*
(m) *Δ is in $\mathbf{S}(\mathscr{Z}(X \times X))$;*
(n) *Δ is in $\mathscr{Z}(X \times X)$;*
(o) *Δ is a $\mathscr{G}_\delta$-subset of $X \times X$;*
(p) *Δ is in $\mathbf{S}(\mathscr{G}(X \times X))$.*

Proof We shall prove the following implications:

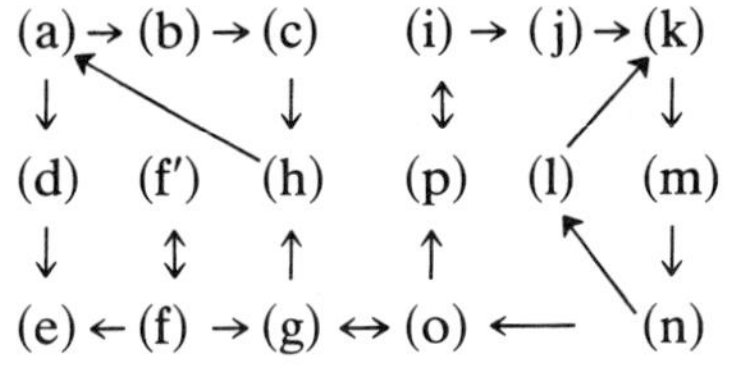

(a) implies (b). Immediate.

(b) implies (c). If $f : M \to X$ is a continuous map of a separable metric space M onto X, then $(m_1, m_2) \to (f(m_1), f(m_2))$ is a continuous map of $M \times M$ onto $X \times X$. Since the separable metric space $M \times M$ is hereditarily

Lindelöf, and the continuous image of a Lindelöf space is Lindelöf, it follows that $X \times X$ is hereditarily Lindelöf.

(c) implies (h). If $X \times X$ is hereditarily Lindelöf, then $(X \times X)\backslash \Delta$ is an open Lindelöf subset, where Δ is the diagonal $\{(x, y): x = y\}$. Hence there is a sequence $\{(G_n, G'_n): n \in \mathbb{N}\}$ of pairs of open subsets X such that

$$(X \times X)\backslash \Delta = \bigcup_{n=1}^{\infty} G_n \times G'_n ,$$

and so $A_n = X \backslash G_n$ and $A'_n = X \backslash G'_n$ give the required sequence of pairs.

(g) implies (h). Note that in (c) implies (h) we only used that $(X \times X)\backslash \Delta$ was Lindelöf.

(h) implies (a). Let $\{(A_n, A'_n): n \in \mathbb{N}\}$ be the given sequence, and let $K^0_n : \mathbb{N}^{\mathbb{N}} \to \mathcal{K}(X)$ and $K^1_n : \mathbb{N}^{\mathbb{N}} \to \mathcal{K}(X)$ with $K^0_n(\mathbb{N}^{\mathbb{N}}) = A_n$, $K^1_n(\mathbb{N}^{\mathbb{N}}) = A'_n$ be upper semi-continuous maps. Let $L_n : \{0, 1\} \times \mathbb{N}^{\mathbb{N}} \to \mathcal{K}(X)$ be the upper semi-continuous map defined by

$$L_n(i, \sigma) = K^i_n(\sigma) ,$$

and $L : (\{0, 1\} \times \mathbb{N}^{\mathbb{N}})^{\mathbb{N}} \to \mathcal{K}(X^{\mathbb{N}})$ the upper semi-continuous map defined, for

$$(\mathbf{i}, \boldsymbol{\alpha}) = ((i_1, \alpha_1), (i_2, \alpha_2), \ldots) \in (\{0, 1\}, \mathbb{N}^{\mathbb{N}})^{\mathbb{N}} ,$$

by

$$L(\mathbf{i}, \boldsymbol{\alpha}) = \prod_{n=1}^{\infty} L_n(i_n, \alpha_n) .$$

For each $x \in X$ let $\mathbf{x}$ denote the constant sequence $(x, x, \ldots)$ in $X^{\mathbb{N}}$. Then the diagonal $D = \{\mathbf{x}: x \in X\}$ is a closed subset of $X^{\mathbb{N}}$ and is homeomorphic to X.

Now let F be the closed subset of $(\{0, 1\} \times \mathbb{N}^{\mathbb{N}})^{\mathbb{N}}$ defined by

$$F = \{(\mathbf{i}, \boldsymbol{\alpha}): L(\mathbf{i}, \boldsymbol{\alpha}) \cap \mathrm{D} \neq \emptyset\} .$$

We claim that the map L', defined, for each $(\mathbf{i}, \boldsymbol{\alpha}) \in F$, by

$$L'(\mathbf{i}, \boldsymbol{\alpha}) = L(\mathbf{i}, \boldsymbol{\alpha}) \cap D$$

is a continuous (point-valued) map of F onto D, and that it is one-to-one if each K^i_n is a disjoint upper semi-continuous map. This will complete the proof, since F is a Polish space.

Since for each n, $L_n(\{0, 1\} \times \mathbb{N}^{\mathbb{N}}) = X$, it follows that

$$L((\{0, 1\} \times \mathbb{N}^{\mathbb{N}})^{\mathbb{N}}) = X^{\mathbb{N}} \quad \text{and} \quad L'(F) = 0 .$$

Now to show that $L'(\mathbf{i}, \boldsymbol{\alpha})$ is a single point for each $(\mathbf{i}, \boldsymbol{\alpha})$ in F, suppose $x, y \in L'(\mathbf{i}, \boldsymbol{\alpha})$. If $x \neq y$, then there is an m such that $x \in A_m$ and $y \notin A'_m$. This means that both x and y can not belong to $L_m(i_m, \alpha_m)$. Hence $\mathbf{x}$ and $\mathbf{y}$ can not both belong to $L(\mathbf{i}, \boldsymbol{\alpha})$. Thus we must have $\mathbf{x} = \mathbf{y}$. The injectivity of L, when each map K_n^i is disjoint, follows similarly.

(a) implies (d). If X is analytic, then $X \times X$ is also analytic by the argument used to show that (b) implies (c). Thus let $f : \mathbb{N}^{\mathbb{N}} \to X \times X$ be a continuous map onto $X \times X$. Then $f^{-1}((X \times X)\backslash \Delta)$ is open in $\mathbb{N}^{\mathbb{N}}$ and, consequently, is a Polish space. Likewise $f^{-1}(\Delta)$ is closed in $\mathbb{N}^{\mathbb{N}}$ and so is a Polish space. It follows that $(X \times X)\backslash \Delta$ and Δ are disjoint analytic sets in $X \times X$. Thus by Theorem 5.1.6 there is a set $B \in \mathcal{G}_{(\sigma,\delta)}(X \times X)$ such that

$$\Delta \subset B \subset (X \times X)\backslash((X \times X)\backslash \Delta) = \Delta ;$$

i.e. $B = \Delta$.

(d) implies (e). Immediate.

(e) implies (f). This follows from Corollary 2.3.3.

(f) implies (f'). This follows from Theorem 2.5.4.

(f') implies (f). This follows from Theorem 2.5.2.

(f') implies (g). This follows from Theorem 2.7.1.

(g) implies (o). Since $X \times X$ is completely regular and $(X \times X)\backslash \Delta$ is an open subset, for each point (x, y), with $x \neq y$, there are closed neighbourhoods F_x of x and F_y of y in X such that

$$F_x \times F_y \subset (X \times X)\backslash \Delta .$$

Since $(X \times X)\backslash \Delta$ is Lindelöf, there are points $\{(x_n, y_n) : n \in \mathbb{N}\}$ such that

$$(X \times X)\backslash \Delta = \bigcup_{n=1}^{\infty} F_{x_n} \times F_{y_n} ;$$

i.e.

$$\Delta = \bigcap_{n=1}^{\infty} (X \times X)\backslash(F_{x_n} \times F_{y_n}) .$$

(o) implies (g). If Δ is a $\mathcal{G}_\delta$-set in $X \times X$, then $(X \times X)\backslash \Delta$ is an $\mathcal{F}_\sigma$-set. Since each closed subset of the K-analytic space $X \times X$ is K-analytic, it follows that $(X \times X)\backslash \Delta$ is K-analytic and so Lindelöf.

(o) implies (p). Immediate.

(p) implies (i). Write the diagonal Δ in $X \times X$ as

$$\Delta = \bigcup_{\sigma} \bigcap_{n} G(\sigma|n) , \qquad G(\sigma|n) \in \mathcal{G}(X \times X) .$$

Since Δ is closed in the K-analytic space $X \times X$, Δ is K-analytic and so Lindelöf. Now for each $\sigma|n \in \mathbb{N}^{(\mathbb{N})}$ and $(x, x) \in G(\sigma|n)$ there is an open subset $U(x, \sigma|n)$ of X such that

$$U(x, \sigma|n) \times U(x, \sigma|n) \subset G(\sigma|n) .$$

Since Δ is Lindelöf there is a set $\{x_m \in X: m \in \mathbb{N}\}$ such that for each $n \in \mathbb{N}$

$$\Delta \cap \bigcup_{m=1}^{\infty} [U(x_m, \sigma|n) \times U(x_m, \sigma|n)] = \Delta .$$

Let $G^*(\sigma|n)$ be the union of those sets, in

$$\{U(x_m, \sigma|n) \times U(x_m, \sigma|n): m \in \mathbb{N}\},$$

which are contained in $G(\sigma|n)$. Then

$$\Delta = \bigcup_{\sigma} \bigcap_{n} G^*(\sigma|n),$$

and the countable family of sets $\{U(x_m, \sigma|n): m \in \mathbb{N}, \sigma|n \in \mathbb{N}^{(\mathbb{N})}\}$ have the required property.

(i) implies (p). Let $\{G_n: n \in \mathbb{N}\}$ be the given sequence of distinct open sets, which we may suppose to be closed under finite intersections. Each $x \in X$ has associated a (possibly finite) subsequence $\{G^x_{n_i} = G_{n_i}: x \in G_{n_i}\}$. By discarding some sets, if necessary, and re-indexing we can suppose that

$$G^x_1 \supset G^x_2 \supset G^x_3 \supset \cdots,$$

where these sets are all distinct or eventually contain but a single point. Since X is Lindelöf, there are points $\{x_m: m \in \mathbb{N}\}$ such that

$$X = \bigcup_{m=1}^{\infty} G^{x_m}_1$$

and the sets in $\{G^{x_m}_1: m \in \mathbb{N}\}$ are distinct, with the possible exception of one set occurring from some stage onwards; i.e. $G^{x_n}_1 = G^{x_p}_1$, $n \geqslant p$,

$$G(m) = \begin{cases} G^{x_m}_1, & \text{for each distinct } G^{x_m}_1, \\ \varnothing, & \text{if } G^{x_m}_1 = G^{x_{m-1}}_1 . \end{cases}$$

Now suppose that $G(\sigma|n)$ has been defined for n and for all $\sigma \in \mathbb{N}^{\mathbb{N}}$, so that $G(\sigma|n) = G^x_n$ for some $x \in X$ or is empty, and

$$X = \bigcup_{\sigma_n=1}^{\infty} G(\sigma|n-1, \sigma_n).$$

As before, since X is Lindelöf there are points $\{y_m: m \in \mathbb{N}\}$ such that

$$X = \bigcup_{m=1}^{\infty} G^{y_m}_{n+1},$$

and the sets $\{G^{y_m}_{n+1}: m \in \mathbb{N}\}$ are distinct, with the possible exception of one set occurring from some stage onwards. Define

$$G(\sigma|n, m) = \begin{cases} G^{y_m}_{n+1}, & \text{for each distinct } G^{y_m}_{n+1}, \\ \varnothing, & \text{if } G^{y_m}_{n+1} = G^{y_{m-1}}_{n+1} . \end{cases}$$

Then we have

$$X = \bigcup_{\sigma} \bigcap_{n} G(\sigma|n) ,$$

with $\bigcap_{n} G(\sigma|n)$ empty or a single point. Consequently

$$\Delta = \bigcup_{\sigma} \bigcap_{n} [G(\sigma|n) \times G(\sigma|n)] .$$

(i) implies (j). We have shown that (i) and (p) are equivalent. In the proof that (p) implies (i) the complete regularity of X implies that we can take the sets $U(x, \sigma|n)$, and hence the sets $G^*(\sigma|n)$, to be cozero sets. Thus, if $f_{\sigma|n} : X \to \mathbb{R}$ are continuous functions such that

$$X \backslash Z(f_{\sigma|n}) = G^*(\sigma|n) ,$$

we have that this countable set of functions $\{f_{\sigma|n} : \sigma|n \in \mathbb{N}^{(\mathbb{N})}\}$ separate the points.

(j) implies (k). If $\{f_n : n \in \mathbb{N}\}$ are continuous real-valued functions which separate the points of X, then the function $f : X \times X \to \mathbb{R}^{\mathbb{N}}$ defined by

$$f(x, y) = (f_1(x), f_1(y), f_2(x), f_2(y), \ldots)$$

has the required properties.

(k) implies (m). Since $X \times X$ is K-analytic and Δ is closed in $X \times X$, Δ is also K-analytic. Thus $f(\Delta)$ is an analytic subset of $\mathbb{R}^{\mathbb{N}}$ (by the equivalence of (a) and (b)). Since $\mathbb{R}^{\mathbb{N}}$ is metrizable, $f(\Delta) \in \mathbf{S}(\mathscr{Z}(\mathbb{R}^{\mathbb{N}}))$. Hence $\Delta = f^{-1} \circ f(\Delta) \in \mathbf{S}(\mathscr{Z}(X \times X))$.

(m) implies (n). Since X is K-analytic, $X \times X$ is K-analytic. Let $K : \mathbb{N}^{\mathbb{N}} \to \mathscr{K}(X \times X)$, $K(\mathbb{N}^{\mathbb{N}}) = X \times X$, be an upper semi-continuous map. Let $f : X \times X \to \mathbb{R}^{\mathbb{N}}$ be a continuous map such that $f^{-1} \circ f(\Delta) = \Delta$, let $\overline{f(X \times X)}$ be a metrizable compactification of $f(X \times X)$, and let $\hat{f} : \beta(X \times X) \to \overline{f(X \times X)}$ be the continuous extension of f to the Stone–Čech compactification of $X \times X$. Then

$$\hat{f}^{-1} \circ \hat{f}((X \times X) \backslash \Delta) \cap (X \times X) = (X \times X) \backslash \Delta$$

and $\hat{f}^{-1} : \overline{f(X \times X)} \to \mathscr{K}(\beta(X \times X))$ is an upper semi-continuous map. Hence the map

$$F : \mathbb{N}^{\mathbb{N}} \times \overline{f(X \times X)} \to \mathscr{K}(\beta(X \times X))$$

defined by

$$F(\sigma, y) = K(\sigma) \cap \hat{f}^{-1}(y)$$

is an upper semi-continuous map, and

$$F(\mathbb{N}^{\mathbb{N}} \times f((X \times X)\backslash \Delta)) = (X \times X)\backslash \Delta .$$

Consequently, since $\mathbb{N}^{\mathbb{N}} \times f((X \times X)\backslash \Delta)$ is Lindelöf, $(X \times X)\backslash \Delta$ is Lindelöf.

Now, since $X \times X$ is completely regular, each point in $(X \times X)\backslash \Delta$ has a closed neighbourhood disjoint from Δ. As $(X \times X)\backslash \Delta$ is Lindelöf, it is the union of countably many of these closed neighbourhoods, say

$$(X \times X)\backslash \Delta = \bigcup_{n=1}^{\infty} F_n .$$

Since $X \times X$ is Lindelöf and regular, it is completely regular. Thus there are continuous functions $f_n : X \times X \to [0, 1]$, such that $f_n(\Delta) = 0$ and $f_n(F_n) = 1$, by Urysohn's lemma. The continuous function $f : X \times X \to \mathbb{R}$ defined by

$$f(x, y) = \sum_{n=1}^{\infty} \frac{1}{2^n} f_n(x, y)$$

has $\{(x, y) : f(x, y) = 0\} = \Delta$; i.e. $\Delta \in \mathscr{Z}(X \times X)$.

(n) implies (l). If $\Delta = Z(f)$, then

$$(X \times X)\backslash \Delta = \bigcup_{n=1}^{\infty} \left\{(x, y) : |f(x, y)| \geq \frac{1}{n}\right\} ;$$

i.e. $(X \times X)\backslash \Delta \in \mathscr{Z}_\sigma(X \times X) \subset S(\mathscr{Z}(X \times X))$.

(l) implies (k). If

$$(X \times X)\backslash \Delta = \bigcup_{\sigma} \bigcap_{n} Z(f_{\sigma|n}) ,$$

then, re-indexing $\{f_{\sigma|n} : \sigma|n \in \mathbb{N}^{(\mathbb{N})}\}$ by $\{f_n : n \in \mathbb{N}\}$, the map f defined by $f(x, y) = (f_1(x, y), f_2(x, y), \ldots)$ has the required properties.

(n) implies (o). If $\Delta = Z(f)$, then

$$\Delta = \bigcap_{n=1}^{\infty} \left\{(x, y) : |f(x, y)| < \frac{1}{n}\right\} ;$$

i.e. Δ is a $\mathscr{G}_\delta$-subset of $X \times X$. This completes the proof of Theorem 5.5.1.

Much of Theorem 5.5.1 does not depend on X being K-analytic, but only on its being K-countably determined. The above proof gives

THEOREM 5.5.2 *Let X be a K-countably determined space. The following statements are equivalent*:

(a) *X is the continuous image of a separable metric space;*
(b) *$X \times X$ is hereditarily Lindelöf;*
(c) *$(X \times X)\backslash \Delta$ is K-countably determined;*
(d) *$(X \times X)\backslash \Delta$ is Lindelöf;*
(e) *there is a sequence $\{(A_n, A'_n): n \in \mathbb{N}\}$ of pairs of K-countably determined subsets of X such that*
  (1) *$A_n \cup A'_n = X$ for each $n \in \mathbb{N}$, and*
  (2) *for each $x \neq x'$ in X, there is an $n \in \mathbb{N}$ such that $x \in A_n$ and $x' \notin A'_n$.*

If, furthermore, X is regular, then (a)–(e) *are equivalent to the following:*

(f) *there is a sequence $\{G_n: n \in \mathbb{N}\}$ of open sets in X such that, if $x \neq y$ in X, then there is an $n \in \mathbb{N}$ such that $x \in G_n$ and $y \notin G_n$;*
(g) *there is a sequence $\{f_n: n \in \mathbb{N}\}$ of continuous real-valued functions which separate the points of X;*
(h) *there is a continuous map $f: X \times X \to \mathbb{R}^{\mathbb{N}}$ such that $f^{-1} \circ f(\Delta) = \Delta$;*
(i) *$(X \times X)\backslash \Delta$ is in $\mathbf{S}(\mathscr{Z}(X \times X))$;*
(j) *Δ is in $\mathbf{S}(\mathscr{Z}(X \times X))$;*
(k) *Δ is in $\mathscr{Z}(X \times X)$;*
(l) *Δ is a $\mathscr{G}_\delta$-subset of $X \times X$;*
(m) *Δ is a $\mathbf{S}(\mathscr{G}(X \times X))$.*

The next theorem depends on X being K-analytic. It does not hold for K-countably determined spaces; for example, part (c) would not imply part (d).

THEOREM 5.5.3 (Fremlin, 1977) *The following statements are equivalent. They are implied by Martin's axiom and the negation of the continuum hypothesis. Their negation is implied by the continuum hypothesis.*

(a, b) *A K-analytic (K-Lusin) space, in which every compact subset is metrizable, is analytic (Lusin).*
(c) *If $E \subset \mathbb{N}^{\mathbb{N}}$ is uncountable, then there is a compact set $K \subset \mathbb{N}^{\mathbb{N}}$ such that $E \cap K$ is uncountable.*
(d, e) *If X is a K-analytic (K-Lusin) space and $E \subset X$ is uncountable, then there is a compact set $K \subset X$ such that $E \cap K$ is uncountable.*
(f, g) *A K-analytic (K-Lusin) space, in which every compact subset is hereditarily Lindelöf, is hereditarily Lindelöf.*
(h, i) *A K-analytic (K-Lusin) space, in which every compact subset is hereditarily separable, is hereditarily separable.*

Proof We first prove the following implications:

$$\begin{array}{ccccc} (\mathrm{f}) & \longleftarrow & (\mathrm{d}) & \rightarrow & (\mathrm{h}) \\ \downarrow & & \uparrow & \swarrow & \\ (\mathrm{a}) \rightarrow & (\mathrm{b}) & \rightarrow (\mathrm{c}) & & \\ & \uparrow & \downarrow & \nwarrow & \\ (\mathrm{g}) \leftarrow & (\mathrm{e}) & \rightarrow & (\mathrm{i}) & \end{array}$$

(a) implies (b). Immediate.

(b) implies (c). Suppose that there is an uncountable set $E \subset \mathbb{N}^{\mathbb{N}}$ that intersects each compact set in a countable set. Let Y be the set $\{0, 1\} \times [0, 1]$ with the topology for which every point in $\{1\} \times [0, 1]$ is open and each point $(0, x)$ has as a neighbourhood base all sets of the form $G \cup G'$, where G is open neighbourhood of x in the usual topology of $[0, 1]$ and G' is the projection of $G \backslash \{x\}$ on $\{1\} \times [0, 1]$. With this topology Y is a first-countable compact Hausdorff space. We identify $\mathbb{N}^{\mathbb{N}}$ with the set of irrational numbers in $[0, 1]$. Then

$$K = (\{0\} \times [0, 1]) \cup (\{1\} \times E)$$

is a compact subset of Y, and

$$X = (\{0\} \times \mathbb{N}^{\mathbb{N}}) \cup (\{1\} \times E) = \bigcap_{n=1}^{\infty} K \cap (Y \backslash \{0, q_n\}),$$

where $\{q_n : n \in \mathbb{N}\}$ are the rational numbers in $[0, 1]$. Note that $K \cap (Y \backslash \{(0, q_n)\})$ is a cozero subset of K, and so X is a Baire set and hence a K-Lusin set in K.

Let C be a compact subset of X. Then, since $C \cap (\{0\} \times \mathbb{N}^{\mathbb{N}})$ is compact, we have by hypothesis that $C \cap (\{1\} \times E)$ is at most countable. Thus C is the union of a compact metrizable set and a countable set and so is metrizable. But X is not an analytic space, since, for example, $\{1\} \times E$ is an uncountable discrete subspace and so X is not hereditarily Lindelöf.

(c) implies (d). Let $E \subset X$ be uncountable and let $F : \mathbb{N}^{\mathbb{N}} \to \mathscr{K}(X)$, $F(\mathbb{N}^{\mathbb{N}}) = X$, be an upper semi-continuous map. If $E' = \{\sigma \in \mathbb{N}^{\mathbb{N}} : F(\sigma) \cap E \neq \varnothing\}$ is countable, then one of the compact sets $F(\sigma)$, $\sigma \in E'$, will suffice. If E' is uncountable and $K \subset \mathbb{N}^{\mathbb{N}}$ is a compact set with $K \cap E'$ uncountable, then $F(K)$ is a compact set in X with $F(K) \cap E$ uncountable.

(d, e) implies (f, g). Let X be a K-analytic (K-Lusin) space in which every compact subset is hereditarily Lindelöf. Suppose X is not hereditarily Lindelöf. Then there is a family $\mathscr{G}$ of open subsets of X with $\bigcup \mathscr{G} \neq \bigcup \mathscr{G}'$ for every countable subfamily $\mathscr{G}'$ of G. Choose inductively $E=\{x_\alpha: 0<\alpha<\omega_1\}$ in X and $\{G_\alpha: 0<\alpha<\omega_1\}$ in $\mathscr{G}$, by first choosing

$$x_\alpha \in \cup \mathscr{G} \setminus \bigcup_{\beta<\alpha} G_\beta ,$$

and then choosing G_β in $\mathscr{G}$ with $x_\alpha \in G_\alpha$. Let $K \subset X$ be a compact set with $E \cap K$ uncountable. By hypothesis K is hereditarily Lindelöf. Thus there is a $\gamma<\omega_1$ such that

$$K \cap \bigcup_{\alpha<\gamma} G_\alpha = K \cap \bigcup_{\alpha<\omega_1} G_\alpha .$$

But, since $E \cap K$ is uncountable, there is a $\beta \geqslant \gamma$ such that

$$x_\beta \in K \cap (G_\beta \setminus \bigcup_{\alpha<\beta} G_\alpha) ,$$

which is a contradiction.

(f) implies (a). If X is a K-analytic space in which every compact subset is metrizable, then $X\times X$ is a K-analytic space with the same property. Since a compact metrizable set is hereditarily Lindelöf, $X\times X$ is hereditarily Lindelöf. Thus, by Theorem 5.5.1, X is analytic.

(d) implies (h). Let X be a K-analytic space in which every compact subset is hereditarily separable. Suppose X is not hereditarily separable. Then there is a set $E=\{x_\alpha: 0<\alpha<\omega_1\}$ such that, for each $\alpha<\omega_1$,

$$x_\alpha \in X \setminus \mathrm{cl}\,\{x_\beta : \beta<\alpha\} .$$

No uncountable subset of E is separable, and so no compact (hereditarily separable) subset can intersect E in an uncountable set, contrary to (d).

(h, i) implies (c). If (c) is false, then the example in the proof that (b) implies (c) shows that (h, i) are false.

We have now completed the proof that all of the statements are equivalent. Now suppose that the continuum hypothesis is true. Let $\{K_\alpha: 0<\alpha<\omega_1\}$ be an enumeration of all of the compact subsets of $\mathbb{N}^{\mathbb{N}}$. Since $\mathbb{N}^{\mathbb{N}}$ is not σ-compact, for each $\alpha<\omega_1$ we can choose

$$x_\alpha \in \mathbb{N}^{\mathbb{N}} \setminus \bigcup_{\beta<\alpha} (K_\beta \cup \{x_\beta\}) .$$

Then $E=\{x_\alpha: 0<\alpha<\omega_1\}$ is an uncountable set such that $E\cap E_\alpha$ is countable for each $\alpha<\omega_1$.

Now suppose Martin's axiom is true and that the continuum hypothesis is false. Let $E \subset \mathbb{N}^{\mathbb{N}}$ be uncountable. Let E' be a subset of E with cardinality $\aleph_1$.

Then there is a $\sigma \in \mathbb{N}^{\mathbb{N}}$ such that $\{n \in \mathbb{N}: \sigma_n < \tau_n\}$ is finite for each $\tau = (\tau_1, \tau_2, \ldots) \in E'$ (Rudin, 1977, Corollary 11). Thus

$$E' \subset \bigcup_{n=1}^{\infty} K_n,$$

where K_n is the compact set

$$\{\tau \in \mathbb{N}^{\mathbb{N}}: \tau_i \leq \max(n, \sigma_i) \text{ for all } i \in \mathbb{N}\},$$

and $E \cap K_n \supset E' \cap K_n$ is uncountable for some $n \in \mathbb{N}$. This completes the proof of the theorem.

5.6 Metrizability

In the last section (Section 5.5) we observed that metrizable K-analytic (K-Lusin) spaces are analytic (Lusin). The converse is, of course, not true. Example 5.2.1 is perhaps the most elementary example of a regular Lusin space which is not metrizable. Every infinite-dimensional separable Banach space with its weak topology is an example of a non-metrizable regular Lusin space. This section is devoted to characterizing metrizable K-analytic and K-Lusin spaces.

THEOREM 5.6.1 *If X is a proper K-countably determined space satisfying the equivalent statements in Theorem* 5.5.2, *then X is metrizable.*

Proof Suppose X satisfies (g) of Theorem 5.5.2; i.e. that there is a sequence of continuous functions into $[0, 1]$ that separate the points of X. Let $g_n : \beta X \to [0, 1]$ be the continuous extensions of these functions. By hypothesis X admits a proper map onto a subset of $\mathbb{R}^{\mathbb{N}}$, which we can suppose to be contained in $[0, 1]^{\mathbb{N}}$. Let $f : \beta X \to [0, 1]^{\mathbb{N}}$ be its continuous extension. By Proposition 5.2.3, $f^{-1} \circ f(X) = X$.

Let $\{G_m : m \in \mathbb{N}\}$ be a countable open base of $[0, 1]$ and $\{H_m : m \in \mathbb{N}\}$ a countable open base of $[0, 1]^{\mathbb{N}}$. Define

$$\mathcal{H} = \{g_n^{-1}(G_m): n, m \in \mathbb{N}\} \cup \{f^{-1}(H_m): m \in \mathbb{N}\}.$$

To prove that X is metrizable it will suffice to show that the countable set

$$\mathcal{H} \cap X = \{H \cap X : H \in \mathcal{H}\}$$

is a subbase of X.

Let $x \in X$ and U be an open neighbourhood of x in βX. For each $y \in \beta X \setminus U$ there are $H_y, H'_y \in \mathcal{H}$ such that

$$x \in H_y, \qquad y \in H'_y, \qquad H_y \cap H'_y = \varnothing.$$

Since $\beta X \setminus U$ is compact, there are $y_1, \ldots, y_k \in \beta X \setminus U$ such that $\bigcup_{i=1}^k H'_{y_i}$ covers $\beta X \setminus U$. Thus

$$x \in \bigcap_{i=1}^{k} H_{y_i} \subset U,$$

and so $\mathscr{H} \cap X$ is a subbase of X.

From this theorem, Theorem 5.5.1 and the discussion in Section 5.2 we get

THEOREM 5.6.2 *A K-analytic space is metrizable, if, and only if, it is proper K-countably determined and satisfies the equivalent statements in Theorem* 5.5.1. *In particular, a regular analytic (resp. Lusin) space X is metrizable, if, and only if,* $X \in \mathbf{S}(\mathscr{Z}(\beta X))$ *(resp.* $\mathscr{Z}_{(\sigma,\delta)}(\beta X)$*).*

We can now add two equivalent statements to those in Theorem 5.5.3.

THEOREM 5.6.3 *The following statements are equivalent to those in Theorem* 5.5.3, *and are implied by Martin's axiom and the negation of the continuum hypothesis, while their negation is implied by the continuum hypothesis:*

(j, k) *a proper K-analytic (proper K-Lusin) space, in which every compact subset is metrizable, is metrizable.*

Proof We will relate these statements to (a) and (c) of Theorem 5.5.3. We show that (a) implies (j) implies (k) implies (c).

(a) implies (j). Let X be a proper K-analytic space with metrizable compact subsets. Then (a) implies that X is analytic, and Theorem 5.6.2 then implies that X is metrizable.

(j) implies (k). Immediate.

(k) implies (c). If (c) does not hold, then the space constructed in the proof that (b) implies (c) is a proper K-Lusin space, with metrizable compact subsets, which is not metrizable. Hence (k) fails.

5.7 Disjoint countable unions

It is advantageous, especially in measure theory, to be able to work with countable disjoint unions of sets, rather than arbitrary countable unions. In this section we will examine to what extent this is possible in the theory of K-analytic and K-Lusin spaces.

As before, if $\mathscr{H}$ is a family of subsets of a set X, then $\mathscr{H}_\sigma$, $\mathscr{H}_+$, $\mathscr{H}_\delta$, and $\mathscr{H}_c$ will denote the families of, respectively, countable unions, countable disjoint unions, countable intersections, and complements of sets in $\mathscr{H}$. A combination of these symbols refers to their successive application. We denote by $\mathscr{H}_{(+,\delta)}$, for example, the smallest family $\mathscr{H}^*$ of sets containing $\mathscr{H}$ such that $\mathscr{H}^* = \mathscr{H}^*_+ = \mathscr{H}^*_\delta$.

The result upon which this section rests is

THEOREM 5.7.1 (Sierpiński, 1925) *For the real line* $\mathbb{R}$ *we have*

$$\mathscr{Z}_c(\mathbb{R}) \subset \mathscr{Z}_{+\delta+}(\mathbb{R}) .$$

Consequently,

$$\mathscr{Z}_{(+,\delta)}(\mathbb{R}) = \mathscr{Z}_{(\sigma,\delta)}(\mathbb{R}) .$$

Proof Let $a, b \in \mathbb{R}$, $a < b$. First we show that the open interval (a, b) belongs to $\mathscr{Z}_{+\delta+}(\mathbb{R})$. Let $\{I_n : n \in \mathbb{N}\}$ be closed intervals in $\mathbb{R}$, with union dense in (a, b), and let

$$B = (a, b) \Big\backslash \left(\bigcup_{n=1}^{\infty} I_n \right) .$$

Then $\bar{B}$ (the closure of B in $\mathbb{R}$) is a nowhere dense closed subset of the closed interval $[a, b]$, and $\bar{B} \backslash B$ is a countable set $\{x_m : m \in \mathbb{N}\}$ consisting of a and b and the end points of the intervals (I_n). Since (a, b) is the countable disjoint union

$$(a, b) = B \cup I_1 \cup I_2 \cup \cdots ,$$

and

$$B = \bigcap_{m=1}^{\infty} (\bar{B} \backslash \{x_1, \ldots, x_m\}) ,$$

it will suffice to show that $\bar{B} \backslash \{x_1, \ldots, x_m\}$ is the countable disjoint union of closed sets.

Since $\bar{B}$ is nowhere dense, it is 0-dimensional. Thus $\bar{B} \backslash \{x_1, \ldots, x_m\}$ is the countable union of relatively clopen subsets of $\bar{B}$; i.e.

$$\begin{aligned} \bar{B} \backslash \{x_1, \ldots, x_m\} &= \bigcup_{k=1}^{\infty} C_k , \qquad C_k \text{ clopen} \\ &= C_1 \cup (C_2 \backslash C_1) \cup (C_3 \backslash C_1 \cup C_2) \cup \cdots , \end{aligned}$$

which is a disjoint union of closed subsets of $\bar{B}$. Consequently we have

$$(a, b) \in \mathscr{Z}_{+\delta+}(\mathbb{R}) .$$

Now, since each open subset of $\mathbb{R}$ is the countable union of disjoint open intervals, we have

$$\mathscr{Z}_c(\mathbb{R}) \subset \mathscr{Z}_{+\delta+}(\mathbb{R}) .$$

To show that

$$\mathscr{Z}_{(+,\delta)}(\mathbb{R}) = \mathscr{Z}_{(\sigma,\delta)}(\mathbb{R})$$

it now suffices to show that the family $\mathscr{H}$ of sets in $\mathscr{Z}_{(+,\delta)}(\mathbb{R})$, whose complements also belong to $\mathscr{Z}_{(+,\delta)}(\mathbb{R})$, is closed under countable disjoint unions and countable intersections. If the sets $(E_n) \subset \mathscr{H}$ are disjoint, then

$$\bigcup_n E_n \in \mathscr{Z}_{(+,\delta)}(\mathbb{R})$$

and also

$$\mathbb{R} \Big\backslash \Big(\bigcup_n E_n \Big) = \bigcap_n (\mathbb{R} \backslash E_n) \in \mathscr{Z}_{(+,\delta)}(\mathbb{R}) .$$

Thus $\mathscr{H}$ is closed under countable disjoint unions. Finally, if $(E_n) \subset \mathscr{H}$, then

$$\bigcap_n E_n \in \mathscr{Z}_{(+,\delta)}(\mathbb{R})$$

and

$$\mathbb{R} \Big\backslash \Big(\bigcap_n E_n \Big) = \bigcup_n (\mathbb{R} \backslash E_n) = \bigcup_n \Big[(\mathbb{R} \backslash E_n) \cap \bigcap_{m<n} E_m \Big] \in \mathscr{Z}_{(+,\delta)}(\mathbb{R}) .$$

Hence $\mathscr{H}$ is closed under countable intersections, and the proof is complete.

THEOREM 5.7.2 *For any topological space X we have*

$$\mathscr{Z}_c(X) \subset \mathscr{Z}_{+\delta+}(X)$$

and

$$\mathscr{Z}_{(+,\delta)}(X) = \mathscr{Z}_{(\sigma,\delta)}(X) .$$

Proof Let $U \in \mathscr{Z}_c(X)$ amd $f : X \to [0, 1)$ a continuous function such that

$$U = \{x \in X : f(x) \in (0, 1)\} .$$

Then, since $(0, 1) \in \mathscr{Z}_{+\delta+}(\mathbb{R})$ by Theorem 5.7.1,

$$U = f^{-1}[(0, 1)] \in \mathscr{Z}_{+\delta+}(X) .$$

As in the proof of Theorem 5.7.1 we obtain

$$\mathscr{Z}_{(+,\delta)}(X) = \mathscr{Z}_{(\sigma,\delta)}(X) .$$

We will need the fact that

$$\bigcup_\sigma \bigcap_n H(\sigma|n) = \bigcap_n \bigcup_\sigma H(\sigma|n),$$

if for each $\sigma|n$,

$$H(\sigma|n+1) \subset H(\sigma|n)$$

and

$$H(\sigma|n) \cap H(\tau|n) = \varnothing,$$

whenever $\sigma|n \neq \tau|n$. We prove a stronger result.

PROPOSITION 5.7.3 (Kunugui, 1938) *Let H be a family of subsets of a set X and $H(\sigma|n) \in \mathscr{H}$ for each $\sigma \in \mathbb{N}^{\mathbb{N}}$ and $n \in \mathbb{N}$. Then*

$$\bigcup_\sigma \bigcap_n H(\sigma|n) = \bigcap_n \bigcup_\sigma H(\sigma|n),$$

if for each $\sigma|n$

$$H(\sigma|n+1) \subset H(\sigma|n)$$

and, for each

$$x \in \bigcup_\sigma \bigcap_n H(\sigma|n)$$

and each $n \in \mathbb{N}$, the set

$$\{\sigma|n : \sigma \in \mathbb{N}^{\mathbb{N}}, x \in H(\sigma|n)\}$$

is finite.

Proof It suffices to show that

$$\bigcap_n \bigcup_\sigma H(\sigma|n) \subset \bigcup_\sigma \bigcap_n H(\sigma|n).$$

Let

$$x \in \bigcap_n \bigcup_\sigma H(\sigma|n)$$

and define $\tau = (\tau_1, \tau_2, \ldots) \in \mathbb{N}^{\mathbb{N}}$ by

$$\tau_n = \max\{\sigma_n : x \in H(\sigma_1, \sigma_2, \ldots, \sigma_n)\}$$

Let

$$A = \{\sigma \in \mathbb{N}^{\mathbb{N}} : \sigma_1 \leq \tau_i \text{ for all } i \in \mathbb{N}\},$$

and

$$B_n = \{\sigma \in \mathbb{N}^{\mathbb{N}} : x \in H(\sigma|n)\}.$$

Then A is compact and B_n is closed in $\mathbb{N}^{\mathbb{N}}$, and so

$$A \cap B_n$$

is compact. If $x \in H(\sigma|n)$, then

$$(\sigma_1, \sigma_2, \ldots, \sigma_n, 1, 1, \ldots) \in A \cap B_n.$$

Thus $A \cap B_n \neq \varnothing$ and, since

$$A \cap B_1 \supset A \cap B_2 \supset \cdots,$$

there is a

$$\gamma \in \bigcap_n A \cap B_n.$$

Clearly

$$x \in \bigcap_n H(\gamma|n) \subset \bigcup_\sigma \bigcap_n H(\sigma|n).$$

The next theorem is derived from Theorem 5.1.6 in the same way that Theorem 3.3.3 is obtained from Theorem 3.3.1. Theorem 5.7.2 is used in part (c)

THEOREM 5.7.4 (Multiple first separation principle) *Let X be a Hausdorff space and (A_n) a disjoint sequence of K-analytic subsets of X.*

(a) (Theorem 3.3.2) There is a disjoint sequence (B_n) *such that for each* $n \in \mathbb{N}$, $A_n \subset B_n$.
(b) *If X is regular, then there is a disjoint sequence $(B_n) \subset \mathscr{F}_{(\sigma,\delta)}(X)$ such that for each $n \in \mathbb{N}$, $A_n \subset B_n$.*
(c) *If X is completely regular, then there is a disjoint sequence $(B_n) \subset \mathscr{Z}_{(+,\delta)}(X)$ such that for each $n \in \mathbb{N}$, $A_n \subset B_n$.*

Recall that $\mathscr{F}_{(+,\delta)}(X)$ denotes the smallest class of sets containing the closed sets $\mathscr{F}(X)$ of X and closed under countable disjoint unions and countable intersections. We always have

$$\mathscr{F}_{(+,\delta)}(X) \subset \mathscr{F}_{(\sigma,\delta)}(X) \subset \mathscr{F}_{(\sigma,c)}(X).$$

The proof of Theorem 3.6.1 derives the next theorem from Theorem 5.7.4. We give the proof of part (d), the first part of which states that in a completely regular space X each K-Lusin subset is contained in $\mathscr{F}_{(+,\delta)}(X)$.

THEOREM 5.7.5 *Let X be a Hausdorff space and B a K-Lusin subset of X.*

(a) (Theorem 3.6.1) *B is a Borel subset of X.*
(b) *If each open set in X is in* $\mathbf{S}(\mathscr{F}(X))$, *then* $B \in \mathscr{G}_{(\sigma,\delta)}(X)$. *Furthermore, if X is K-analytic and each open set in X is in* $\mathbf{S}(\mathscr{F}(X))$, *then* $B \in \mathscr{F}_{(\sigma,\delta)}(X)$.
(c) *If X is regular, then* $B \in \mathscr{F}_{(\sigma,\delta)}(X)$.
(d) *If X is completely regular, then*
   (i) (Jayne, 1977) $B \in \mathscr{F}_{(+,\delta)}(X)$;
   (ii) (Frolík, 1967) $B \in \mathscr{Z}_{(+,\delta)}(X)$, *if, and only if, there is a continuous map* $f: X \to \mathbb{R}^{\mathbb{N}}$ *such that* $f^{-1} \circ f(X) = X$;
   (iii) *if each open set in X is in* $\mathbf{S}(\mathscr{F}(X))$, *then*

$$B \in \mathscr{Z}_{(+,\delta)}(X)\,.$$

Proof (d)(i). Let B be a K-Lusin subset of a completely regular space X. Let

$$F: \mathbb{N}^{\mathbb{N}} \to \mathscr{K}(X)$$

be an upper semi-continuous map with

$$B = \bigcup_{\sigma \in \mathbb{N}^{\mathbb{N}}} F(\sigma)$$

and

$$F(\sigma) \cap F(\tau) = \varnothing$$

whenever $\sigma \neq \tau$.

We have (as in the proof of Theorem 2.5.2)

$$B = \bigcup_{\sigma} \bigcap_{n} \operatorname{cl} F(I(\sigma|n))\,,$$

where $I(\sigma|n) = \{\tau \in \mathbb{N}^{\mathbb{N}} : \tau|n = \sigma|n\}$ and cl denotes closure in X, and

$$\bigcap_{n} \operatorname{cl} F(I(\sigma|n)) \cap \bigcap_{m} \operatorname{cl} F(I(\tau|m)) = \varnothing$$

whenever $\sigma \neq \tau$.

For each n-tuple $\sigma|n$ define

$$B(\sigma|n) = \bigcup_{\tau \in I(\sigma|n)} \bigcap_{m=1}^{\infty} \operatorname{cl} F(I(\tau|m)) \in \mathbf{S}(\mathscr{F}(X))\,.$$

Since for each $\sigma|n \neq \tau|n$,

$$B(\sigma|n) \cap B(\tau|n) = \varnothing,$$

we have that for each $n \in \mathbb{N}$

$$\{B(\sigma|n): \sigma|n \in \mathbb{N}^{(\mathbb{N})}\}$$

is a countable family of disjoint analytic sets. Thus by Theorem 5.7.4(c) there is, for each n, a disjoint family

$$\{C(\sigma|n): \sigma \in \mathbb{N}^{\mathbb{N}}\} \subset \mathscr{Z}_{(+,\delta)}(X)$$

such that for each $\sigma \in \mathbb{N}^{\mathbb{N}}$

$$B(\sigma|n) \subset C(\sigma|n).$$

Note that, since for each $\sigma|n$ we have $B(\sigma|n+1) \subset B(\sigma|n)$, we can choose $C(\sigma|n)$ such that $C(\sigma|n+1) \subset C(\sigma|n)$.

We clearly have

$$B = \bigcup_{\sigma} \bigcap_{n} B(\sigma|n) \subset \bigcup_{\sigma} \bigcap_{n} \operatorname{cl} B(\sigma|n),$$

and, since $B(\sigma|n) \subset \operatorname{cl} F(I(\sigma|n))$, we have

$$\bigcup_{\sigma} \bigcap_{n} \operatorname{cl} B(\sigma|n) \subset \bigcup_{\sigma} \bigcap_{n} \operatorname{cl} F(I(\sigma|n)) = B.$$

Hence

$$\begin{aligned} B &= \bigcup_{\sigma} \bigcap_{n} \operatorname{cl} B(\sigma|n) \\ &= \bigcup_{\sigma} \bigcap_{n} \{(\operatorname{cl} B(\sigma|n)) \cap C(\sigma|n)\} \\ &= \bigcap_{n} \bigcup_{\sigma} \{(\operatorname{cl} B(\sigma|n)) \cap C(\sigma|n)\}, \end{aligned}$$

by Proposition 5.7.3. Thus we get

$$B \in \mathscr{F}_{(+,\delta)}(X).$$

(d)(ii). If $B \in \mathscr{Z}_{(+,\delta)}(X)$, then B can be obtained by iterating countable unions and countable intersections a countable number of times starting from a countable set of zero sets $Z(f_1), Z(f_2), \ldots$, where the $f_n : X \to \mathbb{R}$ are continuous functions. Then the continuous map $f: X \to \mathbb{R}^{\mathbb{N}}$, defined by

$$f(x) = (f_1(x), f_2(x), \ldots),$$

has the property that $f^{-1} \circ f(B) = B$.

Now suppose $F:\mathbb{N}^{\mathbb{N}}\to\mathcal{K}(X)$ is a disjoint upper semi-continuous map with $F(\mathbb{N}^{\mathbb{N}})=B$. Let $h_1:X\to\mathbb{R}^{\mathbb{N}}$ such that $h_1^{-1}\circ h_1(B)=B$. By Theorem 5.7.4(c) there are sets $B(\sigma|n)\in\mathcal{Z}_{(+,\delta)}(X)$ such that

$$F(I(\sigma|n))\subset B(\sigma|n)\,,\qquad \sigma|n\in\mathbb{N}^{(\mathbb{N})}\,,$$

and

$$B(\sigma|n)\cap B(\tau|n)=\varnothing\,,\qquad \sigma|n\neq\tau|n\,.$$

Since the family $\{B(\sigma|n)\colon \sigma|n\in\mathbb{N}^{(\mathbb{N})}\}$ is countable we can obtain, as in the first paragraph of this proof, a continuous map $h_2:X\to\mathbb{R}^{\mathbb{N}}$ such that

$$h_2^{-1}\circ h_2(B(\sigma|n))=B(\sigma|n)\,,\qquad \sigma|n\in\mathbb{N}^{(\mathbb{N})}\,.$$

Now the map $h:X\to\mathbb{R}^{\mathbb{N}}(\approx\mathbb{R}^{\mathbb{N}}\times\mathbb{R}^{\mathbb{N}})$ defined by

$$h(x)=(h_1(x),\,h_2(x))$$

is continuous and satisfies:

(1) $h\circ F(\sigma)\cap h\circ F(\tau)=\varnothing\,,$ if $\sigma\neq\tau$, and
(2) $h^{-1}\circ h(B)=B.$

By (1) and part (d)(i) above, we have that $h(B)=h\circ F(\mathbb{N}^{\mathbb{N}})$ is in $\mathcal{Z}_{(+,\delta)}(\mathbb{R}^{\mathbb{N}})$. By (2) and the continuity of h we have found that $B\in\mathcal{Z}_{(+,\delta)}(X)$.

(d)(iii). Now suppose furthermore that each open set in X is in $\mathbf{S}(\mathcal{F}(X))$. Then for each $\sigma|n\in\mathbb{N}^{(\mathbb{N})}$, by Theorem 5.1.6(c), there is a set $C'(\sigma|n)\in\mathcal{Z}_{(+,\sigma)}(X)$ such that

$$B(\sigma|n)\subset C'(\sigma|n)\subset\operatorname{cl}B(\sigma|n)\,.$$

Thus, taking $C''(\sigma|n)=C(\sigma|n)\cap C'(\sigma|n)$, we get

$$B=\bigcap_n\bigcup_\sigma C''(\sigma|n)\,.$$

Hence

$$B\in\mathcal{Z}_{(+,\delta)}(X)\,.$$

5.8 Disjoint Souslin operation

Recall that if $\mathcal{H}$ is a family of subsets of a set X, then $\mathbf{S}_d(\mathcal{H})$ denotes the family of sets in X with disjoint Souslin representations by means of sets in $\mathcal{H}$: i.e. $B\in\mathbf{S}_d(\mathcal{H})$, if, and only if, we can write

$$B=\bigcup_\sigma\bigcap_n H(\sigma|n)\,,\qquad H(\sigma|n)\in\mathcal{H}\,,$$

with

$$\bigcap_n H(\sigma|n) \cap \bigcap_n H(\tau|n) = \varnothing, \qquad \sigma \neq \tau.$$

As we will see in this section, the families $\mathscr{H}_{(+,\delta)}$ and $\mathbf{S}_d(\mathscr{H})$ are often closely related. For example, they coincide when H is either the family of closed subsets or zero sets of a regular K-analytic space (Theorem 5.8.5).

The following form of Kuratowski's reduction theorem for coanalytic subsets of Polish spaces will be used in the proof of Theorem 5.8.3(a) (and nowhere else in this section). For two sets it is due to K. Kunugui (1935) and for a sequence of sets to C. A. Rogers (1973). The classical version is treated by Dellacherie (see Part 2, Chapitre III, Paragraph 28 in the present volume) and by Kechris and Martin (see Part 4, Theorem 7.3.1 in this volume).

THEOREM 5.8.1 (Reduction theorem) *Let $\mathscr{H}$ be a family of subsets of a set X such that $\mathscr{H}_c \subset \mathbf{S}(\mathscr{H})$. If (H_n) is a sequence of sets in $\mathbf{S}(\mathscr{H})_c$, then there is a disjoint sequence (C_n) in $\mathbf{S}(\mathscr{H})_c$ such that for each $n \in \mathbb{N}$*

$$C_n \subset H_n, \qquad \bigcup_n C_n = \bigcup_n H_n.$$

From this theorem for two sets of H_1 and H_2 we obtain the following corollary in the same way as Theorem 3.3.3 is obtained from Theorem 3.3.1.

COROLLARY 5.8.2 *Let $\mathscr{H}$ be a family of subsets of a set X such that $\mathscr{H}_c \subset \mathbf{S}(\mathscr{H})$. If (H_n) is a disjoint sequence in $\mathbf{S}(\mathscr{H})$, then there is a disjoint sequence (B_n) in $\mathbf{S}(\mathscr{H}) \cap \mathbf{S}(\mathscr{H})_c$ such that for each $n \in \mathbb{N}$, $H_n \subset B_n$.*

THEOREM 5.8.3 *Let $\mathscr{H}$ be a family of subsets of a set X.*

(a) (Kunugui, 1935) *If $\mathscr{H}_c \subset \mathbf{S}(\mathscr{H})$, then $\mathbf{S}_d(\mathscr{H})_c \subset \mathbf{S}(\mathscr{H})$.*
(b) (Rogers, 1970a) *If $\mathscr{H}_c \subset \mathscr{H}_{(\sigma,\delta)}$, then $\mathscr{H}_{(\sigma,\delta)} = \mathscr{H}_{(\sigma,c)} \subset \mathbf{S}_d(\mathscr{H})$.*

Proof (a) Let $B \in \mathbf{S}_d(\mathscr{H})$. Then

$$B = \bigcup_\sigma \bigcap_n B(\sigma|n), \qquad B(\sigma|n) \in \mathscr{H},$$

and

$$\bigcap_n B(\sigma|n) \cap \bigcap_n B(\tau|n) = \varnothing, \qquad \sigma \neq \tau.$$

Letting

$$F(\sigma|n) = \bigcap_{m=1}^{n} B(\sigma|m)$$

we get

$$B = \bigcup_{\sigma} \bigcap_{n} F(\sigma|n), \qquad F(\sigma|n+1) \subset F(\sigma|n), \qquad F(\sigma|n) \in \mathbf{S}(\mathscr{H}),$$

and

$$\bigcap_{n} F(\sigma|n) \cap \bigcap_{n} F(\tau|n) = \varnothing, \qquad \sigma \neq \tau.$$

Now defining

$$F^*(\sigma|n) = \bigcup_{\tau \in I(\sigma|n)} \bigcap_{m} F(\tau|m) \in \mathbf{S}(\mathscr{H}),$$

we have

$$B = \bigcup_{\sigma} \bigcap_{n} F^*(\sigma|n), \qquad F^*(\sigma|n+1) \subset F^*(\sigma|n),$$

and

$$F^*(\sigma|n) \cap F^*(\tau|n) = \varnothing, \qquad \sigma|n \neq \tau|n.$$

Thus $\{F^*(\sigma|n) \colon \sigma \in \mathbb{N}^{\mathbb{N}}\}$ is a countable family of disjoint sets in $\mathbf{S}(\mathscr{H})$. Hence by Corollary 5.8.2 there is, for each n, a disjoint family

$$\{C(\sigma|n) \colon \sigma \in \mathbb{N}^{\mathbb{N}}\} \in \mathbf{S}(\mathscr{H}) \cap \mathbf{S}(\mathscr{H})_c,$$

such that for each $\sigma \in \mathbb{N}^{\mathbb{N}}$

$$F^*(\sigma|n) \subset C(\sigma|n).$$

Since $F^*(\sigma|n+1) \subset F^*(\sigma|n)$, we can choose

$$C^*(\sigma|n) = \bigcap_{m=1}^{n} C(\sigma|m),$$

so that

$$C^*(\sigma|n+1) \subset C^*(\sigma|n), \qquad F^*(\sigma|n) \subset C^*(\sigma|n),$$

and

$$C^*(\sigma|n) \in \mathbf{S}(\mathscr{H}) \cap \mathbf{S}(\mathscr{H})_c.$$

We now have

$$B = \bigcup_{\sigma}\bigcap_{n} (F(\sigma|n) \cap C^*(\sigma|n))$$

$$= \bigcap_{n}\bigcup_{\sigma} (F(\sigma|n) \cap C^*(\sigma|n)) ,$$

by Proposition 5.7.3. Thus

$$X \backslash B = \bigcup_{n}\bigcap_{\sigma} [X \backslash (F(\sigma|n) \cap C^*(\sigma|n))] \in \mathbf{S}(\mathcal{H}) .$$

(b) First note that

$$\mathcal{H}_c \subset \mathcal{H}_{(\sigma,\delta)} \text{ implies } \mathcal{H}_{(\sigma,\delta)} = \mathcal{H}_{(\sigma,c)} .$$

Now let $\mathcal{H}^*$ denote the class of sets B with representations.

$$B = \bigcup_{\sigma}\bigcap_{n} H(\sigma|n) , \qquad H(\sigma|n) \in \mathcal{H}$$

such that for each n-tuple $\sigma|n$ we have

$$\bigcup_{\tau \in I(\sigma|n)} \bigcap_{m} H(\tau|m) \in \mathcal{H}_{(\sigma,\delta)} .$$

A modification of the proofs of Theorem 2.3.1 and Corollary 2.3.3 shows that

$$\mathcal{H}^* = \mathcal{H}^*_{(\sigma,\delta)} .$$

We have $\mathcal{H} \subset \mathcal{H}^*$, since for $H \in \mathcal{H}$ we can write

$$H = \bigcup_{\sigma}\bigcap_{n} H(\sigma|n) ,$$

where $H(\sigma|n) = H$ for each $\sigma|n$. Thus

$$\mathcal{H}_{(\sigma,c)} = \mathcal{H}_{(\sigma,\delta)} \subset \mathcal{H}^*_{(\sigma,\delta)} = \mathcal{H}^* .$$

We now show that $\mathcal{H}^* \subset \mathbf{S}_d(\mathcal{H})$. In particular, we show that each set $B \in \mathcal{H}^*$ has a representation

$$B = \bigcup_{\sigma}\bigcap_{n} H(\sigma|n) ,$$

such that

$$H(\sigma|n) \in \mathcal{H} , \qquad \bigcup_{\tau \in I(\sigma|n)} \bigcap_{m} H(\tau|m) \in \mathcal{H}_{(\sigma,\delta)} ,$$

and

$$\bigcap_n H(\sigma|n) \cap \bigcap_m H(\tau|m) = \varnothing\,, \qquad \sigma \neq \tau\,.$$

By hypothesis, $B \in \mathcal{H}^*$ has a representation

$$B = \bigcup_\sigma \bigcap_n F(\sigma|n)\,,$$

such that

$$F(\sigma|n) \in \mathcal{H} \quad \text{and} \quad \bigcup_{\tau \in I(\sigma|n)} \bigcap_m F(\tau|m) \in \mathcal{H}_{(\sigma,\delta)}\,.$$

Writing

$$F^*(\sigma|n) = \bigcap_{m=1}^{n} F(\sigma|m)\,,$$

we have

$$\bigcap_n F(\sigma|n) = \bigcap_n F^*(\sigma|n)$$

and consequently

$$B = \bigcup_\sigma \bigcap_n F^*(\sigma|n)\,, \qquad F^*(\sigma|n+1) \subset F^*(\sigma|n)\,,$$

and

$$B(\sigma|n) \equiv \bigcup_{\tau \in I(\sigma|n)} \bigcap_m F^*(\tau|m) \in \mathcal{H}_{(\sigma,\delta)}\,.$$

We now partition the sequence of positive integers into an infinite sequence of infinite sequences

$$\begin{gathered} 1, 3, 5, 7, \ldots, \\ 2, 3\times 2, 5\times 2, 7\times 2, \ldots, \\ 4, 3\times 4, 5\times 4, 7\times 4, \ldots. \end{gathered}$$

For each positive integer n there are unique integers $h(n) \geq 0$, $t(n) \geq 1$, $q(n) \geq 1$ satisfying the diophantine equations

$$n = t(n)[2q(n)-1]\,, \qquad t(n) = 2^{h(n)}\,.$$

Then n will lie in the $q(n)$th position in the $(h(n)+1)$th sequence. We will use $\mathbf{j}_t(\sigma)$ to denote the vector

$$\mathbf{j}_t(\sigma)=\sigma_t, \sigma_{3t}, \sigma_{5t}, \sigma_{7t}, \ldots .$$

We now inductively define sets

$$H(\sigma|n), \qquad C(\sigma|n)\in\mathscr{H}, \qquad n\geq 1, \qquad \sigma\in\mathbb{N}^{\mathbb{N}},$$

and

$$L_{h+1}(\sigma|h;\tau|q)\in\mathscr{H}, \qquad D_{h+1}(\sigma|h;\tau|q)\in\mathscr{H}_{(\sigma,\delta)}, \qquad h\geq 0, q\geq 1, \sigma, \tau\in\mathbb{N}^{\mathbb{N}}.$$

The general relations between the sets will be as follows:

$$H(\sigma|n)=L_{h(n+1)}(\sigma|h(n);\mathbf{j}_{t(n)}(\sigma)|q(n)), \tag{30}$$

$$C(\sigma|n)=C(\sigma|n-1)\cap\Big[D_{h(n)+1}(\sigma|h(n);\mathbf{j}_{t(n)}(\sigma)|q(n))$$
$$\Big\backslash\Big\{\bigcup_{k<\sigma_n} D_{h(n)+1}(\sigma|h(n);\mathbf{j}_{t(n)}(\sigma)|q(n)-1,k)\Big\}\Big], \tag{31}$$

where $C(\varnothing)=X$ (recall that $\varnothing$ denotes the empty sequence of zero length),

$$C(\sigma|h)=\bigcup_{\tau}\bigcap_{q} L_{h+1}(\sigma|h;\tau|q), \tag{32}$$

and

$$D_{h+1}(\sigma|h;\tau|q)=\bigcup_{\nu\in I(\tau|q)}\bigcap_{p} L_{h+1}(\sigma|h;\nu|p)\in\mathscr{H}_{(\sigma,\delta)}. \tag{33}$$

We need to show that the inductive definition can be initiated and then carried out using these formulae in the right order. We start the process by taking

$$L_1(\varnothing;\tau|q)=F^*(\tau|q),$$

$$D_1(\varnothing;\tau|q)=\bigcup_{\nu\in I(\tau|q)}\bigcap_{p} L_1(\varnothing;\nu|p)$$

$$=\bigcup_{\nu\in I(\tau|q)}\bigcap_{p} F^*(\nu|p)$$

$$=B(\tau|q).$$

Then $L_1(\varnothing;\tau|q)\in\mathscr{H}$ and $D_1(\varnothing;\tau|q)\in\mathscr{H}_{(\sigma,\delta)}$. The next step is to define the closed sets H_{σ_1}, $\sigma_1>1$, by taking

$$H(\sigma_1)=L_1(\varnothing;\sigma_1)=F^*(\sigma_1),$$

in accordance with the case $n=1$ of (30). So the set

$$D_1(\varnothing;\sigma_1)=B(\sigma_1)$$

is a subset of $H(\sigma_1)$. Now we define $C(\sigma_1)$ by the formula

$$C(\sigma_1)=D_1(\varnothing;\sigma_1)\Big\backslash\Big\{\bigcup_{k<\sigma_1} D_1(\varnothing;k)\Big\},$$

which is the case $n=1$ of (31). Then

$$C(\sigma_1)=B(\sigma_1)\Big\backslash\Big\{\bigcup_{k<\sigma_1} B(k)\Big\}$$

and the sets $C(\sigma_1)$, for $\sigma_1\geq 1$, are disjoint with

$$H(\sigma_1)\supset C(\sigma_1)\in\mathscr{H}_{(\sigma,\delta)}=\mathscr{H}_{(\sigma,c)}.$$

Now suppose that, for some integer $n\geq 2$, we have reached the stage when the sets

$$H(\sigma|h),\qquad C(\sigma|h),$$
$$L_{h+1}(\sigma|h;\tau|q),\qquad D_{h+1}(\sigma|h;\tau|q)$$

have been defined for all $h<n$. As $h(n)<n$, we can define $H(\sigma|n)$ by the formula (30), and $C(\sigma|n)$ by the formula (31). Then $C(\sigma|n)$ is in $\mathscr{H}_{(\sigma,\delta)}$, and we can choose closed sets $L_{n+1}(\sigma|n;\tau|q)\in\mathscr{H}$ so that $C(\sigma|n)$ has the representation

$$C(\sigma|n)=\bigcup_{\tau}\bigcap_{q} L_{n+1}(\sigma|n;\tau|q),$$

in accordance with (32), such that we can define

$$D_{n+1}(\sigma|n;\tau|q)=\bigcup_{\nu\in I(\tau|q)}\bigcap_{p} L_{n+1}(\sigma|n;\nu|p)\in\mathscr{H}_{(\sigma,\delta)},$$

in accordance with (33). Thus the four families of sets can be constructed inductively in this way.

We now write

$$H(\sigma)=\bigcap_{n} H(\sigma|n),$$

and study the set C with the representation

$$C=\bigcup_{\sigma}\bigcap_{n} H(\sigma|n).$$

If c is any point of C, there is a $\sigma^* \in \mathbb{N}^{\mathbb{N}}$ with

$$\begin{aligned} c \in H(\sigma^*) &= \bigcap_n H(\sigma^*|n) \\ &\subset \bigcap_q H(\sigma^*|2q-1) \\ &= \bigcap_q F^*(\mathbf{j}_1(\sigma^*)|q) \\ &= F^*(\mathbf{j}_1(\sigma^*)) \subset B\ ; \end{aligned} \tag{34}$$

and so $C \subset B$.

Similarly, for any $\sigma \in \mathbb{N}^{\mathbb{N}}$ and any $n \geqslant 1$,

$$\begin{aligned} H(\sigma) &= \bigcap_m H(\sigma|m) \\ &\subset \bigcap_q H(\sigma|2^n(2q-1)) \\ &= \bigcap_q L_{n+1}(\sigma|n\,; \mathbf{j}_{2^n}(\sigma)|q) \\ &\subset C(\sigma|n)\,. \end{aligned} \tag{35}$$

Now, if σ, τ are different vectors of $\mathbb{N}^{\mathbb{N}}$, we can choose $n > 1$ so that

$$\sigma|n \neq \tau|n \quad \text{and} \quad \sigma|r = \tau|r \quad \text{if } 1 \leqslant r < n\,.$$

But, under these conditions, the formula (31) ensures that the corresponding sets $C(\sigma|n)$ and $C(\tau|n)$ are disjoint. Hence, by (35), the sets $H(\sigma)$ and $H(\tau)$ are disjoint. Thus the set C has a disjoint Souslin representation.

Now consider any point $b \in B$. We prove by induction that there is an infinite sequence of integers

$$\sigma^* = \sigma_1^*, \sigma_2^*, \ldots,$$

with

$$b \in H(\sigma^*|n), \qquad n \geqslant 1\,.$$

Since

$$b \in B = \bigcup_{\sigma_1=1}^{\infty} B(\sigma_1)\,,$$

we can choose an integer σ_1 so that

$$b \in B(\sigma_1) = D_1(\varnothing\,; \sigma_1)\,.$$

So there will be a smallest positive integer σ_1^* with

$$b \in D_1(\emptyset; \sigma_1^*).$$

Then, by (31),

$$b \in C(\sigma_1^*).$$

Writing

$$E(\sigma|n) = D_{h(n)+1}(\sigma|h(n); \mathbf{j}_{t(n)}(\sigma)|q(n)), \tag{36}$$

we have, in particular,

$$b \in E(\sigma_1^*) \cap C(\sigma_1^*).$$

We show that the components $\sigma_1^*, \sigma_2^*, \ldots$ can be chosen inductively so that, for $n > 1$,

$$b \in E(\sigma^*|n) \cap C(\sigma^*|n).$$

We suppose that, for some $n > 2$, $\sigma_1^*, \sigma_2^*, \ldots, \sigma_{n-1}^*$ have been chosen so that

$$b \in E(\sigma^*|r) \cap C(\sigma^*|r), \qquad 1 \leq r < n.$$

Now, if n is an exact power of 2, we have $n = t(n)$ and

$$b \in C(\sigma^*|h(n)),$$

since $h(n) < n$. So, in this case,

$$\begin{aligned} b &\in C(\sigma^*|h(n)) \\ &= \bigcup_\tau \bigcap_q L_{h(n)+1}(\sigma^*|h(n); \tau|q) \\ &= \bigcup_{j_1=1}^{\infty} D_{h(n)+1}(\sigma^*|h(n); j_1). \end{aligned}$$

But, if n is not an exact power of 2, we have $n = t(n)\{2q(n) - 1\}$ with $q(n) > 2$ and

$$\begin{aligned} b &\in E(\sigma^*|t(n)\{2q(n) - 3\}) \\ &= D_{h(n)+1}(\sigma^*|h(n); \sigma_{t(n)}^*, \sigma_{3t(n)}^*, \ldots, \sigma_{\{2q(n)-3\}t(n)}^*) \\ &= \bigcup_{\sigma_n=1}^{\infty} D_{h(n)+1}(\sigma^*|h(n); \mathbf{j}_{t(n)}(\sigma^*)|\{q(n) - 1\}, \sigma_n). \end{aligned}$$

Thus, in either case we have

$$b \in C(\sigma^*|n-1) \cap \bigcup_{\sigma_n=1}^{\infty} D_{h(n)+1}(\sigma^*|h(n); \mathbf{j}_{t(n)}(\sigma^*)|\{q(n) - 1\}, \sigma_n).$$

So we can choose σ_n^* to be the smallest σ_n with

$$b \in D_{h(n)+1}(\sigma^*|h(n); \mathbf{j}_{t(n)}(\sigma^*)|\{q(n)-1\}, \sigma_n).$$

Then, by (31) and (36), we have

$$b \in E(\sigma^*|n) \cap C(\sigma^*|n).$$

Thus the components of the vector σ^* can be chosen by induction. Then

$$\begin{aligned} b &\in \bigcap_n E(\sigma^*|n) \\ &= \bigcap_n D_{h(n)+1}(\sigma^*|h(n); \mathbf{j}_{t(n)}(\sigma^*)|q(n)) \\ &\subset \bigcap_n L_{h(n)+1}(\sigma^*|h(n); \mathbf{j}_{t(n)}(\sigma^*)|q(n)) \\ &= \bigcap_n H(\sigma^*|n) \\ &\subset C. \end{aligned}$$

Consequently $B \subset C$. Hence $B = C$ and $B \in \mathbf{S}_d(\mathscr{H})$. Finally, since

$$\bigcap_m H(\sigma|m) \subset C(\sigma|n) \quad \text{and} \quad C(\sigma|n) \cap C(\tau|n) = \varnothing$$

when $\sigma|n \neq \tau|n$, it follows that

$$\bigcup_{\tau \in I(\sigma|n)} \bigcap_m H(\tau|m) = C(\sigma|n) \in \mathscr{H}_{(\sigma,\delta)}.$$

COROLLARY 5.8.4 (Rogers, 1970) *Let $\mathscr{H}$ be a family of subsets of a set $\mathbf{X}$ and consider the following conditions.*

(1) $\mathscr{H}_c \subset \mathscr{H}_{(\sigma,\delta)}$.
(2) *Each set $B \in \mathscr{H}_c$ has a representation*

$$B = \bigcup_\sigma \bigcap_n H(\sigma|n), \qquad H(\sigma|n) \in \mathscr{H},$$

such that for each n-tuple $\sigma|n$

$$\bigcup\left\{\bigcap_{m=1}^{\infty} H(\tau|m) \colon \tau \in \mathbb{N}^{\mathbb{N}} \text{ and } \tau|n = \sigma|n\right\} \in \mathscr{H}_{(\sigma,\delta)}.$$

(3) *Each set* $B \in \mathcal{H}_c$ *has a representation*

$$B = \bigcup_{\sigma} \bigcap_{n} H(\sigma|n), \qquad H(\sigma|n) \in \mathcal{H},$$

such that for each n-tuple $\sigma|n$

$$\bigcup \left\{ \bigcap_{m=1}^{\infty} H(\tau|m) \colon \tau \in \mathbb{N}^{\mathbb{N}} \text{ and } \tau|n = \sigma|n \right\} \in \mathbf{S}_d(\mathcal{H}).$$

(4) $\mathcal{H}_c \subset \mathbf{S}_d(\mathcal{H})$.
(5) $\mathcal{H}_c \subset \mathbf{S}(\mathcal{H})$.

We always have

$$(1) \Rightarrow (2) \Rightarrow (3) \Rightarrow (4) \Rightarrow (5).$$

If X is a K-analytic Hausdorff space and $\mathcal{H} = \mathcal{F}(X)$, *then we also have*

$$(5) \Rightarrow (1).$$

Proof The only part that is neither trivial nor contained in the proof of Theorem 5.8.3(b) is (5) implies (1) when X is K-analytic. This follows from Theorem 3.3.1.

THEOREM 5.8.5 (Jayne, 1977) *If X is a completely regular K-analytic space then*

$$\mathcal{F}_{(+,\delta)}(X) = \mathbf{S}_d(\mathcal{F}(X)),$$

and (even if X is not completely regular)

$$\mathcal{Z}_{(+,\delta)}(X) = \mathbf{S}_d(\mathcal{Z}(X)).$$

Proof By Corollary 2.3.3 we have $\mathcal{F}_{(+,\delta)} \subset \mathbf{S}_d(\mathcal{F}(X))$. Let $B \in \mathbf{S}_d(\mathcal{F}(X))$. Then

$$B = \bigcup_{\sigma} \bigcap_{n} F(\sigma|n), \qquad F(\sigma|n) \in \mathcal{F}(X),$$

$$\bigcap_{n} F(\sigma|n) \cap \bigcap_{n} F(\tau|n) = \varnothing,$$

whenever $\sigma \neq \tau$. For each $\sigma|n \in \mathbb{N}^{(\mathbb{N})}$ define

$$B(\sigma|n) = \bigcup_{\tau \in I(\sigma|n)} \bigcap_{m} F(\tau|m) \in \mathbf{S}_d(\mathcal{F}(X)).$$

Thus each $B(\sigma|n)$ is K-analytic (Theorem 2.5.3) and we have

$$B(\sigma|n)\cap B(\tau|n)=\emptyset\,, \qquad \sigma|n\neq\tau|n\,.$$

Thus for each $n\in\mathbb{N}$ Theorem 5.7.4(c) gives a disjoint sequence of sets $\{C(\sigma|n)\colon \sigma\in\mathbb{N}^{\mathbb{N}}\}\subset\mathscr{Z}_{(+,\delta)}(X)$ such that

$$B(\sigma|n)\subset C(\sigma|n)\,.$$

We have, as in the proof of Theorem 5.7.5(d), that

$$B=\bigcap_n\bigcup_\sigma(\{\mathrm{cl}\,B(\sigma|n)\}\cap C(\sigma|n))\in\mathscr{F}_{(+,\delta)}(X)\,,$$

where $\mathrm{cl}\,B(\sigma|n)$ denotes the closure in X.

By Corollary 2.3.3 we also have $\mathscr{Z}_{(+,\delta)}\subset\mathbf{S}_d(\mathscr{Z}(X))$. Let $B\in\mathbf{S}_d(\mathscr{Z}(X))$. Then by Theorem 5.8.3(a)

$$X\backslash B\in\mathbf{S}(\mathscr{Z}(X))\,.$$

Thus, by Theorem 5.1.6(c), $B\in\mathscr{Z}_{(+,\delta)}(X)$.

We can also prove that $B\in\mathscr{Z}_{(+,\delta)}(X)$, without recourse to Theorem 5.8.3(a), by using the first part of this theorem. We have

$$B=\bigcup_\sigma\bigcap_n Z(f_{\sigma|n})\,,$$

$$\bigcap_n Z(f_{\sigma|n})\cap\bigcap_n Z(f_{\tau|n})=\emptyset\,,$$

whenever $\sigma\neq\tau$, where each $f_{\sigma|n}:X\to R$ is continuous. Re-index $\{f_{\sigma|n}\colon \sigma|n\in\mathbb{N}^{(\mathbb{N})}\}$ by $\{f_n\colon n\in\mathbb{N}\}$ and define a continuous map $f:X\to\mathbb{R}^{\mathbb{N}}$ by

$$f(x)=(f_1(x),f_2(x),\ldots)\,.$$

Then $f(x)$ is an analytic subset of the separable metric space $\mathbb{R}^{\mathbb{N}}$, $f^{-1}\circ f(Z(f_n))=Z(f_n)$ for each $n\in\mathbb{N}$, $f^{-1}\circ f(B)=B$, and

$$f(B)=\bigcup_\sigma\bigcap_n f(Z(f_{\sigma|n}))\in\mathbf{S}_d(\mathscr{Z}(f(X)))\,.$$

By the first part of this theorem,

$$\mathbf{S}_d(\mathscr{Z}(f(X)))=\mathscr{Z}_{(+,\delta)}(f(X))\,,$$

since $\mathscr{Z}_{(+,\delta)}(f(X))=\mathscr{F}_{(+,\delta)}(f(X))$. Consequently,

$$B=f^{-1}\circ f(B)\in\mathscr{Z}_{(+,\delta)}(X)\,.$$

The next two theorems are now easy consequences of Theorems 5.7.5(c), 5.7.5(d) and 5.8.5.

THEOREM 5.8.6 (Jayne, 1977) *The following conditions are equivalent for a completely regular Hausdorff space X:*

(1) *X is K-Lusin;*
(2) *$X \in \mathscr{F}_{(+,\delta)}(Y)$ for any completely regular Hausdorff space Y containing X;*
(3) *$X \in \mathscr{F}_{(+,\delta)}(\beta X)$, where βX denotes the Stone-Čech compactification of X;*
(4) *$X \in \mathbf{S}_d(\mathscr{F}(Y))$ for any completely regular Hausdorff space Y containing X;*
(5) *$X \in \mathbf{S}_d(\mathscr{F}(\beta X))$;*
(6) *$X \in \mathbf{S}_d(\mathscr{F}(K))$ for some compact Hausdorff space K.*

The equivalence of the first three statements in the next theorem to each other and to X being a Baire set in βX is due to Frolík (1961*c*). The remaining parts are due to Jayne (1973, 1977).

THEOREM 5.8.7 *The following conditions are equivalent for a completely regular Hausdorff space X:*

(1) *X is K-Lusin and proper K-countably determined;*
(2) *X and $(\beta X)\backslash X$ are K-analytic;*
(3) *X is a proper K-Lusin space;*
(4) *$X \in \mathscr{Z}_{(+,\delta)}(\beta X)$;*
(5) *$X \in \mathscr{Z}_{(+,\delta)}(Y)$ for any completely regular Hausdorff space Y containing X as a dense subspace;*
(6) *$X \in \mathbf{S}_d(\mathscr{Z}(\beta X))$;*
(7) *$X \in \mathbf{S}_d(\mathscr{Z}(Y))$ for any completely regular Hausdorff space Y containing X as a dense subspace;*
(8) *X is homeomorphic to a closed subset of a product space $K \times B$ for some compact space K and a Borel subset B of some Polish space.*
(9) *$X \in \mathbf{S}_d(\mathscr{Z}(K))$ for some compact Hausdorff space K.*

THEOREM 5.8.8 *The following conditions are equivalent for a completely regular Hausdorff space X:*

(1) *X is K-analytic;*
(2) *$X \in \mathbf{S}(\mathscr{F}(Y))$ for any completely regular Hausdorff space Y containing X;*
(3) *$X \in \mathbf{S}(\mathscr{F}(\beta X))$;*
(4) *$X \in \mathbf{S}(\mathscr{F}(K))$ for some compact Hausdorff space K.*

Proof (1) implies (2) by Theorem 2.5.2. Clearly (2) implies (3) and (3) implies (4); and (4) implies (1) by Theorem 2.5.3 since every compact space is trivially K-analytic.

Most of the proof of the next theorem has been given in Section 5.2.

THEOREM 5.8.9 *The following conditions are equivalent for a completely regular Hausdorff space X:*

(1) *X is K-analytic and proper K-countably determined;*
(2) *X is a proper K-analytic space;*
(3) $X \in \mathbf{S}(\mathscr{Z}(\beta X))$;
(4) $X \in \mathbf{S}(\mathscr{Z}(Y))$ *for any completely regular Hausdorff space Y containing X as a dense subspace;*
(5) *X is homeomorphic to a closed subset of a product space $K \times A$ for some compact Hausdorff space and an analytic subset of A of some Polish space;*
(6) $X \in \mathbf{S}(\mathscr{Z}(K))$ *for some compact Hausdorff space K.*

THEOREM 5.8.10 *Let X be a K-Lusin space and let $\mathscr{L}$ denote the family of K-Lusin subsets of X.*

(a) (Rogers, 1970*a*) *If every open set in X is in* $\mathbf{S}(\mathscr{F}(X))$, *then*

$$\mathscr{L} = \mathscr{F}_{(\sigma,\delta)}(X) = \mathscr{F}_{(c,\delta)}(X) = \mathbf{S}_d(\mathscr{F}(X)) .$$

(b) (Jayne, 1977) *If X is completely regular, then*

$$\mathscr{L} = \mathscr{F}_{(+,\delta)}(X) = \mathbf{S}_d(\mathscr{F}(X)) .$$

Proof (a) By Corollary 5.8.4 we have

$$\mathscr{F}_c(X) \subset \mathscr{F}_{(\sigma,\delta)}(X) ,$$

and so by Theorem 5.8.3(b), we have

$$\mathscr{F}_{(\sigma,c)}(X) = \mathscr{F}_{(\sigma,\delta)}(X) \subset \mathbf{S}_d(\mathscr{F}(X)) .$$

Since X is a K-Lusin space, it follows that

$$\mathscr{F}(X) \subset \mathscr{L}$$

and so, by Theorem 5.1.4,

$$\mathbf{S}_d(\mathscr{F}(X)) \subset \mathbf{S}_d(\mathscr{L}) = \mathscr{L} .$$

Finally, we have

$$\mathscr{L} \subset \mathscr{F}_{(\sigma,c)}(X) ,$$

by Theorem 3.6.1.

(b) As in part (a), since $\mathscr{F}(X)\subset\mathscr{L}$, we have by Theorem 5.1.4 and Corollary 2.3.3 that

$$\mathscr{F}_{(+,\delta)}(X)\subset \mathbf{S}_d(\mathscr{F}(X))\subset \mathbf{S}_d(\mathscr{L})=\mathscr{L}\,.$$

Finally, by Theorem 5.7.5(d) we have

$$\mathscr{L}=\mathscr{F}_{(+,\delta)}(X)\,.$$

We conclude this section with two conditions on an upper semi-continuous map $K:\mathbb{N}^{\mathbb{N}}\to\mathscr{K}(X)$ which imply, when each open set in X is in $\mathbf{S}(\mathscr{F}(X))$, that X is a K-Lusin space. We will only give the proof of the first condition, which is closely related to the proof of Theorem 5.8.3(b).

THEOREM 5.8.11 *Let X be a Hausdorff space with the property that each open set in X is in $\mathbf{S}(\mathscr{F}(X))$.*

(a) (Rogers, 1970*a*) *X is a K-Lusin space, if, and only if, there is an upper semi-continuous map $K:\mathbb{N}^{\mathbb{N}}\to\mathscr{K}(X)$, $X=K(\mathbb{N}^{\mathbb{N}})$, with*

$$K(I(\sigma|n))\in\mathscr{F}_{(\sigma,\,\delta)}(X)$$

for all $\sigma|n\in\mathbb{N}^{(\mathbb{N})}$.

(b) (Rogers, 1965) *X is a K-Lusin space if there is an upper semi-continuous map $K:\mathbb{N}^{\mathbb{N}}\to\mathscr{K}(X)$, $X=K(\mathbb{N}^{\mathbb{N}})$, such that each non-empty perfect compact subset of X contains a point x such that the set $\{\sigma\in\mathbb{N}^{\mathbb{N}}: x\in K(\sigma)\}$ is finite or countable.*

Proof of (*a*) Suppose X is a K-Lusin space and $K:\mathbb{N}^{\mathbb{N}}\to\mathscr{K}(X)$, $X=K(\mathbb{N}^{\mathbb{N}})$, is a disjoint u.s.c. map. Then, since each $I(\sigma|n)$ is homeomorphic to $\mathbb{N}^{\mathbb{N}}$ and K restricted to $I(\sigma|n)$ is a disjoint u.s.c. map, Theorem 5.2.5(b) implies that

$$K(I(\sigma|n))\in\mathscr{F}_{(\sigma,\,\delta)}(X)\,.$$

Now suppose X is a K-analytic space and that $K:\mathbb{N}^{\mathbb{N}}\to\mathscr{K}(X)$, $X=K(\mathbb{N}^{\mathbb{N}})$, is an u.s.c. map such for each $\sigma|n\in\mathbb{N}^{(\mathbb{N})}$

$$K(I(\sigma|n))\in\mathscr{F}_{(\sigma,\,\delta)}(X)\,.$$

For each $\sigma\in\mathbb{N}^{\mathbb{N}}$ and $n\in\mathbb{N}$ define $F(\sigma|n)$ to be the closure of $K(I(\sigma|n))$ in X. Write

$$F(\sigma)=\bigcap_{n=1}^{\infty}F(\sigma|n)\,.$$

Clearly

$$F(\sigma)=\bigcap_n F(\sigma|n)\supset\bigcap_n K(I(\sigma|n))\supset K(\sigma)\,.$$

Now, if $x \in F(\sigma)$ and $x \notin K(\sigma)$, using the fact that X is Hausdorff and $K(\sigma)$ is compact we can choose disjoint open sets U_1 and U_2 with

$$x \in U_1\,; \qquad K(\sigma) \subset U_2\,.$$

Then for sufficiently large n

$$K(I(\sigma|n)) \subset U_2 \subset X \backslash U_1\,.$$

As $X\backslash U_1$ is closed this implies that

$$x \in F(\sigma) \subset F(\sigma|n) \subset X \backslash U_1\,,$$

for all sufficiently large n. This contradiction shows that $F(\sigma) \subset K(\sigma)$. Thus $F(\sigma) = K(\sigma)$ for all $\sigma \in \mathbb{N}^{\mathbb{N}}$ and X has the Souslin representation

$$X = \bigcup_{\sigma} \bigcap_{n} F(\sigma|n)$$

with the sets

$$B(\sigma|n) \equiv \bigcup_{\tau \in I(\sigma|n)} \bigcap_{m} F(\tau|m) \in \mathscr{F}_{(\sigma,\,\delta)}(X)\,.$$

We now suppose that the whole construction used in the proof of Theorem 5.8.3(b) has been carried through, with B put equal to X, and we use the notation and results of that proof. We obtain

$$X = B = \bigcup_{\sigma} \bigcap_{n} H(\sigma|n)\,, \qquad H(\sigma|n) \in \mathscr{F}(X)\,,$$

$$\bigcap_{n} H(\sigma|n) \cap \bigcap_{n} H(\tau|n) = \varnothing\,, \qquad \sigma \neq \tau\,,$$

and for each n-tuple $\sigma|n$

$$\bigcup_{\tau \in I(\sigma|n)} \bigcap_{m} H(\tau|n) \in \mathscr{F}_{(\sigma,\,\delta)}(X)\,.$$

So it will suffice to show that $H : \mathbb{N}^{\mathbb{N}} \to \mathscr{K}(X)$ is an upper semi-continuous map, where

$$H(\sigma) \equiv \bigcap_{n} H(\sigma|n)\,.$$

First we show that $H(\sigma) \in \mathscr{K}(X)$. We have

$$H(\sigma) = \bigcap_n H(\sigma|n) = \bigcap_{h=0}^{\infty} \bigcap_{q=1}^{\infty} H(\sigma|2^h(2q-1))$$

$$= \bigcap_{h=0}^{\infty} \bigcap_{q=1}^{\infty} L_{h+1}(\sigma|h; \mathbf{j}_{2^h}(\sigma)|q)$$

$$= \left[\bigcap_{q=1}^{\infty} F(\mathbf{j}_1(\sigma)|q)\right] \cap \left[\bigcap_{h=1}^{\infty} \bigcap_{q=1}^{\infty} L_{h+1}(\sigma|h; \mathbf{j}_{2^h}(\sigma)|q)\right]$$

$$= F(\mathbf{j}_1(\sigma)) \cap \left[\bigcap_{h=1}^{\infty} \left\{\bigcap_{q=1}^{\infty} L_{h+1}(\sigma|h; \mathbf{j}_{2^h}(\sigma)|q)\right\}\right].$$

Now

$$F(\mathbf{j}_1(\sigma)) = K(\mathbf{j}_1(\sigma))$$

and we write

$$L_{h+1}(\sigma|h; \tau) = \bigcap_{q=1}^{\infty} L_{h+1}(\sigma|h; \tau|q).$$

Thus we have

$$H(\sigma) = K(\mathbf{j}_1(\sigma)) \cap \left[\bigcap_{h=1}^{\infty} L_{h+1}(\sigma|h; \mathbf{j}_{2^h}(\sigma))\right].$$

As $K(\mathbf{j}_1(\sigma))$ is compact and the sets $L_1(\sigma|h; \tau|q)$ are closed, the sets $H(\sigma)$ are compact.

Now, to establish the upper semi-continuity of H, consider any $\sigma^* \in \mathbb{N}^{\mathbb{N}}$ and any open set G in X with

$$H(\sigma^*) \subset G.$$

If $K(\mathbf{j}_1(\sigma^*)) \backslash G \neq \emptyset$, consider a point

$$x \in K(\mathbf{j}_1(\sigma^*)) \backslash G.$$

Then

$$x \notin H(\sigma^*) = K(\mathbf{j}_1(\sigma^*)) \cap \left[\bigcap_{h=1}^{\infty} L_{h+1}(\sigma^*|h; \mathbf{j}_{2^h}(\sigma^*))\right].$$

So we may choose $h(x) \geq 1$ and then $q(x) \geq 1$ so that

$$x \notin L_{h(x)+1}(\sigma^*|h(x); \mathbf{j}_{2^{h(x)}}(\sigma^*)|q(x)).$$

Now the open set

$$U(x) = X \backslash L_{h(x)+1}(\sigma^*|h(x); \mathbf{j}_{2^{h(x)}}(\sigma^*)|q(x))$$

contains x and

$$U(x)\cap H(\sigma^*)=\varnothing\,.$$

Indeed, for all τ^* with

$$\tau^*|2^{h(x)}(2q(x)-1)=\sigma^*|2^{h(x)}(2q(x)-1)\,,$$

we have

$$U(x)\cap H(\tau^*)=\varnothing\,.$$

Since $K(\mathbf{j}_1(\sigma^*))\backslash G$ is compact, we can choose a finite set of points $x_1,\ldots,x_m$ so that

$$K(\mathbf{j}_1(\sigma^*)\backslash G\subset\bigcup_{i=1}^{m}(x_i)\,.$$

Thus

$$K(\mathbf{j}_1(\sigma^*))\subset G\cup\bigcup_{i=1}^{m}U(x_i)\,.$$

By the upper semi-continuity of $K:\mathbb{N}^{\mathbb{N}}\to\mathcal{K}(X)$, we can choose q^* so that

$$K(I(\mathbf{j}_1(\sigma^*)|q^*))\subset G\cup\bigcup_{i=1}^{m}U(x_i)\,.$$

Take

$$n^*=\max\{2q^*-1,2^{h(x_1)}(2q(x_1)-1),\ldots,2^{h(x_m)}(2q(x_m)-1)\}\,.$$

Then for all $\sigma\in I(\sigma^*|n^*)$ we obtain

$$\begin{aligned}H(\sigma)\subset&\left[G\cup\left\{\bigcup_{i=1}^{m}U(x_i)\right\}\right]\cap\left[\bigcap_{i=1}^{m}L_{h(x_i)+1}(\sigma^*|h(x_i);\,\mathbf{j}_{2^{h(x_i)}}(\sigma^*)|q(x_i))\right]\\ \subset&\, G\,.\end{aligned}$$

Thus we have the required upper semi-continuity of H.

5.9 Continuous images of K-Lusin spaces

The continuous image of a K-Lusin space is, of course, a K-analytic space, but, as the examples in Section 4.5 show, not necessarily a K-Lusin space. Thus the problem arises, "What conditions on the continuous function and on the spaces ensure that the image will be a K-Lusin space?" For example, if X and Y are metrizable, Y is K-Lusin and f: $Y\to X$ is a continuous surjection such that for each x in X the set $f^{-1}(x)$ is countable (see

Dellacherie, Part 2, Chapitre II, Section 4 in this volume) or, more generally, σ-compact (see Dellacherie, Chapitre IV, Section 4), then X is a K-Lusin space. In the absence of metrizability this result no longer holds, although the injective continuous image of a K-Lusin space is clearly always a K-Lusin space (see Theorem 5.1.3).

The following example is a completely regular K-Lusin space Y (in fact, $Y \in \mathscr{L}_\sigma(\beta Y)$) with a closed (i.e. maps closed sets to closed sets) continuous map of Y onto a completely regular space X, which is injective except for one x in X for which $f^{-1}(x)$ is countable, and with X not a K-Lusin space.

Example 5.9.1 As in Example 5.2.2, let Y be the discrete union of countably many copies of the one-point compactification of an uncountable discrete set. Let X be the set Y with the countable set of compactifying points identified. Give X the quotient topology. Then the quotient map is closed, continuous, and injective except that the countable set of compactifying points map onto a single point. As shown in Example 5.2.2, X is not a K-Lusin space.

If $f: Y \to X$ is a proper surjection and X is a K-Lusin space, then Y is a K-Lusin space. This is because $f^{-1}: X \to \mathscr{K}(Y)$ is a disjoint upper semi-continuous map and so Theorem 5.1.3 applies. If X and Y are metrizable, and Y is a K-Lusin space, then X is a K-Lusin space, since $f^{-1}(x)$ is compact for each x in X (see Dellacherie, Chapitre II, Section 3). More generally, if X and Y are Hausdorff spaces and f: $Y \to X$ is a proper surjection, then X is a proper K-Lusin space, if, and only if, Y is a proper K-Lusin space. This follows from Propositions 5.2.2 and 5.2.3, Theorem 5.1.6(c), and the remarks after Proposition 5.2.3.

The next example is a proper surjection of a completely regular K-Lusin space onto a completely regular space which is not a K-Lusin space.

Example 5.9.2 For each $n \in \mathbb{N}$ let D_n be the one-point compactification of an uncountable discrete set D, with compactifying point $y_n \in D_n$. Let Y be the disjoint union of the sets D_n, $n \in \mathbb{N}$. Define the topology on Y by taking each $y \in Y \backslash \{y_n: n \in \mathbb{N}\}$ to be open; a neighbourhood base of each y_n, $n \geqslant 2$, to be its neighbourhood base in D_n; and a neighbourhood base of y_1 to be all sets of the form

$$N(y_1) = U_1 \cup \bigcup_{n=m}^{\infty} U_n,$$

for some $m \in \mathbb{N}$, where U_n, $n = 1, m, m+1, \ldots$ are neighbourhoods of x_n in D_n. As in the last example, let X be the set Y with the points $\{y_n: n \in \mathbb{N}\}$ identified, and give X the quotient topology. Then Y is a completely regular

K-Lusin space, since it is the countable disjoint union of compact sets, although unlike the last example, $Y \notin \mathscr{Z}_{(+,\delta)}(\beta Y)$. The quotient map is proper and, as shown in Example 5.2.2, X is not a K-Lusin space.

Instead of asking when a continuous surjection f of Y into X preserves some property we can ask when the projection of the graph of f on X preserves the property. Thus for the property of being a K-Lusin space we can ask when the projection on X of a K-Lusin set in $X \times Y$ is a K-Lusin set. This reformulation allows us to introduce conditions on the placement of the K-Lusin set in the product (e.g. contained in $\mathscr{Z}_{(+,\delta)}(X \times Y)$), and leads to positive results for non-metrizable spaces.

In the last two examples the graph of the quotient map f of Y onto X may be regarded as a closed K-Lusin subset of $X \times \beta Y$ or as a K-Lusin subset of $\beta X \times \beta Y$. In each case the graph of f is not a Baire subset of the product.

In the case of metrizable spaces X and Y, if certain hypotheses allow one to conclude that the projection on X of each K-Lusin set B in $X \times Y$, satisfying the hypothesis, is K-Lusin, then one can usually (e.g. when $B^x = \{y: (x, y) \in B\}$ is σ-compact for each x in X) also conclude that B can be uniformized by a K-Lusin set B_0; that is, there is a K-Lusin set $B_0 \subset B$ such that $\pi_x(B) = \pi_x(B_0)$ and $\pi_x | B_0$ is injective. This situation no longer persists in the absence of metrizability, not even when X is a compact metric space and Y is a perfectly normal compact space (cf. Example 5.9.3).

If X and Y are completely regular K-Lusin spaces and $B \in \mathscr{Z}_{(+,\delta)}(X \times Y)$ with $B^x = \{y: (x, y) \in B\}$ at most countable for each x in X, then, by Theorem 5.9.6, $\pi_x(B) \in \mathscr{Z}_{(+,\delta)}(X)$ and hence, by Theorem 5.1.4, $\pi_x(B)$ is a K-Lusin set. The next paragraph gives an example of a perfectly normal compact space Y and, for $X = [0, 1]$, a compact set $B \in \mathscr{Z}(X \times Y) = \mathscr{F}(X \times Y)$ with the cardinality of $B^x = \{y: (x, y) \in B\}$ at most two for each x in X, such that there is no K-analytic set which uniformizes B.

Example 5.9.3 We take Y to be the well-known double arrow topology on two half open unit intervals:

$$A: [\!-\!\!\underset{a}{\bullet}\!\!-\!),$$

$$A': (\!-\!\!\underset{a'}{\bullet}\!\!-\!],$$

$$Y = A \cup A',$$

where a neighbourhood base of a point $a \in A$ is given by sets of the form $[a, a+\varepsilon) \cup (a', a'+\varepsilon)$, where a' is the projection of a on A', and of a point $a' \in A'$ by sets of the form $(a-\varepsilon, a) \cup (a'-\varepsilon, a']$. Then Y is a compact perfectly normal non-metrizable space.

Define $f: Y \to [0, 1]$ by $f(a) = f(a') = a$, $f(0) = 0$ and $f(1') = 1$.

Suppose, to obtain a contradiction, that C is a K-analytic subset of Y which uniformizes f; i.e. $f(C) = [0, 1]$ and $f|C$ is injective. Since Y is perfectly normal,

$$C \in S(\mathscr{Z}(X)) = S(\mathscr{F}(X)) .$$

Thus, by Theorem 5.8.9, C is a proper K-analytic space. Since $f|C$ is a continuous real-valued function that separates the points of C, it follows from Theorem 5.6.1 that C is metrizable. Hence, by Theorem 3.5.1, C contains an uncountable compact subset. Now C must have an uncountable number of points in the top or bottom segment of Y. We can suppose that $C \cap A$ is uncountable.

Let $\mathscr{B}$ be an open base for the relative topology on C. For each point x in $C \cap A$, $[x, 1) \cup (x', 1)$ must contain an open neighbourhood of x in $\mathscr{B}$. Since $C \cap A$ is uncountable, this implies that $\mathscr{B}$ is uncountable. This contradicts the fact that C is metrizable and Lindelöf, and consequently has a countable base.

The graph of f is a zero set, in $[0, 1] \times Y$, since it is a closed set and the product of a perfectly normal compact space and a compact metric space is always perfectly normal.

In the last example we can define a function $g: [0, 1] \to Y$ by

$$g(x) = \begin{cases} x & \text{in } A \quad \text{for } x \neq 1 , \\ 1' & \text{in } A' \quad \text{for } x = 1 . \end{cases}$$

Then $f \circ g(x) = x$ for all x in $[0, 1]$, and $g^{-1}(B) \in \mathscr{Z}_{(+,\delta)}([0, 1])$ for each $B \in \mathscr{Z}_{(+,\delta)}(Y)$; i.e. g is a Baire measurable selection for f (whose graph is necessarily not a Baire, nor even K-analytic, subset of $[0, 1] \times Y$).

The existence of such a Baire-measurable selection in this example is not so much dependent on the function as on the space Y—at least supposing the continuum hypothesis. For it is not difficult to establish that if Y is a compact Hausdorff space with weight at most $\aleph_1$ (i.e. Y has an open base of cardinality at most $\aleph_1$) and f is a continuous map of Y onto a Hausdorff space X, then there is a map $g: X \to Y$ such that $f \circ g(x) = x$ for each x in X and $g^{-1}(B) \in \mathscr{Z}_{(+,\delta)}(X)$ for each $B \in \mathscr{Z}_{(+,\delta)}(Y)$.

We will now describe two methods of generalizing results for metrizable analytic and Lusin spaces to K-analytic and K-Lusin spaces. The first entails taking a classical theorem and analysing its proof in the light of the more general definitions. The following two theorems illustrate this approach. The first of them generalizes Lusin's theorem that the projection of a Borel set, in the product of Polish spaces, with countable sections is a Borel set. The second theorem generalizes the Kunugui–Arsenin theorem that the projection of a Borel set, in the product of Polish spaces, with σ-compact sections is a Borel set, and also Ščegol'kov's theorem that such a Borel set can be uniformized by a Borel set.

THEOREM 5.9.1 (Rogers, 1965) *Let Y be a K-Lusin space and let X be a Hausdorff space with the property that each open set in X is in $\mathbf{S}(\mathcal{F}(X))$. If B is a K-Lusin subset of $X \times Y$ such that each non-empty perfect subset of $\pi_X(B)$ contains a point x such that the section*

$$B^x = \{y \in Y: (x, y) \in B\}$$

is at most countable, then $\pi_X(B)$ is a K-Lusin set.

THEOREM 5.9.2 (Rogers) *Let Y be a compact metric space and let X be a Hausdorff space with the property that each open set in X is in $\mathbf{S}(\mathcal{F}(X))$. If B is a K-Lusin set in $X \times Y$ such that the section B^x is σ-compact for each x in X, then B can be uniformized by a K-Lusin set and $\pi_X(B)$ is a K-Lusin set.*

The second method of obtaining results for K-analytic spaces is to reduce to the metrizable case. Suppose we are given a statement about the K-analytic subsets of a completely regular Hausdorff space X. The idea is to define, in some way, a continuous map f of X into $\mathbb{R}^{\mathbb{N}}$ (or equivalently, define a separable, continuous pseudo-metric on X), so that the validity of the statement for the analytic subsets of $f(x) \subset \mathbb{R}^{\mathbb{N}}$ implies its validity in X. As we will see, it is very often possible to design theorems so that such a continuous map into $\mathbb{R}^{\mathbb{N}}$ can be defined.

One limitation of this approach is that not many relevant properties are preserved both by a continuous map f and by its inverse. If f is a proper map, then compactness, σ-compactness, and in certain contexts, Baire classes are preserved. These are the properties exploited in the next theorems.

Another limitation of this approach is that it does not seem to give theorems like Theorems 5.9.1 and 5.9.2.

The following theorem illustrates this approach. Its proof is a reduction to Chapter III, Theorem 32 and a special case of Chapter IV, Theorem 51 in Part 2 of this volume. In the context of Polish spaces, the fact that C_1 is coanalytic is due to Kunugui and the fact that C_2 is coanalytic is due to Arsenin.

THEOREM 5.9.2 *Let X be a completely regular Hausdorff space, Y be a proper K-Lusin space, and let $B \subset X \times Y$ be such that both B and its complement belong to $\mathbf{S}(\mathcal{Z}(X) \times \mathcal{Z}(Y))$. The sets*

$$C_1 = \{x \in \pi_X(B): B^x \text{ is compact}\}$$

and

$$C_2=\{x\in\pi_X(B)\colon B^x \text{ is } \sigma\text{-compact}\}$$

are both complements of sets in $\mathbf{S}(\mathscr{Z}(X))$. *Consequently, if* B^x *is* σ*-compact for each* x *in* X, *then both* $\pi_X(B)$ *and its complement belong to* $\mathbf{S}(\mathscr{Z}(X))$.

Proof By hypothesis we can write

$$B=\bigcup_\sigma\bigcap_n Z(f_{\sigma|n})\times Z(g_{\sigma|n}),$$

$$(X\times Y)\backslash B=\bigcup_\sigma\bigcap_n Z(f'_{\sigma|n})\times Z(g'_{\sigma|n}),$$

where $f_{\sigma|n}, f'_{\sigma|n}: X\to\mathbb{R}$ and $g_{\sigma|n}, g'_{\sigma|n}: Y\to\mathbb{R}$ are continuous functions. Re-index the sets $\{f_{\sigma|n}, f'_{\sigma|n}\colon \sigma|n\in\mathbb{N}^{(\mathbb{N})}\}$ and $\{g_{\sigma|n}, g'_{\sigma|n}\colon \sigma|n\in\mathbb{N}^{(\mathbb{N})}\}$ by $\{f_n\colon n\in\mathbb{N}\}$ and $\{g_n\colon n\in\mathbb{N}\}$. Since Y is a proper K-Lusin space, there is a proper map $g_o: Y\to g_0(Y)\subset\mathbb{R}^{\mathbb{N}}$.

Define continuous functions $f: X\to\mathbb{R}^{\mathbb{N}}$ and $g: Y\to\mathbb{R}^{\mathbb{N}}$ by

$$f(x)=(f_1(x), f_2(x), \ldots),$$

$$g(y)=(g_0(y), g_1(y), g_2(y), \ldots).$$

Since g_0 is a proper map, it follows that $g: Y\to g(Y)$ is a proper map.

Now define a continuous function

$$\Phi\colon X\times Y\to\Phi(X\times Y)\subset\mathbb{R}^{\mathbb{N}}\times\mathbb{R}^{\mathbb{N}}$$

by

$$\Phi(x, y)=(f(x), g(y)).$$

The map Φ has the following properties:

(i) the sets $f(Z(f_{\sigma|n}))$, $f(Z(f'_{\sigma|n}))$, $g(Z(g_{\sigma|n}))$, and $g(Z(g'_{\sigma|n}))$ are closed in $f(X)$ and $g(Y)$ respectively;

(ii) $\Phi(B)=\bigcup_\sigma\bigcap_n f(Z(f_{\sigma|n}))\times g(Z(g_{\sigma|n}))$ and $\Phi((X\times Y)\backslash B)=\bigcup_\sigma\bigcap_n f(Z(f'_{\sigma|n}))\times g(Z(g'_{\sigma|n}))$;

(iii) $\Phi^{-1}\circ\Phi(B)=B$ and so $\Phi((X\times Y)\backslash B)=(X\times Y)\backslash\Phi(B)$;

(iv) $g(B^x)=\Phi(B)^{f(x)}$ for each x in X;

(v) $g: g^{-1}(\Phi(B)^{f(x)})\to\Phi(B)^{f(x)}$ is a proper map;

(vi) $C_1=f^{-1}(\{f(x)\in\pi_{f(X)}(f(B))\colon \Phi(B)^{f(x)} \text{ is compact}\})$;

(vii) $C_2=f^{-1}(\{f(x)\in\pi_{f(X)}(f(B))\colon \Phi(B)^{f(x)} \text{ is } \sigma\text{-compact}\})$.

Now from (i), (ii) and (iii) it follows that $\Phi(B)$ and $(f(X)\times g(Y))\backslash\Phi(B)$ are contained in $\mathbf{S}(\mathscr{Z}(f(X))\times\mathscr{Z}(g(Y)))$. Since g is a proper map, $g(Y)$ is a Lusin space (which is all that is required in Dellacherie's Theorem 32 in Chapter

III). It now follows from Dellacherie's Theorems 32 in Chapter III and 51 (and his Remarks 52(d)) in Chapter IV that the sets

$$\{f(x) \in \pi_{f(X)}(\Phi(B)) \colon \Phi(B)^{f(x)} \text{ is compact}\},$$
$$\{f(x) \in \pi_{f(X)}(\Phi(B)) \colon \Phi(B)^{f(x)} \text{ is } \sigma\text{-compact}\}$$

are complements of sets in $\mathbf{S}(\mathscr{Z}(f(X)))$. The first part of the theorem now follows from (vi) and (vii).

The second part of the theorem, when B^x is σ-compact for each x in X, follows from the first part and Theorem 2.6.5.

To obtain a neater form of the last theorem (and subsequent theorems) we need the following lemma:

LEMMA 5.9.4 *If X and Y are completely regular Hausdorff spaces and $X \times Y$ is Lindelöf, then*

$$(\mathscr{Z}_{(+,\delta)}(X) \times \mathscr{Z}_{(+,\delta)}(X \times Y))_{(\sigma,\delta)} = \mathscr{Z}_{(+,\delta)}(X \times Y)\,. \tag{37}$$

Consequently, if $A \in \mathbf{S}(\mathscr{Z}(X \times Y))$, then $\pi_X(A) \in \mathbf{S}(\mathscr{Z}(X))$.

Proof Suppose $Z(f) \in \mathscr{Z}(X)$ and $Z(g) \in \mathscr{Z}(Y)$. Then

$$Z(f) \times Z(g) = Z(f \circ \pi_X) \cap Z(g \circ \pi_X)$$

is a zero set in $X \times Y$. Hence the left-hand side of (37) is contained in the right—recalling that $\mathscr{Z}_{(+,\delta)}(X \times Y) = \mathscr{Z}_{(\sigma,\delta)}(X \times Y)$ by Theorem 5.7.2.

Suppose $Z(h) \in \mathscr{Z}(X \times Y)$. Then

$$(X \times Y) \backslash Z(h) = \bigcup_{n=1}^{\infty} \left\{(x, y) \colon |h(x, y)| \geqslant \frac{1}{n}\right\},$$

and consequently $(X \times Y) \backslash Z(h)$ is Lindelöf. Thus $(X \times Y) \backslash Z(h)$ is the countable union of products of cozero sets in X and Y, since X and Y are completely regular. It now follows that the right-hand side of (37) is contained in the left.

If $A \in \mathbf{S}(\mathscr{Z}(X \times Y))$, then, by the first part of this theorem and by Theorem 2.6.5, $\pi_X(A) \in \mathbf{S}(\mathscr{Z}(X))$.

From Theorem 5.9.3, Lemma 5.9.4 and the first separation theorem (Theorem 5.1.6(c)), we now obtain

THEOREM 5.9.5 *Let X be a regular K-analytic space, Y is a proper K-Lusin space, and let $B \in \mathscr{Z}_{(+,\delta)}(X \times Y)$.*

The sets

$$C_1 = \{x \in \pi_X(B) \colon B^x \text{ is compact}\}$$

and

$$C_2 = \{x \in \pi_X(B) \colon B^x \text{ is } \sigma\text{-compact}\}$$

are complements of sets in $\mathbf{S}(\mathscr{Z}(X))$. *Consequently, if* B^x *is* σ*-compact for each* x *in* X, *then*

$$\pi_X(B) \in \mathscr{Z}_{(+,\delta)}(X)\,.$$

From this theorem, Lemma 5.9.4 and the fact that every regular K-Lusin space is Baire-isomorphic to a proper K-Lusin space (Theorem 5.1.7) we now obtain

THEOREM 5.9.6 *If* X *is a regular* K*-analytic space,* Y *a regular* K*-Lusin space, and* B *a set in* $\mathscr{Z}_{(+,\delta)}(X \times Y)$ *such that* B^x *is at most countable for each* x *in* X, *then* $\pi_X(B) \in \mathscr{Z}_{(+,\delta)}(X)$.

Saint-Raymond (1976*a*) has shown that a Borel set $B \subset X \times Y$, where X and Y are compact metric spaces, such that B^x is σ-compact for each x in X, can be expressed as the countable union of Borel sets (B_n) with B_n^x compact for each $n \in \mathbb{N}$ and each x in X. Dellacherie has given two forms of this theorem in his Corollaries 47 and 48 in Chapter IV. The next two theorems are obtained from these corollaries using the method of the proof of Theorem 5.9.3.

THEOREM 5.9.7 *If* X *is a regular* K*-analytic space,* Y *a proper* K*-analytic space, and* B *is a set in* $\mathscr{Z}_{(+,\delta)}(X \times Y)$ *such that* B^x *is an* $\mathscr{F}_\sigma$*-set for each* x, *then* B *is the countable union of sets* (B_n) *in* $\mathscr{Z}_{(+,\delta)}(X \times Y)$ *such that* B_n^x *is a zero set for each* $n \in \mathbb{N}$ *and each* $x \in X$.

THEOREM 5.9.8 *Let* X *be a regular* K*-analytic space,* Y *a proper* K*-Lusin space, and* B *a set in* $\mathbf{S}(\mathscr{Z}(X \times Y))$ *(resp.* $\mathscr{Z}_{(+,\delta)}(X \times Y)$*) such that* B^x *is contained in some* σ*-compact subset, depending on* x, *of* Y *for each* $x \in X$.

(a) *B is the countable union of sets* (B_n) *in* $\mathbf{S}(\mathscr{Z}(X \times Y))$ *(resp.* $\mathscr{Z}_{(+,\delta)}(X \times Y)$*) such that* B_n^x *is relatively compact for each* $n \in \mathbb{N}$ *and each* $x \in X$.

(b) *If* $B \in \mathscr{Z}_{(+,\delta)}(X \times Y)$ *and* B^x *is* σ*-compact for each* x *in* X, *then* B *is the countable union of sets* (B_n) *in* $\mathscr{Z}_{(+,\delta)}(X \times Y)$ *such that* B_n^x *is compact for each* $n \in \mathbb{N}$ *and each* $x \in X$.

A. Louveau (1977*a*) and J. Bourgain (198?*b*) have independently proved that if B is a Borel set in the product $X \times Y$ of two compact metric spaces such that B^x is a $\mathcal{K}_{\sigma\delta}$-set for each x in X, then

$$B = \bigcap_{m=1}^{\infty} \bigcup_{n=1}^{\infty} B_{mn},$$

where each B_{mn} is a Borel set with B^x_{mn} compact for each x in X. In order to obtain this result in compact Hausdorff spaces we need the highly non-trivial fact that if B is a $\mathscr{Z}_{\sigma\delta}$-subset of some compact Hausdorff space and f is a proper map of B onto a subset of a compact metric space, then $f(B)$ is a $\mathcal{K}_{\sigma\delta}$-set. This is contained in Theorem 5.9.13. Hence we can obtain the next theorem by reducing to the above result.

THEOREM 5.9.9 *Let X and Y be compact Hausdorff spaces and let B be a set in $\mathscr{Z}_{(+,\delta)}(X \times Y)$ such that $B^x \in Z_2(Y)$ for each x in X. Then there are sets B_{mn} in $\mathscr{Z}_{(+,\delta)}(X \times Y)$, with B_{mn} compact for each x in X and each $m, n \in \mathbb{N}$ such that*

$$B = \bigcap_{m=1}^{\infty} \bigcup_{n=1}^{\infty} B_{mn}.$$

Kondô (1939) has proved that every coanalytic subset of the product of two Polish spaces can be uniformized by a coanalytic set. This result is treated by Kechris and Martin (see Part 4, Theorem 8.2.1) and by Dellacherie (Corollary 39, Chapter IV). Rogers and Willmott (1968) have obtained a form of Kondô's theorem in the context of K-analytic spaces by the method of proof analysis already mentioned. The following theorem is obtained by reducing to Dellacherie's Corollary 39, Chapter IV, using the method of the proof of Theorem 5.9.3.

THEOREM 5.9.10 *Let X be a completely regular Hausdorff space, Y a proper K-Lusin space, and let C be the complement of a set in $\mathbf{S}(\mathscr{Z}(X) \times \mathscr{Z}(Y))$. There is a set $U \subset C$, which is the complement of a set in $\mathbf{S}(\mathscr{Z}(X) \times \mathscr{Z}(Y))$, such that $\pi_X(U) = \pi_X(C)$ and U^x is compact for each x in X.*

From this theorem and Lemma 5.9.4 we get

COROLLARY 5.9.11 *Let X be a regular K-analytic space, Y a proper K-Lusin space, and let C be the complement of a set in $\mathbf{S}(\mathscr{Z}(X \times Y))$. There is a set $U \subset C$, which is the complement of a set in $\mathbf{S}(\mathscr{Z}(X \times Y))$, such that $\pi_X(U) = \pi_X(C)$ and U^x is compact for each x in X.*

The Baire sets of multiplicative class α, denoted $Z_\alpha(X)$, and of additive class α, denoted $CZ_\alpha(X)$ $(0 \leqslant \alpha \leqslant \omega_1)$ of a completely regular Hausdorff space X can be inductively defined by

$$Z_0(X) = \mathscr{Z}(X), \qquad CZ_0(X) = \{X \backslash F : F \in Z_0(X)\};$$

$$Z_{\alpha+1}(X) = (CZ_\alpha(X))_\delta,$$

$$CZ_{\alpha+1}(X) = (Z_\alpha(X))_\sigma,$$

and for each limited ordinal λ by

$$Z_\lambda(X) = \Big(\bigcup_{\xi<\lambda} CZ_\xi(X)\Big)_{\sigma\delta},$$

$$CZ_\lambda(X) = \Big(\bigcup_{\xi<\lambda} Z_\xi(X)\Big)_{\delta\sigma}.$$

This classification is described in more detail at the beginning of Section 6; e.g. $CZ_\alpha(X) = \{B : X \backslash B \in Z_\alpha(X)\}$ and

$$\mathscr{Z}_{(+,\delta)}(X) = Z_{\omega_1}(X) = \bigcup_{\alpha<\omega_1} Z_\alpha(X).$$

Starting from any family $\mathscr{H}$ of subsets of X, this inductive definition gives a family $\mathscr{H}_\alpha$ for each $\alpha \leqslant \omega_1$. We use this notation in the proof of Theorem 5.9.13.

Using our method of proof reduction, we are now going to develop and build upon the following theorem:

THEOREM 5.9.12 (Saint-Raymond, 1976*b*; Taĭmanov, 1960, for $\alpha \geqslant \omega_0$) *Let X and Y be compact metric spaces, f a continuous map of Y onto X, and let B be a subset of X. Then for each α, $0 \leqslant \alpha < \omega_1$, $B \in Z_\alpha(X)$ (resp. $CZ_\alpha(X)$), if, and only if, $f^{-1}(B) \in Z_\alpha(Y)$ (resp. $CZ_\alpha(Y)$).*

In Theorem 5.9.12 (and Theorem 5.9.13) it suffices to consider the class $Z_\alpha(X)$, since $CZ_\alpha(X) = \{X \backslash B : B \in Z_\alpha(X)\}$. If $B \in Z_\alpha(X)$, then it is immediate that $f^{-1}(B) \in Z_\alpha(Y)$. Also, if $F^{-1}(B) \in Z_\alpha(Y)$, then $B \in Z_{\omega_1}(X) = \mathscr{Z}_{(+,\delta)}(X)$ by the first separation theorem (Theorem 5.1.6). The difficult part is proving that $B \in Z_\alpha(X)$. After minor modification, the proof of (c) implies (d) in Theorem 5.4.3 shows that $B \in Z_\alpha(X)$ for $\alpha = 0, 1$ and $\alpha \geqslant \omega_0$, and $B \in Z_{\alpha+1}(X)$ for $2 \leqslant \alpha < \omega_0$.

The first step in our development is to extend Theorem 5.9.12 to compact Hausdorff spaces.

THEOREM 5.9.13 *Let X and Y be compact Hausdorff spaces, f a continuous map of Y into X, and let B be a subset of X. Then for each α, $0 \leqslant \alpha < \omega_1$, $B \in Z_\alpha(X)$ (resp. $CZ_\alpha(X)$), if, and only if $f^{-1}(B) \in Z_\alpha(Y)$ (resp. $CZ_\alpha(Y)$).*

Proof As remarked above, it is only necessary to show that $B \in Z_\alpha(X)$, when $f^{-1}(B) \in Z_\alpha(Y)$. The first separation theorem (Theorem 5.1.6) implies that

$$B \in \bigcup_{\alpha < \omega_1} Z_\alpha(X) = \mathscr{Z}_{(+,\delta)}(X) .$$

Compactness considerations give the result for $\alpha = 0, 1$. Suppose $\alpha \geqslant 2$ and that $f^{-1}(B) \in Z_\alpha(Y)$. Also suppose, to get a contradiction, that $B \in Z_\beta(X)$. Let $\beta > \alpha$ be the least ordinal such that $B \in Z_\beta(X)$. Let $\{Z(g_n): n \in \mathbb{N}\}$ be zero sets of continuous functions $g_n : X \to \mathbb{R}$ which represents B as a set in $Z_\beta(X)$; i.e. $B \in \{Z(g_n): n \in \mathbb{N}\}_\beta$. Define a continuous function $g : X \to \mathbb{R}^{\mathbb{N}}$ by

$$g(x) = (g_1(x), g_2(x), \ldots) .$$

Then g has the properties

(1) $g^{-1} \circ g(Z(g_n)) = Z(g_n)$, for each $n \in \mathbb{N}$, and $g^{-1} \circ g(B) = B$;
(2) $g(B) \in \{g(Z(g_n)): n \in \mathbb{N}\}_\beta \subset Z_\beta(g(X))$;
(3) $g(B) \notin \bigcup_{\xi<\beta} Z_\xi(g(X))$, since $g^{-1} \circ g(B) = B \notin \bigcup_{\xi<\beta} Z_\xi(X)$.

Now let $\{Z(h_n): n \in \mathbb{N}\}$ be zero sets of continuous functions $h_n : Y \to \mathbb{R}$ such that $f^{-1}(B) \in \{Z(h_n): n \in \mathbb{N}\}_\alpha$. Define a continuous function $h : Y \to \mathbb{R}^{\mathbb{N}}$ by

$$h(x) = (h_1(x), h_2(x), \ldots) .$$

Then, as with the function g, $h \circ f^{-1}(B) \in Z_\alpha(h(Y))$.

Now we define a continuous map $\Phi : Y \to \mathbb{R}^{\mathbb{N}} \times \mathbb{R}^{\mathbb{N}}$ by

$$\Phi(y) = (g \circ f(y), h(y)) .$$

Then Φ has the properties

(4) $\Phi^{-1} \circ \Phi(Z(g_n)) = Z(g_n)$, for each $n \in \mathbb{N}$, and $\Phi^{-1} \circ \Phi \circ f^{-1}(B) = f^{-1}(B)$;
(5) $\Phi \circ f^{-1}(B) \in \{\Phi(Z(g_n)): n \in \mathbb{N}\}_\alpha \subset Z_\alpha(\Phi(Y))$;
(6) Letting π denote the projection of $\Phi(Y)$ onto $g \circ f(Y)$, we have

$$\pi^{-1} \circ \pi \circ \Phi \circ f^{-1}(B) = \Phi \circ f^{-1}(B) ,$$

$$\pi \circ \Phi \circ f^{-1}(B) = g \circ f \circ f^{-1}(B) = g(B) ,$$

and consequently

$$\pi^{-1} \circ g(B) = \Phi \circ f^{-1}(B) .$$

Now, by (3), $g(B) \notin Z_\alpha(g(X)) = Z_\alpha(g \circ f(Y))$, and so, by Theorem 5.9.12,

$$\pi^{-1} \circ g(B) \notin Z_\alpha(\Phi(Y)) .$$

By (6), $\pi^{-1} \circ g(B) = \Phi \circ f^{-1}(B)$, and so, by (5),

$$\pi^{-1} \circ g(B) \in Z_\alpha(\Phi(Y)) ,$$

which is a contradiction. This completes the proof.

If $B \in Z_\alpha(X)$, $\alpha > 0$, for some complete metric space X, and B is homeomorphic to a subset B' of a complete metric space Y, then $B \in Z_\alpha(Y)$. This follows from Lavrentiev's theorem, which implies that if $f\colon B \to B'$ is such a homeomorphism, then there exist $\mathscr{G}_\delta$-sets $C \supset B$ and $C' \supset B'$ such that f extends to a homeomorphism of C onto C'. Although we do not have a form of Lavrentiev's theorem for compact Hausdorff spaces, we can still obtain the invariance of Baire classes under homeomorphic embeddings. We must, however, require that the embedded set B' be a Baire set in Y, or, what is equivalent (Theorem 5.7.5(d)(ii)), that there is a continuous map $f\colon Y \to \mathbb{R}^{\mathbb{N}}$ such that $f^{-1} \circ f(B') = B'$, since even compact subsets need not be Baire subsets.

THEOREM 5.9.14 *Let X and Y be compact Hausdorff spaces, and let $B \in Z_\alpha(X)$. If B is homeomorphic to a set B' in $\mathscr{Z}_{(+,\delta)}(Y)$, then $B' \in Z_\alpha(Y)$.*

Proof Let $f: B \to B'$ be a homeomorphism. Let $i: \beta B \to \mathrm{cl}\, B$, where $\mathrm{cl}\, B$ is the closure of B in X, be the continuous extension of the identity map on B. By Proposition 5.2.3 $i^{-1} \circ i(B) = B$, and so $B \in Z_\alpha(\beta B)$ (since $B \in Z_\alpha(\mathrm{cl}\, B)$).

Since $B' \in \mathscr{Z}_{(+,\delta)}$, there is a continuous map $g: Y \to \mathbb{R}^{\mathbb{N}}$ such that $g^{-1} \circ g(B') = B'$. Let $h: \beta B \to \mathrm{cl}\, g(B')$, where $\mathrm{cl}\, g(B')$ is the closure of $g(B')$ in $\mathbb{R}^{\mathbb{N}}$, be the continuous extension of $g \circ f: B \to g(B')$. Then, by Proposition 5.2.3, $h^{-1} \circ g(B') = B$, and so, by Theorem 5.9.13,

$$g(B') \in Z_\alpha(\mathrm{cl}\, g(B')) \subset Z_\alpha(g(Y))\,.$$

Hence, again by Theorem 5.9.13,

$$g^{-1} \circ g(B') = B' \in Z_\alpha(Y)\,,$$

which completes the proof.

Let B and B' be separable metric spaces, X and Y metrizable compactifications of B and B', respectively, and let f be a proper map of B onto B'. Then $B \in Z_\alpha(X)$, if, and only if, $B' \in Z_\alpha(Y)$. To see this, note that B is homeomorphic to the graph of f, denoted by $\mathrm{gr}\, f$, in $X \times Y$. The closure of $\mathrm{gr}\, f$ in $X \times Y$, denoted by $\mathrm{cl}\,(\mathrm{gr}\, f)$, is a compactification of $\mathrm{gr}\, f$ and

$$\pi_Y: \mathrm{cl}\,(\mathrm{gr}\, f) \to Y$$

is a continuous extension of f, necessarily with $\pi_Y^{-1}(B') = B$. Thus, by Theorem 5.9.12, $B' \in Z_\alpha(Y)$, if, and only if, $\mathrm{gr}\, f \in Z_\alpha(\mathrm{cl}\,(\mathrm{gr}\, f))$, and, by Theorem 5.9.12, $\mathrm{gr}\, f \in Z_\alpha(\mathrm{cl}\,(\mathrm{gr}\, f))$, if, and only if, $B \in Z_\alpha(X)$.

Suppose the continuous map above is only closed, rather than proper. If $B \in Z_\alpha(X)$, then $B' \in Z_\alpha(Y)$. There are two ways to see this. One is that there is a closed subset F of B such that $f(F) = B'$ and $f: F \to B'$ is proper.

The other is that $f^{-1}(y)$ is, in fact, compact for all except at most countably many y in B' (Taĭmanov, 1955). The first follows from the results of Vaĭnšteĭn (1947) that the boundary of $f^{-1}(y)$, denoted $\mathrm{Bd}\, f^{-1}(y)$, is compact for each $y \in B'$. To see this let $F_y = \mathrm{Bd}\, f^{-1}(y)$, if this set is non-empty, and let F_y be an arbitrary point in $f^{-1}(y)$ otherwise. Then $F = \bigcup\{F_y \colon y \in B)$ is closed, since $B \backslash F$ is the union of the interiors of $f^{-1}(y)$ with one point removed when $\mathrm{Bd}\, f^{-1}(y) = \varnothing$, and $f \colon F \to B'$ is proper. So from both points of view it follows from the previous paragraph that $B' \in Z_\alpha(Y)$.

The first approach in the last paragraph combined with Theorem 5.9.13 leads to part (a) of the next theorem. The second approach combined with Theorem 5.9.13 leads to part (b).

THEOREM 5.9.15 (a) (Čoban, 1970*a*, for $\alpha \geqslant \omega_0$) *Let X and Y be complete metric spaces and, for $\alpha \geqslant 1$, let $B \in Z_\alpha(X)$ (resp. $CZ_\alpha(X)$). If $f \colon B \to B'$ is a closed continuous map onto a subset B' of Y, then $B' \in Z_\alpha(Y)$ (resp. $CZ_\alpha(Y)$).*

(b) *Let X and Y be compact Hausdorff spaces and, for $\alpha \geqslant 1$, let $B \in Z_\alpha(X)$ (resp. $CZ_\alpha(X)$). If $f \colon B \to B'$ is a closed continuous map onto a set $B' \in \mathbf{S}(\mathscr{Z}(Y))$, then $B' \in Z_\alpha(Y)$ (resp. $CZ_\alpha(Y)$).*

In (b) of the last theorem, if f is also an open map, then the hypothesis "$B' \in \mathbf{S}(\mathscr{Z}(Y))$" may be replaced by "$B'$ be a dense subset of Y". For open continuous maps, we have the following theorem.

THEOREM 5.9.16 (a) (Čoban, 1970*a*, for $\alpha \geqslant \omega_0$) *Let X and Y be complete metric spaces and, for $\alpha \geqslant 1$, let $B \in Z_\alpha(X)$ (resp. $CZ_\alpha(X)$). If $f \colon B \to B'$ is an open continuous map onto a subset B' of Y such that, for some metric on B, $f^{-1}(y)$ is complete for each $y \in B'$, then $B' \in Z_\alpha(Y)$ (resp. $CZ_\alpha(Y)$).*

(b) *Let X and Y be compact Hausdorff spaces and let $B \in Z_\alpha(X \times Y)$. (resp. $CZ_\alpha(X \times Y)$). If B^x is compact for each $x \in X$ and $\pi_X \colon B \to \pi_X(B)$ is an open map, then $\pi_X(B) \in Z_\alpha(X)$ (resp. $CZ_\alpha(Y)$).*

The proof of Theorem 5.9.5 together with Theorem 5.9.13 allows us to generalize a recent result of Louveau (1977*b*), which for multiplicative class 0 was proved by Novikov, for additive class 1 by Arsenin (1940), and for multiplicative class 2 by J. Bourgain (198?*a*).

THEOREM 5.9.17 (Louveau, 1977*b*, for compact metric spaces) *Let X and Y be compact Hausdorff spaces and let $B \in \mathscr{Z}_{(+,\delta)}(X \times Y)$. For each α, $0 \leqslant \alpha < \omega_1$, the sets*

$$\{x \in X \colon B^x \in Z_\alpha(Y)\}$$

and

$$\{x \in X : B^x \in CZ_\alpha(Y)\}$$

are complements of sets in $\mathbf{S}(\mathscr{Z}(X))$.

Let $\mathscr{H}$ be a family of subsets of a set X and let bi-$\mathbf{S}(\mathscr{H})$ denote the set $\{B \subset X : B, X\backslash B \in \mathbf{S}(\mathscr{H})\}$. Let Y be a completely regular Hausdorff space and for each α, $-1 \leqslant \alpha < \omega_1$, inductively define

$$\begin{aligned}\mathscr{J}_{-1} &= (\text{bi-}\mathbf{S}(\mathscr{H})) \times CZ_0(Y) \\ &= \{B \times C : B \in \text{bi-}\mathbf{S}(\mathscr{H}),\ C \in CZ_0(Y)\},\end{aligned}$$

$$C\mathscr{J}_\alpha = \begin{cases} \left(\bigcup_{\xi<\alpha} \mathscr{J}_\xi\right)_\sigma, & \text{if } \alpha \text{ is not a limit ordinal}, \\ \left(\bigcup_{\xi<\alpha} \mathscr{J}_\xi\right)_{\sigma c \sigma}, & \text{if } \alpha \text{ is a limit ordinal}, \end{cases}$$

$$\mathscr{J}_\alpha = \{B : (X \times Y)\backslash B \in C\mathscr{J}_\alpha\}.$$

Now we define $\mathscr{L}_{-1} = \mathscr{J}_{-1}$ and for each α, $0 \leqslant \alpha < \omega_1$,

$$C\mathscr{L}_\alpha = \{B \in \text{bi-}\mathbf{S}(\mathscr{H} \times CZ_0(Y)) : B^x \in CZ_\alpha(Y) \text{ for all } x \in X\},$$

$$\mathscr{L}_\alpha = \{B : (X \times Y)\backslash B \in C\mathscr{L}_\alpha\}.$$

When Y is a Polish space, Louveau (198?) has proved that, for all α, $0 \leqslant \alpha < \omega_1$, $\mathscr{J}_\alpha = \mathscr{L}_\alpha$ and $C\mathscr{J}_\alpha = C\mathscr{L}_\alpha$.

The method of the proof of Theorem 5.9.3, together with Theorem 5.9.13, gives the next theorem.

THEOREM 5.9.18 *Let X and Y be completely regular Hausdorff spaces and suppose that* $Y \in Z_1(\beta Y)$. *Taking* $\mathscr{H} = \mathscr{Z}(X)$ *in the above definitions we have, for* $0 \leqslant \alpha < \omega_1$, $\mathscr{J}_\alpha = \mathscr{L}_\alpha$ *and* $C\mathscr{J}_\alpha = C\mathscr{L}_\alpha$.

This theorem, together with the first separation theorem (Theorem 5.1.6(c)) and Lemma 5.9.4, gives

THEOREM 5.9.19 *Let X and Y be regular K-analytic spaces and suppose that* $Y \in Z_1(\beta Y)$. *Taking* $\mathscr{H} = \mathscr{Z}(X)$ *in the above definitions we have, for* $0 \leqslant \alpha < \omega_1$,

$$\mathscr{L}_{-1} = \mathscr{J}_{-1} = \mathscr{Z}_{(+,\delta)}(X) \times CZ_0(Y),$$

$$C\mathscr{J}_\alpha = C\mathscr{L}_\alpha = \{B \in \mathscr{Z}_{(+,\delta)}(X \times Y) : B^x \in CZ_\alpha(Y) \text{ for all } x \in X\},$$

$$\mathscr{J}_\alpha = \mathscr{L}_\alpha.$$

It follows from Theorem 2.8.1 that every K-analytic space is the continuous image of some proper K-Lusin space. The next theorem gives two conditions for the continuous image of a proper K-analytic space to be proper K-analytic.

THEOREM 5.9.20 *Let X and Y be compact Hausdorff space and let $A \in \mathbf{S}(\mathscr{Z}(X))$.*

(a) *If f is an open and closed continuous map of A onto a dense subset A' of Y, then $A' \in \mathbf{S}(\mathscr{Z}(Y))$.*
(b) *If f is an open continuous map of A onto a dense subset A' of Y such that $f^{-1}(y)$ is compact for each $y \in A'$, then $A' \in \mathbf{S}(\mathscr{Z}(Y))$.*

The final theorem in this section considers the preservation of the property of having a disjoint Souslin representation under injective projections. We will only give the proof of part (b). Part (a) is related to Theorem 2.6.7.

THEOREM 5.9.21 (a) *Let X be a completely regular Hausdorff space, Y a regular K-analytic space, and let $\mathscr{Z}^*$ be the family of Lindelöf zero sets in $X \times Y$. If $B \in \mathbf{S}_d(\mathscr{Z}^*)$, and $\pi_X|B$ is injective, then $\pi_X(B) \in \mathbf{S}_d(\mathscr{Z}(X))$.*

(b) *Let X be a Hausdorff space and Y a regular K-Lusin space. If $B \in \mathbf{S}_d(\mathscr{F}(X \times Y))$ and $\pi_X|B$ is injective, then $\pi_X(B) \in \mathbf{S}_d(\mathscr{F}(X))$.*

Proof (b) By hypothesis we have

$$B = \bigcup_\sigma \bigcap_n F(\sigma|n)\,, \qquad F(\sigma|n) \in \mathscr{F}(X \times Y)\,;$$

$$F(\sigma|n+1) \subset F(\sigma|n)\,, \qquad \sigma|n \in \mathbb{N}^{(\mathbb{N})}\,;$$

$$\bigcap_n F(\sigma|n) \cap \bigcap_n F(\tau|n) = \varnothing\,, \qquad \sigma \neq \tau\,.$$

Let $K : \mathbb{N}^{\mathbb{N}} \to \mathscr{K}(Y)$, $K(\mathbb{N}^{\mathbb{N}}) = Y$, be an upper semi-continuous map with

$$K(\sigma) \cap K(\tau) = \varnothing\,, \qquad \sigma \neq \tau\,.$$

Then we obtain

$$B = \bigcup_\sigma \bigcap_n F'(\sigma|n)\,,$$

where

$$F'(\sigma|n) = \begin{cases} X \times \operatorname{cl} K(I(\sigma_1)), & \text{if } n = 1\,, \\ [X \times \operatorname{cl} K(I(\sigma_1, \sigma_3, \sigma_5, \ldots, \sigma_n))] \cap F'(\sigma|n-1)\,, & \text{if } n > 1 \text{ is odd} \\ F(\sigma_2, \sigma_4, \ldots, \sigma_n) \cap F'(\sigma|n-1)\,, & \text{if } n \text{ is even}\,, \end{cases}$$

where, as usual the cl denotes closure in Y and

$$I(\sigma_1) = \{\tau \in \mathbb{N}^{\mathbb{N}}: \tau_1 = \sigma_1\}.$$

Then we have

$$\bigcap_n F'(\sigma|n) \cap \bigcap_n F'(\tau|n) = \varnothing, \qquad \sigma \neq \tau. \tag{38}$$

Now, since $\bigcap_n F'(\sigma|n)$ is a closed set contained in

$$X \times \bigcap_n \operatorname{cl} K(I(\sigma_1, \sigma_3, \sigma_5, \ldots, \sigma_{2n+1})),$$

and since $\bigcap_n \operatorname{cl} K(I(\sigma_1, \sigma_3, \sigma_5, \ldots, \sigma_{2n+1}))$ is compact, the projection

$$P(\sigma) = \pi_X\Big(\bigcap_n F'(\sigma|n)\Big)$$

is closed in X. Further,

$$\begin{aligned}\pi_X(B) &= \pi_X\Big(\bigcup_\sigma \bigcap_n F'(\sigma|n)\Big)\\ &= \bigcup_\sigma P(\sigma).\end{aligned}$$

Suppose that for some $\sigma, \tau \in \mathbb{N}^{\mathbb{N}}$, $\sigma \neq \tau$, we had

$$x \in P(\sigma) \cap P(\tau).$$

By hypothesis $\pi_X|B$ is injective and so $(x, t) \in B$ for at most one $y \in Y$. Hence for some $y \in Y$, we have

$$(x, y) \in \Big(\bigcap_n F'(\sigma|n)\Big) \cap \Big(\bigcap_n F'(\tau|n)\Big),$$

contrary to (38).

Now write

$$P(\sigma|n) = \operatorname{cl} \bigcup \{P(\tau): \tau \in I(\sigma|n)\}.$$

To show that $\pi_X(B)$ has the disjoint Souslin representation

$$\bigcup_\sigma \bigcap_n P(\sigma|n), \qquad P(\sigma|n) \in \mathscr{F}(X),$$

it remains only to verify that for each $\sigma \in \mathbb{N}^{\mathbb{N}}$ we have

$$P(\sigma) = \bigcap_n P(\sigma|n).$$

Clearly

$$P(\sigma) \subset \bigcap_n P(\sigma|n).$$

If $x \notin P(\sigma)$, then

$$(\{x\} \times Y) \cap \left(\bigcap_n F'(\sigma|n)\right) = \varnothing ,$$

and, in particular,

$$(\{x\} \times \bigcap_n \operatorname{cl} K(\sigma_1, \sigma_3, \sigma_5, \ldots, \sigma_{2n+1})) \cap \left(\bigcap_n F'(\sigma|n)\right) = \varnothing .$$

Now, since $\{x\} \times \bigcap_n \operatorname{cl} K(\sigma_1, \sigma_3, \sigma_5, \ldots, \sigma_{2n+1})$ is compact and each $F'(\sigma|n)$ is closed, there will be an m^* so that

$$\left(\{x\} \times \bigcap_n K(\sigma_1, \sigma_3, \sigma_5, \ldots, \sigma_{2n+1})\right) \cap F'(\sigma|m^*) = \varnothing .$$

As $\bigcap_n \operatorname{cl} K(\sigma_1, \sigma_3, \sigma_5, \ldots, \sigma_{2n+1})$ is compact, there are sets U in X and V in Y such that

$$x \in U, \qquad \bigcap_n \operatorname{cl} K(\sigma_1, \sigma_3, \sigma_5, \ldots, \sigma_{2n+1}) \subset V,$$

and

$$(U \times V) \cap F'(\sigma|m^*) = \varnothing .$$

Then we can choose $l^* > m^*$ so that

$$\operatorname{cl} K(\sigma_1, \sigma_3, \sigma_5, \ldots, \sigma_{l^*}) \subset V.$$

This ensures that, when $n^* \geq l^*$, we have

$$F'(\sigma|n^*) \subset X \times V,$$

and, consequently,

$$(U \times Y) \cap F'(\sigma|n^*) = \varnothing .$$

This in turn ensures that

$$U \cap P(\tau) = \varnothing , \qquad \tau|n^* = \sigma|n^* .$$

Thus, since $x \in U$, we have

$$x \notin P(\sigma|n^*) \supset \bigcap_n P(\sigma|n) .$$

5.10 Compact convex sets

In this section K will denote a compact convex subset of a Hausdorff locally convex real topological vector space and $\mathscr{E}(K)$ its set of extreme points. When K is metrizable $\mathscr{E}(K)$ is a $\mathscr{G}_\delta$-subset (hence $\mathscr{K}_{\sigma\delta}$-subset) of K. This follows from the fact that

$$K \backslash \mathscr{E}(K) = \phi((K \times K) \backslash \Delta)$$

is σ-compact, where Δ is the diagonal in $K \times K$ and $\varphi: K \times K \to K$ is the continuous function defined by $\varphi(x, y) = \frac{1}{2}(x + y)$. Furthermore, every Polish space is homeomorphic to $\mathscr{E}(K)$ for some metrizable compact complex set K (in fact, a simplex)—see Haydon (1975).

Since $\mathscr{G}_\delta$-subsets of complete metric spaces are completely metrizable, it follows that $\mathscr{E}(K)$ is a Baire space (i.e. the countable intersection of dense open subsets of $\mathscr{E}(K)$ is dense). Although $\mathscr{E}(K)$ is not necessarily a $\mathscr{G}_\delta$-subset of K when K is not metrizable, we still have

THEOREM 5.10.1 (Choquet, 1969, p. 146) *For any compact convex set K, $\mathscr{E}(K)$ is a Baire space.*

In spite of Theorem 5.10.1 $\mathscr{E}(K)$ can in general be topologically quite pathological. At the same time rather mild conditions on $\mathscr{E}(K)$ imply that K is metrizable, as the next three theorems show.

THEOREM 5.10.2 (Haydon, 1976; Corson, 1970, for X a Polish space) *Let K be a compact convex set. If $\mathscr{E}(K)$ is the continuous image of a separable metric space X, then K is metrizable.*

We can now obtain the next theorem immediately from Theorem 5.6.1 and Theorem 5.10.2.

THEOREM 5.10.3 *Let K be a compact convex set. If $\mathscr{E}(K)$ is a K-countably determined space satisfying the equivalent statements* (a)–(m) *in Theorem* 5.5.2, *then K is metrizable.*

THEOREM 5.10.4 (Jayne, 1978; MacGibbon, 1972, for $\mathscr{E}(K) \in \mathbf{S}(\mathscr{Z}(K))$). *Let K be a compact convex set. If the following three equivalent statements are satisfied, then K is metrizable.*

(a) *There exists a continuous map $f: K \to \mathbb{R}^{\mathbb{N}}$ such that*

$$f^{-1} \circ f(\mathscr{E}(K)) = \mathscr{E}(K) .$$

(b) *There is a sequence $\{Z_n : n \in \mathbb{N}\} \subset \mathscr{Z}(K)$ such that for each $x \in \mathscr{E}(K)$ and $y \in K \setminus \mathscr{E}(K)$ there is an $m \in \mathbb{N}$ such that*

$$x \in Z_m \quad \text{and} \quad y \notin Z_m .$$

(c) *There is a subset S of $\mathbb{N}^{\mathbb{N}}$ and sets $\{Z(\sigma|n): \sigma \in S\} \subset \mathscr{Z}(K)$ such that*

$$\mathscr{E}(K) = \bigcup_{\sigma \in S} \bigcap_n Z(\sigma|n) .$$

Proof For the equivalence of (a)–(c) see Section 5.2.

Let f be the continuous map given by (a). Let $\varphi : K \times K \to K$ be the continuous map defined by

$$\varphi(x, y) = \tfrac{1}{2}(x + y).$$

Then we have for the diagonal Δ in $K \times K$ that

$$((K \times K)\backslash \Delta) = K \backslash \mathscr{E}(K).$$

Define $\Phi : K \times K \to \mathbb{R}^{\mathbb{N}} \times \mathbb{R}^{\mathbb{N}} \times \mathbb{R}^{\mathbb{N}}$ (which is homeomorphic to $\mathbb{R}^{\mathbb{N}}$) by

$$\Phi(x, y) = (f(x), f(y), f \circ \varphi(x, y)).$$

Then Φ is a continuous map and, writing $\Delta = \{(x, y) \in K \times K : x = y\}$, satisfies

$$\Phi^{-1} \circ \Phi[(\mathscr{E}(K) \times \mathscr{E}(K))\backslash \Delta] = (\mathscr{E}(K) \times \mathscr{E}(K))\backslash \Delta.$$

Thus, as the inverse of the Lindelöf space

$$\Phi[(\mathscr{E}(K) \times \mathscr{E}(K))\backslash \Delta]$$

under the proper map Φ, $(\mathscr{E}(K) \times \mathscr{E}(K))\backslash \Delta$ is Lindelöf. Therefore, since f restricted to $\mathscr{E}(K)$ is a proper map, Theorem 5.6.1 (using Theorem 5.5.2(a)$\Leftrightarrow$(d)) implies that $\mathscr{E}(K)$ is metrizable. $\mathscr{E}(K)$ is separable, since it is Lindelöf. Finally, the metrizability of K now follows from Theorem 5.10.2.

There are compact convex sets K such that $\mathscr{E}(K)$ is σ-compact, but not compact, and such that $\mathscr{E}(K)$ is a $\mathscr{K}_{\sigma\delta}$-set in K, but not σ-compact. There are no known examples of compact convex sets K such that $\mathscr{E}(K)$ is K-analytic without $\mathscr{E}(K)$ being a $\mathscr{K}_{\sigma\delta}$-subset of K.

THEOREM 5.10.5 (Talagrand, 1979*a*) *Let K be a compact convex set.*

(a) *If $\mathscr{E}(K)$ is K-countably determined (in particular, K-analytic), then it is possible to write*

$$\mathscr{E}(K) = \bigcap_n (F_n \cup V_n),$$

where, for each n, F_n is closed in K, V_n is open in K, and $F_n \cap V_n = \varnothing$.

(b) *If $\mathscr{E}(K)$ is proper K-countably determined (in particular, proper K-analytic, then it is possible to write*

$$\mathscr{E}(K) = \bigcap_n (F_n \cup O_n),$$

where, for each n, F_n is closed in K, O_n is an open σ-compact set in K, and $F_n \cap O_n = \varnothing$. Consequently, $\mathscr{E}(K)$ is a $\mathscr{K}_{\sigma\delta}$-set in K.

Remarks (1) In (b) of Theorem 5.10.5 we can also conclude that $\mathscr{E}(K)$ is a K-Lusin space. This follows from Theorem 5.8.10(b), since an open σ-compact set in K is necessarily a cozero set.

(2) For closed (not necessarily compact) convex subsets K of separable Banach spaces the situation is quite different. Although $K\backslash\mathscr{E}(K)$ is always analytic, $\mathscr{E}(K)$ need not be analytic (see Jayne and Rogers, 1977).

6 Baire functions and isomorphisms

In this section we outline some of the basic facts concerning Baire functions and isomorphisms and give some of the more accessible proofs.

6.1 Baire functions

Let X be a completely regular Hausdorff space and $C(X)$ the space of continuous real-valued functions on X. Let $B_0(X)=C(X)$, and inductively define $B_\alpha(X)$ for each ordinal $\alpha\leqslant\omega_1$ (ω_1 denotes the first uncountable ordinal) to be the space of pointwise limits of sequences of functions in $\bigcup_{\xi<\alpha} B_\xi(X)$. Let $B_\alpha^*(X)$ be the space of bounded functions in $B_\alpha(X)$. With the pointwise operations $B_\alpha(X)$ and $B_\alpha^*(X)$ are lattice-ordered algebras. With the supremum norm $B_\alpha^*(X)$ is a Banach algebra (see Hausdorff, 1957, § 41). The functions in

$$B_{\omega_1}(X)=\bigcup_{\alpha<\omega_1} B_\alpha(X)$$

are called Baire functions or, if X is metrizable, Borel functions.

The Baire sets of X of multiplicative class α, denoted by $Z_\alpha(X)$, are defined to be the zero sets of functions in $B_\alpha^*(X)$. Those of additive class α, denoted by $CZ_\alpha(X)$, are defined as the complements of sets in $Z_\alpha(X)$. Finally, those of ambiguous class α, denoted by $A_\alpha(X)$, are the sets which are simultaneously in $Z_\alpha(X)$ and $CZ_\alpha(X)$. With the set-theoretic operations of unions and intersection, $A_\alpha(X)$ is a Boolean algebra for each $\alpha\leqslant\omega_1$. Note that with this notation we have $Z_{\omega_1}(X)=\mathscr{Z}_{(+,\delta)}(X)$.

For each $\alpha\leqslant\omega_1$, $Z_\alpha(X)$ is closed under countable intersections and finite unions; and $CZ_\alpha(X)$ is closed under countable unions and finite intersections. Furthermore, we have

THEOREM 6.1.1 (Lebesgue–Hausdorff) *For each* $\alpha<\omega_1$

$$Z_{\alpha+1}(X)=(CZ_\alpha(X))_\delta,$$

and

$$CZ_{\alpha+1}(X)=(Z_\alpha(X))_\sigma;$$

and, if λ *is a limit ordinal, then*

$$Z_\lambda(X)=\Big(\bigcup_{\xi<\lambda} CZ_\xi(X)\Big)_{\sigma\delta},$$

and

$$CZ_\lambda(X) = \left(\bigcup_{\xi<\lambda} Z_\xi(X)\right)_{\delta\sigma}.$$

Remark Note that for $0 \leqslant \alpha < \omega_1$,

$$\boldsymbol{\Sigma}^0_{\alpha+1} = CZ_\alpha(\mathbb{N}^\mathbb{N}),$$
$$\boldsymbol{\Pi}^0_{\alpha+1} = Z_\alpha(\mathbb{N}^\mathbb{N}),$$
$$\boldsymbol{\Delta}^0_{\alpha+1} = A_\alpha(\mathbb{N}^\mathbb{N}).$$

The basic existence results are given in the next theorem.

THEOREM 6.1.2 (a) (Lebesgue, 1905) *If X is a complete metric space containing a non-empty perfect subset, then for all $\alpha < \omega_1$,*

$$B_{\alpha+1}(X) \backslash B_\alpha(X) \neq \varnothing.$$

(b) (Čoban, 1970*a*; Jayne, 1970) *If X is a compact Hausdorff space containing a non-empty perfect subset, then for all $\alpha < \omega_1$,*

$$B_{\alpha+1}(X) \backslash B_\alpha(X) \neq \varnothing.$$

(c) (P. R. Meyer, 1966) *If X is a compact Hausdorff space which contains no non-empty perfect subsets, then $B_1(X) = B_2(X)$.*

Proof (a) A non-empty perfect subset of a complete metric space will contain a separable non-empty perfect subset which will contain, by Corollary 3.5.2, a copy of the Cantor set. The result now follows from Theorems 4.3.2 and 6.1.1.

(b) A compact Hausdorff space X with a non-empty perfect subset admits a continuous map f onto the unit interval by Proposition 5.4.1(a). It follows from Theorems 5.9.13 and 6.1.1 that for each $\alpha < \omega_1$ the map

$$T_f : B_{\omega_1}([0, 1]) \to B_{\omega_1}(X)$$

defined by

$$T_f(h)(x) = h \circ f(x)$$

has the property that

$$T_f[B_{\alpha+1}([0, 1]) \backslash B_\alpha([0, 1])] = T_f B_{\alpha+1}([0, 1]) \backslash T_f B_\alpha([0, 1]) \subset B_{\alpha+1}(X) \backslash B_\alpha(X).$$

By (a) of this theorem, $B_{\alpha+1}([0,1])\backslash B_\alpha([0,1]) \neq \varnothing$. Hence $B_{\alpha+1}(X)\backslash B_\alpha(X) \neq \varnothing$.

(c) It will suffice, in view of Theorem 6.1.1, to show that the range of every real-valued Baire function on a compact space X without non-empty perfect subsets is countable. So suppose, to get a contradiction, that $f \in B_{\omega_1}(X)$ and that $f(X)$ is uncountable. Let $f_n : X \to \mathbb{R}$, $n \in \mathbb{N}$, be continuous functions such that f is contained in the pointwise sequential closure of $\{f_n : n \in \mathbb{N}\}$. Define

$$\Phi : X \to \mathbb{R}^{\mathbb{N}}$$

by

$$\Phi(x) = (f_1(x), f_2(x), \ldots).$$

Then, if $f(X)$ is uncountable, $\Phi(X)$ must be uncountable.

Since $\Phi(X)$ is a compact metric space, there is a copy of the Cantor set $C \subset \Phi(X)$. Let X' be the intersection of all compact subsets Y of X, such that $f(Y) = C$. By Hypothesis X' has an isolated point x. Write $X'' = X'\backslash\{x\}$. Then X'' is compact and $f(X'')$ is a proper compact subset of C. Thus $f(x)$ is an isolated point of C, which is impossible. Thus $\Phi(X)$ and, consequently, $f(X)$ must be countable sets.

6.2 Baire isomorphisms

A bijection $f: X \to Y$ between completely regular spaces is called a Baire isomorphism if $f(\mathscr{Z}_{(+,\delta)}(X)) = \mathscr{Z}_{(+,\delta)}(Y)$; and is said to be of class α, β if

$$f^{-1}(Z_0(Y)) \subset Z_\alpha(X) \quad \text{and} \quad f(Z_0(X)) \subset Z_\beta(Y).$$

If X and Y are metrizable, then f is usually called a Borel isomorphism. In this section we state a number of classical theorems of Kuratowski and Hausdorff (see Kuratowski, 1966), which serve to introduce the recent results of Jayne and Rogers contained in the next section. The first is a generalization of Lavrentiev's theorem on the extension of homeomorphisms, and one of its consequences.

THEOREM 6.2.1 (Kuratowski, 1934) (a) *Every Borel isomorphism of class α, β between two subsets of $\mathbb{R}^{\mathbb{N}}$ can be extended to a class α, β isomorphism from a set in $Z_{\alpha+\beta+1}(\mathbb{R}^{\mathbb{N}})$ onto a set in $Z_{\beta+\alpha+1}(\mathbb{R}^{\mathbb{N}})$.*

(b) *If $X \in Z_\alpha(\mathbb{R}^{\mathbb{N}})$, $\alpha > 0$, and $f: X \to f(X) \subset \mathbb{R}^{\mathbb{N}}$ is a class α, β isomorphism, then $f(X) \in Z_{\beta+\alpha}(\mathbb{R}^{\mathbb{N}})$.*

THEOREM 6.2.2 (Kuratowski, 1934) *For each pair of countable ordinals α, β there exists a class α, β Borel isomorphism f of $\mathbb{N}^{\mathbb{N}}$ onto itself such that f is not of class less than α and f^{-1} is not of class less than β.*

THEOREM 6.2.3 (a) (Kuratowski, 1934)

$$B \in \begin{cases} Z_{\alpha+1}(\mathbb{R}^{\mathbb{N}}), & \text{if } 0 < \alpha < \omega_0, \\ Z_\alpha(\mathbb{R}^{\mathbb{N}}), & \text{if } \alpha \geq \omega_0, \end{cases}$$

if, and only if, there is a closed set $F \subset \mathbb{N}^{\mathbb{N}}$ *and a class* $0, \alpha$ *Borel isomorphism* $f: F \to B$.

(b) (Hausdorff, 1937) *Let B be a subset of* $\mathbb{R}^{\mathbb{N}}$ *such that for each open set U in* $\mathbb{R}^{\mathbb{N}}$ *the set* $U \cap B$ *is uncountable. Then*

$$B \in \begin{cases} Z_{\alpha+1}(\mathbb{R}^{\mathbb{N}}), & \text{if } 0 < \alpha < \omega_0, \\ Z_\alpha(\mathbb{R}^{\mathbb{N}}), & \text{if } \alpha \geq \omega_0, \end{cases}$$

if, and only if, there is a class $0, \alpha$ *Borel isomorphism* $f: \mathbb{N}^{\mathbb{N}} \to B$.

Remark Theorem 2.4.1 is a special case of Theorem 6.2.3(b).

COROLLARY 6.2.4 (a) (Kuratowski, 1934) *If* $X \in Z_{\alpha+1}(\mathbb{R}^{\mathbb{N}})$, $\alpha + 1 > 2$, *and* $Y \in Z_{\beta+1}(\mathbb{R}^{\mathbb{N}})$, $\beta + 1 > 2$, *then there exists a class* α, β *Borel isomorphism of X onto Y (provided, of course, that both X and Y are uncountable).*

(b) (Kuratowski, 1934) *Any two uncountable sets in* $Z_1(\mathbb{R}^{\mathbb{N}})$ *are class* $1, 1$ *Borel isomorphic.*

(c) (Hausdorff, 1937) *An uncountable subset of* $\mathbb{R}^{\mathbb{N}}$ *is class* $1, 1$ *Borel isomorphic to* $\mathbb{N}^{\mathbb{N}}$ *if, and only if, it belongs to* $Z_2(\mathbb{R}^{\mathbb{N}})$.

Proof Part (a) follows from Theorem 6.2.3(a). Part (b) follows essentially from Corollary 2.4.2, which is a special case of Theorem 6.2.3(a). Similarly, part (c) follows from Theorem 6.2.3(b) and Theorem 6.2.1(b).

6.3 First-level Baire isomorphisms

A bijection $f: X \to Y$ is called an αth-level Baire isomorphism ($0 \leq \alpha \leq \omega_1$) if $f(Z_\alpha(X)) = Z_\alpha(Y)$. As usual, the name "$\alpha$th-level Borel isomorphism" is used in the context of metric spaces. This definition is motivated by

THEOREM 6.3.1 (Jayne, 1974*a*) *If X and Y are both realcompact spaces (i.e. homeomorphic to closed subsets of products of real lines) or both metric spaces, then the following are equivalent:*

(a) *X and Y are* α*th-level Baire isomorphic;*
(b) $B_\alpha^*(X)$ *and* $B_\alpha^*(Y)$ *are isometrically isomorphic as Banach spaces;*
(c) $B_\alpha(X)$ *and* $B_\alpha(Y)$ *are ring isomorphic.*

Outline proof of (b) implies (a) for realcompact X and Y Consider the compact set, denoted by X_α, of non-zero multiplicative linear functionals on $B_\alpha^*(X)$ with the weak-star topology. It follows, from the fact that, for each pair of disjoint sets B_1, B_2 in $Z_\alpha(X)$, there is a

$$C \in A_\alpha(X) = Z_\alpha(X) \cap CZ_\alpha(X),$$

with

$$B_1 \subset C \subset X \backslash B_2,$$

that X_α has a base of clopen sets. Since the Boolean algebra of clopen sets of X is isomorphic to $A_\alpha(X)$, the Stone space of $A_\alpha(X)$ is homeomorphic to X_α.

Now, since $B_\alpha^*(X)$ and $B_\alpha^*(Y)$ are isomorphically isometric with $C(X_\alpha)$ and $C(Y_\alpha)$, respectively, it follows that if $B_\alpha^*(X)$ and $B_\alpha^*(Y)$ are isomorphically isometric, then X_α and Y_α are homeomorphic.

The canonical embedding of X into X_α, which assigns a point in X to the evaluational functional at that point, maps X onto a dense subset of X_α. The induced topology on X from X_α is discrete, if, and only if, every point in X is a $\mathscr{G}_\delta$. The space X_α may thus be considered as a compactification of X with the topology having $CZ_\alpha(X)$ as an open base. From this point of view each $f \in B_\alpha^*(X)$ has a unique extension to an $\hat{f} \in C(X_\alpha)$ and the map

$$T: B_\alpha^*(X) \to C(X_\alpha)$$

defined by $T(f) = \hat{f}$ is a norm preserving algebra isomorphism.

A filter $\mathscr{F}$ of sets in $A_\alpha(X)$(resp. $Z_\alpha(X)$) is said to have the CIP (countable intersection property) if for each countable family $\{C_n\} \subset \mathscr{F}$ there is a $C \in A_\alpha(X)$ (resp. $Z_\alpha(X)$) such that

$$C \subset \bigcap_{n=1}^{\infty} C_n .$$

A filter $\mathscr{F}$ of sets in $A_\alpha(X)$(resp. $Z_\alpha(X)$) is said to be fixed if

$$\bigcap \{F : F \in \mathscr{F}\} \neq \varnothing .$$

For any completely regular spaces X and Y, if f is a Boolean algebra isomorphism of $A_\alpha(X)$ onto $A_\alpha(Y)$ and $\mathscr{M}$ is a maximal filter in $A_\alpha(X)$ with the CIP, then $f(\mathscr{M})$ is a maximal filter in $A_\alpha(Y)$ with the CIP.

If X is realcompact, then every maximal filter with the CIP in $A_\alpha(X)$, $\alpha \geqslant 1$ is fixed. To see this let $\mathscr{M} \subset A_\alpha(X)$ be a maximal filter with the CIP. Then, since each set in $Z_\alpha(X)$ is the countable intersection of sets $A_\alpha(X)$, $\mathscr{M}_\delta$ is a maximal filter with the CIP in $Z_\alpha(X)$. Thus, since X is realcompact and

$$Z_\alpha(X) \subset \mathbf{S}(Z_0(X)),$$

$\mathcal{M}_\delta$ is fixed. Thus $\mathcal{M}$ is fixed. Here we have used the following result of Frolík (1971): let $\mathcal{H}_1$ and $\mathcal{H}_2$ be families of subsets of a set X which are closed under countable intersections and let $\mathbf{M}(\mathcal{H}_i)$, $i = 1, 2$, denote the set of free maximal filters in $\mathcal{H}_i$ with the CIP. If

$$\mathcal{H}_1 \subset \mathcal{H}_2 \quad \text{and} \quad \mathcal{H}_2 \subset \mathbf{S}(\mathcal{H}_1)\,,$$

then the map

$$\mathcal{M} \to \mathcal{M} \cap \mathcal{H}_1\,, \qquad \mathcal{M} \in \mathbf{M}(\mathcal{H}_2)$$

is one-to-one onto $\mathbf{M}(\mathcal{H}_1)$.

From this it follows that any homeomorphism f of X_α onto Y_α maps X onto Y. We also have that

$$B \in A_\alpha(X) \Leftrightarrow f(B) \in A_\alpha(Y)\,.$$

This completes the outline of the proof.

We might define inter-level isomorphisms. For example we could define a Baire isomorphism $f: X \to Y$ to be of level α, β

$$f(Z_\beta(X)) = Z_\alpha(Y)\,.$$

As in Theorem 6.3.1, the existence of such a map is equivalent to the existence of an isometric isomorphism between $B_\beta^*(X)$ and $B_\alpha^*(Y)$. But the following theorem shows that such maps do not exist, when $\alpha \neq \beta$, in perhaps the most interesting spaces. For metrizable compact spaces the next theorem is due to Dashiell (1974).

THEOREM 6.3.2 (Jayne, 1974*a*) (a) *Let X and Y be compact Hausdorff spaces and suppose that either X or Y contains a non-empty perfect subset. For $0 \leq \alpha < \beta \leq \omega_1$, $B_\alpha^*(X)$ is not linearly isometric to $B_\beta^*(Y)$.*

(b) *If X and Y are infinite and contain no non-empty perfect subsets, then $B_0^*(X)$ is not linearly isometric to $B_1^*(Y)$.*

Remark Note that $B_1^*(X) = B_2^*(X)$ in (b), by Theorem 6.1.2.

The next two theorems indicate that first-level isomorphisms lead to a classification of Polish spaces that is both manageable and fine enough to distinguish many different classes. Finally, Theorem 6.3.5 shows that most uncountable Polish spaces, in particular finite-dimensional spaces, are second-level isomorphic.

THEOREM 6.3.3 (Jayne and Rogers, 1979) *If X and Y are first-level Borel isomorphic separable metric spaces and X is σ-compact, then*

(a) *Y is σ-compact,*
(b) $\dim X = \dim Y$.

Proof We will only give the proof of part (a). Let $\phi : X \to Y$ be a first-level Borel isomorphism. Then Y is a Borel subset of its completion $\bar{Y}$ under a metric d on Y. To see this, define a map $\psi: X \times Y \to \mathbb{R}$ by

$$\psi(x, y) = d(y, \varphi(x))\,.$$

Then ψ is Borel measurable and the graph of φ, denoted by gr φ, is equal to $\psi^{-1}(0)$. Thus gr φ is a Borel subset of $X \times \bar{Y}$, and, since $\pi_Y|\text{gr }\varphi$ is injective,

$$\pi_Y(\text{gr }\varphi) = Y\,.$$

is a Borel subset of $\bar{Y}$.

Suppose that Y is not σ-compact. Then by a theorem of Hurewicz (see Hoffmann–Jørgensen and Topsøe, Part 3, Corollary 1.3.5) Y contains a closed set that is homeomorphic to $\mathbb{N}^{\mathbb{N}}$, which we will identify with $\mathbb{N}^{\mathbb{N}}$. We will construct in a compact subset of $\varphi^{-1}(\mathbb{N}^{\mathbb{N}})$ a countable dense-in-itself set whose image under φ is a closed discrete subset of $\mathbb{N}^{\mathbb{N}}$. This will give a contradiction.

Since $\varphi^{-1}(\mathbb{N}^{\mathbb{N}})$ is an $\mathscr{F}_\sigma$-set in the σ-compact space X, it is σ-compact. Since $\mathbb{N}^{\mathbb{N}}$ is not σ-compact, there must be a compact subset F of $\varphi^{-1}(\mathbb{N}^{\mathbb{N}})$ such that $\varphi(F)$ is not σ-compact.

For $l \geq 1$, let

$$\mathbf{h}^{(l,1)}, \mathbf{h}^{(l,2)}, \mathbf{h}^{(l,3)}, \ldots$$

be an enumeration of the finite sequence of positive integers of length l. Suppose that for each $l \geq 1$ the set

$$I(\mathbf{h}^{(l,s)}) \cap \varphi(F)$$

is σ-compact, except for only finitely many $s \geq 1$. By rearranging the enumeration we can choose $S(l) \geq 0$ such that

$$I(\mathbf{h}^{(l,s)}) \cap \varphi(F)$$

is not σ-compact when $1 \leq s \leq S(l)$, but is σ-compact when $s > S(l)$. Now

$$\bigcap_{l=1}^{\infty} \bigcup_{1 \leq s \leq S(l)} I(\mathbf{h}^{(l,s)})$$

is compact and $\varphi(F)$ is an $\mathscr{F}_\sigma$-set in Y. Hence

$$\varphi(F) = \left[\varphi(F) \cap \bigcap_{l=1}^{\infty} \bigcup_{1 \leq s \leq S(l)} I(\mathbf{h}^{(l,s)})\right] \cup \left[\bigcup_{l=1}^{\infty} \bigcup_{s=S(l)+1}^{\infty} \{I(\mathbf{h}^{(l,s)}) \cap \varphi(F)\}\right]$$

is σ-compact, contrary to the choice of F. Thus for some $l \geq 1$ the set $I(\mathbf{h}^{(l,s)}) \cap \varphi(F)$ is not σ-compact for infinitely many values of s. So we can choose an infinite sequence $\{\mathbf{g}^{(r)}: r \in \mathbb{N}\}$ of distinct finite sequences of positive integers of length l with

$$I(\mathbf{g}^{(r)}) \cap F$$

not σ-compact for each $r \in \mathbb{N}$.

For each r the set

$$\varphi^{-1}(I(\mathbf{g}^{(r)})) \cap F$$

is σ-compact. Thus for each $r \in \mathbb{N}$ we can choose a compact set H_r such that $\varphi(H_r)$ is not σ-compact,

$$H_r \subset F \quad \text{and} \quad g(H_r) \subset I(\mathbf{g}^{(r)}) .$$

For a fixed r suppose that each point $x \in H_r$ has a closed neighbourhood $N(x)$ such that $\varphi(H_r \cap N(x))$ is σ-compact. Since H_r is compact this would imply that $\varphi(H_r)$ is σ-compact. Hence we can choose a point $y_r \in H_r$ such that $\varphi(H_r \cap N)$ is not σ-compact for each closed neighbourhood N of y_r.

Now, since F is compact, (y_r) has a subsequence, also denoted (y_r), which converges to a point $x \in F$. As the sequences $\mathbf{g}^{(r)}$ are distinct, the sets

$$\varphi(H_r) \subset I(\mathbf{g}^{(r)}), \qquad r \in \mathbb{N},$$

are disjoint, and the points $\{y_r: r \in \mathbb{N}\}$ are distinct. Now $\varphi(x)$ belongs to some Baire interval $I(\mathbf{g})$ corresponding to a sequence $\mathbf{g}$ of length l and $\mathbf{g}$ can coincide with at most one member of the sequence $\{\mathbf{g}^{(r)}: r \in \mathbb{N}\}$. Omitting one term from the sequence and re-indexing again, if necessary, we may suppose that $\mathbf{g}, \mathbf{g}^{(1)}, \mathbf{g}^{(2)}, \ldots$ are all distinct.

Let $\{G_r: r \in \mathbb{N}\}$ be open sets in X with

$$F \cap \left(\bigcap_{r=1}^{\infty} G_r \right) = \{x\} \quad \text{and} \quad y_r \in G_r .$$

As the points $x, y_1, y_2, \ldots$ are distinct we can choose for each $r \geq 1$ a closed neighbourhood N_r of y_r such that

$$N_r \cap N_s = \varnothing , \qquad 1 \leq r \leq s ;$$

$$N_r \subset G_r , \qquad r \geq 1 .$$

Write $F_r = H_r \cap N_r$ for each $r \in \mathbb{N}$. Then the sets $\{x\}, F_1, F_2, \ldots$ are disjoint compact subsets of F, and there are open sets $G_1, G_2, \ldots$, such that for each $r \in \mathbb{N}$

$$F_r \subset G_r \quad \text{and} \quad F \cap \bigcap_{r=1}^{\infty} G_r = \{x\} .$$

There is a sequence $\mathbf{g}, \mathbf{g}^{(1)}, \mathbf{g}^{(2)}, \ldots$ of distinct finite sequences of positive integers of length l with

$$\varphi(x) \in I(\mathbf{g}),$$

$$g(F_r) \subset I(\mathbf{g}^{(r)}), \qquad r \geq 1,$$

and

$$\varphi(F_r) \text{ is not } \sigma\text{-compact}.$$

We write $\tau = \varphi(x)$, and have $\tau | l = \mathbf{g}$.

We apply the above construction to each F_r, $r \geq 1$, and then iterate this procedure. Thus we obtain a system of compact sets

$$F, F(\sigma_1), F(\sigma_1\sigma_2), F(\sigma_1\sigma_2\sigma_3), \ldots, \qquad \sigma_i \geq 1,$$

of points

$$x, x_{\sigma_1}, x_{\sigma_1\sigma_2}, x_{\sigma_1\sigma_2\sigma_3}, \ldots, \qquad \sigma_i \geq 1,$$

of positive integers

$$l, l_{\sigma_1}, l_{\sigma_1\sigma_2}, l_{\sigma_1\sigma_2\sigma_3}, \ldots, \qquad \sigma_i \geq 1,$$

and points of $\mathbb{N}^{\mathbb{N}}$,

$$\tau, \tau^{\sigma_1}, \tau^{\sigma_1\sigma_2}, \tau^{\sigma_1\sigma_2\sigma_3}, \ldots, \qquad \sigma_i \geq 1,$$

with the following properties:

$$x \in F, \qquad x_{\sigma|n} \in F(\sigma|n),$$

$$F(\sigma|n, m) \to x_{\sigma|n} \quad \text{as} \quad m \to \infty,$$

the sets

$$\{x_{\sigma|n}\}, \qquad F(\sigma|n, m), \qquad m \geq 1,$$

are disjoint for each fixed $\sigma|n$;

$$F(\sigma|n+1) \subset F(\sigma|n),$$

$$l_{\sigma|n+1} > l_{\sigma|n},$$

$$\varphi(x_{\sigma|n}) = \tau^{\sigma|n},$$

$$\tau^{\sigma|n+1}|l_{\sigma|n} \neq \tau^{\sigma|n}|l_{\sigma|n} \quad \text{and} \quad \tau^{\sigma|n,m}|l_{\sigma|n} \neq \tau^{\sigma|n,p}|l_{\sigma|n}, \qquad p > m \geq 1;$$

and finally $\varphi(F(\sigma|n+1))$ is not σ-compact and

$$\varphi(F(\sigma|n+1)) \subset I(\tau^{\sigma|n+1}|l_{\sigma|n}).$$

Now each point $\tau^{\sigma|n}$ is in the Baire interval $I(\tau^{\sigma|n}|l_{\sigma|n})$, which contains no other point of the system $\{\tau^{\sigma|n}: \sigma|n \in \mathbb{N}^{(\mathbb{N})}\}$. Thus $\{\tau^{\sigma|n}: \sigma|n \in \mathbb{N}^{(\mathbb{N})}\}$ is a countable discrete closed subset of $\mathbb{N}^{\mathbb{N}}$. But the set

$$\{x_{\sigma|n}: \sigma|n \in \mathbb{N}^{(\mathbb{N})}\} = \{\varphi^{-1}(\tau^{\sigma|n}): \sigma|n \in \mathbb{N}^{(\mathbb{N})}\}$$

is a dense-in-itself subset of the compact set F. Being a countable dense-in-itself set, this set must be homeomorphic to the rational numbers (see Hoffmann–Jørgensen and Topsøe, Part 3, Section 1.1), which has no complete metric and so cannot be a $\mathscr{G}_\delta$-subset of F and so not of X. This completes the proof.

Remark Theorem 6.3.3(a) does not hold for non-metrizable compact Hausdorff spaces. The space $\{p\} \cup \mathbb{N}^{\mathbb{N}}$ in Example 5.2.3 is a non-σ-compact completely regular K-analytic space whose Baire subsets are precisely the countable subsets which do not contain the point p. If $\{\infty\} \cup \mathbb{N}^{\mathbb{N}}$ denotes the one point compactification of the set $\mathbb{N}^{\mathbb{N}}$ with the discrete topology, then the Baire subsets of $\{\infty\} \cup \mathbb{N}^{\mathbb{N}}$ are likewise the countable subsets which do not contain the point $\{\infty\}$. Thus the bijection of $\{\infty\} \cup \mathbb{N}^{\mathbb{N}}$ onto $\{p\} \cup \mathbb{N}^{\mathbb{N}}$, which is the identity of $\mathbb{N}^{\mathbb{N}}$, is a first-level Baire isomorphism of a compact Hausdorff space onto a non-σ-compact completely regular K-analytic space. Theorem 6.3.3(a) does, however, hold under certain conditions—for example, when the spaces are perfectly normal (see Jayne and Rogers, 1979).

THEOREM 6.3.4 (Jayne and Rogers, 1979) (a) *If X is an uncountable $\mathscr{G}_\delta$-subset of $\mathbb{R}$, then X is first-level Borel isomorphic either to* $\mathbb{N}^{\mathbb{N}}$, $2^{\mathbb{N}}$, $[0, 1]$ *or* $\mathbb{N}^{\mathbb{N}} \cup [0, 1]$; *and these four spaces are not first-level isomorphic.*

(b) *An uncountable $\mathscr{G}_\delta$-subset X of $\mathbb{R}^n$, $1 \leq n < \infty$, is first-level isomorphic to $\mathbb{R}^n$, if, and only if, X is σ-compact and* $\dim X = n$.

Proof We will only give the proof of part (a). It is immediately clear from Theorem 6.3.3 that these four spaces are not first-level isomorphic. To prove that any $\mathscr{G}_\delta$-subset of $\mathbb{R}$ is first-level isomorphic to one of them, we prove a Cantor–Bernstein type lemma: if A and C are metric spaces, each homeomorphic to a closed subset of the other, then A and C are first-level isomorphic.

To prove this last statement, suppose A is a closed subset of C and that φ is a homeomorphism of C onto A. Let

$$H = (C \backslash A) \cup \varphi(C \backslash A) \cup \varphi \circ \varphi(C \backslash A) \cup \cdots \cup \varphi^n(C \backslash A) \cup \cdots.$$

Then

$$C \backslash H = \left[\bigcap_{n=1}^{\infty} \varphi^n(A)\right] \cup (A \backslash \varphi(C)) \cup \bigcup_{n=1}^{\infty} [\varphi^n(A) \backslash \varphi^{n+1}(C)].$$

Hence both H and $C\backslash H$ are $\mathscr{F}_\sigma$-subsets of C.

Define $f: C \to A$ by

$$f(x) = \begin{cases} x, & \text{if } x \in C\backslash H, \\ \varphi(x), & \text{if } x \in H. \end{cases}$$

Then f is injective and, since

$$(C\backslash A) \cap \varphi(H) = \varnothing \quad \text{and} \quad \varphi(H) \subset A,$$

we have $C\backslash A = H\backslash\varphi(H)$ and

$$\varphi(H) \cup (C\backslash H) = C\backslash(H\backslash\varphi(H)) = C\backslash(C\backslash A) = A.$$

Hence

$$f(C) = f(H) \cup f(C\backslash H) = \varphi(H) \cup (C\backslash H) = A$$

and f is surjective.

Since $f|H$ and $f|C\backslash H$ are homeomorphisms and H and $C\backslash H$ are $\mathscr{F}_\sigma$-subsets of C, f is a first-level isomorphism.

Now suppose X is an uncountable $\mathscr{G}_\delta$-set in $\mathbb{R}$. If $\dim X = 0$ and X is not σ-compact, then X contains a closed subset homeomorphic to $\mathbb{N}^\mathbb{N}$ (see Hoffmann–Jørgensen and Topsøe, Part 3, Section 1.3) and so X is first-level isomorphic to $\mathbb{N}^\mathbb{N}$, as every 0-dimensional Polish space is homeomorphic to a closed subset of $\mathbb{N}^\mathbb{N}$.

If $\dim X = 0$ and X is compact, then it contains a copy of $2^\mathbb{N}$ and is homeomorphic to a (necessarily) closed subset of $2^\mathbb{N}$. Hence X is first-level isomorphic to $2^\mathbb{N}$. If X is σ-compact, then, since it is 0-dimensional, we can write it as the disjoint union of 0-dimensional compact sets, say K_n. Pick $x \in K_1$ and define a topology on $\bigcup_n K_n$ by taking sets of the form

$$U_1 \cup \bigcup_{n=m}^{\infty} K_m$$

for some $m \geq 1$, where U_1 is an open neighbourhood of x_1 in K_1. Then $\bigcup_n K_n$ is a compact 0-dimensional metric space and the identity map is a first-level isomorphism to X. Hence X is first-level isomorphic to $\mathbb{N}^\mathbb{N}$.

If $\dim X = 1$ and X is compact, then X must contain a closed interval, and, of course, is contained in an interval. Thus X is first-level isomorphic to $[0, 1]$. If $X = (a, b)$, let $x_n = (b - a)/n$, $n \geq 2$, and define a map $f: (a, b) \to a, b$ by $f(x_2) = a$, $f(x_3) = b$, $f(x_4) = x_2$, $f(x_5) = x_3, \ldots, f(x_{n+2}) = x_n, \ldots$, and $f(x) = x$ otherwise. Then f is a first-level isomorphism. If X is σ-compact, then X is the disjoint union of an 0-dimensional compact set and a finite or countable union of open intervals. Piecing together isomorphisms and using, if necessary, the convergence of a sequence of sets technique used in the 0-dimensional case, we get that X is first-level isomorphic to $[0, 1]$.

Finally, suppose $\dim X = 1$ and X is not σ-compact. Since $\dim X = 1$, the union of all σ-compact open subsets of X will have an uncountable number of points and hence be first-level isomorphic to $[0, 1]$ by the results of the previous paragraph. Since X is not σ-compact, the remainder will be an uncountable 0-dimensional $\mathcal{G}_\delta$-set that is not σ-compact, and hence be first-level isomorphic to $\mathbb{N}^{\mathbb{N}}$. Piecing these two isomorphisms together completes the proof.

THEOREM 6.3.5 (Jayne and Rogers, 1979) *Any two Polish spaces which are the countable unions of finite-dimensional spaces are second-level Borel isomorphic.*

Proof Let X be a space satisfying the hypothesis of the theorem. We will show that X is second-level isomorphic to $\mathbb{N}^{\mathbb{N}}$. First note that $\mathbb{N}^{\mathbb{N}}$ and $2^{\mathbb{N}}$ are second-level isomorphic. Define $C \subset \mathbb{N}^{\mathbb{N}}$ by

$$C = \{(n, 1, 1, \ldots): n \in \mathbb{N}\}$$

and $D \subset 2^{\mathbb{N}}$ by

$$D = \{\sigma \in 2^{\mathbb{N}}: \text{for some } m, \sigma_n = 0 \text{ for all } n \geq m\}.$$

Then $\mathbb{N}^{\mathbb{N}} \backslash C$ and $2^{\mathbb{N}} \backslash D$ are homeomorphic to $\mathbb{N}^{\mathbb{N}}$, and if $f_1: C \to D$ is any bijection and $f_2: \mathbb{N}^{\mathbb{N}} \backslash C \to 2^{\mathbb{N}} \backslash D$ is any homeomorphism, then $f: \mathbb{N}^{\mathbb{N}} \to 2^{\mathbb{N}}$, defined by

$$f(x) = \begin{cases} f_1(x), & \text{if } x \in C, \\ f_2(x), & \text{if } x \in \mathbb{N}^{\mathbb{N}} \backslash C, \end{cases}$$

is a second-level isomorphism.

There is no loss in generality in supposing that X is the countable disjoint union of uncountable 0-dimensional $\mathcal{G}_\delta$-subsets. Each of these $\mathcal{G}_\delta$-subsets is first-level isomorphic to either $\mathbb{N}^{\mathbb{N}}$ or $2^{\mathbb{N}}$ by Theorem 6.3.4(a) and these are second-level isomorphic. Thus piecing these isomorphisms together we get that X is second-level isomorphic to $\mathbb{N}^{\mathbb{N}}$.

6.4 First level Baire functions

A function $f: X \to Y$ is called an αth-level Baire function ($0 \leq \alpha < \omega_1$), if $f^{-1}(CZ_\alpha(Y)) \subset CZ_\alpha(X)$. The name "$\alpha$th-level Borel function" is used in the context of metric spaces. A function $f: X \to Y$ between metric spaces is called piece-wise continuous, if there exist closed sets $\{F_n\}$ in X such that $X = \bigcup_{n=1}^{\infty} F_n$ and f restricted to F_n is continuous for each n. It is not difficult to check that, if a bijection $f: X \to Y$ and its inverse are piece-wise continuous, then the sets $\{F_n\}$ may be chosen such that, for each n, $f(F_n)$ is closed in Y and f restricted to F_n is a homeomorphism onto $f(F_n)$. We call such a bijection a piece-wise homeomorphic map.

THEOREM 6.4.1 (Jayne and Rogers, 198?*a*) *Let X and Y be complete metric spaces, and $A \in \mathbf{S}(\mathcal{F}(X))$. A function f from A into Y is a first level Borel function, if, and only if, it is piece-wise continuous. If f is injective and $F(A) \in \mathbf{S}(\mathcal{F}(Y))$, then F is a first level Borel isomorphism onto $f(A)$, if, and only if, f is a piece-wise homeomorphic map onto $f(A)$.*

Remark Theorem 6.3.3 is an immediate corollary of this theorem.

A first level Baire function $f: X \to Y$ between metric spaces is called an $\mathcal{F}_\sigma$-map, if the image of every countable union of closed sets in X is the countable union of closed sets in Y. A function $f: X \to Y$ between metric spaces is called piece-wise closed, if there exist closed sets $\{F_n\}$ in X such that $X = \bigcup_{n=1}^{\infty} F_n$ and f restricted to each F_n is a closed continuous map into Y (in particular, $f(F_n)$ is closed in Y).

THEOREM 6.4.2 (Jayne and Rogers, 198?*b*) *Let X and Y be complete metric spaces, and let $A \in \mathbf{S}(\mathcal{F}(X))$ and $B \in \mathbf{S}(\mathcal{F}(Y))$. A function f from A onto B is an $\mathcal{F}_\sigma$-map, if, and only if, f is piece-wise closed.*

References

Alexandrov, P. S.
1916 Sur la puissance des ensembles mesurables *B*, *C.R. Acad. Sci., Paris*, **162**, 323–325.

Arhangel'skiĭ, A. V.
1969 On the cardinality of bicompacta satisfying the first axiom of countability, *Dokl. Akad. Nauk SSSR*, **187**, 967–970 (= *Soviet Math. Dokl.*, **10**, 151–155).

Arsenin, V. Ya.
1940 Sur les projections de certains ensembles mesurables *B*, *Dokl. Akad. Nauk SSSR*, **27**, 107–109.

Baire, R.
1900 Sur les fonctions de variables réelles, *Annali di Math., Serie 3*, **3**, 1–124.

Bourgain, J.
198?*a* $F_{\sigma\delta}$-sections of Borel sets, *Fund. Math.*, in press.
198?*b* Borel sets with $F_{\sigma\delta}$-sections, *Fund. Math.*, in press.

Choquet, G.
1951 Ensembles boréliens et analytiques dans les espaces topologiques, *C.R. Acad. Sci., Paris*, **232**, 2174–2176.

1953 Theory of capacities, *Ann. Inst. Fourier, Grenoble*, **5**, 131–295.
1969 *Lectures on Analysis*, Vol. II, Benjamin, New York.

Čoban, M. M.
1970*a* Baire sets in complete topological spaces, *Ukrain. Mat. Ž.*, **22**, 330–342.
1970*b* Many-valued mappings and Borel sets I, *Trudy Moskov. Mat. Obšč.* **22**, 229–250 (= *Trans. Moscow Math. Soc.*, **22**, 258–280).

Corson, H. H.
1970 Metrizability of compact convex sets, *Trans. Amer. Math. Soc.* **151**, 589–596.

Dashiell, F. K.
1974 Isomorphism problems for the Baire classes, *Pacific J. Math.*, **52**, 29–43.

Davies, R. O.
1952 Subsets of finite measure in analytic sets, *Indag. Math.*, **14**, 488–489.
1956 Non-σ-finite closed subsets of analytic sets, *Proc. Camb. Phil. Soc.*, **52**, 174–177.
1968 A theorem on the existence of non-σ-finite subsets, *Mathematika*, **15**, 60–62.
1970 Increasing sequences of sets and Hausdorff measure. *Proc. London Math. Soc. (3)*, **20**, 222–236.

Davies, R. O. and Rogers, C. A.
1969 The problem of subsets of finite positive measure, *Bull. London Math. Soc.*, **1**, 47–54.

Davies, R. O., Jayne, J. E., Ostaszewski, A. J., and Rogers, C. A.
1977 Theorems of Novikov type, *Mathematika*, **24**, 97–114.

Dellacherie, C. and Meyer, P. A.
1975 Ensembles analytiques et temps d'arrêt, in *Seminaire de Probabilités IX*, Lecture Notes in Mathematics no. 465, Springer-Verlag, Berlin, Heidelberg and New York, pp. 373–405.

De Wilde, M.
1978 *Closed Graph Theorems and Webbed Spaces*, Pitman, London.

Diestel, J.
1975 *Geometry of Banach Spaces—Selected Topics*, Lecture Notes in Mathematics no. 485, Springer Verlag, Berlin, Heidelberg and New York.

Engelking, R.
1968 *Outline of General Topology*, North-Holland, Amsterdam.
1977 *General Topology*, PWN, Warsaw.

Fremlin, D. H.
1977 *K*-analytic spaces with metrizable compacta, *Mathematika*, **24**, 257–261.

Frolík, Z.
1961*a* On analytic spaces, *Bull. Acad. Polon. Sci.*, **8**, 747–750.
1961*b* On borelian and bianalytic spaces, *Czech. Math. J.*, 629–631.
1961*c* A contribution to the descriptive theory of sets and spaces, in *General Topology and its Relations to Modern Analysis and Algebra*, Academic Press, New York and London, 157–173.
1967 Baire sets that are Borelian subspaces, *Proc. Roy. Soc. A*, **299**, 287–290.
1968 On the Souslin-graph theorem, *Comment. Math. Univ. Carolinae*, **9**, 243–249.
1970*a* A survey of the separable descriptive theory of sets and spaces, *Czech Math. J.*, **20**, 406–467.
1970*b* Analytic and Borelian sets in general spaces, *Proc. London Math. Soc.* (*3*), **31**, 674–692.

1970*c* Projective limits of measure spaces, *Proceedings of the 6th Berkeley Symposium on Mathematical Statistics and Probability*, Vol. II, Berkeley, pp. 67–80.
1971 Realcompactness is a Baire measurable property, *Bull. Acad. Polon. Sci.*, **19**, 617–621.

Gödel, K.
1938 The consistency of the axiom of choice and of the generalized continuum hypothesis, *Proc. Nat. Acad. Sci., U.S.A.*, **24**, 556–557.
1940 The consistency of the continuum hypothesis, *Ann. of Math. Studies*, **3**, Princeton University Press, Princeton, N.J.

Hausdorff, F.
1916 Die Mächtigkeit der Borelschen Mengen, *Math. Ann.*, **77**, 430–437.
1937 Die schlichten stetigen Bilder des Nullraums, *Fund. Math.*, **29**, 151–158.
1957 *Set Theory*, Chelsea, New York.

Haydon, R.
1975 A new proof that every Polish space is the extreme boundary of a simplex, *Bull. London Math. Soc.*, **7**, 97–100.
1976 An extreme point criterion for separability of a dual Banach space, and a new proof of a theorem of Corson, *Quart. J. Math. Oxford* (*3*), **27**, 379–385.

Holický, P.
1974 The convex generation of convex Borel sets in locally convex spaces, *Mathematika*, **21**, 207–215

Hurewicz, W.
1928 Relativ perfekte Teile von Punktmengen und Mengen (A) *Fund. Math.*, **12**, 78–109.
1930 Zur Theorie der analytischen Mengen, *Fund. Math.*, **15**, 4–17.

Jayne, J. E.
1970 Descriptive set theory in compact spaces, *Notices Amer. Math. Soc.*, **17**, 268.
1973 Characterizations and metrization of proper analytic spaces, *Inventiones Mathematicae*, **22**, 51–62.
1974*a* The space of class α Baire functions, *Bull. Amer. Math. Soc.*, **80**, 1151–1156.
1974*b* Spaces of Baire functions I, *Ann. Inst. Fourier, Grenoble*, **24**, 47–76.
1974*c* Continuous images of proper analytic and proper Borel spaces, *Proc. Camb. Phil. Soc.*, **75**, 185–191.
1976 Structure of analytic Hausdorff spaces, *Mathematika*, **23**, 208–211.
1977 Generation of σ-algebras, Baire sets and descriptive Borel sets, *Mathematika*, **24**, 241–256.
1978 Metrization of compact convex sets, *Math. Ann*, **234**, 109–115.

Jayne, J. E. and Rogers, C. A.
1977 The extremal structure of convex sets, *J. Functional Anal.*, **26**, 251–288.
1979 Borel isomorphisms at the first level, *Mathematika*, **26**, 125–179, and a corrigendum to appear in *Mathematika*, **27**.
1980 The potency of Polish spaces: retractions: restrictions, *Proc. London Math. Soc.* (3), to appear.
198?*a* First level Borel functions and isomorphisms, to appear.
198?*b* Piece-wise closed maps, to appear.

Kelley, J. L.
1955 *General Topology*, Van Nostrand, New York.

Klee, V.
1951 Convex sets in linear spaces, *Duke Math. J.*, **18**, 443–466.
1953 Convex sets in linear spaces, III, *Duke Math. J.* **20**, 875–883.
Kondô, M.
1939 Sur l'uniformisation des complémentaires analytiques et les ensembles projectifs de la seconde classe, *Japan J. Math.*, **15**, 197–230.
Kunugui, K.
1935 La théorie des ensembles analytiques et les especes abstracts, *J. Fac. Sci. Hokkaido Imp. Univ., Ser. 1, Math.*, **4**, 1–40.
1938 Contribution à la théorie des ensembles Boreliens et analytiques I, *J. Fac. Sci. Hokkaido Imp. Univ., Ser. 1, Math.*, **7**, 161–189.
1940 Sur un problem de M. E. Szpilrajn, *Proc. Imp. Acad. Tokyo*, **16**, 73–78.
Kuratowski, K.
1933 Sur le prolongement de l'homéomorphie, *C. R. Acad. Sci., Paris*, **197**, 1090
1934 Sur une généralisation de la notion d'homéomorphie, *Fund. Math.*, **22**, 306–220.
1935 Quelques problèmes concernant les espaces metriques non-separables, *Fund. Math.*, **25**, 534–545.
1936 Sur les théorèmes de séparation dans la théorie des ensembles, *Fund. Math.*, **26**, 183–191.
1966 *Topology*, Vol. 1, Academic Press, New York and London.
Kuratowski, K. and Ryll-Nardzewski, C.
1961 A general theorem on selectors, *Bull. Acad. Polon. Sci.*, **13**, 397–403.
Larman, D. G.
1967*a* On Hausdorff measure in finite-dimensional compact metric spaces, *Proc. London Math. Soc.* (*3*), **17**, 193–306.
1967*b* On selection of non-σ-finite subsets, *Mathematatika*, **14**, 161–164.
1971 The convex Borel sets in $\mathbb{R}^3$ are convexly generated, *J. London Math. Soc.*, **2**, 5–14.
Lebesgue, H.
1905 Sur les fonctions représentables analytiquement, *J. Math. Pures Appl.*, **1**, 139–216.
Louveau, A.
1977*a* Boréliens à coupes $K_{\sigma\delta}$, *C.R. Acad. Sci., Paris*, **285**, 309–312.
1977*b* La hiérarchie borélienne des ensembles Δ_1^1, *C.R. Acad. Sci., Paris*, **285**, 601–604.
1980 A separation theorem for Σ_1^1 sets, *Trans. Amer. Math. Soc.*, **260**, 363–378.
Lusin, N. N.
1917 Sur la classification de M. Baire, *C.R. Acad. Sci., Paris*, **164**, 91–94.
1925 Sur les ensembles non mesurable *B* et l'emploi de la diagonale de Cantor, *C. R. Acad. Sci., Paris*, **181**, 95–96.
1927 Sur les ensembles analytiques, *Fund. Math.*, **10**, 1–95.
1930 *Les Ensembles Analytiques*, Gauthier-Villars, Paris.
Lusin, N. N. and Novikov, P. S.
1935 Choix éffectif d'un point dans un complémentaire analytique arbitraire, donne pas un crible, *Fund. Math.* **25**, 559–560.
Lusin, N. N. and Sierpiński, W.
1918 Sur quelques propriétés des ensembles (*A*), *Bull. Acad. Cracovie*, **4**, 35–48.

Mac Gibbon, B.
1972 A criterion for metrizability of a compact convex set in terms of the set of extreme points, *J. Functional Anal.*, **11**, 385–392.
Mackey, G. W.
1957*a* Borel structure in groups and their duals, *Trans. Amer. Math. Soc.*, **85**, 134–165.
1957*b* Les ensembles boréliens et les extensions des groupes, *J. Math. Pures Appl.*, **36**, 171–178.
Maitra, A. and Ryal-Nardzewski, C.
1970 On the existence of two analytic non-Borel sets which are not isomorphic, *Bull. Acad. Polon. Sci.*, **18**, 177–178.
Marczewski, E. (*see* Szpilrajn-Marczewski, E.)
Mazurkiewicz, S.
1936 Über die Menge der differenzierbaren Funktionen, *Fund. Math.*, **27**, 244–249.
Meyer, P. A.
1966 *Probabilités et Potential*, Herman, Paris.
1966 The Baire order problem for compact spaces, *Duke Math. J.*, **33**, 33–40.
Mickle, E. J.
1958 On a closure property of measurable sets, *Proc. Amer. Math. Soc.*, **9**, 688–689.
Mokobodzki, G.
1976 Démonstration élémentaire d'un théorème de Novikov, in *Séminaire de Probabilités X*, Lecture Notes in Math. no. 511, Springer-Verlag, Berlin, Heidelberg and New York.
Montgomery, D.
1934 Non-separable metric spaces, *Fund. Math.*, **25**, 517–533.
Neumann, J. von
1949 On rings of operators. Reduction theory. *Ann. of Math.*, **50**, 401–485.
Nikodym, O.
1925 Sur une propriété de l'opération *A*, *Fund. Math.*, **7**, 149–154.
Novikov, P. S.
1934 Sur la séparabilité-*B* dénombrable des ensembles analytiques, *Dokl. Akad. Nauk SSSR*, **2**, 145–148.
1951 On the consistency of some propositions of the descriptive theory of sets, *Trudy Mat. Inst. Steklov*, **38**, 279–316 (= *Amer. Math. Soc. Trans.* (*2*), **29** (1963), 51–89).
Preiss, D.
1973 The convex generation of convex Borel sets in Banach spaces, *Mathematika*, **20**, 1–3.
Rogers, C. A.
1965 Descriptive Borel sets, *Proc. Roy. Soc. A*, **286**, 455–478.
1970*a* Spaces with good Borel structures, *J. London Math. Soc.*, **2**, 372–384.
1970*b* *Hausdorff Measures*, Cambridge University Press, Cambridge.
1973 Lusin's second separation theorem, *J. London Math. Soc.*, **6**, 491–503.
1977 Universal properties of certain analytic sets, *J. London Math. Soc.* (*2*), **16**, 177–183.
Rogers, C. A. and Willmott, R. C.
1966 On the projection of Souslin sets, *Mathematika*, **13**, 147–150.

1968 On the uniformization of sets in topological spaces, *Acta Math.*, **120**, 1–52.

Rudin, M. E.

1977 Martin's axiom, in *Handbook of Mathematical Logic* (J. Barwise, ed.), North-Holland, Amsterdam, Sect. B6.

Saint-Pierre, J.

1977 Séparation simultanée d'un ensemble analytique de suites d'ensembles $\mathscr{F}$-sousliniens, *C.R. Acad. Sci., Paris*, **285**, 933–936.

Saint-Raymond, J.

1975 Approximation des sous-ensembles analytiques par l'intérieur, *C.R. Acad. Sci., Paris*, **281**, 85(87.

1976*a* Boréliens a coupes K_σ, *Bull. Soc. Math. France*, **104**, 389–400.

1976*b* Fonctions boréliennes sur un quotient, *Bull. Sci. Math.* (*2*), **100**, 141–147.

Saks, S.

1937 *Theory of the Integral*, 2nd edn, Hafner, New York, p. 347.

Ščgol'kov, E. A.

1948 On the uniformization of certain *B*-sets. *Dokl. Akad. Nauk SSSR*, **59**, 1065–1068.

Schwartz, L.

1966 Sur le théorème de graphe fermé, *C.R. Acad. Sci., Paris*, **263**, 602–605.

Sierpiński, W.

1925 *Funkeje Przedstawialne Analityncznie* (in Polish), Lwow.

1929 Sur l'éxistence de diverses classes d'ensembles, *Fund. Math.*, **14**, 82–91.

1933 Le théorème d'unicité de M. Lusin pour les espaces abstraits, *Fund. Math.*, **21**, 250–275.

1950 *Les Ensembles Projectifs et Analytiques*, Memorial de Science Mathematique, no. 112, Gauthier-Villars, Paris.

1952 *General Topology* (translated by C. C. Krieger), University of Toronto Press, Toronto.

Sion, M.

1960*a* On analytic sets in topological spaces, *Trans. Amer. Math. Soc.*, **96**, 341–354.

1960*b* Topological and measure theoretic properties of analytic sets, *Proc. Amer. Math. Soc.*, **11**, 769–776.

1960*c* On the uniformization of sets in topological spaces, *Trans. Amer. Math. Soc.*, **96**, 237–245.

1961 Continuous images of Borel sets, *Proc. Amer. Math. Soc.*, **12**, 385–391.

1963 On capacitability and measurability, *Ann. Inst. Fourier, Grenoble*, **13**, 83–99.

Šneĭder, V. E.

1945*a* Continuous images of Souslin and Borel sets: metrization theorems (in Russian), *Dokl. Akad. Nauk SSSR*, **50**, 77–79.

1945*b* Descriptive theory of sets in topological spaces (in Russian), *Dokl. Akad. Nauk SSSR*, **50**, 81–83.

1948 Descriptive theory of sets in topological spaces (in Russian), *Ucenye Zapiski Moskov, Gos. Univ., Matematika*, **135**, 37–85.

Souslin, M.

1917 Sur une définition des ensembles mesurables B sans nombres transfinis, C.R. Acad. Sci., Paris, **164**, 88–91.

1923 Sur un corp non dénombrable de nombres réels, (edited by K. Kuratowski), *Fund. Math.*, **4**, 311–315.

Szpilrajn-Marczewski, E.

1929 O mierzalnosci i warunku Baire'a, *C.R. du I Congres des Math. des Pays Staves, Varsovie.*

1933 Sur certain invariants de l'operation (*A*), *Fund. Math.*, **21**, 229–235.

Taĭmanov, A. D.

1955 On closed mappings I, *Mat. Sb.*, **36**, 349–352.

1960 On closed mappings II, *Mat. Sb.*, **52**, 579–588.

Talagrand, M.

1975 Sur une conjecture de H. H. Corson, *Bull. Sci. Math.*, **99**, 211–212.

1979 Sur les convexes compacts dont l'ensemble des points extremaux est K-analytique, *Bull. Soc. Math. France*, **107**, 49–53.

1979*a* Deux generalisations d'un théorèm de I. Namioka, *Pacific J. Math.*, **81**, 239–251.

1979*b* Sur la *K*-analyticité de certains espaces d'operateurs, *Israel J. Math.*, **32**, 124–130.

198? Espaces de Banach faiblement *K*-analytique, to appear.

Treves, F.

1967 *Topological Vector Spaces, Distributions and Kernels*, Academic Press, New York and London.

Vaĭnšteĭn, I. A.

1947 On closed mappings of metric spaces, *Dokl. Akad. Nauk SSSR*, **57**, 319–321.

Vašak, L.

1976 On one generalization of weakly compactly generated Banach spaces. Preprint.

Part 2

Un cours sur les ensembles analytiques

C. DELLACHERIE

Dédié au moussaillon qui a souqué ferme
dans la tempête

Préface

Les notes qui suivent sont une rédaction, amplifiée, des sept conférences que j'ai faites à l'Ecole d'été de Londres, en juillet 1979, sur la théorie des ensembles analytiques. Deux thèmes principaux, constituant à mon avis les deux pôles fondamentaux de la théorie, sont développés: d'une part, la conjonction "capacités et ensembles analytiques", qui fournit le premier niveau de la théorie (définition plutôt topologique, stabilité, premier théorème de séparation, capacitabilité, etc.), et d'autre part la conjonction "relations bien-fondées et ensembles analytiques", qui fournit le second niveau de la théorie (définition plutôt combinatoire, constituants, deuxième théorème de séparation, théorème de la borne, etc.).

Avant de donner quelques "conseils pratiques" sur l'utilisation de ce cours, je vais donner les grandes lignes architecturales de son développement (d'autres indications peuvent être trouvées dans les "commentaires", placés à la fin du cours).

Le premier thème est exposé au Chapitre I, sans doute de manière un peu trop abrupte (j'ai songé un moment à écrire un Chapitre 0). Mais la majeure partie de l'auditoire était composée de "connaisseurs", et j'ai voulu tout de suite les intéresser à mon thème favori, encore bien méconnu malgré la vingtaine d'années écoulée depuis les travaux fondamentaux de Choquet. Cependant, plutôt que de continuer tout de suite à exposer le premier thème jusqu'en ses derniers retranchements, j'ai préféré rejeter cela en appendice et montrer tout de suite comment la conjonction "capacités et ensembles analytiques" permettait de traiter le premier niveau de la théorie.

Le Chapitre II est une espèce de pause: on y expose des résultats classiques, soit en y introduisant des thèmes secondaires (comme la structure mesurable d'Effros), soit en y effleurant le second thème.

Au Chapitre III, j'ai voulu, tout en introduisant le langage des temps d'arrêt, exposer les résultats classiques du deuxième niveau (allant jusqu'au théorème de Kondô) en me limitant à une exposition simple du second thème.

C'est au Chapitre IV que nous avons développé la conjonction "relations bien-fondées et ensembles analytiques" sous une forme qui n'est pas la plus générale, mais qui est à la fois agréable et utile. Enfin, premier thème et deuxième thème se retrouvent dans le "décorticage" auquel est soumis le théorème de Saint-Raymond à la fin de ce chapitre.

Un troisième thème—très important—n'a pas du tout été abordé dans ce cours: la conjonction "jeux et ensembles analytiques". Je le regrette un peu, mais comme cela est un des thèmes principaux des notes de Martin et Kechris, le lecteur n'y perd rien, tandis que j'y gagne à ne pas devoir affronter deux formidables joueurs.

On trouvera, juste devant cette préface, la table des matières indiquant: quatre chapitres, un appendice, des commentaires, un index de notations, un autre terminologique, et une bibliographie. Les "commentaires" précisent en certains points la structure du cours, mais sont essentiellement consacrés à quelques évocations historiques (d'un passé relativement récent, le plus souvent). L'index de notations a aussi été traité comme un lexique, auquel on peut se reporter non seulement pour se remémorer la signification d'une notation, mais aussi pour pallier l'insuffisance de l'index terminologique. La bibliographie est relativement succincte: en dehors de mes travaux (charité bien ordonnée commençant par soi-même!), on n'y trouvera catalogués que les travaux cités dans le cours, la plupart récents et parfois non encore publiés; les auteurs non cités se trouvent en très bonne compagnie (Hausdorff, Kuratowski, Lusin, Sierpiński, . . .).

Quand elles ne sont pas expliquées dans le texte, les notations employées sont, je l'espère, des notations courantes. Je me contenterai ici de signaler que "ssi" est une abréviation pour "si et seulement si", que j'ai aussi employé à la place de "$\Leftrightarrow$". Pour le lecteur qui n'est pas de langue française, je rappelle que "croissant" (resp. "positif", etc.) est l'équivalent, en anglais, de "non-decreasing" (resp. "non-negative", etc.); on ajoute l'adverbe "strictement" quand on veut parler de l'autre notion voisine.

Le système interne de référence n'a rien à voir avec la pagination. Chaque chapitre (et l'appendice) a été découpé en "numéros" si bien qu'une évocation, au cours du Chapitre III, de 23 (ou du no. 23) renvoie au no. 23 de ce chapitre, tandis qu'une évocation de I–23 (resp. A-23) renvoie au no. 23 du Chapitre I (resp. de l'appendice).

Il y a, dans le texte, un certain nombre d'exercices laissés au lecteur: en règle générale (qui souffre quelques exceptions, le plus souvent par inadvertance), leur matière n'est point utilisée dans le texte principal. Leur raison d'être est donc soit d'aider à la compréhension du texte, soit d'apporter un complément intéressant mais non développé (parfois par paresse).

Le texte, tel qu'il est, comporte quelques imperfections dont je suis conscient (et de nombreuses autres!), mais je n'ai pas voulu abuser outre mesure de la patience de C. A. Rogers en lui demandant d'attendre une seconde rédaction . . .. J'ai en particulier eu trop tendance à employer, en variant parfois les graphismes, les lettres I et J pour des notions à la fois importantes et diverses: j'espère que le lecteur n'aboutira pas à la conclusion que $\aleph_1$ est une capacité ayant une infinité dénombrable d'éléments! J'ai essayé de corriger ce défaut en rappelant, dans l'index des notations, l'emploi des lettres I et J pour chaque chapitre (en omettant d'y signaler que I désigne aussi le premier chapitre!).

Les deux premiers chapitres ont été essentiellement écrits, en trois semaines, dans le beau cadre du Mathematisches Forschunginstitut de Oberwolfach, avant que se tienne l'Ecole d'été. Le reste, qui a pris beaucoup plus de temps, a été écrit après l'Ecole d'été, alors que j'étais aussi membre de l'Institute for Advanced Study de Princeton, où j'ai bénéficié, entre autres, d'une bourse de la National Science Foundation.

I Capacités et ensembles analytiques

Les lettres E et F (avec ou sans indices) désignent des espaces topologiques séparés. Un ensemble X est une partie d'un espace topologique séparé, espace "ambiant" souvent sous-entendu s'il n'y a pas risque de confusion; la lettre A sera réservée aux ensembles analytiques (quand ils seront définis), la lettre B aux boréliens (i.e. aux éléments de la tribu engendrée par les ouverts), la lettre K aux compacts et la lettre U aux ouverts. On note $\mathscr{A}(E)$, $\mathscr{B}(E)$, $\mathscr{K}(E)$ et $\mathscr{U}(E)$ les classes correspondantes de parties de E.

L'ensemble des parties de E est noté $\mathscr{P}(E)$. On confondra généralement une partie X de E avec sa fonction indicatrice 1_X; en particulier, on notera souvent la réunion "sup" et l'intersection "inf".

1 Capacités et multicapacités

1 Soit f une application continue de E dans F et soit m une mesure sur F (on entendra toujours par mesure une mesure $m \geq 0$, bornée, sur $\mathscr{B}(F)$, telle que $m(Y) = \inf m(U)$, $U \supseteq Y$, pour tout fermé Y de F). Définissons une fonction I sur $\mathscr{P}(E)$ en posant

$$I(X) = m[f(X)]$$

(on supposera toujours que m est prolongée à $\mathscr{P}(F)$ par la mesure extérieure: $m(Y) = \inf m(B)$, $B \supseteq Y$). Par rapport à m, la fonction I a beaucoup perdu en "additivité" et en "continuité"; elle garde néanmoins les propriétés suivantes:

(a) elle est *croissante*: si $X \subseteq Y$, alors $I(X) \leq I(Y)$;
(b) elle est *montante*: si $X^n \uparrow X$ (i.e. si (X^n) est une suite croissante et $X = \sup_n X^n$), alors $I(X^n) \uparrow I(X)$;
(c) elle est *descendante sur les compacts*: si $K^n \downarrow K$, alors $I(K^n) \downarrow I(K)$;

(d) elle est *continue à droite*: pour tout $\varepsilon > 0$, il existe $U \supseteq X$ tel que $I(U) < I(X) + \varepsilon$;

(e) elle est *fortement sous-additive*:

$$I(X_1 \cup X_2) + I(X_1 \cap X_2) \leq I(X_1) + I(X_2) .$$

En fait, la descente sur les compacts résulte de la continuité à droite sur les compacts (et de la croissance): si $K^n \downarrow K \subseteq U$ et si $I(K) \leq I(U) < I(K) + \varepsilon$, alors K^n est inclus dans U pour n grand et donc $I(K^n) < I(K) + \varepsilon$; réciproquement, si E est métrisable compact, la descente sur les compacts entraine la continuité à droite sur les compacts: si (U^n) est une suite décroissante d'ouverts telle que $\bar{U}^n \downarrow K$, on a alors $I(U_n) < I(K) + \varepsilon$ pour n grand.

Oubliant la forte sous-additivité (e) (que nous évoquerons cependant de temps à autre; elle joue en particulier un grand rôle dans la construction de capacités, dont nous ne parlerons pas ici) et ne retenant de (d) que la continuité à droite sur les compacts (la continuité à droite générale a partie liée avec la forte sous-additivité), nous obtenons alors la notion de capacité:

2 DÉFINITION Une *capacité sur E* est une fonction I de $\mathcal{P}(E)$ dans $\bar{\mathbb{R}}_+$, croissante, montante, finie et continue à droite sur les compacts.

Remarques. (a) Le terme "capacité" ayant parfois une signification différente, en particulier en théorie du potentiel, il conviendrait plutôt de dire "capacité de Choquet" et même de rajouter "continue à droite" si on veut insister sur le fait qu'on demande plus que la descente sur les compacts.

(b) C'est à Choquet que l'on doit les fondements de la théorie des capacités, mais c'est Sion qui a montré que la continuité à droite sur les compacts est la bonne hypothèse à faire dans le cadre topologique "général", plutôt que la descente sur les compacts. Il reste cependant que c'est cette dernière propriété qui peut être adaptée aisément dans le cadre "abstrait", dont nous ne parlerons pas.

(c) Toujours au sujet de la continuité à droite sur les compacts, il faut prendre garde de bien écrire

$$\forall t > I(K) \qquad \exists U \supseteq K \qquad I(U) < t$$

et non la propriété plus faible

$$\forall t > I(K) \qquad \exists U \supseteq K \qquad \forall K' \subseteq U \qquad I(K') < t$$

même si cette dernière est souvent pourvue du même nom.

Il est rare qu'une fonction d'ensembles soit d'emblée définie sur tout $\mathcal{P}(E)$. Aussi est il important de connaitre des méthodes pour construire des capacités. Nous n'en parlerons pas cependant ici (le lecteur pourra se reporter à Dellacherie (1972*b*)), faute de temps. Signalons quand même que c'est la montée de la fonction d'ensembles qui est difficile à obtenir, ou à vérifier.

Voici néanmoins quelques exemples "tout construits".

3 *Exemples* (a) D'abord un exemple trivial, mais fondamental. Soit L un compact de E et posons $I_L(X)=0$ si $L\cap X=\varnothing$ et $I_L(X)=1$ sinon. Il est clair que I_L est une capacité (même si L est seulement fermé).

(b) Ensuite l'exemple le plus familier, que nous avons déjà vu: toute mesure (extériorisée) est une capacité.

(c) Un moyen pour fabriquer une capacité à partir d'une autre: si φ est un homéomorphisme croissant de $\bar{\mathbb{R}}_+$ sur un intervalle de $\bar{\mathbb{R}}_+$, alors $\varphi(I)$ est une capacité si I en est une.

(d) Un autre moyen, beaucoup plus intéressant: si f est une application continue de E dans F et si I est une capacité sur F, alors J définie par $J(X)=I[f(X)]$ est une capacité sur E. Remarquons que cela résulte du fait que $X\to f(X)$ est croissante, montante (i.e. si $X^n\uparrow X$, alors $f(X^n)\uparrow f(X)$) et continue à droite sur les compacts (i.e. si K est compact, $f(K)$ l'est aussi, et, pour V ouvert contenant $f(K)$, il existe U ouvert contenant K tel que $f(U)\subseteq V$).

(e) Un dernier exemple, généralisant (a) et (b). Supposons E polonais (ce qui, en particulier, nous évite de parler de mesures de Radon), et désignons par $\mathcal{M}^1(E)$ l'ensemble des mesures de masse $\leqslant 1$ sur E. Nous supposerons toujours que $\mathcal{M}^1(E)$ est muni de la topologie de la convergence étroite, i.e. de la topologie la moins fine rendant continues les applications $m\to m(f)=\int f\,dm$ où f parcourt l'ensemble $\mathscr{C}_b(E)$ des fonctions réelles, continues et bornées, sur E. On sait que $\mathcal{M}^1(E)$ est alors une espace polonais (et est métrisable compact si E l'est). Soit maintenant L un compact de $\mathcal{M}^1(E)$ et posons, pour toute partie X de E,

$$I_L(X)=\sup m(X)\,,\qquad m\in L\,.$$

On définit ainsi une capacité I_L sur E (seule la continuité à droite n'est pas évidente: pour $\varepsilon>0$ et $m\in L$, prendre un ouvert U_m contenant K tel que $m(U_m)<m(K)+\varepsilon$, et utiliser le fait que $m\to m(U)$ est semicontinue inférieurement—en abrégé, s.c.i.). On retrouve (a) si L est constitué de mesures de Dirac, et (b) si L est réduit à un point.

4 Etant donnée une capacité I sur E, une partie X de E est dite *I-capacitable* si l'on a

$$I(X)=\sup I(K)\,,\qquad K\subseteq X\,,$$

et *universellement capacitable* si elle est capacitable pour toute capacité (sur E). Noter que la classe $\mathscr{C}$ des ensembles universellement capacitable (E variant) est stable pour les images directes par les applications continues: c'est à peu près tout ce qu'on sait dire sur sa stabilité pour des opérations ensemblistes (avec aussi les images réciproques par les applications continues si, par exemples, les espaces ambiants sont métrisables compacts). Revenons à notre capacité I sur E; posons, pour tout X,

$$J(X) = \inf I(B), \qquad B \supseteq X$$

(remarquer que l'inf est atteint sur $\mathscr{B}(E)$). On a $I(B) = J(B)$ pour tout B et donc, en particulier, $I(K) = J(K)$ pour tout K et $I(U) = J(U)$ pour tout U. Il est facile de vérifier que J est aussi une capacité et donc, si X est universellement capacitable, on a

$$I(X) = \sup\{I(K), K \subseteq X\} = \sup\{J(K), K \subseteq X\} = J(X)$$

si bien que l'on a $I(X) = \inf I(B)$, $B \supseteq X$. Autrement dit, l'approximation intérieure par des compacts entraine l'approximation extérieure par des boréliens:[1] nous verrons un peu plus loin combien cette idée est importante.

Revenons maintenant à l'Exemple 3(a): on voit que s'y introduit naturellement une fonction d'ensembles $I(X_1, X_2) = I_{X_1}(X_2)$ à deux arguments, qui vaut 0 si $X_1 \cap X_2 = \varnothing$ et vaut 1 sinon. Et cette fonction est une "bicapacité": elle est croissante, montante, en chacun de ses arguments, et continue à droite du couple pour les compacts (i.e. pour tout $\varepsilon > 0$ il existe $U_1 \supseteq K_1$ et $U_2 \supseteq K_2$ tels que l'on ait $I(U_1, U_2) < I(K_1, K_2) + \varepsilon$; pour notre I, cela exprime simplement le fait que deux compacts disjoints sont séparables par des ouverts disjoints). Nous allons définir, de manière générale, la notion de multicapacité, à un nombre fini ou infini dénombrable d'arguments (nous dirons "dénombrable" pour "fini ou infini dénombrable").

5 Nous désignerons désormais par $\mathscr{I}$, ou $\mathscr{J}$ (avec ou sans indices), un ensemble dénombrable d'indices (variable) que, le plus souvent, nous identifierons tacitement à un segment initial de $\mathbb{N} = 1, 2, \ldots$ ou de $\mathbb{N}^\circ = 0, 1, 2, \ldots$. Pour des raisons typographiques évidentes, nous serons souvent amenés à contracter " . . . " en "." (qui ne désignera donc pas une variable muette).

DÉFINITION Une *multicapacité* I *sur* $\prod_{i \in \mathscr{I}} E_i$ est une application de $\prod_i \mathscr{P}(E_i)$ dans $\bar{\mathbb{R}}_+$ vérifiant les conditions suivantes:

[1] Si I est fortement sous-additive, on a une approximation extérieure par des ouverts: cela est crucial en théorie du potentiel.

(1) elle est *globalement* croissante:

$$X_1 \subseteq Y_1, ., X_i \subseteq Y_i, \ldots \Rightarrow I(X_1, ., X_i, \ldots) \leq I(Y_1, ., Y_i, \ldots);$$

(2) elle est *séparément* montante: si tous les arguments X_i sont fixés sauf pour $i = j$, on a

$$X_j^n \uparrow X_j \Rightarrow I(X_1, ., X_{j-1}, X_j^n, X_{j+1}, \ldots) \uparrow I(X_1, ., X_{j-1}, X_j, X_{j+1}, \ldots);$$

(3) elle est *globalement* continue à droite sur les compacts: I est finie sur les compacts[1] et, pour tout $t > I(K_1, ., K_i, \ldots)$, il existe pour tout i un ouvert $U_i \supseteq K_i$ tel que $U_i = E_i$ sauf pour un nombre fini d'indices, et qu'on ait $I(U_1, ., U_i, \ldots) < t$.

Remarques (a) Lorsque $\mathscr{I}$ est fini, la montée séparée est évidemment équivalente à la montée globale. Nous verrons à l'occasion de l'exemple 6(b) que ce n'est pas le cas si $\mathscr{I}$ est infini. D'autre part, la considération d'ensembles d'indices infinis (dénombrables) n'est pas un luxe: elle est inévitable si l'on veut traiter commodément la théorie des ensembles analytiques à l'aide des capacités.

(b) La continuité à droite globale sur les compacts entraine (avec la croissance globale) la descente globale sur les compacts: si, pour tout i, $K_i^n \downarrow K_i$, alors $I(K_1^n, ., K_i^n, \ldots) \downarrow I(K_1, ., K_i, \ldots)$. Si les espaces E_i sont métrisables compacts, on a aussi l'implication dans l'autre sens, et donc une équivalence.

Nous donnons maintenant deux exemples; nous en verrons d'autres au cours de la Section 2.

6 *Exemples* (a) (à deux arguments) Reprenons l'exemple 3(e), en faisant cette fois varier le "paramètre" L. On obtient alors la bicapacité I sur $\mathcal{M}^1(E) \times E$ suivante:

$$I(X_1, X_2) = \sup m(X_2), \qquad m \in X_1.$$

(b) (à une infinité d'arguments) Nous généralisons cette fois l'exemple 3(a). Prenons $E_i = E$ pour tout i et posons

$$I(X_1, ., X_i, \ldots) = 0 \quad \text{si } \inf_i X_i = \varnothing \text{ et} = 1 \text{ sinon.}$$

Il est facile de voir que I est bien une multicapacité (la continuité à droite n'est pas tout à fait évidente: nous laissons ce petit exercice de topologie au lecteur). Remarquons, à cette occasion, que la montée séparée n'entraine pas la montée globale: si $X_i^n = \varnothing$ pour $i \leq n$ et $X_i^n = E$ pour $i > n$, on a $X_i^n \uparrow E$ pour tout i et $I(X_1^n, ., X_i^n, \ldots) = 0$ pour tout n alors que $I(E, ., E, \ldots) = 1$.

(1) Condition bien naturelle, mais qui n'est pas essentielle.

7 Disons, sans nous illusionner sur l'intérêt de cette notion, qu'une famille $(X_i)_{i\in\mathcal{I}}$ est *universellement capacitable* si l'on a

$$I(X_1, ., X_i, \ldots) = \sup I(K_1, ., K_i, \ldots), \qquad K_i \subseteq X_i,$$

pour toute multicapacité I sur $\prod_i E_i$. Le point faible de cette définition est qu'elle fait intervenir tous les membres à la fois de la famille alors qu'on aimerait avoir une propriété individuelle de chacun des X_i. Cette notion nous fournit cependant l'occasion de démontrer deux lemmes qui motiveront notre définition des ensembles analytiques, et seront très utiles dans leur étude.

Dans le premier lemme, nous déduisons de l'approximation par l'intérieur fournie par 7 une approximation par l'extérieur:

8 LEMME *Si I est une multicapacité sur $\prod E_i$ et si (X_i) est une famille universellement capacitable, alors on a*

$$I(X_1, ., X_i, \ldots) = \inf I(B_1, ., B_i, \ldots), \qquad B_i \supseteq X_i,$$

et l'inf est atteint.

Démonstration Posons, pour toute famille (X_i) d'arguments pour I,

$$J(X_1, ., X_i, \ldots) = \inf I(B_1, ., B_i, \ldots), \qquad B_i \supseteq X_i.$$

Comme la classe des boréliens est stable pour les intersections dénombrables et que I est globalement croissante, il est clair que l'inf est atteint dans la définition de J. Nous allons montrer que J est aussi une multicapacité sur $\prod E_i$: comme I et J coincident sur les boréliens et donc, en particulier, sur les compacts, le lemme résultera aussitôt de 7. La croissance globale de J et sa continuité globale sur les compacts sont triviales. Reste à vérifier la montée séparée. Supposons par exemple $X_2, ., X_i, \ldots$ fixés et soit $X_1^n \uparrow X_1$. Désignons, pour tout n, par Y_1^n, $Y_2^n, ., Y_i^n, \ldots$ des boréliens tels que $X_1^n \subseteq Y_1^n$, $X_i \subseteq Y_i^n$ pour $i \geqslant 2$ et que l'on ait

$$J(X_1^n, X_2, ., X_i, \ldots) = I(Y_1^n, Y_2^n, ., Y_i^n, \ldots).$$

Quitte à remplacer, pour $i \geqslant 2$, Y_i^n par $\inf_n Y_i^n$, on peut supposer que les Y_i^n ne dépendent pas de n pour $i \geqslant 2$, et donc supprimer l'indice n pour $i \geqslant 2$. D'autre part, pour n fixé, on peut remplacer Y_1^n par $\inf_{m\geqslant n} Y_1^m$ et donc supposer que la suite (Y_1^n) est croissante. Si on désigne par Y_1 la réunion de cette suite, on a alors

$$J(X_1^n, X_2, ., X_i, .) = I(Y_1^n, Y_2, ., Y_i, .) \uparrow I(Y_1, Y_2, ., Y_i, .)$$
$$\geqslant J(X_1, X_2, ., X_i, .).$$

D'où, finalement, la montée de J.

Remarque On peut remplacer $\mathscr{B}(E_i)$ par toute classe $\mathscr{C}(E_i)$ stable pour les intersections dénombrables et les réunions de suites monotones, et vérifiant la condition suivante: pour K, U dans E_i tels que $K \subseteq U$, il existe un ouvert V dans $\mathscr{C}$ tel que $K \subseteq V \subseteq U$. En particulier, si E_i est complètement régulier, on peut prendre pour $\mathscr{C}$ la tribu de Baire (i.e. la tribu engendrée par les fonctions réelles continues sur E_i).

Appliquons ce lemme à l'Exemple 6(b): on obtient un théorème de séparation qui nous fournira immédiatement, plus loin, le théorème de séparation de Novikov pour les ensembles analytiques:

9 COROLLAIRE *Soit (X_i) une famille universellement capacitable telle que $\inf_i X_i = \varnothing$. Il existe alors une famille (B_i) de boréliens telle que $X_i \subseteq B_i$ pour tout i et que $\inf_i B_i = \varnothing$. En particulier, si le couple $(X, E-X)$ est universellement capacitable, alors X est borélien.*

Notre but est maintenant de trouver une classe $\mathscr{C}$, aussi large et aussi "stable" que possible, telle que (X_i) soit universellement capacitable dès que chaque X_i appartient à $\mathscr{C}$. Il est tout à fait trivial, et cela ne mène pas loin, que l'on peut prendre $\mathscr{C} = \mathscr{K}$. Pour des familles finies (X_i), il est clair, vu la montée séparée d'une multicapacité, que l'on peut prendre $\mathscr{C} = \mathscr{K}_\sigma$ (i.e. la classe des réunions dénombrables de compacts); mais, la montée n'étant que séparée, il n'est pas évident que l'on peut prendre $\mathscr{C} = \mathscr{K}_\sigma$ pour des familles infinies (dénombrables). Cela nous est cependant assuré par le lemme suivant:

10 LEMME *Si, pour chaque i, X_i est un $\mathscr{K}_\sigma$ de E_i, alors (X_i) est une famille universellement capacitable.*

Démonstration Ecrivons $X_i = \sup_n K_i^n$ où, pour chaque i, (K_i^n) est une suite *croissante* de compacts. Soit alors I une multicapacité sur $\prod E_i$ et soit $t < I(X_1, ., X_i, \ldots)$; nous allons montrer qu'il existe une suite $i \to n_i$ d'entiers tels que l'on ait

$$t < I(K_1^{n_1}, ., K_i^{n_i}, \ldots)$$

ce qui prouvera le lemme. D'abord, à l'aide d'un raisonnement par récurrence, on établit aisément, grâce à la montée séparée de I, l'existence d'une suite $j \to n_j$ d'entiers telle que, pour tout i,

$$t < I(K_1^{n_1}, ., K_i^{n_i}, X_{i+1}, ., X_{i+k}, \ldots).$$

Maintenant, grâce à la croissance globale, on a aussi pour tout i

$$t < I(K_1^{n_1}, ., K_i^{n_i}, E_{i+1}, ., E_{i+k}, \ldots).$$

Par ailleurs, comme I est continue à droite sur les compacts, pour $\varepsilon > 0$ fixé, on a, pour i suffisamment grand,

$$I(K_1^{n_1}, . , K_i^{n_i}, E_{i+1}, . , E_{i+k}, .) < \varepsilon + I(K_1^{n_1}, . , K_i^{n_i}, K_{i+1}^{n_{i+1}}, . , K_{i+k}^{n_{i+k}}, .)$$

(on n'a utilisé en fait qu'une faible partie de la continuité à droite). Il est alors clair qu'on a $t < I(K_1^{n_1}, . , K_i^{n_i}, \ldots)$.

Remarque La démonstration est plus riche que l'énoncé: elle contient en germe la définition des schémas de Souslin. Nous reprendrons cela en détail au moment opportun.

On pourrait ensuite faire un pas élémentaire de plus en démontrant, "à la main", que l'on peut prendre $\mathscr{C} = \mathscr{K}_{\sigma\delta}$ (i.e. la classe des intersections dénombrables d'éléments de $\mathscr{K}_\sigma$); puis, profitant du fait que l'on peut faire le "changement de variables" $X_i = f_i(X_i')$, où les f_i sont des applications continues, dans une multicapacité (cf l'Exemple 3(d), qui passe aux multicapacités), on pourrait finalement prendre pour $\mathscr{C}$ la "classe image" de $\mathscr{K}_{\sigma\delta}$ par les applications continues: on retrouve là la démarche de Choquet pour définir les ensembles analytiques.

Nous ferons les choses plus globalement—ce qui trivialisera les premiers pas de la théorie des ensembles analytiques (définition, stabilité, capacitabilité, séparation): nous allons regarder tout de suite tous les "bons changements de variables" pour les multicapacités. Ce seront, en gros (cf. Exemple 3(d)), les opérations ensemblistes—avec un nombre fini ou infini dénombrable d'arguments—croissantes, montantes, et continues à droite sur les compacts. Puis, nous prendrons pour classe $\mathscr{C}$ la "classe image" de $\mathscr{K}_\sigma$ pour ces opérations, et appellerons ensembles analytiques ses éléments; nous montrerons aussi que l'on obtient ainsi la même notion d'ensemble analytique que celle de Choquet, Sion, Frolík et Rogers, ce qui justifiera leur appellation.

2 Opérations capacitaires et ensembles analytiques

Comme nous le verrons bientôt, la classe d'opérations ensemblistes que nous allons définir maintenant est stable par composition et comprend toutes les "bonnes" opérations usuelles ne faisant pas intervenir la complémentation.

11 DÉFINITION Une *opération capacitaire* I sur $\prod_{i \in \mathscr{I}} E_i$, à valeur dans F, est une application de $\prod_i \mathscr{P}(E_i)$ dans $\mathscr{P}(F)$ vérifiant les conditions suivantes:

(1) elle est *globalement* croissante:

$$X_1 \subseteq Y_1, . , X_i \subseteq Y_i, \ldots \Rightarrow I(X_1, . , X_i, \ldots) \subseteq I(Y_1, . , Y_i, \ldots) ;$$

(2) elle est *séparément* montante: si tous les arguments X_i sont fixés sauf pour $i=j$, on a

$$X_j^n \uparrow X_j \Rightarrow I(X_1,.,X_{j-1},X_j^n,X_{j+1},\ldots)\uparrow I(X_1,.,X_{j-1},X_j,X_{j+1},\ldots);$$

(3) elle est *globalement* continue à droite sur les compacts: si les arguments K_i sont compacts, alors $I(K_1,.,K_i,\ldots)$ est compact, et, pour tout ouvert $V\supseteq I(K_1,.,K_i,\ldots)$, il existe, pour toute i, un ouvert $U_i\supseteq K_i$ tel que $U_i=E_i$ sauf pour un nombre fini d'indices et qu'on ait $I(U_1,.,U_i,\ldots)\subseteq V$.

Remarques (a) Nous avons inclus dans la définition de la continuité à droite sur les compacts le fait qu'une opération capacitaire est à valeurs compactes sur les compacts: cette condition est nécessaire pour assurer la stabilité de notre notion pour la composition des applications.

(b) Il y a bien plus qu'une analogie entre les Définitions 5 et 11. Ainsi, si l'on note $I_y(\cdots)$ la valeur de l'indicatrice de $I(\cdots)$ au point $y\in F$, on voit que, pour y fixé, la fonction $I_y(\cdots)$ est une multicapacité à valeurs dans $\{0, 1\}$ si I est une opération capacitaire. Il est possible de chapeauter les notions d'opération capacitaire et de multicapacité par la notion de "multinoyau capacitaire" qui, en gros, est obtenue en permettant, dans 11, que arguments et valeurs puissent être des fonctions dans $\bar{\mathbb{R}}_+$. Voir l'appendice.

(c) Ici encore, la continuité à droite sur les compacts entraine (avec la croissance globale) la descente globale sur les compacts. Et la réciproque est vraie si les espaces E_i sont métrisables compacts (et si, bien entendu, I est à valeurs compactes sur les compacts).

Nous énonçons tout de suite, avant de donner des exemples, un résultat éminemment important[(1)] ... et à peu près trivial: la stabilité par composition de la classe des opérations capacitaires. La difficulté de la démonstration étant essentiellement d'ordre dactylographique, nous laissons au lecteur le soin de l'écrire. Nous ferons cependant deux remarques à son sujet, après l'énoncé.

12 THÉORÈME *Soit d'une part I une opération capacitaire (resp. une multicapacité) sur $\prod_{i\in\mathcal{I}} E_i$ et à valeur dans F (resp. $\bar{\mathbb{R}}_+$) et soient d'autre part, pour tout $i\in\mathcal{I}$, une opération capacitaire J_i sur $\prod_{j\in\mathcal{I}^i} E_j^i$ et à valeur dans E_i. Alors l'application J obtenue par composition de I et des J_i, i.e. définie par*

$$J(\ldots,X_j^i,\ldots)=I[J_1(X_1^1,.,X_j^1,.),\ldots,J_i(X_1^i,.,X_j^i,.),\ldots]$$

est une opération capacitaire (resp. une multicapacité) sur $\prod_{i,j} E_j^i$ à valeur dans F (resp. $\bar{\mathbb{R}}_+$).

(1) Ce résultat joue, dans notre cadre, le rôle de l'idempotence de l'opération A de Souslin en théorie classique.

Remarques (a) On prendra garde, lorsque certains espaces E_j^i sont identiques, de ne pas identifier pour cela les arguments. La montée n'étant que séparée, il se peut que $I(X_1, ., X_i, \ldots)$ soit une opération capacitaire à une infinité d'arguments pris dans le même espace et que l'opération $J(X) = I(X, ., X, \ldots)$ ne soit pas capacitaire. Ce type d'opération est étudié dans l'appendice.

(b) C'est essentiellement dans la démonstration de ce théorème qu'on utilise la continuité à droite sur les compacts, dans toute sa force. Plus loin, dans la démonstration du théorème de capacitabilité des ensembles analytiques, nous "n'exhiberons" que le fait que notre I est à valeur compact sur les compacts et vérifie

$$I(K_1, ., K_i, E_{i+1}, ., E_{i+k}, \ldots) \downarrow I(K_1, ., K_i, K_{i+1}, ., K_{i+k}, \ldots),$$

quand i tend vers l'infini (cf. la démonstration du Lemme 10), si bien que la nécessité de la continuité à droite, dans toute sa force, sera cachée dans l'évocation du théorème de composition.

Voici maintenant quelques exemples importants d'opérations capacitaires. Afin de ne pas alourdir l'exposé, nous avons laissé au lecteur le soin de la vérification du caractère capacitaire des opérations lorsqu'elle est "sans problème", et différons cette vérification (cf. 30) quand elle est plus délicate. En fait, nous n'aurons vraiment besoin pour la partie fondamentale de la suite que des cas "sans problème".

13 *Exemples* (a) (à un argument): image directe par une application continue.

(b) (à un argument): les espaces ambiants étant métrisables compacts, image réciproque par une application continue.

(c) (à deux arguments): E et F étant deux espaces métrisables compacts, on prend $E_1 = E$ et $E_2 = E \times F$ et on désigne par π la projection de $E \times F$ sur F. On pose

$$I(X_1, X_2) = \pi[(X_1 \times F) \cap X_2].$$

L'opération capacitaire I de $E_1 \times E_2$ dans F ainsi définie (c'est en fait la composée des trois opérations élémentaires $X \to \pi(X)$, $X \to X \times F$ et $(X, Y) \to X \cap Y$) permet d'obtenir l'opération d'image directe par une application f de E dans F et celle d'image réciproque par une application g de F dans E: en effet, si X_2 est le graphe de l'application f (resp. g), alors $I(X_1, X_2)$ est égal à $f(X_1)$ (resp. $g^{-1}(X_1)$). Noter que X_2 est borélien si f (resp. g) est borélienne.

(d) (à un nombre fini ou infini d'arguments):

$$I(X_1, ., X_i, \ldots) = \prod_i X_i.$$

(e) (à un nombre fini ou infini d'arguments):

$$I(X_1, ., X_i, \dots) = \inf_i X_i .$$

On sous-entend bien évidemment, pour que I ait un sens, que les arguments X_i sont pris dans le même espace ambiant.

(f) (à un nombre fini ou infini d'arguments): nous venons de voir qu'une intersection dénombrable est une opération capacitaire. Il est clair qu'une réunion finie en est une aussi, mais ce n'est pas le cas pour une réunion infinie (dénombrable). Il n'en reste pas moins que l'on peut obtenir la réunion dénombrable à l'aide d'une opération capacitaire en procédant comme suit. On prend $\mathcal{I} = \mathbb{N}^\circ = 0, 1, 2, \dots$ puis $E_0 = \mathbb{N} = 1, 2, \dots$ et $E_i = F$ pour $i > 0$ (i.e. l'espace ambiant dans lequel on fait une réunion). On pose alors

$$I(X_0, X_1, ., X_i, \dots) = \sup_{i \in X_0} X_i .$$

On définit bien ainsi une opération capacitaire I, et l'on a

$$\sup_{i>0} X_i = I(\mathbb{N}, X_1, ., X_i \dots) .$$

Noter que $\mathbb{N}$ est un $\mathcal{K}_\sigma$.

(g) (à une infinité d'arguments):[1] on se propose ici de montrer comment on peut obtenir l'opération A de Souslin à l'aide d'une opération capacitaire. Notre ensemble d'indices $\mathcal{I}$ sera ici l'ensemble dénombrable S des suites finies d'éléments de $\mathbb{N}$, augmenté d'un élément noté 0. On prend $E_0 = \mathbb{N}^{\mathbb{N}}$ et $E_i = F$ pour $i \neq 0$ (i.e. l'espace dans lequel on considère des schémas de Souslin). On pose alors, pour X_0 dans E_0 et pour X_s dans F, $s \in S$,

$$I(X_0, \dots, X_s, \dots) = \sup_{\sigma \in X_0} \inf_{s \dashv \sigma} X_s ,$$

où "$s \dashv \sigma$" signifie que la suite infinie σ commence par la suite finie s. On définit ainsi une opération capacitaire I (démonstration reportée au no. 30) et l'on a

$$\sup_{\sigma \in \mathbb{N}^{\mathbb{N}}} \inf_{s \dashv \sigma} X_s = I(\mathbb{N}^{\mathbb{N}}, \dots, X_s, \dots) ,$$

le premier membre étant, par définition, le noyau du schéma de Souslin $(X_s)_{s \in S}$. Noter que $\mathbb{N}^{\mathbb{N}}$ est un $\mathcal{K}_{\sigma\delta}$ de $\bar{\mathbb{N}}^{\mathbb{N}}$, où $\bar{\mathbb{N}} = \mathbb{N} \cup \{\infty\}$.[2]

[1] La compréhension de cet exemple n'est pas indispensable pour la suite.

[2] $\bar{\mathbb{N}}$ est évidemment le compactifié d'Alexandrov de $\mathbb{N}$, et $\bar{\mathbb{N}}^{\mathbb{N}}$ est muni de la topologie produit.

(h) (à un argument): on se propose ici de montrer qu'une application semicontinue supérieurement (en abrégé, s.c.s.) Ψ de $\mathbb{N}^{\mathbb{N}}$ dans $\mathcal{K}(F)$ donne naissance à une opération capacitaire. Rappelons que Ψ est dite s.c.s. si, pour tout ouvert $V \supseteq \Psi(\sigma)$, il existe $s \dashv \sigma$ tel que l'on ait $\Psi(\tau) \subseteq V$ pour tout τ tel que $s \dashv \tau$. Posons tout simplement, pour X partie de $\mathbb{N}^{\mathbb{N}}$,

$$I(X) = \sup_{\sigma \in X} \Psi(\sigma) .$$

On vérifie aisément que l'on définit ainsi une opération capacitaire (le seul point non trivial—$I(K)$ est compact pour K compact—est un petit exercice de topologie laissé au lecteur). Notons au passage le fait bien connu que tout schéma de Souslin $(K_s)_{s \in S}$ sur les compacts donne naissance à une telle application s.c.s. Ψ: il suffit de poser $\Psi(\sigma) = \inf_{s \dashv \sigma} K_s$.

Voici, enfin, notre définition des ensembles analytiques:

14 DÉFINITION Un ensemble A est dit *analytique* s'il existe une opération capacitaire I et des ensembles $H_1, ., H_i, \ldots$ qui sont des arguments $\mathcal{K}_\sigma$ pour I, tels que l'on ait $A = I(H_1, ., H_i, \ldots)$. On peut choisir I de sorte que $H_i = \mathbb{N}$ pour tout i (exercice!).

Remarque La définition est volontairement ambigüe: on n'y précise pas si I est une opération capacitaire de $\prod H_i$ dans A (en fait, sur A), ou si I en est une de $\prod E_i$ dans F, où E_i est un espace ambiant pour H_i (le fait pour H_i d'être un $\mathcal{K}_\sigma$ ne depend pas de E_i) et F un espace ambiant pour A. Un moment de réflexion convainc que cela revient au même. Autrement dit, la notion d'ensemble analytique est absolue: le fait pour un espace topologique séparé d'être analytique est une propriété topologique, ne dépendant pas de son plongement éventuel dans un autre espace séparé.

Etant donné le théorème de composition 12, on a immédiatement le théorème d'invariance suivant

15 THÉORÈME *Soit I une opération capacitaire et soient $A_1, ., A_i, \ldots$ des arguments pour I. Si les A_i sont analytiques, alors l'image $I(A_1, ., A_i, \ldots)$ est aussi analytique.*

Passant en revue les exemples d'opérations capacitaires de 13, on obtient les propriétés de stabilité classiques. Nous oublions pour le moment 13(h), qui sera utilisé plus loin.

16 COROLLAIRE (1) *La classe $\mathcal{A}$ des ensembles analytiques est stable pour les réunions et intersections dénombrables, pour les produits dénombrables, et pour l'opération de Souslin.*

(2) *La classe $\mathscr{A}$ contient la classe stabilisée de $\mathscr{K}$ pour les réunions et intersections dénombrables (i.e. la classe des $\mathscr{K}$-boréliens, comme l'on dit parfois).*

(3) *La classe $\mathscr{A}$ est stable pour les images directes par des applications continues. Si on se restreint à la catégorie des espaces ambiants métrisables séparables, la classe $\mathscr{A}$ est alors stable pour les images directes par des applications boréliennes, et pour les images réciproques si l'espace ambiant de départ est analytique.*

Démonstration Remarquons d'abord qu'un ensemble compact K est analytique (prendre l'opération capacitaire $I(\cdot\cdot\cdot)$ constante et égale à K). Ceci dit, (1) résulte immédiatement du théorème appliqué aux Exemples 13(f), (e), (d) et (g) (pour l'Exemple (g), il faut remarquer d'abord qu'un $\mathscr{K}_{\sigma\delta}$ est analytique d'après (f) et (e)); le point (2) résulte immédiatement de (1), et la première partie du point (3) de l'Exemple 13(a). Lorsque les espaces ambiants sont métrisables compacts, la deuxième partie du point (3) résulte immédiatement de l'Exemple 13(c), une fois remarqué que, d'après (2), les boréliens d'un espace métrisable compact sont analytiques. Dans le cas où les espaces ambiants sont métrisables séparables, on se ramène au cas des espaces métrisables compacts en utilisant le lemme suivant (qui nous sera encore bien utile plus loin).

17 LEMME (1) *Tout espace métrisable séparable peut être plongé dans un espace métrisable compact (plus précisément, dans $[0, 1]^{\mathbb{N}}$).*

(2) *Soient E, F des espaces plongés respectivement dans E', F' où F' est un espace métrisable compact, et soit f une application borélienne de E dans F. Il existe alors une application borélienne f' de E' dans F' telle que f soit égale à la restriction de f' à E, F.*

Demonstration Nous ne démontrons pas (1), qui est bien classique. Pour le point (2), nous remarquons d'abord qu'une application borélienne de E dans F est encore borélienne de E dans F'; donc, quitte à plonger F' dans $[0, 1]^{\mathbb{N}}$, on peut supposer que $F = F' = [0, 1]^{\mathbb{N}}$. La donnée de f borélienne revient alors à la donnée d'une suite (f^n) de fonctions boréliennes de E dans $[0, 1]$, si bien que nous nous sommes finalement ramenés au cas où $F = F' = [0, 1]$. Maintenant, comme $\mathscr{B}(E) = \mathscr{B}(E')|_E$, il est clair que (2) est vrai si f est une fonction étagée (i.e. une combinaison linéaire finie d'indicatrices). Dans le cas général, f est limite simple (et même uniforme) d'une suite (f_m) de fonctions étagées boréliennes, et on peut alors prolonger f à E' en posant $f' = \liminf f'_m$ où, pour chaque m, f'_m est une extension à E' de f_m.

Voici encore un appendice très important de 16

18 COROLLAIRE *Tout borélien d'un espace analytique métrisable séparable est analytique. En particulier, tout borélien d'un espace polonais est analytique.*

Démonstration Il est classique que tout espace polonais plongé dans un compact métrisable y est un $\mathcal{G}_\delta$ (i.e. une intersection dénombrable d'ouverts) et donc un $\mathcal{K}_{\sigma\delta}$: par conséquent, tout espace polonais est analytique. Ceci dit, tout borélien d'un espace métrisable compact est analytique d'après 16(2), et donc tout borélien d'un espace analytique métrisable séparable, par plongement.

Nous allons maintenant voir les théorèmes d'approximation intérieure et extérieure. D'abord, le théorème de capacitabilité.

19 THÉORÈME *Soit I une opération capacitaire, ou une multicapacité, et soient $A_1, ., A_i, \ldots$ des arguments analytiques pour I. On a alors*

$$I(A_1, ., A_i, \ldots) = \sup I(K_1, ., K_i, \ldots) \qquad (\circ)$$

où K_i parcourt les compacts contenus dans A_i.

Démonstration Si I est une opération capacitaire à valeur dans F, alors, pour $y \in F$ fixé, la valeur $I_y(\cdots)$ de l'indicatrice de $I(\cdots)$ en y est une multicapacité. Il suffit donc de démontrer le théorème lorsque I est une multicapacité. Maintenant, d'après le lemme 10, on sait que (∘) est vrai si les A_i sont des $\mathcal{K}_\sigma$. Dans le cas général, il existe, par définition, pour tout i, une opération capacitaire J_i et des arguments $\mathcal{K}_\sigma$ pour J_i, $H_1^i, ., H_j^i, \ldots$ tels que

$$A_i = J_i(H_1^i, ., H_j^i, \ldots) .$$

Désignons par J la multicapacité obtenue en composant I avec les J_i et appliquons lui (∘), pour les arguments H_j^i; il vient

$$\begin{aligned} I(A_1, ., A_i, \ldots) &= J(\ldots, H_j^i, \ldots) \\ &= \sup J(\ldots, K_j^i, \ldots) , \qquad K_j^i \subseteq H_j^i \\ &= \sup I(\ldots, J_i(\,., K_j^i, .), \ldots) , \end{aligned}$$

et il n'y a plus qu'à remarquer que $J_i(\ldots, K_j^i, \ldots)$ est un compact contenu dans A_i si, pour tout j, K_j^i est un compact contenu dans H_j^i.

Nous donnerons, plus loin, une forme beaucoup plus précise de ce théorème, où nous verrons apparaitre, de manière naturelle, les schémas de Souslin et les applications s.c.s. sur $\mathbb{N}^\mathbb{N}$ à valeurs dans les compacts. Nous passons maintenant à l'extérieur.

20 THÉORÈME *Soit I une opération capacitaire, ou une multicapacité, et soient* $A_1, ., A_i, \ldots$ *des arguments analytiques pour I. On a alors*

$$I(A_1, ., A_i, \ldots) = \inf I(B_1, ., B_i, \ldots)\,, \qquad (\circ\circ)$$

où B_i *parcourt les boréliens contenant* A_i.

De plus, si I est une multicapacité, l'inf est atteint.

Démonstration On se ramène, comme précédemment, au cas où I est une multicapacité. Alors ($\circ\circ$) est conséquence immédiate du théorème précédent et du Lemme 8.

Remarque D'après la remarque suivant le Lemme 8, on peut remplacer les boréliens par les éléments du stabilisé de la classe des ouverts (de Baire si les espaces ambiants sont complètement réguliers) pour les lim inf de suites.

21 COROLLAIRE *Soit I une opération capacitaire, ou une multicapacité, et soient* $A_1, ., A_i, \ldots$ *des arguments analytiques pour I. Si on a*

$$I(A_1, ., A_i, \ldots) = \varnothing (\text{ou } 0)\,,$$

alors il existe, pour tout i, un borélien $B_i \supseteq A_i$ *tel que*

$$I(B_1, ., B_i, \ldots) = \varnothing (\text{ou } 0)\,.$$

Démonstration Appliquer le théorème à I si I est une multicapacité et, si I est une opération capacitaire, à la multicapacité $J[I(\cdots)]$, où $J(X) = 0$ si $X = \varnothing$ et $= 1$ sinon.

Appliquons le corollaire à l'Exemple 6(b); on obtient le théorème de séparation de Novikov:

22 COROLLAIRE *Si* (A_i) *est une suite de parties analytiques de E et si* $\inf_i A_i = \varnothing$, *alors il existe, pour tout i, un borélien* $B_i \supseteq A_i$ *tel que* $\inf_i B_i = \varnothing$.

Et, corollaire du corollaire du corollaire, le théorème bien classique de Souslin:

23 COROLLAIRE *Soit A une partie analytique de E. Si* $A^c = E - A$ *est analytique, alors A est une partie borélienne.*

24 *Remarques* (a) Plus généralement, si (A_i) est une partition dénombrable de E en parties analytiques, alors chacun des A_i est borélien (appliquer 23 à A_j et $\sup_{i \neq j} A_i$).

(b) Disons qu'un espace métrisable séparable est *coanalytique* si, plongé dans un espace polonais, c'est le complémentaire d'une partie analytique (il résulte facilement de 17(2) et 16(3) que cela ne dépend pas du plongement choisi). Nous verrons au Chapitre III qu'il existe un couple de parties coanalytiques disjointes d'un espace métrisable compact, non séparables par des boréliens disjoints. Si C_1, C_2 est un tel couple, alors $C = C_1 \times C_2$, qui est encore coanalytique, n'est pas capacitable pour la capacité $J(X)$ sur l'espace ambiant $E \times E$ qui vaut 0 si les projections $\pi_1(X)$, $\pi_2(X)$ sont séparables par des boréliens, et 1 sinon (exercice!).

(c) On obtient une généralisation, due à Liapunov, du théorème de séparation de Novikov en appliquant 21 à l'Exemple 13(g), i.e. aux schémas de Souslin. Voir à ce sujet les commentaires.

25 Nous revenons maintenant au théorème de capacitabilité 19, dont nous allons préciser l'énoncé. Nous rappelons d'abord quelques notations introduites à l'occasion des Exemples 16(g) et (h), et en ajoutons quelques autres. Nous désignons par Σ l'espace polonais $\mathbb{N}^{\mathbb{N}}$, par S l'ensemble dénombrable des suites finies d'entiers; la notation $s \dashv t$, $s \dashv \sigma$ pour $s, t \in S$ et $\sigma \in \Sigma$ signifie que la suite t, σ commence par la suite s (pour t, on suppose de plus $s \neq t$). La topologie de Σ admet pour base d'ensembles ouverts (et fermés) les *ilots*, où l'ilot I_s d'indice $s \in S$ est l'ensemble $\{\sigma\colon s \dashv \sigma\}$. Un *schéma de Souslin* sur une classe $\mathscr{C}(E)$ de parties de E est une famille $\mathbf{X} = (X_s)_{s \in S}$ d'éléments de $\mathscr{C}(E)$ indexée par S; le schéma est dit *régulier* (ou monotone) si $s \dashv t \Rightarrow X_s \supseteq X_t$. Le *noyau* du schéma de Souslin $\mathbf{X}$ est l'ensemble

$$\mathbf{N}[\mathbf{X}] = \sup_{\sigma \in \Sigma} \inf_{s \dashv \sigma} X_s = \sup_{\sigma \in \Sigma} \inf_{n} X_{\sigma|n},$$

où l'on a noté $\sigma|n$ la suite finie des n premiers éléments de σ. Enfin au schéma de Souslin $\mathbf{X}$ nous associons l'application $\Psi_{\mathbf{X}}$ de Σ dans $\mathscr{P}(E)$ définie par

$$\Psi_{\mathbf{X}}(\sigma) = \inf_{n} X_{\sigma|n},$$

si bien que le noyau $\mathbf{N}[\mathbf{X}]$ est égal à $\sup_{\sigma \in \Sigma} \Psi_{\mathbf{X}}(\sigma)$. Si $\Psi_{\mathbf{X}}$ est à valeur dans $\mathscr{K}(E)$ et si, pour tout $U \supseteq \Psi_{\mathbf{X}}(\sigma)$, il existe $t \dashv \sigma$ tel que $\inf_{s \dashv t} X_s$ soit contenu dans U, alors l'application $\Psi_{\mathbf{X}}$ est s.c.s. (rappelons que cela signifie que, pour tout ouvert U contenant le compact $\Psi_{\mathbf{X}}(\sigma)$, il existe un ilot I_s contenant σ tel que $\sup_{\tau \in I_s} \Psi_{\mathbf{X}}(\tau)$ soit contenu dans U); c'est en particulier le cas si $\mathbf{X}$ est un schéma de Souslin sur $\mathscr{K}(E)$. Réciproquement, si Ψ est une application s.c.s. de Σ dans $\mathscr{K}(E)$ et si on pose $X_s = \sup_{\sigma \in I_s} \Psi(\sigma)$ alors $\mathbf{X} = (X_s)$ est un schéma de Souslin régulier tel que $\Psi = \Psi_{\mathbf{X}}$ et pour tout $U \supseteq \Psi(\sigma)$, il existe $t \dashv \sigma$ tel que $X_t \subseteq U$.

Nous commençons par tirer toute la substance de la démonstration du lemme 10.

26 LEMME *Soit I une opération capacitaire de $\prod E_i$ dans F et soit, pour tout i, (K_i^n) une suite croissante de compacts de E_i. Identifiant l'ensemble d'indices $\mathscr{I}$ à un segment initial de $\mathbb{N}$, posons, pour $s = (s_1, \ldots, s_i) \in S$ de longueur i,*

$$L_s = I(K_1^{s_1}, ., K_i^{s_i}, E_{i+1}, ., E_{i+k}, \ldots),$$

et pour $\sigma = (\sigma_1, \ldots) \in \Sigma$,

$$K_\sigma = I(K_1^{\sigma_1}, ., K_i^{\sigma_i}, K_{i+1}^{\sigma_{i+1}}, ., K_{i+k}^{\sigma_{i+k}}, \ldots).$$

Alors, pour $\sigma \in \Sigma$, $(L_{\sigma|n})$ est une suite décroissante de parties de F; on a

$$K_\sigma = \inf_n L_{\sigma|n},$$

et, pour tout ouvert U contenant K_σ, $L_{\sigma|n}$ est inclus dans U pour n suffisamment grand. Enfin, si on pose, pour tout i, $H_i = \lim \uparrow K_i^n$, on a

$$I(H_1, ., H_i, \ldots) = \sup_{\sigma\in\Sigma} K_\sigma = \sup_{\sigma\in\Sigma} \inf_n L_{\sigma|n}.$$

Démonstration Ne reculant devant aucun sacrifice dactylographique, nous recopions la démonstration du Lemme 10, en l'adaptant au cas d'une opération capacitaire. Il est clair d'abord que, pour σ fixé, K_σ est un compact de F et $(L_{\sigma|n})$ est une suite décroissante. De plus, il résulte immédiatement de la continuité à droite de I sur les compacts que, pour $U \supseteq K_\sigma$, on a $L_{\sigma|n} \subseteq U$ pour n suffisamment grand; on a donc $K_\sigma = \inf_n L_{\sigma|n}$ et $\sup_\sigma K_\sigma = \sup_\sigma \inf_n L_{\sigma|n}$. D'autre part, il est clair que $I(H_1, ., H_i, \ldots)$ contient $\sup_\sigma K_\sigma$; réciproquement, et on retrouve là un cas particulier du Lemme 10, pour $y \in I(H_1, ., H_i, \ldots)$ fixé, considérons la multicapacité I_y, à valeur dans $\{0, 1\}$. Etant donnée la montée séparée, on construit aisément, par récurrence, une suite infinie σ telle que, pour tout i, le point y appartienne à $I(K_1^{\sigma_1}, ., K_i^{\sigma_i}, H_{i+1}, ., H_{i+k}, \ldots)$: y appartient alors a fortiori à $L_{\sigma|n}$ pour tout n et donc à K_σ.

Nous déduisons d'abord de ce lemme que la notion d'ensemble analytique que nous avons définie coincide avec celle de Choquet, Sion, Frolík et Rogers (cf. Jayne (1976) ou les notes de Jayne et Rogers, Part 1, Section 2.8).

27 THEOREME *Une partie A de E est analytique ssi il existe une application s.c.s. Ψ de Σ dans $\mathscr{K}(E)$ telle que $A = \sup_{\sigma\in\Sigma} \Psi(\sigma)$. Nous dirons que Ψ*

représente A (noter que cette notion ne dépend pas en fait de l'espace ambiant E).

Demonstration La suffisance de la condition résulte de 13(h) et de 15, l'espace Σ étant analytique. Etant donnée la Definition 14, la nécessité résulte immédiatement du lemme précédent: avec les notations de ce dernier, il suffit de poser $\Psi(\sigma) = K_\sigma$.

Nous allons voir maintenant que l'on peut tirer beaucoup mieux du Lemme 26.

28 THÉORÈME *Il existe un choix $A \to \Psi_A$ associant à chaque ensemble analytique A une application Ψ_A le représentant de sorte que la propriété suivante soit satisfaite: pour toute opération capacitaire (ou multicapacité) I et pour toute famille $A_1, ., A_i, \ldots$ d'arguments analytiques pour I, on a*

$$I(A_1, ., A_i, \ldots) = \sup_{\sigma \in \Sigma} \cdots \sup_{\tau \in \Sigma} \cdots I(\Psi_{A_1}(\sigma), ., \Psi_{A_i}(\tau), \ldots) .$$

Autrement dit, puisque $A_i = \sup_\tau \Psi_{A_i}(\tau)$, les opérations I et $\sup_{\sigma \in \Sigma}$ commutent.

Démonstration Nous prenons pour Ψ_A l'application représentant A définie au cours de la démonstration précédente. Ainsi si, par définition, $A_i = J_i(H_1^i, ., H_j^i, \ldots)$ (cf. 14; on choisit ici une définition de A_i, avec une représentation explicite $H_j^i = \lim \uparrow K_j^{i,n}$), alors, par définition de Ψ_{A_i}, on a $A_i = \sup_\sigma \Psi_{A_i}(\sigma)$. Si, maintenant, on compose l'opération (ou la multicapacité) I de l'énoncé 28 avec les opérations J_i, on obtient l'opération capacitaire (ou la multicapacité) J; écrivant que $I(A_1, ., A_i, \ldots) = J(\ldots, H_j^i, \ldots)$, il ne reste plus qu'à appliquer le Lemme 26 à J et aux suites croissantes de compacts $(K_j^{i,n})_{n \in \mathbb{N}}$ pour obtenir la formule de l'énoncé (le Lemme 26 étant valable, *mutatis mutandis*, pour une multicapacité).

Nous passons maintenant aux schémas de Souslin, en nous limitant à quelques résultats que nous affinerons au fur et à mesure des besoins dans les chapitres ultérieurs.

29 THÉORÈME (1) *Une partie analytique A de E est noyau d'un schéma de Souslin sur les fermés de E.*

(2) *Réciproquement, si E est un espace analytique métrisable séparable, tout noyau A d'un schéma de Souslin sur les fermés de E est une partie analytique de E.*

Démonstration Tout compact étant égal à l'intersection de ses voisinages fermés, le point (1) résulte immédiatement de 27: si Ψ représente A, alors A

est noyau de $(\bar{X}_s)$ avec $X_s = \sup_{\sigma\in I_s} \Psi(\sigma)$. Pour le point (2), on voit aisément que, par plongement, on peut se ramener au cas où E est métrisable compact. On conclut alors par 27, tout schéma de Souslin sur les compacts engendrant une "bonne" application Ψ.

Nous terminons ce paragraphe en montrant, comme promis (cf. 13(g)), que l'opération de Souslin provient d'une opération capacitaire.

30 Rappelons d'abord la définition de l'opération J que nous avions associée à l'opération A de Souslin: nous avons ajouté un élément noté 0 à S et posé pour $X_0 \in \mathcal{P}(\Sigma)$ et $X_s \in \mathcal{P}(E)$, $s \in S$,

$$J(X_0, \ldots, X_s, \ldots) = \sup_{\sigma\in X_0} \inf_n X_{\sigma|n}\,,$$

si bien que, si $\mathbf{X} = (X_s)$, on a

$$\mathbf{N}[\mathbf{X}] = J(\Sigma, \ldots, X_s, \ldots)\,.$$

Nous allons ramener, bien classiquement, l'étude de J à celle d'un "schéma de projection", ce qui nous permettra d'écrire J comme composition d'opérations capacitaires élémentaires. Désignant par π la projection de $E \times \Sigma$ sur E, nous remarquons que l'on a

$$\inf_n X_{\sigma|n} = \inf_n \pi[(X_{\sigma|n} \times I_{\sigma|n})] = \pi\left[\inf_n (X_{\sigma|n} \times I_{\sigma|n})\right],$$

où I_s est l'ilot d'indice s. On a donc

$$J(X_0, \ldots, X_s, \ldots) = \pi\left[\sup_{\sigma\in X_0} \inf_n (X_{\sigma|n} \times I_{\sigma|n})\right].$$

Mais, deux ilots distincts sont ou emboités ou disjoints suivant que leurs indices sont "compatibles" ou non, si bien qu'en permutant "sup" et "inf" dans le crochet, on obtient

$$J(X_0, \ldots, X_s, \ldots) = \pi\left[\inf_n \sup_{s\in X_0^n} (X_s \times I_s)\right],$$

où $X_0^n = X_0|n$ désigne l'ensemble des $s \in S$ de longueur n telles que $I_s \cap X_0 \neq \varnothing$. On a ainsi presqu'écrit J comme composition d'opérations capacitaires élémentaires; le "presque" venant du fait que, les ilots n'étant pas compacts, le produit $X_s \times I_s$ s'interprète mal. Nous remédions à cela en remplaçant l'ilot I_s de Σ par l'ilot correspondant I_s^+ de $\Sigma^+ = \bar{\mathbb{N}}^{\mathbb{N}}$: I_s^+ est l'ensemble des suites infinies d'éléments de $\bar{\mathbb{N}} = \mathbb{N} \cup \{\infty\}$ commençant par $s \in S$. Ainsi, π désignant maintenant la projection de $E \times \Sigma^+$ sur E, on a encore

$$J(X_0, \ldots, X_s, \ldots) = \pi\left[\inf_n \sup_{s \in X_0^n} (X_s \times I_s^+)\right].$$

Nous laissons alors au lecteur le soin patient de décortiquer le second membre en opérations capacitaires élémentaires.

3 Espaces de Blackwell. Théorème de Souslin et Lusin

Nous donnons ici quelques applications importantes du théorème de séparation 22 à la théorie abstraite des espaces mesurables. C'est aussi, pour nous, une première occasion de montrer comment ramener une situation "abstraite" à une situation "topologique" très classique. D'abord, un peu de terminologie.

31 Un *espace mesurable* est un couple $(\Omega, \mathscr{F})$ où Ω est un ensemble et $\mathscr{F}$ une tribu de parties de Ω. L'espace mesurable $(\Omega, \mathscr{F})$ est dit *séparé* si ses *atomes* (i.e. les classes d'équivalence pour la relation d'équivalence $\forall H \in \mathscr{F}$ $\omega \in H \Leftrightarrow w \in H$) sont réduits aux singletons; il est dit *séparable* (ce qui n'a rien à voir avec "séparé"!) si il existe une suite (G_n) d'éléments de $\mathscr{F}$ engendrant la tribu $\mathscr{F}$.

Soit $(\Omega, \mathscr{F})$ un espace mesurable séparable, mais non nécessairement séparé, et soit (G_n) une suite de générateurs de $\mathscr{F}$. La relation d'équivalence R définissant les atomes étant équivalente à la relation $\forall n\ \omega \in G_n \Leftrightarrow w \in G_n$, les atomes de $(\Omega, \mathscr{F})$ appartiennent alors à la tribu $\mathscr{F}$. Désignons par $\bar{\Omega}$ le quotient Ω/R, que nous munissons de la tribu $\bar{\mathscr{F}}$ constituée par les $\bar{H} = H/R$ quand H parcourt $\mathscr{F}$ (il est clair que les éléments de $\mathscr{F}$ sont saturés pour R): le lecteur se convaincra sans peine que, du point de vue de la structure mesurable, il n'y a pratiquement aucune différence entre $(\Omega, \mathscr{F})$ et son *séparé* $(\bar{\Omega}, \bar{\mathscr{F}})$ (qui est encore séparable). En particulier, nous confondrons toute v.a. sur Ω (i.e. toute fonction $\mathscr{F}$-mesurable) avec la v.a. qu'elle induit sur $\bar{\Omega}$.

Le théorème suivant, dû à Marczewski (du temps où il s'appelait encore Szpilrajn), mériterait d'être mieux connu: il montre que, du point de vue de la structure mesurable, il n'y a pas de différence entre un espace mesurable séparable et un espace métrisable séparable muni de sa tribu borélienne.

32 THÉORÈME *Soit $(\Omega, \mathscr{F})$ un espace mesurable séparable, et soit (G_n) une suite de générateurs de $\mathscr{F}$. Posons, pour tout $\omega \in \Omega$,*

$$\Phi(\omega) = \sum_n 2 \times 3^{-n} 1_{G_n}(\omega).$$

La fonction Φ ainsi définie de Ω dans l'ensemble triadique de Cantor, que nous appellerons l'indicatrice de Marczewski de (G_n), *a les propriétés suivantes: si on pose $X = \Phi(\Omega)$, on a*

(1) *pour tout $B \in \mathscr{B}(X)$, $\Phi^{-1}(B)$ appartient à $\mathscr{F}$;*
(2) *pour tout $x \in X$, $\Phi^{-1}(\{x\})$ est un atome de $(\Omega, \mathscr{F})$;*
(3) *pour tout $H \in F$, $\Phi(H)$ appartient à $\mathscr{B}(X)$.*

Ainsi Φ est un isomorphisme de $(\bar{\Omega}, \bar{\mathscr{F}})$ sur $(X, \mathscr{B}(X))$.

Démonstration Il est clair que Φ est une bijection mesurable de $\bar{\Omega}$ sur X. Pour démontrer (3), qui est le seul point qui n'est pas tout à fait évident, il suffit de prouver que le tribu engendrée par Φ, qui est une sous-tribu de $\bar{\mathscr{F}}$, est en fait égale à $\bar{\mathscr{F}}$. Et cela résulte immédiatement du fait que, pour tout n, on a $G_n = \Phi^{-1}(B_n)$ où B_n est l'ensemble des x dont le n-ième terme du développement en base 3 est égal à 2.

Voici maintenant une notion d'espace mesurable très utile en probabilités:

33 DÉFINITION Un espace mesurable séparable $(\Omega, \mathscr{F})$ est appelé *espace de Blackwell* si, pour toute v.a. f sur Ω, l'ensemble $f(\Omega)$ est une partie analytique de $\mathbb{R}$.

En fait, un espace mesurable est de Blackwell ssi, quitte à considérer son séparé, il est isomorphe à une partie analytique de $[0, 1]$ munie de sa tribu borélienne. Plus précisément,

34 THÉORÈME (1) *Si $(\Omega, \mathscr{F})$ est un espace de Blackwell, alors $(\bar{\Omega}, \bar{\mathscr{F}})$ est isomorphe à $(X, \mathscr{B}(X))$, où X est une partie analytique de* $[0, 1]$.

(2) *Réciproquement, si X est un espace analytique métrisable séparable, alors $(X, \mathscr{B}(X))$ est un espace de Blackwell.*

Démonstration Le point (1) résulte immédiatement de 32 et le point (2) de 16(3), la tribu $\mathscr{B}(X)$ étant séparable.

Remarque Disons que E est un espace *souslinien* s'il est image continue d'un espace analytique métrisable séparable X (nous verrons plus loin que X et donc E est image continue de $\mathbb{N}^{\mathbb{N}}$ si bien que E est souslinien au sens de Bourbaki). Un tel espace est analytique, séparable, mais non nécessairement métrisable. Nous verrons cependant au no. 36 que $\mathscr{B}(E)$ est séparable, et $(E, \mathscr{B}(E))$ de Blackwell.

La conséquence très simple suivante du théorème de séparation fournit un outil précieux aux probabilistes

35 THÉORÈME *Soit* $(\Omega, \mathscr{F})$ *un espace de Blackwell et soit* $\mathscr{G}$ *une soustribu séparable de* $\mathscr{F}$. *Pour qu'une v.a.* f *soit* $\mathscr{G}$*-mesurable, il faut et il suffit que* f *soit constante sur chaque atome de* $(\Omega, \mathscr{G})$.

Démonstration La nécessité est triviale. Pour démontrer la suffisance, il suffit de montrer que la tribu engendrée par f est contenue dans $\mathscr{G}$: on se ramène donc au cas où f est l'indicatrice d'un élément H de $\mathscr{F}$. Désignons par Φ l'indicatrice de Marczewski d'une suite de générateurs de $\mathscr{G}$. Comme H est, par hypothèse, saturé pour la relation d'équivalence définissant les atomes de $\mathscr{G}$, les ensemble $\Phi(H)$ et $\Phi(\Omega - H)$ sont des parties disjointes de $[0, 1]$. Et ce sont des parties analytiques d'après 34(1) et 18; on peut donc les séparer par des boréliens disjoints B et B' de $[0, 1]$ (cf. 22, pour deux ensembles analytiques). Mais alors, on a nécessairement $H = \Phi^{-1}(B)$, $\Omega - H = \Phi^{-1}(B')$, et donc H appartient à $\mathscr{G}$.

36 COROLLAIRE *Soient* $(\Omega, \mathscr{F})$ *et* $(W, \mathscr{G})$ *deux espaces mesurables, séparables et séparés. Si* $(\Omega, \mathscr{F})$ *est un espace de Blackwell et si* f *est une bijection mesurable de* Ω *sur* W, *alors* f *est un isomorphisme de* $(\Omega, \mathscr{F})$ *sur* $(W, \mathscr{G})$.

Démonstration La tribu $f^{-1}(\mathscr{G})$ est une sous-tribu séparable de $\mathscr{F}$, ayant les mêmes atomes que $\mathscr{F}$ (en fait, les points de Ω). Il résulte alors du théorème appliqué aux indicatrices des éléments de $\mathscr{F}$ que $\mathscr{F} = f^{-1}(\mathscr{G})$; d'où f est un isomorphisme.

Remarque Soit E un espace souslinien, image d'un espace analytique métrisable séparable X par une fonction continue f. Comme $(X, \mathscr{B}(X))$ est un espace de Blackwell, il est clair que, pour toute sous-tribu séparable $\mathscr{E}$ de $\mathscr{B}(E)$, $(E, \mathscr{E})$ est aussi de Blackwell. Et, d'après le corollaire, il nous suffit, pour démontrer que $(E, \mathscr{B}(E))$ est de Blackwell, de trouver une sous-tribu séparable $\mathscr{E}$ dont les atomes sont les singletons (appliquer 36 à la tribu séparable engendrée par $\mathscr{E}$ et $H \in \mathscr{B}(E)$ et à la tribu $\mathscr{E}$). Or, soit (U_n) une base dénombrable de la topologie de X: E étant séparé, il est facile de voir que la tribu $\mathscr{E}$ engendrée par les $f(U_n)$ convient.

Nous démontrons maintenant, à l'aide de 36 et du lemme 17 une version abstraite du célèbre théorème sur les images injectives de boréliens (théorème dû à Souslin et Lusin, et attribué parfois, érronément, à Kuratowski)

37 THÉORÈME *Soient* $(\Omega, \mathscr{F})$ *et* $(W, \mathscr{G})$ *deux espaces mesurables, séparables et séparés. Si* $(\Omega, \mathscr{F})$ *est isomorphe à* $(X, \mathscr{B}(X))$, *où* X *est un* borélien *d'un espace polonais* P, *et si* f *est une application mesurable* injective *de* Ω *dans* W, *alors on a* $f(H) \in \mathscr{G}$ *pour tout* $H \in \mathscr{F}$.

Démonstration Notons d'abord que $(\Omega, \mathscr{F})$ est un espace de Blackwell d'un type particulier; par ailleurs, quitte à plonger P dans $[0, 1]^{\mathbb{N}}$ (P y est alors un $\mathscr{G}_\delta$ et donc un borélien), on peut supposer que $P = [0, 1]^{\mathbb{N}}$. D'autre part, grâce à l'indicatrice de Marczewski, on peut supposer que W est une partie de $[0, 1]$ et que $\mathscr{G} = \mathscr{B}(W)$. Posons $Y = f(\Omega)$: comme $(\Omega, \mathscr{F})$ est de Blackwell, Y est une partie analytique de $[0, 1]$ et, d'après 36, f est un isomorphisme de $(\Omega, \mathscr{F})$ sur $(Y, \mathscr{B}(Y))$. Comme Y est une partie de W, il nous suffit de montrer, pour conclure, que Y est un borélien de $[0, 1]$. Or, d'après le Lemme 17, il existe une fonction borélienne f' de $[0, 1]^{\mathbb{N}}$ dans $[0, 1]$ telle que $f = f'|_\Omega$ et, g désignant l'application inverse de f allant donc de Y dans $[0, 1]^{\mathbb{N}}$, il existe une application borélienne g' de $[0, 1]$ dans $[0, 1]^{\mathbb{N}}$ telle que $g = g'|_Y$ (appliquer 17 aux composantes de g). Posons $\Omega' = \{x \in [0, 1]^{\mathbb{N}}: g'(f'(x)) = x\}$ et d'autre part $Y' = \{y \in [0, 1]: f'(g(y)) = y\}$: Ω' est un borélien de $[0, 1]^{\mathbb{N}}$ contenant Ω, Y' un borélien de $[0, 1]$ contenant Y. Maintenant, $(\Omega', \mathscr{B}(\Omega'))$ et $(Y', \mathscr{B}(Y'))$ sont isomorphes, $f'|_{\Omega'}$ et $g'|_{Y'}$ formant un couple d'isomorphismes réciproques. Comme Ω est un borélien de $[0, 1]^{\mathbb{N}}$ et donc de Ω', on en déduit que $Y = f'(\Omega)$ est un borélien de Y' et donc de $[0, 1]$.

Remarque Disons que E est un espace *lusinien* s'il est image continue et injective d'un borélien X d'un espace polonais P par une application f (on montre ci-dessous que X et donc E est image continue et injective d'un fermé de $\mathbb{N}^{\mathbb{N}}$ si bien que E est lusinien au sens de Bourbaki). D'après la remarque de 36, $(E, \mathscr{B}(E))$ est de Blackwell et donc, d'après 36, f est un isomorphisme de $(X, \mathscr{B}(X))$ sur $(E, \mathscr{B}(E))$. Ceci dit, soit g une application continue et injective de E dans F: $\mathscr{B}(F)$ n'étant pas nécessairement séparable, on ne peut conclure directement du théorème que $g(E)$ est un borélien de F. On pourra cependant conclure qu'il en est ainsi si l'on peut trouver une sous-tribu séparable $\mathscr{G}$ de $\mathscr{B}(F)$ telle que tout singleton de $g(E)$ soit un atome de $\mathscr{G}$ (appliquer le théorème à $(\bar{F}, \bar{\mathscr{G}})$). Or si, comme plus haut, on désigne par (U_n) une base dénombrable de la topologie de X, il est facile de voir que la tribu $\mathscr{G}$ engendrée par les $\overline{g(f(U_n))}$ convient.

Le théorème suivant, bien classique, est élémentaire quoiqu'un peu longuet à démontrer

38 THÉORÈME *Tout borélien d'un espace polonais P est image continue et injective d'un fermé de* $\Sigma = \mathbb{N}^{\mathbb{N}}$.

Démonstration D'abord, quitte à plonger P dans $[0, \frac{1}{2}]^{\mathbb{N}}$, on se ramène au cas où $P = [0, 1[^{\mathbb{N}}$. Appelons alors $\mathscr{C}$ la classe des parties de P de la forme $\prod_n [a_n, b_n[$ où $[a_n, b_n[= [0, 1[$ sauf pour un nombre fini de coordonnées: le complémentaire d'un élément de $\mathscr{C}$ étant une réunion finie d'éléments disjoints de $\mathscr{C}$, on vérifie aisément que $\mathscr{B}(P)$ est le stabilisé de $\mathscr{C}$ pour les intersections dénombrables et les réunions dénombrables "disjointes". Par ailleurs, tout élément de $\mathscr{C}$ a la propriété **P** voulue, i.e. est image continue et injective d'un fermé de Σ. En effet, $\Sigma^{\mathbb{N}}$ étant homéomorphe à Σ (écrire que $(\mathbb{N}^{\mathbb{N}})^{\mathbb{N}} = \mathbb{N}^{(\mathbb{N}\times\mathbb{N})}$), il suffit de montrer qu'un intervalle $[a_n, b_n[$, et donc simplement $[0, 1[$, est image de Σ par une bijection continue f; pour $[0, 1[$, il suffit de poser

$$f(\sigma) = 1 - \sum_n 2^{-\|\sigma|n\|}$$

où $\|s\|$ désigne la somme des termes de $s \in S$. Ceci fait, il ne reste plus qu'à montrer que l'ensemble des parties de $P = [0, 1[^{\mathbb{N}}$ ayant la propriété **P** est stable pour les intersections dénombrables et les réunions dénombrables "disjointes". Soient donc (X_n) une suite de parties de P et, pour chaque n, F_n un fermé de Σ et f_n une bijection continue de F_n sur X_n. On a alors

(a) pour les intersections dénombrables: on identifie d'abord P à la diagonale de $P^{\mathbb{N}}$ si bien que $\inf_n X_n$ est identifié à la trace de $\prod_n X_n$ sur cette diagonale. Posant $f = \prod f_n$, on remarque que f est une bijection continue de $\prod F_n$ sur $\prod X_n$ et que $\inf X_n$ est alors identifié à l'image par f du fermé $f^{-1}(P)$. On conclut en identifiant $f^{-1}(P)$ à un fermé de Σ, $\Sigma^{\mathbb{N}}$ étant homéomorphe à Σ.

(b) pour les réunions dénombrables "disjointes": les X_n sont cette fois disjoints. Tout ilot de Σ étant homéomorphe à Σ, on peut suppose F_n contenu dans l'ilot I_n (n étant ici la suite s n'ayant qu'un élément, à savoir n). Alors $F = \sup_n F_n$ est fermé dans Σ et $\sup_n X_n$ est image de F par l'application continue f obtenue en recollant les f_n.

Comme un fermé de Σ est polonais pour la topologie induite, on a, en combinant ce théorème avec 36 et la remarque de 37

39 THÉORÈME *Soit E un espace lusinien. Il existe sur E une topologie polonaise plus fine, et toute topologie polonaise plus fine a pour tribu borélienne celle de la topologie initiale.*

Le théorème suivant, qui n'est pas assez connu, rend parfois de grands services: il montre que, fréquemment, un problème portant sur une suite d'applications boréliennes peut se ramener à un problème portant sur une suite d'applications continues.

40 THÉORÈME *Soit E un espace lusinien et soient, pour tout n, d'une part f_n une application borélienne de E dans un espace à base dénombrable F_n et, d'autre part, g_n une application borélienne de E dans E. Il existe alors une topologie polonaise sur E, plus fine que l'initiale, telle que les f_n et g_n soient continues.*

Démonstration D'abord, d'après le théorème précédent, on peut supposer E polonais. Nous allons d'abord renforcer la topologie de E en une nouvelle topologie polonaise de sorte que les f_n deviennent continues: comme cela ne changera pas la tribu borélienne de E, les g_n ne perdent rien pour attendre. Nous désignons par (U_m) une base dénombrable de la topologie de E et, pour tout n, par (U_m^n) une telle base pour F_n. Nous écrivons en une seule suite (V_k) tous les ensembles U_m et ceux de la forme $f_n^{-1}(U_m^n)$ (qui sont boréliens). Puis, nous définissons une application ouverte, borélienne et injective $h=(h_k)$ de E dans $\{0, 1\}^{\mathbb{N}}$ et posant $h_k(x)= 1_{V_k}(x)$ pour $x \in E$. D'après 37, $h(E)$ est un borélien de l'espace métrisable compact $\{0, 1\}^{\mathbb{N}}$ et il est clair que, identifiant l'ensemble E à $h(E)$, la topologie de $h(E)$ est plus fine que celle de E et rend continues les applications f_n. Nous renforçons encore la topologie de $E=h(E)$ à l'aide de 39 pour avoir une topologie polonaise et nous considérons, pour la suite, cette dernière comme étant la topologie initiale de E. Nous passons maintenant aux g_n: c'est un peu plus délicat car on modifie à la fois la topologie de l'espace de départ et celle de l'espace d'arrivée des applications. Nous posons $E_0=E$, les indices allant noter des topologies polonaises de plus en plus fines sur E. Si E_m est défini, nous désignons par E_{m+1} l'ensemble E muni d'une topologie polonaise plus fine que celle de E_m de sorte que les applications g_n de E_{m+1} dans E_m soient continues (ce qui est possible d'après ce qui précède). Enfin, nous désignons par E_∞ l'ensemble E muni de la topologie polonaise qui le rend homéomorphe à la diagonale de $\prod_m E_m$: il est alors clair que les g_n sont des applications continues de E_∞ dans E_∞.

Remarques (a) Il n'est pas possible, en général, de rendre continue une application borélienne de $E \times E$ dans E tout en conservant la structure produit de $E \times E$ (exercice!).

(b) Harrington a découvert (en passant par le forcing, mais il existe une démonstration élémentaire cf Louveau (1976/77) que, si (A_n) est une suite de parties analytiques de E, on peut renforcer la topologie de E en une topologie à base dénombrable telle que E soit un espace de Baire et que chacun des A_n soit ouvert. Cela fournit un nouvel outil puissant (voir en particulier les travaux récents de Louveau (1980)).

(c) Si, dans l'énoncé de 40, on suppose seulement que E est souslinien métrisable, il existe alors une topologie souslinienne métrisable plus fine sur E telle que les f_n et g_n soient continues. La démonstration est analogue.

II Quelques théorèmes de section et de sélection

Il s'agit en fait d'un chapitre d'applications du chapitrc I: on y utilise essentiellement la notion d'espace analytique métrisable séparable et le théorème de séparation I-22 (soit, plus précisément, ce qu'on appelle le premier théorème de séparation, car nous en verrons un second, au chapitre III, qui fait appel à des techniques très différentes de celles du Chapitre I).

Logiquement, les paragraphes de ce chapitre auraient donc pu se trouver dans le Chapitre I, à la suite des autres. Cependant, comme nous allons changer quelque peu nos conventions et notre terminologie, nous avons préféré marquer une coupure afin d'éviter tout risque de confusion. *Désormais*—cela vaut pour les chapitres ultérieurs—, *tous les espaces topologiques considérés seront métrisables séparables.* A ceci près, nos conventions, quant à l'emploi des lettres, restent celles du chapitre I; E et F sont des espaces métrisables séparables (variables), A un ensemble analytique (voir cependant l'alinéa suivant), B un borélien, K un compact, U un ouvert, etc. La lettre $\mathscr{F}$ nous servant à noter une tribu, nous nous sommes résolu à écrire $\mathbf{F}(E)$ la classe de fermés de E, qui interviendra souvent dans ce chapitre, et donc $\mathbf{G}(E)$ la classe des ouverts: d'où les notations, par ailleurs classiques, $\mathbf{F}_\sigma$, $\mathbf{G}_\delta$, etc.

1 Soit E plongé dans F analytique; nous dirons qu'une partie H de E est *E-analytique* s'il existe $A' \in \mathscr{A}(F)$ tel que $H = A' \cap E$, et nous nous permettrons de noter A une partie E-analytique de E. Etant donné I-29(2), il est clair que A est E-analytique ssi A est noyau d'un schéma de Souslin sur $\mathbf{F}(E)$ (ou encore sur $\mathscr{B}(E)$ d'après I-16(1) et I-18): la notion de partie E-analytique de E ne dépend donc pas du plongement dans F; elle est cependant relative à E (sauf si E est lui-même un espace analytique). Nous dirons qu'une partie C de E est *E-coanalytique* si $C^c = E - C$ est *E-analytique* (si E est un espace polonais, cette notion est absolue—cf. I-17(2), I-24(b)—et nous omettrons alors le préfixe "E"); nous dirons que C est *E-bianalytique* si C et $E - C$ cont E-analytiques (si E est un espace analytique, C est alors borélien d'après I-23). Enfin, nous noterons $\mathbf{A}(E)$, $\mathbf{CA}(E)$, $\mathbf{BA}(E)$ la classe des parties respectivement E-analytiques, E-coanalytiques et E-bianalytiques: $\mathbf{A}(E)$ et $\mathbf{CA}(E)$ sont stables pour les réunions et intersections dénombrables d'après I-16(1); donc $\mathbf{BA}(E)$ est une tribu, qui contient $\mathscr{B}(E)$ d'après I-18 (et lui est égale si E est un espace analytique).

Remarque Nous aurions pu, comme de nombreux auteurs, réserver le qualificatif "analytique" à la notion absolue, et attribuer le qualificatif "soulinien" à la notion relative. Mais, comme d'autres auteurs (par exemple Bourbaki) emploient le qualificatif "souslinien" dans un sens absolu (cf. I-34), nous nous sommes résolu à utiliser la terminologie exposée ci-dessus, proche de celle que nous avions utilisée ailleurs.

2 Venons en aux problèmes abordés dans ce chapitre. D'abord, nous nous poserons des problèmes de *section* (ou, comme l'on dit aussi, plus classiquement d'ailleurs, d'*uniformisation*): étant donnés un espace produit $E \times F$ muni des projections π_E, π_F, et une partie H de $E \times F$, choisir, pour tout $y \in \pi_F(H)$, un point $h(y)$ de la *coupe* $H(y) = \pi_E[H \cap (E \times \{y\})]$ de sorte que l'application $y \to h(y)$, dont le graphe fournit une *section* de H, soit une "bonne" application. Le problème de *sélection* est un problème voisin: on se donne une relation d'équivalence R sur E et il s'agit de choisir, pour chaque $x \in E$, un représentant $r(x)$ de la classe d'équivalence $[x]$ de sorte que l'on ait $xRy \Rightarrow r(x) = r(y)$ et que l'application $x \to r(x)$ soit une "bonne" *sélection*.

Chemin faisant, nous étudierons la structure des boréliens à coupes fermées ou dénombrables.

1 Le théorème de section de Jankov et von Neumann

Nous allons démontrer ici le théorème suivant (que nous metrons en forme un peu plus loin): si $H(\omega)$ est une partie analytique d'une espace analytique E, dépendant mesurablement du paramètre ω (lequel appartient à un espace mesurable $(\Omega, \mathscr{F})$), alors il existe une application universellement mesurable h de Ω dans E telle que l'on ait $h(\omega) \in H(\omega)$ chaque fois que $H(\omega)$ est non vide.

La méthode standard pour choisir un point dans un ensemble X est la suivante: mettre un "bon" ordre sur X (le mot "bon" n'ayant pas ici son sens technique; on suppose cependant que X a un plus petit élément pour cet ordre), et choisir le plus petit élément de X pour cet ordre. Maintenant, si X dépend d'un paramètre ω et si l'ordre "construit" dépend "convenablement" de ω (ce qui exclut, pour X non dénombrable, que l'ordre soit un bon ordre au sens technique du terme; voir à ce sujet le Chapitre IV), on peut espérer obtenir un "bon" choix en fonction du paramètre ω.

Nous allons, en premier lieu, établir un théorème très simple et très important: il existe un "bon" choix d'un point dans un fermé (non vide) de

$\Sigma = \mathbb{N}^{\mathbb{N}}$. Ayant muni $\mathbb{N}^{\mathbb{N}}$ de l'ordre lexicographique (nous l'expliciterons plus loin), nous choisirons tout simplement, dans un fermé non vide, le plus petit point pour cet ordre. Mais, pour donner tout son sens au mot "bon", nous allons avant tout mettre une topologie polonaise naturelle sur $\mathbf{F}(\Sigma)$. Que le lecteur qui connait bien le traité de topologie de Kuratowski prenne garde: ce ne sera ni la topologie exponentielle ni la topologie induite par la distance "naturelle" de Hausdorff, topologies qui ne sont pas polonaises (la première n'est pas métrisable, la seconde n'est pas séparable).

3 Nous munirons $\mathbf{F}(\Sigma)$ de la topologie la moins fine telle que les ensembles de la forme

$$\{X : X \cap I_s = \varnothing\}, \qquad \{X : X \cap I_s \neq \varnothing\},$$

où X parcourt $\mathbf{F}(\Sigma)$ et I_s les ilots (ilot Σ, d'indice vide, compris), soient ouverts (et donc aussi fermés). Une base dénombrable pour cette topologie, formée d'ensembles à la fois ouverts et fermés, est consituée, par $\{\varnothing\}$ (i.e. le fermé vide est un point isolé) et par les ensembles de la forme

$$\{X : X \cap V = \varnothing\} \cap \{X : X \cap U_1 \neq \varnothing\} \cap \cdots \cap \{X : X \cap U_n \neq \varnothing\},$$

où chacun des $n+1$ ensembles $V, U_1, \ldots, U_n$ est une réunion finie d'ilots. Nous laissons au lecteur le soin de vérifier que la topologie de $\mathbf{F}(\Sigma)$ est définie par la distance d suivante: pour X_1 et X_2 fermés de Σ, on a $d(X_1, X_2) = 1/n$, où n est le plus petit entier k pour lequel existe une suite s, dont la somme des termes notée $\|s\|$ est égale à k, telle que l'ilot I_S différencie X_1 et X_2, i.e. vérifiie la condition

$$(X_1 \cap I_s = \varnothing \text{ et } X_2 \cap I_2 \neq \varnothing) \quad \text{ou} \quad (X_1 \cap I_s \neq \varnothing \text{ et } X_2 \cap I_s = \varnothing).$$

L'espace $\mathbf{T}(\Sigma)$ étant à base dénombrable, le lemme suivant établit que *sa topologie est polonaise*

LEMME $\mathbf{F}(\Sigma)$ *est complete pour la distance d.*

Démonstration Soit (X_n) une suite de Cauchy pour d. Pour $s \in S$ fixé soit N un entier assez grand de sorte que l'on ait $d(X_p, X_q) < 1/\|s\|$ pour $p, q \geqslant N$: on a alors, pour $n \geqslant N$, $I_s \cap X_n = \varnothing$ ou $\neq \varnothing$ suivant que $I_s \cap X_N = \varnothing$ ou $\neq \varnothing$. Désignons, pour n entier, par S_n l'ensemble fini $\{s \in S : \|s\| = n$ et $\forall m \geqslant n\ I_s \cap X_m \neq \varnothing\}$: il est facile de voir que (X_n) converge vers le fermé

$$X = \inf_n \sup_{s \in S_n} I_s.$$

4 Nous munissons maintenant Σ de l'*ordre lexicographique*, noté $\leqslant$: pour $\sigma, \tau \in \Sigma$, on a $\sigma < \tau$ ssi il existe un entier n tel que l'on ait $\sigma|n-1 = \tau|n-1$ et

$\sigma(n) < \tau(n)$ (où $\sigma(n)$ désigne le n-ième terme de σ). Il est clair que $\leqslant$ est un ordre total et que $\{(\sigma, \tau) : \sigma < \tau\}$ est ouvert dans $\Sigma \times \Sigma$ (et donc $\leqslant$ est une relation d'ordre fermée). Pour toute suite τ, nous désignerons par J_τ l'ensemble $\{\sigma : \sigma < \tau\}$, qui est un ouvert de Σ. Notons le fait important que *tout ilot I_s est de la forme $J_\tau - J_\sigma$*: en effet, si $s = s(1), \ldots, s(n-1), s(n)$, il suffit de prendre $\tau = s(1), \ldots, s(n-1), s(n)+1, 1, .., 1, \ldots$ et d'autre part $\sigma = s(1), .., s(n-1), s(n), 1, .., 1, \ldots$. En particulier, les J_τ engendrent la tribu borélienne $\mathscr{B}(\Sigma)$. Nous dirons qu'une application f de E dans Σ *est s.c.s.* si $f^{-1}(J_\tau)$ est ouvert pour toute suite τ; une telle application est alors borélienne, et même, vu l'expression des ilots ci-dessus, de lère classe de Baire (i.e. l'image réciproque d'un ouvert est un $\mathbf{F}_\sigma$).

Nous pouvons enfin formuler le "bon choix" tant attendu

5 THÉORÈME (1) *Tout fermé non vide X de Σ a un plus petit élément $c(X)$ pour l'ordre lexicographique.*

(2) *L'application $X \to c(X)$ de $\mathbf{F}(\Sigma) - \{\varnothing\}$ dans Σ est s.c.s.*

Démonstration Le point (1) est à peu près évident. Le point (2) résulte aisément de l'équivalence logique

$$c(X) < \tau \Leftrightarrow \exists n \ \exists \sigma \ \sigma < \tau \quad \text{et} \quad I_{\sigma|n} \cap X \neq \varnothing .$$

Le théorème suivant, classique (le point (2) sert classiquement à définir les ensembles analytiques), permet souvent de ramener un problème sur un espace polonais ou analytique à un problème sur Σ.

6 THÉORÈME (1) *Tout espace polonais P est image continue et ouverte de $\Sigma = \mathbb{N}^{\mathbb{N}}$.*

(2) *Tout espace analytique E est image continue de Σ.*

Démonstration Soit d une distance sur P, définissant sa topologie et telle que P soit complet pour d; nous noterons $d(X)$ le diamètre de $X \in \mathscr{P}(E)$ pour cette distance d. Désignant par S_n l'ensemble des $s \in S$ de longueur n, nous définissons, par récurrence sur la longueur, un schéma de Souslin $(U_s)_{s \in S}$ sur $\mathbf{G}(P)$ de la manière suivante: $(U_s)_{s \in S}$ est une base dénombrable de la topologie de P; si U_s est défini pour $s \in S_n$, nous prenons pour (U_t), t parcourant les suites de S_{n+1} telles que $s \dashv t$, une base dénombrable de la topologie (induite) de U_s telle que l'adhérence $\bar{U}_t$ de U_t dans P soit contenue dans U_s et que l'on ait $d(U_t) < 1/n$. Ceci fait, nous définissons une application f de Σ dans P en prenant pour $f(\sigma)$ l'unique élément du singleton $\inf_n U_{\sigma|n}$: comme on a $f(I_s) = U_s$, il est clair que f est une application continue et ouverte de P sur Σ. Nous passons maintenant au point (2). Nous plongeons d'abord E dans un espace compact F; puis nous désignons par Ψ une application s.c.s. de Σ dans $\mathscr{K}(F)$ représentant E (cf. I-27). Le grand tort de Ψ étant de ne pas être forcément à valeurs dans les singletons, nous modifions Ψ comme suit: f étant une application continue de Σ sur F (point (1)), nous posons pour toute suite σ,

$$\Phi(\sigma) = \Psi(\sigma_1) \cap f(\{\sigma_2\}),$$

où $\sigma \to (\sigma_1, \sigma_2)$ est un homéomorphisme de Σ sur $\Sigma \times \Sigma$ (prendre, par exemple, pour σ_1 (resp. σ_2) la suite des termes de rang impair (resp. pair) de σ). L'application Φ ainsi définie, qui est à valeurs dans les compacts comportant au plus un point, est encore s.c.s. et représente encore E. Nous définissons finalement une application continue g de Σ sur E en choisissant un point arbitraire $x_0 \in E$ et en posant $g(\sigma) = x_0$ si $\Phi(\sigma)$ est vide et $g(\sigma) =$ le seul point de $\Phi(\sigma)$ sinon. Le lecteur qui n'aime pas l'arbitraire pourra choisir x_0 comme suit: prendre pour x_0 l'unique point de $\Phi(\sigma_0)$ où σ_0 est la plus petite suite σ de Σ (pour l'ordre lexicographique) telle que $\Phi(\sigma)$ ne soit pas vide.

Remarque La réciproque du point (1) est vraie: toute image continue et ouverte d'un espace polonais est polonaise.

Voici maintenant, d'abord sous sa forme topologique, le théorème de section de Jankov et Von Neumann:

7 THÉORÈME *Soit H une partie $(E \times F)$-analytique d'un produit $E \times F$ où E est un espace analytique. Il existe une application f de F dans E, mesurable lorsque E est muni de la tribu $\mathscr{B}(E)$ et F de la tribu engendrée par $\mathbf{A}(F)$, et telle que l'on ait $f(y) \in H(y)$ pour tout $y \in F$ tel que $H(y)$ soit non vide.*

Démonstration Nous traitons d'abord le cas où F est lui-même un espace analytique, si bien que H est aussi un espace analytique. D'après le théorème précédent, il existe alors une application continue g de Σ dans $E \times F$ telle que $H = g(\Sigma)$; nous écrivons g sous la forme d'un couple (g_E, g_F) d'applications, en composant g avec les projections π_E, π_F. Désignons par X l'application qui à $y \in F$ associe le fermé $X(y) = \overset{-1}{g_F}(\{y\})$ de Σ si bien que $g_E[X(y)] = H(y)$, coupe de H selon y, puis par $\hat{c}$ la fonction de choix définie en 5 prolongée à l'espace $\mathbf{F}(\Sigma)$ tout entier en posant $\hat{c}(\emptyset) = \sigma_0$ où σ_0 est un point arbitrairement choisi dans Σ. Nous définissons alors une application f de F dans E en posant, pour tout $y \in F$,

$$f(y) = g_E[\hat{c}(X(y))],$$

il est clair que l'on a $f(y) \in H(y)$ si $H(y)$ est non vide. En ce qui concerne la measurabilité de f, on pourrait la déduire directement de l'égalité

$X^{-1}[c^{-1}(J_\tau)] = g_E(J_\tau)$ (cf. 4 pour la définition de J_τ) et faire ainsi l'économie de la définition de la topologie de $\mathbf{F}(\Sigma)$, mais nous trouvons plus instructif de compliquer les choses. Nous remarquons que g_E étant continue et $\hat{c}$ borélienne d'après 5, il nous suffit, pour étudier la mesurabilité de f, de regarder celle de X. Or, on a l'équivalence, y parcourant F et s parcourant S,

$$X(y) \cap I_s \neq \varnothing \Leftrightarrow g_E(I_s) \cap H(y) \neq \varnothing .$$

Les ensembles de la forme $\{Y: Y \cap I_s \neq \varnothing\}$ engendrant la tribu borélienne de $\mathbf{F}(\Sigma)$ (cf. 3), on voit que X est mesurable, pour cette tribu sur $\mathbf{F}(\Sigma)$, lorsque F est muni d'une tribu contenant les ensembles de la forme

$$A_s = \{y: g_E(I_s) \cap H(y) \neq \varnothing\} = \pi_F[(g_E(I_s) \times F) \cap H] .$$

Il est clair que les A_s sont analytiques, d'où la mesurabilité de X pour la tribu sur F engendrée par $\mathbf{A}(F) = \mathscr{A}(F)$. Noter que la mesurabilité pour $\mathscr{B}(F)$ semble difficile à atteindre car il y a deux images directes dans l'écriture de A_s; si H est polonais, on peut avoir $g_E(I_s)$ ouvert d'après 6(1), mais reste $\pi_F[\cdots]$. Nous arrêtons là cette discussion qui sera reprise plus loin, et passons au cas où F est arbitraire (sous-entendu: mais métrisable séparable!). On plonge alors F dans un espace compact F' puis, H étant $(E \times F)$-analytique, on trouve H' analytique dans $E \times F'$ tel que $H = H' \cap (E \times F)$, et, ayant obtenu une "bonne" application f' pour H' d'après ce qui précède, on prend pour f la restriction de f' à F.

Remarque Supposons F analytique et E compact. Si H est un borélien de $E \times F$, on ne peut espérer avoir une application borélienne (ou même seulement de graphe analytique) f sectionnant H que si $\pi_F(H)$ est un borélien de F. En effet, Graphe $(f) \cap H$ serait alors un ensemble analytique. Or si G est une section analytique de H (i.e. un ensemble analytique contenu dans h tel que, pour tout $y \in \pi_F(H)$, $G(y)$ soit un singleton), la restriction de π_F à G est un isomorphisme borélien de G sur $\pi_F(G) = \pi_F(H)$ d'après I-36. Donc l'application de $\pi_F(G)$ dans E ayant G pour graphe est borélienne et, de I-17, il résulte l'existence d'une application borélienne g de F dans E telle que G soit contenu dans le graphe de g. Mais alors $F - \pi_F(H)$, qui est F-coanalytique et égal à $F - \pi_F(G)$, est aussi égal à $\pi_F[\text{graphe}(g) - H]$, et donc est analytique si H est borélien. D'où $\pi_F(H)$ est borélien d'après I-23.

L'application f de l'énoncé 7 est universellement mesurable (i.e. mesurable pour toute mesure—positive, bornée—sur $(F, \mathscr{B}(F))$). Cela résulte du théorème de capacitabilité, comme nous allons le voir à l'occasion de la version "abstraite" de 7 que nous voyons maintenant.

8 THÉORÈME *Soient $(\Omega, \mathscr{F})$ un espace mesurable, E un espace analytique et H un noyau de schéma de Souslin sur la tribu produit $\mathscr{B}(E)\times\mathscr{F}$. Il existe alors une application universellement mesurable f de $(\Omega, \mathscr{F})$ dans $(E, \mathscr{B}(E))$ telle que l'on ait $f(\omega)\in H(\omega)$ pour tout $\omega\in\Omega$ tel que $H(\omega)$ soit non vide.*

Démonstration Par définition, H admet une représentation de la forme

$$H = \sup_{\sigma\in\Sigma} \inf_{s\dashv\sigma} L_s ,$$

où (L_s) est une famille d'éléments de $\mathscr{B}(E)\times F$ indexée par S (le lecteur qui aimerait s'épargner les schémas de Souslin pourra se borner à considérer le cas où H appartient à $\mathscr{B}(E)\times\mathscr{F}$). La famille (L_s) étant dénombrable, il existe une sous-tribu *séparable* $\mathscr{G}$ de $\mathscr{F}$ telle que les L_s appartiennent à $\mathscr{B}(E)\times\mathscr{G}$. Mais alors, quitte à remplacer $(\Omega, \mathscr{G})$ par son séparé $(\bar{\Omega}, \bar{\mathscr{G}})$, on peut supposer, d'après I-32, que Ω est une partie F de $[0, 1]$ et $\mathscr{G}$ sa tribu borélienne, et H est alors $(E\times F)$-analytique (en effet, H est alors trace sur $E\times F$ d'un noyau de schéma de Souslin H' sur $\mathscr{B}(E\times[0, 1])$ et H' est analytique d'après I-18 et I-16(1)). Cela nous permet d'appliquer 7 qui nous fournit une application f de F dans E sectionnant H et mesurable pour la tribu $\mathscr{B}(E)$ sur E et la tribu sur F engendrée par $\mathbf{A}(F)$. Maintenant, si m est une mesure sur $(\Omega, \mathscr{F})$, elle induit une mesure m' sur $(\Omega, \mathscr{G})$, et donc une mesure notée encore m' sur $(\bar{\Omega}, \bar{\mathscr{G}}) = (F, \mathscr{B}(F))$; d'où finalement une mesure $\hat{m}$ sur $[0, 1]$, image de m' par l'injection de F dans $[0, 1]$. Or d'après le théorème de capacitabilité I-19 appliqué à $\hat{m}$ extériorisée, toute partie analytique de $[0, 1]$ est $\hat{m}$-mesurable, et donc aussi tout élément de la tribu engendrée par $\mathscr{A}([0, 1])$. Comme la trace de cette tribu sur F est égale à la tribu engendrée par $\mathbf{A}(F)$, on en déduit que tout élément de cette dernière est m'-mesurable. Il est alors clair que f est m-mesurable.

Remarque Lorsqu'on ne désire obtenir qu'une application sectionnante f m-mesurable pour m fixée sur $(\Omega, \mathscr{F})$, il existe une démonstration plus courte (celle de 8 commence au moins en 4), familière aux probabilistes. Supposons que $(\Omega, \mathscr{F}) = ([0, 1]), \mathscr{B}([0, 1])$—sinon, on s'y ramène comme ci-dessus, ou l'on fait de la théorie "abstraite"—et plongeons aussi mesurablement E dans $[0, 1]$ à l'aide de I-32. Ceci fait, considérons la capacité I sur $[0, 1]\times\Omega$ définie par $I(X) = m[\pi_\Omega(X)]$. L'ensemble H étant analytique, d'après le théorème de capacitabilité, il existe, pour $\varepsilon < I(H)/2$ (supposé non nul) un compact K inclus dans H tel que $I(H)\leq I(K)+\varepsilon < 2I(K)$. On définit alors une fonction s.c.i. f sur $\pi_\Omega(K)$ en posant

$$f(\omega) = \inf\{x : x\in K(\omega)\}$$

si bien que f sectionne H "à ε près". On réapplique alors le procédé à $H_1 = H - ([0, 1]\times\pi_\Omega(K))$, et ainsi de suite, jusqu'à épuisement (i.e. jusqu'à ce que l'ensemble résiduel H' ait une projection négligeable), ce qui est certainement le cas au bout d'une infinité dénombrable de pas; on recolle alors les fonctions f obtenues en une seule fonction sectionnant H "à un ensemble négligeable près".

2 Le théorème de section de Kuratowski et Ryll–Nardzewski

Nous ne verrons en fait qu'un cas particulier de ce théorème, qui nous précisera la "bonté" possible d'une application sectionnante de H dans $E\times\Omega$ lorsque E est polonais et H à coupes fermées. Pour ce faire, nous allons commencer par mettre une bonne structure mesurable sur $\mathbf{F}(E)$ pour E analytique; pour $E=\Sigma$, on retrouvera la tribu borélienne de $\mathbf{F}(\Sigma)$.

9 DÉFINITION On appelle *tribu d'Effros* sur l'ensemble $\mathbf{F}(E)$ des fermés d'un espace E la tribu, notée $\mathscr{F}(E)$, engendrée par les ensembles de la forme $\{X: X\cap U\neq\varnothing\}$ où U est un ouvert de E.

Remarques (a) Comme E est à base dénombrable, il est clair que la tribu $\mathscr{F}(E)$ est séparable; elle est aussi séparée.

(b) Bien entendu, si Y est un fermé de E, l'ensemble $\{X: X\subseteq Y\}$ appartient à $\mathscr{F}(E)$. Par contre, en général, $\{X: X\cap Y\neq\varnothing\}$ *n'y appartient pas*, même si $E=\Sigma$. C'est cependant le cas si Y est compact (exercice!).

(c) Si E est compact, alors $\mathscr{F}(E)$ coincide avec la tribu borélienne de $\mathbf{F}(E)=\mathscr{K}(E)$ muni de la topologie exponentielle (ou de Hausdorff) qui, dans ce cas, est métrisable compacte. Rappelons que cette topologie est la topologie la moins fine sur $\mathscr{K}(E)$ telle que tout ensemble de la forme $\{K: K\cap U\neq\varnothing\}$ soit ouvert et tout ensemble de la forme $\{K: J\cap U\neq\varnothing\}$ fermé, et que, si d est une distance sur E définissant sa topologie, alors la distance de Hausdorff d sur $\mathscr{K}(E)$ définie par

$$d(K_1,K_2)=\inf\{\varepsilon>0: J_1\subseteq K_2^\varepsilon \text{ et } K_2\subseteq K_1^\varepsilon\}$$

K_i^ε étant le voisinage d'ordre ε de K_i, définit la topologie de $\mathscr{K}(E)$.

(d) Revenons au cas général. Si E est plongé dans E' et si on identifie $X\in\mathbf{F}(E)$ à son adhérence $\bar{X}$ dans E', ce qui identifie $\mathbf{F}(E)$ au sous-ensemble de $\mathbf{F}(E')$ constitué par les X' tels que $X'=\overline{E\cap X'}$, il est facile de voir que $\mathscr{F}(E)$ est alors la trace de $\mathscr{F}(E')$ sur $\mathbf{F}(E)$ (considérer un générateur de la forme $\{X: X\subseteq Y\}$, Y fermé). En particulier, si l'on plonge E dans un compact E', on voit que $\mathscr{F}(E)$ n'est autre que la tribu borélienne de $\mathbf{F}(E)$ muni de la topologie induite par celle de $\mathscr{K}(E')$. Mais, s'il existe parfois une topologie[(1)] naturelle sur $\mathbf{F}(E)$ telle que $\mathbf{F}(E)=\mathscr{B}(\mathbf{F}(E))$ (par exemple, si E

[(1)] Métrisable, séparable.

est compact, ou si $E = \Sigma$), ce n'est pas le cas en général; la topologie induite par un plongement dans un compact dépend en général du plongement. Cela ne nous empêchera pas, par la suite, de considérer $\mathscr{F}(E)$ comme une tribu borélienne, et même d'utiliser le vocabulaire topologique au sujet de $\mathbf{F}(E)$ quand ce que l'on considère ne dépend pas du plongement: par exemple, $\{X \in \mathbf{F}(E) : X \cap U \neq \varnothing\}$ est toujours ouvert, etc. En particulier, la notion de partie $\mathbf{F}(E)$-analytique de $\mathbf{F}(E)$ est bien définie car elle coincide avec celle de noyau de schéma de Souslin sur $\mathscr{F}(E)$ d'après I-18 et I-16(1).

Nous sous-entendrons toujours désormais que $\mathbf{F}(E)$ est muni de la tribu d'Effros $\mathscr{F}(E)$. Nous allons bientôt voir que $\mathbf{F}(E)$ est un espace de Blackwell (cf. I-33) quand E est analytique, et est encore meilleur quand E est polonais.

10 THÉORÈME *Soit f une application continue de F dans E et posons, pour Y fermé de F,*

$$\Phi(Y) = \overline{f(Y)} .$$

L'application Φ ainsi définie de $\mathbf{F}(F)$ dans $\mathbf{F}(E)$ est borélienne.

Démonstration Pour U ouvert de E et Y fermé de F, on a

$$\Phi(Y) \cap U \neq \varnothing \Leftrightarrow f(Y) \cap Y \neq \varnothing \Leftrightarrow Y \cap f^{-1}(U) \neq \varnothing .$$

Comme f est continue, $f^{-1}(U)$ est ouvert. Il est alors clair que Φ est borélienne.

Remarque Si F est analytique et f seulement borélienne, on peut montrer (en plongeant par exemple F dans un compact) que Φ est encore mesurable si on remplace $\mathscr{F}(F)$ par la tribu sur $\mathbf{F}(F)$ engendrée par ses parties analytiques. En particulier, Φ est alors universellement mesurable.

Comme, d'après 6, tout espace analytique est image continue de Σ, et que $\mathbf{F}(\Sigma)$ est polonais, on déduit immédiatement du théorème:

11 COROLLAIRE *Si E est analytique,* $\mathbf{F}(E)$ *est un espace de Blackwell.*

Nous passons maintenant aux images réciproques.

12 THÉORÈME *Soit f une application* ouverte *de F dans E et posons, pour X fermé de E,*

$$\Psi(X) = \overline{f^{-1}(X)} .$$

L'application Ψ ainsi définie de $\mathbf{F}(E)$ *dans* $\mathbf{F}(F)$ *est borélienne. Si, de plus, f est continue et surjective, alors Ψ est un isomorphisme borélien de* $\mathbf{F}(E)$ *sur une partie* $\mathbf{G}_\delta$ *de* $\mathbf{F}(F)$.

Démonstration Pour V ouvert de F et X fermé de E, on a

$$\Psi(X) \cap V \neq \varnothing \Leftrightarrow f^{-1}(X) \cap V \neq \varnothing \Leftrightarrow X \cap f(V) \neq \varnothing$$

et donc Ψ est borélienne si f est ouverte. Si de plus f est continue (auquel cas on a $\Psi(X) = f^{-1}(X)$), et surjective (auquel cas Ψ est injective), on a, pour U ouvert de E et Y dans l'image de $\mathbf{F}(E)$ par Ψ,

$$\Psi^{-1}(Y) \cap U \neq \varnothing \Leftrightarrow Y \cap f^{-1}(U) \neq \varnothing$$

si bien que Ψ est un isomorphisme de $\mathbf{F}(E)$ sur son image. Enfin, si (V_n) est une base dénombrable de la topologie de F, on a, pour Y fermé de F,

$$Y \in \text{Image}\ (\Psi) \Leftrightarrow \forall n\ Y \cap V_n \neq \varnothing \text{ ou } Y \cap f^{-1}[f(V_n)] = \varnothing\ ,$$

d'où Image (Ψ) est un $\mathbf{G}_\delta$ de $\mathbf{F}(F)$. Noter que, si $F = \Sigma$, on peut prendre pour (V_n) la collection des ilots si bien que Image (Ψ) est même un fermé de $\mathbf{F}(\Sigma)$ muni de sa topologie naturelle.

Remarque Même si E et F sont polonais, Ψ n'est pas en général borélienne si f est continue mais non ouverte. On a cependant, pour E, F analytiques et f borélienne, la mesurabilité de Ψ si $\mathbf{F}(E)$ est muni de la tribu engendrée par ses parties analytiques.

Comme, d'après 6, tout espace polonais est image continue et ouverte de Σ, on déduit immédiatement du théorème le théorème suivant dû à Effros (d'où le nom donné à la tribu $\mathscr{F}(E)$):

13 COROLLAIRE *Si E est polonais, $\mathscr{F}(E)$ est la tribu borélienne d'une topologie polonaise sur* $\mathbf{F}(E)$.

Remarques (a) Pour E polonais, il n'est pas difficile de voir, en plongeant E dans un compact, que $\mathscr{K}(E)$ est un élément de la tribu d'Effros. Cela est *toujours faux* si E est analytique mais non polonais, même pour E homéomorphe à un borélien de polonais (i.e. lusinien dans la terminologie de Bourbaki) auquel cas—c'est facile à voir—$\mathscr{K}(E)$ est une partie $\mathbf{F}(E)$-coanalytique. C'est là un résultat difficile de Christensen (1974) et Saint-Raymond (1971/73).

(b) Il est facile de voir que l'énoncé 13 est encore vrai si E est un espace $\mathscr{K}_\sigma$. Saint-Raymond (1978) a montré que 13 est vrai ssi E est la réunion d'un espace polonais et d'un espace $\mathscr{K}_\sigma$!

(c) Je répare ici un oubli dans la longue série de remarques de 9. Si E est analytique et A une partie analytique de E, on voit facilement, en plongeant E dans un compact, que l'ensemble $\{X \in \mathbf{F}(E): X \cap A \neq \emptyset\}$ est analytique dans $\mathbf{F}(E)$. On n'a pas mieux en général (cf. le (b) de 9) pour E polonais et A fermé.

Voici maintenant une forme "canonique" du théorème de section annoncé de Kuratowski et Ryll–Nardzewski: c'est le théorème du "bon choix" 5 étendu à tout espace polonais

14 THÉORÈME *Si E est un espace polonais, il est possible de choisir un point $c(X)$ dans $X \in \mathbf{F}(E)$ non vide de sorte que l'application $X \to c(X)$ de $\mathbf{F}(E) - \{\emptyset\}$ dans E soit borélienne.*

Démonstration Soit f une surjection continue et ouverte de Σ sur E. D'après 5 et 12, il suffit de poser, pour X fermé non vide de E, $c(X) = \gamma[f^{-1}(X)]$ où γ est le choix pour Σ donné par 5.

Remarques (a) Saint-Raymond (1977) a montré que, pour E lusinien, il existe un choix borélien ssi E est la réunion d'un espace polonais et d'un espace $\mathcal{K}_\sigma$.

(b) Un espace analytique E étant image continue de Σ, on voit, en remplaçant dans la démonstration "12" par la remarque de 12, qu'il existe alors un choix measurable si $\mathbf{F}(E)$ est muni de la tribu engendrée par ses parties analytiques. Un tel choix est universellement mesurable.

De la forme "canonique", on déduit immédiatement, par composition d'applications mesurables, la version classique du théorème (que l'on aurait pu démontrer beaucoup plus vite, directement, en faisant l'économie de la structure mesurable de $\mathbf{F}(E)$—exercice!).

15 THÉORÈME *Soient $(\Omega, \mathcal{F})$ un espace mesurable, E un espace polonais et H une partie de $E \times \Omega$ vérifiant les conditions suivantes: pour tout $\omega \in \Omega$, la coupe $H(\omega)$ est fermée dans E, et, pour tout ouvert U de E, la projection de $H \cap (U \times \Omega)$ sur Ω appartient à $\mathcal{F}$. Il existe alors une application mesurable f de $(\Omega, \mathcal{F})$ dans $(E, \mathcal{B}(E))$ telle que l'on ait $f(\omega) \in H(\omega)$ pour tout $\omega \in \Omega$ tel que $H(\omega)$ soit non vide.*

Démonstration La condition de l'énoncé équivaut à dire que $\omega \to H(\omega)$ est une application mesurable de $(\Omega, \mathcal{F})$ dans $(\mathbf{F}(E), \mathcal{F}(E))$, d'où la conclusion d'après 14 (prolonger arbitrairement c en $\hat{c}$ sur $\mathbf{F}(E)$).

Remarques (a) Il n'y a pas ici la remarque "universellement mesurable" habituelle: on obtiendrait un énoncé moins général que 8!

(b) La condition de l'énoncé sur H implique que H appartient à la tribu produit $\mathscr{B}(E)\times\mathscr{F}$. En effet, si (U_n) est une base dénombrable de la topologie de E, on a l'équivalence pour $x\in E$ et $\omega\in\Omega$

$$(x,\omega)\in H \Leftrightarrow \forall n\ [x\notin U_n \text{ ou } \omega\in\pi_\Omega[H\cap(U_n\times\Omega)]],$$

où π_Ω est la projection de $E\times\Omega$ sur Ω. Comme la condition sur H implique que $\pi_\Omega(H)$ appartient à $\mathscr{F}$, la réciproque est fausse en général. Nous verrons cependant, au paragraphe suivant, qu'elle est vraie si H élément de $\mathscr{B}(E)\times\mathscr{F}$ a ses coupes compactes et si $(\Omega,\mathscr{F})$ est un espace mesurable de Blackwell.

3 Ensembles à coupes fermées ou compactes

Ce que nous avons fait jusqu'ici, dans ce chapitre, reposait sur des moyens élémentaires: essentiellement sur la possibilité de bien paramétrer nos espaces par $\mathbb{N}^\mathbb{N}$ (rappelons que 6(2) sert très souvent à définir les espaces analytiques). Il n'en sera pas de même pour les résultats principaux de ce paragraphe: nous serons amenés à utiliser le théorème de séparation I-22, dans toute sa force (i.e. pour une suite infinie).

16 Nous commençons par rappeler quelques notations classiques. Un sous-ensemble $\mathscr{E}$ de $\mathscr{P}(E)$ contenant la partie vide sera appelé un *pavage* sur E. Si $\mathscr{E}$ est un pavage, nous désignerons respectivement par $\mathscr{E}_s$, $\mathscr{E}_d$, $\mathscr{E}_\sigma$, $\mathscr{E}_\delta$ le plus petit pavage contenant $\mathscr{E}$ et stable respectivement pour les réunions finies, les intersections finies, les réunions dénombrables, les intersections dénombrables. On note $\mathscr{E}_{sd}$ le pavage $(\mathscr{E}_s)_d$, etc. Si $\mathscr{E}=\mathscr{E}_s$ (resp. $\mathscr{E}=\mathscr{E}_d$), il est facile de voir que $\mathscr{E}_\delta=\mathscr{E}_{\delta s}$ (resp. $\mathscr{E}_\sigma=\mathscr{E}_{\sigma d}$). Enfin, si $\mathscr{E}$ est un pavage sur E et $\mathscr{F}$ un pavage sur F, on appelle *pavage produit* le pavage $\mathscr{E}\times\mathscr{F}$ sur $E\times F$ dont les éléments sont les ensembles de la forme $C\times D$ où C est un élément de $\mathscr{E}$ et D un élément de $\mathscr{F}$.

Voici d'abord un résultat élémentaire et néanmoins utile

17 THÉORÈME *Soient, E, F deux espaces, où E est analytique, et soit H une partie $(E\times F)$-analytique de $E\times F$. Alors, la partie $\bar{H}^E$ de $E\times F$ définie comme suit, où x parcourt E et y parcourt F,*

$$(x,y)\in\bar{H}^E \Leftrightarrow x\in\overline{H(y)},$$

appartient au pavage $(\mathbf{F}(E)\times\mathbf{A}(F))_{s\delta}$. En particulier, $\bar{H}^e$ est une partie $(E\times F)$-analytique de $E\times F$.

Démonstration Soit (U_n) *une base dénombrable de la topologie de E.* Posons, pour tout n, $C_n = \{y \in F : H^c(y) \supseteq U_n\}$. L'ensemble C_n est le complémentaire de l'ensemble F-analytique $\pi_F[H \cap (U_n \times F)]$ et on a

$$(E \times F) - \bar{H}^E = \sup_n (U_n \times C_n) \quad \text{d'où} \quad \bar{H}^E = \inf_n ((U_n^c \times F) \cup (E \times C_n^c)).$$

D'où la conclusion.

Remarques (a) Si H est borélien, $\bar{H}^E$ n'est pas borélien en général, même si E et F sont compacts.

(b) On a vu que le théorème de séparation I-22 de Novikov dérive, par l'intermédiaire de la multicapacité I-6(b), de la possibilité de séparer des compacts disjoints par des ouverts disjoints. On peut montrer (exercice pas facile!) que I-22 est équivalent au résultat suivant: soient E et F deux espaces compacts, H et L deux parties analytiques disjointes de $E \times F$ telles que $H(y)$ et $L(y)$ soient compacts pour tout $y \in F$; il existe alors deux boréliens disjoints H' et L' de $E \times F$ tels que l'on ait $H \subseteq H'$ et $L \subseteq L'$ et que $H'(y)$ et $L'(y)$ soient ouverts pour tout $y \in F$.

Nous arrivons maintenant au résultat-clé de ce paragraphe, essentiellement dû à Kunugui:

18 THÉORÈME *Soient E et F deux espaces analytiques et soit H une partie borélienne de* $E \times F$ *telle que* $H(y)$ *soit fermé pour tout* $y \in F$. *Alors H appartient au pavage* $(\mathbf{F}(E) \times \mathcal{B}(F))_{s\delta}$.

Démonstration Comme on a ici $H^c = E - \bar{H}^E$, on peut écrire, en reprenant la représentation obtenue au cours de la démonstration de 17,

$$H^c = \sup_n (U_n \times C_n) \quad \text{où} \quad C_n = F - \pi_F[H \cap (U_n \times F)].$$

(U_n) étant une base dénombrable de la topologie de E. Les ensembles $L_n = (U_n \times C_n) \cup H$ sont $(E \times F)$-coanalytiques et recouvrent $E \times F$. D'après I-22 (en passant aux complémentaires), il existe alors une suite (M_n) de boréliens de $E \times F$ telle que l'on ait $M_n \subseteq L_n$ pour tout n et que $\sup_n M_n = E \times F$; d'où $M_n - H$ est contenu dans $U_n \times C_n$ pour tout n et $\sup_n (M_n - H) = H^c$. Maintenant, $\pi_F(M_n - H)$ est analytique et disjoint de l'ensemble analytique $F - C_n = \pi_F(H \cap (U_n \times F))$. D'après I-22, appliqué à un couple cette fois, il existe un borélien B_n de F contenant $\pi_F(M_n - H)$ et contenu dans C_n. Mais alors, on a aussi $H^c = \sup_n (U_n \times B_n)$ et c'est gagné: $H = \inf_n ((U_n^c \times F) \cup (E \times B_n^c))$.

Remarques (a) Lorsque F n'est pas analytique, on a un résultat analogue si on remplace partout dans l'énoncé le concept de "borélien" par celui de "partie $(E\times F)$-bianalytique" ou "partie F-bianalytique". Il faut, pour démontrer cela, avoir le théorème général de séparation de Novikov (équivalent au théorème de réduction) qui ne sera démontré qu'au Chapitre III, par des techniques très différentes de celles du Chapitre I.

(b) On peut se demander si la représentation donnée dans l'énoncé s'étend aux classes de Baire plus hautes que celle des fermés, i.e. si un borélien H à coupes de classe $\mathbf{F}_\xi$ avec $\xi=\sigma\delta\ldots$ (opération transfinie indexée par les ordinaux dénombrables) appartient au pavage $(\mathbf{F}(E)\times \mathscr{B}(F))_{s\delta\xi}$. La réponse est positive d'après des travaux récents de Louveau (198?), précédés—en remontant le temps—par ceux de Bourgain (198?*a*, *b*) pour $\xi=\sigma\delta$ et ceux de Saint-Raymond (1976) pour $\xi=\sigma$. Nous démontrerons le résultat de Saint-Raymond au Chapitre IV.

(c) D'après I-40 et sa remarque I-40(c), on peut remplacer la topologie de F par une topologie analytique plus fine (une topologie polonaise si F est lusinien) de sorte que les boréliens B_n intervenant dans la représentation de H à la fin de la démonstration soient ouverts dans F pour sa nouvelle topologie. Alors H devient tout simplement un fermé de $E\times F$! (Remarque de Louveau.)

Nous déduisons de ce théorème un résultat très intéressant de Novikov sur les boréliens à coupes compactes:

19 THÉORÈME *Soient E, F deux espaces où E est polonais et F analytique, et soit H une partie borélienne de $E\times F$ telle que $H(y)$ soit compact pour tout $y\in F$. Alors $\pi_F(H)$ est un borélien de F.*

Démonstration D'après 18, on peut écrire $H=\lim_n \downarrow H_n$ où les H_n appartiennent à $(\mathbf{F}(E)\times\mathscr{B}(F))_s$, pavage stable pour les intersections finies; il est clair que les $\pi_F(H_n)$ sont des boréliens de F. Supposons d'abord E compact: alors, on a $\pi_F(H)=\lim_n\downarrow\pi_F(H_n)$ (ceci parce que $H_n(y)=\varnothing$ pour n grand si $H(y)=\varnothing$), et c'est gagné. Dans le cas général, nous plongeons E dans un compact E': E étant borélien dans E', H est encore un borélien à coupes compactes dans $E'\times F$, et c'est fini.

Remarques (a) Le résultat est encore vrai (avec la même démonstration) si on suppose E seulement lusinien. Il est faux par contre, en général, si E est seulement analytique (exercice!).

(b) Revenons à E polonais (ou lusinien), et supposons que le borélien H est, plus généralement, à coupes $\mathscr{K}_\sigma$; d'après le résultat de Saint-Raymond cité en 18(b), H est réunion dénombrable de boréliens à coupes compactes, et donc $\pi_F(H)$ est encore borélien. C'est le résultat maximal dans ce sens puisque, si $E = \bar{\mathbb{N}}^{\mathbb{N}}$, toute partie analytique A de F est projection d'une partie $\mathbf{G}_\delta$ de $E \times F$ (prendre le graphe d'une application continue de $\mathbb{N}^{\mathbb{N}}$ sur A).

20 COROLLAIRE *Soient $(\Omega, \mathscr{F})$ un espace mesurable de Blackwell, E un espace polonais et H un élément de la tribu produit $\mathscr{B}(E) \times \mathscr{F}$ tel que $H(\omega)$ soit compact pour tout $\omega \in \Omega$. Alors*

(1) *l'application $\omega \to H(\omega)$ est une application mesurable de $(\Omega, \mathscr{F})$ dans $(\mathbf{F}(E), \mathscr{F}(E))$;*
(2) *il existe une application mesurable f de $(\Omega, \mathscr{F})$ dans $(E, \mathscr{B}(E))$ telle qu'on ait $f(\omega) \in H(\omega)$ pour tout $\omega \in \Omega$ tel que $H(\omega) \neq \varnothing$.*

Démonstration Quitte à remplacer $(\Omega, \mathscr{F})$ par son séparé, on peut supposer que Ω est un espace analytique F et que $\mathscr{F} = \mathscr{B}(F)$. Alors, si U est un ouvert de E, on peut écrire $U = \sup_n X_n$ avec X_n fermé pour tout n et donc

$$\pi_\Omega[H \cap (U \times \Omega)] = \sup_n \pi_\Omega[H \cap (X_n \times \Omega)]$$

est borélien d'après 19 car les $H \cap (X_n \times \Omega)$ sont des boréliens à coupes compactes. D'où $\omega \to H(\omega)$ est borélienne. Le point (2) est alors conséquence de 15, mais on peut aussi l'établir plus directement en remplaçant le théorème du "bon choix" 14 par un cas particulier, très simple. En effet, quitte à plonger E, on peut supposer $E = [0, 1]^{\mathbb{N}}$; ayant muni $[0, 1]^{\mathbb{N}}$ de l'ordre lexicographique engendré par l'ordre naturel de $[0, 1]$, on peut alors choisir dans $K \in \mathscr{K}(E)$ non vide le plus petit point pour cet ordre. Remarquer aussi que, si $E = [0, 1]$, on peut alors tout simplement prendre $f(\omega) = \inf\{x : x \in H(\omega)\}$ en convenant que $\inf \varnothing = 1$.

Remarques (a) On peut faire ici une remarque analogue à 18(a).

(b) On continue ici la remarque 19(b). Si, plus généralement, les coupes de H sont des $\mathscr{K}_\sigma$, on peut écrire $H = \sup_n H_n$ où les H_n sont des éléments de $\mathscr{B}(E) \times \mathscr{F}$ à coupes compactes. Posons pour tout n $L_n = H_n - (\sup_{m<n} E \times \pi_\Omega(H_n))$, si bien que les $\pi_\Omega(L_n)$ sont disjoints. D'après 20, les L_n peuvent être sectionnés par des applications mesurables, et il est alors facile, en recollant les sections obtenues, d'obtenir une application mesurable f de Ω dans E sectionnant H. Signalons que le résultat de 19(b) sur la projection de H a été établi (sans faire apel à un théorème de structure du type de celui de Saint-Raymond) par Arsenin et Kunugui, tandis que la possibilité de "bien" sectionner H a été établie par Tchégolkov.

Notons enfin que ce théorème de section borélienne est maximal en un certain sens (cf. la remarque de 7).

4 Ensembles à coupes dénombrables

Nous allons démontrer ici le théorème de clivabilité des boréliens à coupes dénombrables—théorème dû à Novikov et non à Lusin, comme on le trouve écrit par de nombreux auteurs (dont moi-même).

Voici l'énoncé du théorème, à la démonstration duquel sera consacré tout cette section; nous reprenons la démonstration du traité classique de Hausdorff, mais en en court-circuitant une bonne part grâce à quelques "tours de passe-passe".

21 THÉORÈME *Soint E, F deux espaces polonais et soit H un borélien de $E \times F$ tel que $H(y)$ soit (au plus) dénombrable pour tout $y \in F$. Il existe alors une suite (f_n) d'applications boréliennes de F dans E telle que H soit contenu dans la réunion des graphes des f_n.*

Remarque L'énoncé est encore vrai si l'on suppose H analytique. Cette extension, sans doute due à Lusin, sera établie au Chapitre IV où nous affinerons les outils utilisés ici.

Démonstration: premiére partie de 21 Nous allons voir, dans cette première partie, qu'on peut se ramener à démontrer 21 dans le cas où E et F sont compacts, et où H est un compact de $E \times F$ (ne pas se leurrer: c'est la partie "facile").

Se ramener au cas où F est compact est chose aisée: il suffit de plonger F dans un compact F' et de remplacer F par F'. Le reste de notre programme est un peu plus délicat sans être très compliqué.

Soit g la restriction de π_F à H: H est lusinien, g est continue et semi-injective, i.e. pour tout $y \in F$, $g^{-1}(\{y\})$ est (au plus) dénombrable. Supposons démontré qu'il existe une partition dénombrable (H_n) de H en boréliens telle que la restriction g_n de g à H_n soit injective. Alors, revenant dans $E \times F$, on voit que H_n y est un "graphe borélien", i.e. un borélien G tel que $G(y)$ soit ou vide ou un singleton pour tout $y \in F$. D'après I-36, l'application g_n est un isomorphisme borélien de H_n sur $\pi_F(H_n)$, lequel est borélien dans F d'après I-37 (ou II-19!). Il ne resterait alors plus qu'à poser $f_n(y) =$ l'unique point de $H_n(y)$ si $y \in \pi_F(H_n)$ et $f(y) = x_0$ sinon où x_0 est un point choisi dans E.

Nous nous sommes donc ramenés, dans un premier temps, à démontrer que la restriction g de π_F à H est "clivable". Maintenant, quitte à renforcer la topologie de H, on peut supposer que H (que nous ne considérons plus plongé dans $E \times F$) est un fermé de $\mathbb{N}^{\mathbb{N}}$ (cf. I-38). Supposons que nous ayons pu plonger le fermé H de $\mathbb{N}^{\mathbb{N}}$ dans $[0, 1]$ de telle sorte que, $\bar{H}$ étant l'adhérence de H dans $[0, 1]$, l'ensemble $\bar{H} - H$ soit dénombrable. Alors, comme le graphe G de g dans $H \times F$ est fermé (g est continue), l'adhérence $\bar{G}$ de G dans $[0, 1] \times F$ sera un compact (rappelons que F est compact) dont les coupes $\bar{G}(y)$ sont dénombrables (et compactes) pour $y \in F$. Si, par conséquent, on sait que l'énoncé 21 est vrai pour E, F compacts et pour un compact de $E \times F$ à coupes dénombrables, on aura un "clivage" de $\bar{G}$ par les graphes des applications boréliennes fournies par 21, d'où un clivage de G dans $H \times F$ et finalement, en projetant sur H, un clivage de H lui-même (les ensembles obtenus par projection sont boréliens d'après I-37).

Pour achever cette première partie de la démonstration de 21, il ne nous reste plus qu'à montrer qu'un fermé H de $\mathbb{N}^{\mathbb{N}}$ se plonge dans $[0, 1]$ de sorte que $\bar{H} - H$ soite dénombrable. Il suffit évidemment de considérer le cas de $\mathbb{N}^{\mathbb{N}}$ tout entier. On peut alors se référer au fait, bien classique, que $\mathbb{N}^{\mathbb{N}}$ est homéomorphe au sous-espace de $[0, 1]$ constitué par les nombres irrationnels. Nous démontrons, plus généralement, le résultat suivant dû à Mazurkiewicz.

22 THÉORÈME *Tout espace polonais X totalement discontinu, dont toute partie compacte est d'intérieur vide, est homéomorphe à $\mathbb{N}^{\mathbb{N}}$.*

Démonstration Nous munissons X d'une distance d définissant sa topologie, pour laquelle X soit complet. Ceci fait, il suffit évidemment de construire un schéma de Souslin régulier $(U_s)_{s \in S}$ sur les parties ouvertes et fermées, non vides, de X vérifiant la condition: pour s parcourant l'ensemble S_k des suites finies de longueur k, les U_s forment une partition de X par des ensembles de diamètre $\leq 1/k$. On est ainsi finalement ramené au problème de partitionner un ouvert-fermé non vide U de X en une suite infinie de parties ouvertes-fermées non vides de petit diamètre: nous laissons au lecteur le soin de résoudre ce petit problème.

Remarque L'homéomorphisme traditionnel de $\mathbb{N}^{\mathbb{N}}$ sur les irrationnels est donné par la décomposition en fraction continue d'un irrationnel. Si l'on désire seulement identifier $\mathbb{N}^{\mathbb{N}}$ avec une partie à complémentaire dénombrable de $[0, 1]$, il y a bien plus simple: il suffit d'identifier $\sigma = (n_1, n_2, n_3, \ldots)$ au nombre $x \in [0, 1]$ tel que le développement dyadique de $x/2$ commence par n_1 "0", continue par n_2 "1", se poursuit par n_3 "0", etc.

Démonstration: seconde partie de 21 Nous commençons par rappeler un peu de terminologie remontant à Cantor:

23 DÉFINITION Soit X un fermé d'un espace E. On appelle *dérivé* de X le fermé noté X' constitué par les points non isolés de X. On appelle *dérivation de Cantor* l'application $X \to X'$ de $\mathbf{F}(E)$ dans $\mathbf{F}(E)$.

Il est clair que l'on a $X' \subseteq X$, et $X' \subseteq Y'$ si Y est un fermé contenant X. Nous étudierons en détail ce type d'applications au Chapitre IV.

24 Si I désigne l'ensemble des ordinaux dénombrables, on définit (Cantor les a inventés pour çà!) la *suite transfinie des dérivés successifs* $(X^i)_{i \in I}$ du fermé X de E comme suit, par récurrence transfinie: si $j = i+1$ est un ordinal de première espèce, on pose $X^{i+1} = (X^i)'$; si j est de seconde espèce, on pose $X^j = \inf_{i<j} X^i$. La suite transfinie (X^i) est évidemment décroissante. Comme E est à base dénombrable, l'inf de toute famille de fermés est atteint pour une sous-famille dénombrable, et il existe donc un plus petit ordinal dénombrable i_0 (dépendant de X) tel que $X^{i_0} = X^{i_0+1} = \ldots$: X^{i_0} est alors un *parfait* (i.e. un fermé sans points isolés—il n'est pas exclu que $X^{i_0} = \emptyset$), appelé le noyau parfait de X. D'autre part, $X^i - X^{i+1}$ est dénombrable pour tout i puisque discret; on voit alors que le fermé X se décompose en un ensemble parfait et un ensemble dénombrable ne contenant aucun parfait non vide (théorème de Cantor et Bendixson). Supposons maintenant E polonais. Alors, en vertu du théorème de Baire, tout fermé dénombrable de E a au moins un point isolé si bien que tout parfait non vide est non dénombrable (et même a la puissance du continu—exercice!). En particulier, si X est un fermé dénombrable du polonais E, X^i est vide à partir d'un certain ordinal dénombrable.

Nous passons maintenant à l'étude de la mesurabilité de $X \to X'$ lorsque E est compact:

25 THÉORÈME *Si E est un espace compact, la dérivation de Cantor est une application borélienne de $\mathscr{K}(E)$ dans $\mathscr{K}(E)$.*

Démonstration Soient (U_n) une base dénombrable de la topologie de E. Pour U ouvert et K compact de E, on a l'équivalence

$$U \cap K' \neq \emptyset \Leftrightarrow \exists x \; x \in U \text{ et } x \in K \text{ et}$$
$$\forall n \; [x \notin U_n \text{ ou } \exists y \; y \neq x \text{ et } y \in U_n \text{ et } y \in K].$$

Comme la projection d'un $\mathbf{F}_\sigma = \mathscr{K}_\sigma$ de $E \times \mathscr{K}(E)$ sur $\mathscr{K}(E)$ est encore un $\mathscr{K}_\sigma$, on en déduit aisément que $K \to K'$ est borélienne.

Remarques (a) Supposons toujours E compact et désignons par D le sous-ensemble de $\mathscr{K}(E)$ constitué par les compacts L dénombrables.

L'ensemble D est une partie coanalytique de $\mathcal{K}(E)$ puisqu'on a

$$L \in D \Leftrightarrow \forall K \; (K \neq \varnothing \text{ et } K \subseteq L) \Rightarrow K \neq K' .$$

Appelons indice de $L \in D$ le plus petit ordinal dénombrable $i(L)$ tel que son i-ème dérivé L^i soit vide. Nous démontrerons ci-dessous que D vérifie le *théorème de la borne*: pour toute partie analytique A de $\mathcal{K}(E)$ contenue dans D, $\sup_{L \in A} i(L)$ est un ordinal dénombrable. On en déduit que, si E n'est pas dénombrable, D n'est pas un borélien de $\mathcal{K}(E)$ (exercice!), et aussi, si $E = [0, 1]$, que l'ensemble des compacts contenus dans les rationnels de $[0, 1]$ est coanalytique non borélien (exercice!).

(b) Si E est seulement analytique, la démonstration de 25 montre que $\{X \in \mathbf{F}(E) : X' \neq \varnothing\}$ est encore analytique, *mais* il peut ne pas être borélien: c'est déjà le cas pour $E = \mathbb{N}^{\mathbb{N}}$. (Exercice! Identifier $\mathbb{N}^{\mathbb{N}}$ aux irrationnels de $[0, 1]$ et utiliser les exercices précédents.) Donc la dérivation de Cantor n'est pas borélienne en général; mais la relation $X \subseteq Y'$ sur $\mathbf{F}(E)$ reste analytique car elle équivaut à "$X = \varnothing$ ou ($Y' \neq \varnothing$ et ...)" où "..." est la relation borélienne

$$\forall n \; [U_n \cap X = \varnothing \text{ ou } \exists p \; \exists q \; U_p \subseteq U_n \text{ et } U_q \subseteq U_n \text{ et } U_p \cap U_q = \varnothing$$
$$\text{et } U_p \cap Y \neq \varnothing \text{ et } U_q \cap Y \neq \varnothing] .$$

On a simplement écrit que tout ouvert qui coupe X doit contenir au moins deux points distincts de Y. Nous verrons au Chapitre IV que le sous-ensemble D de $\mathbf{F}(E)$ constitué par les fermés dénombrables vérifie encore le théorème de la borne.

Nous revenons maintenant à nous moutons, et rappelons la situation à laquelle nous étions arrivés: E et F compacts, H un compact de $E \times F$ tel que $H(y)$ soit un compact dénombrable pour tout $y \in F$. Et l'on veut trouver une partition dénombrable de H en "graphes boréliens". Nous définissons une suite transfinie (H_i) de parties de $E \times F$, indexée par l'ensemble I des ordinaux dénombrables, par

$$(x, y) \in H_i \Leftrightarrow x \in H^i(y) ,$$

où $H^i(y)$ est le i-ème dérivé du compact $H(y)$. On vérifie aisément, par récurrence transfinie, que chaque H_i est un borélien à coupes compactes de $E \times F$: cela résulte simplement, par composition d'applications boréliennes, de 20 et 25 (on peut aussi raisonner plus élémentairement, sans utiliser 20, en remarquant que, H étant compact, $y \to H(y)$ est borélienne, et en montrant, par récurrence transfinie, que $K \to K^i$ est une application borélienne de $\mathcal{K}(E)$ dans $\mathcal{K}(E)$: on obtient alors le fait que $y \to H_i(y)$ est borélienne). Nous allons montrer, en deux temps, que l'on a dans ces conditions:

(1) pour tout i, $H_i - H_{i+1}$, est contenu dans une réunion dénombrable de graphes boréliens;
(2) il existe un ordinal dénombrable i_0 tel que H_{i_0} soit vide.

Il est clair que nous aurons alors fini de démontrer 21.

LEMME *Soit $y \to K(y)$ une application borélienne de F dans $\mathcal{K}(E)$. L'ensemble $\{(x, y): x \in K(y) - K'(y)\}$ est contenu dans une réunion dénombrable de graphes boréliens.*

Démonstration Désignons par (U_n) une base dénombrable de la topologie de E et définissons, pour chaque n, une application $K \to K_n$ de $\mathcal{K}(E)$ dans $\mathcal{K}(E)$ en posant $K_n = U_n \cap K$ si $U_n \cap K$ est un singleton et $K_n = \varnothing$ sinon. Nous laissons au lecteur le soin de vérifier que ces applications sont boréliennes. Maintenant, l'ensemble $\{(x, y): x \in K(y) - K'(y)\}$ est égal à la réunion des graphes $\{(x, y): x \in K_n(y)\}$, lesquels sont boréliens par composition d'applications boréliennes.

Cela règle le point (1). Pour le point (2), nous remarquons que, $y \to H(y)$ étant une application borélienne, l'image de F par cette application est une partie analytique de $\mathcal{K}(E)$ constituée de compacts dénombrables. Le point (2) résulte donc du théorème de la borne, évoqué plus haut, que nous démontrons maintenant:

26 THÉORÈME *L'espace E étant compact, soit $K \to i(K)$ l'application de l'ensemble D des compacts dénombrables de E dans l'ensemble I des ordinaux dénombrables définie par*

$$i(K) = \inf\{i \in I: K^i = \varnothing\}.$$

Alors, pour toute partie analytique A de $\mathcal{K}(E)$ contenue dans D, $\sup_{K \in A} i(K)$ est un ordinal dénombrable.

Démonstration Nous avons vu, à la Remarque 25(a), que D est coanalytique dans $\mathcal{K}(E)$. Par conséquent, d'après le théorème de séparation, pour un couple, il existe un borélien B de $\mathcal{K}(E)$ contenant A et contenu dans D si bien que, quitte à remplacer A par B, on peut supposer A borélien. Maintenant, A étant borélien, il est facile de voir que l'application $y \to H(y)$ de $\mathcal{K}(E)$ dans E définie par $H(y) = \{x: x \in y\}$ pour $y \in A$ et $H(y) = \varnothing$ pour $y \notin A$ est borélienne (autrement dit, vu 20, $H = \{(x, y): x \in y \text{ et } y \in A\}$ est un borélien à coupes compactes de $E \times \mathcal{K}(E)$) $\cdots$ et nous sommes pratiquement revenus à notre point de départ, à ceci près que nous avons dégagé un énoncé 26 très intéressant, et ramené le problème à sa forme "canonique". Nous reprenons alors les notations introduites plus haut (i.e. $F = \mathcal{K}(E)$;

$(x, y) \in H_i \Leftrightarrow x \in H^i(y)$, etc.) et démontrons, artisanalement (nous verrons au Chapitre IV une méthode bien plus puissante pour démontrer des résultats du type de 26), ce fameux point (2) auquel nous sommes revenus. Nous raisonnons par l'absurde: nous supposons $H_i \neq \emptyset$ pour tout $i \in I$ et démontrons qu'il existe alors $y \in F$ tel que $H(y)$ soit non dénombrable. Nous munissons l'espace compact $E \times F$ d'une distance d définissant sa topologie et allons construire un "schéma dyadique" $(U_s)_{s \in S}$ sur les ouverts de $E \times F$ où, pour une fois, S désigne l'ensemble des suites finies de 0 et de 1, ayant lcs propriétés suivantes:

(a) pour $s \in S^n$, où S^n est l'ensemble des suites de longueur n, le diamètre de U_s est $< 1/n$;
(b) pour $s, t \in S^n$ et $s \neq t$, $U_s \cap U_t = \emptyset$;
(c) pour $s \dashv t$, $U_s \supseteq \bar{U}_t$;
(d) enfin, pour tout $i \in I$, et tout n, l'ensemble

$$L_n^i = \inf_{s \in S^n} \pi_F(U_s \cap H_i)$$

n'est pas vide.

Ceci fait, nous laisserons au lecteur le soin de voir qu'il existera alors un unique point y dans $\inf_{s \in S} \pi_F(U_s)$, et que, pour cet $y \in F$, le compact $H(y)$ a la puissance du continu. Supposons donc U_s défini pour tout $s \in S^n$ (on peut toujours prendre $U = E \times F$ pour s de longueur 0!), et voyons comment construire les U_{s0} et U_{s1}, où $s0$ (resp. $s1$) est la suite de longueur $n+1$ obtenue en ajoutant 0 (resp. 1) à "la droite" de $s \in S^n$. Nous choisissons d'abord, pour tout $S \in S^n$, une suite $(U_s^k)_{k \in \mathbb{N}}$ d'ouverts de $E \times F$ de diamètre $< 1/n$, d'adhérence contenue dans U_s, et formant une base de la topologie de U_s. Pour $i \in I$ fixé, choisissons un point y_i dans L_n^{i+1}: comme $H_{i+1}(y)$ est le dérivé de $H_i(y)$, on est alors assuré que $U_s \cap H_i(y_i)$ est infini pour tout $s \in S^n$; il existe donc deux ouverts disjoints U_s^p et U_s^q de notre base de U_s—p et q dépendant de $s \in S^n$—tels que y_i appartienne encore à

$$\inf_{s \in S^n} [\pi_F(U_s^p \cap H_i) \cap \pi_F(U_s^q \cap H_i)].$$

Mais ce n'est pas fini, car $p = p(i)$ et $q = q(i)$ sont, pour $s \in S^n$ fixé, fonctions de l'ordinal i que nous faisons varier maintenant. Or, l'ensemble I des ordinaux dénombrables n'est pas dénombrable, tandis que la famille (U_s^k), k parcourant $\mathbb{N}$ et s parcourant S^n, est dénombrable: nous sommes donc assurés qu'il existe un sous-ensemble non dénombrable J de I tel que, pour $s \in S^n$ fixé, $p(i)$ et $q(i)$ ne dépendent pas de i sur J. Nous posons alors $U_{s0} = U_s^p$ et $U_{s1} = U_s^q$ où p (resp. q) est la valeur prise par $p(i)$ (resp. $q(i)$) sur J. Il est clair que les U_t, $t \in S^{n+1}$, ainsi définis vérifient (a), (b), (c). D'autre part, vu la définition de $p(i)$ et $q(i)$, on sait que, pour $i \in J$, l'ensemble L^i_{n+1} n'est pas vide. Mais, $(H_i)_{i\in I}$ est une suite transfinie décroissante, et J une partie non dénombrable de I si bien que tout élément de I est majoré par un élément de J: il est alors clair que L^i_{n+1} est non vide pour tout $i \in I$. C'est fini....

5 Relations d'équivalence et sélections

Soit R une relation d'équivalence sur E. Une sélection pour R étant une application r de E dans E telle qu'on ait $xRy \Leftrightarrow r(x) = r(y)$, il est clair que trouver une sélection pour R revient à trouver une application f de E/R dans E sectionnant l'ensemble

$$H = \{(x, y) \in E/R \times E : x \in y\}.$$

Seulement, la structure mesurable naturelle de E/R (i.e. celle engendrée par les boréliens de E saturés pour R) est en général "déplorable", même pour une "bonne" relation d'équivalence sur un "bon" espace, si bien que, contrairement aux problèmes de section, le problème d'une "bonne" sélection ne reçoit, en général, qu'une réponse négative.

Il en est ainsi, par example, dans le cas classique suivant, dû à Vitali: E est le tore $\mathbb{R}/\mathbb{Z}$ à une dimension, R la relation d'équivalence: xRy ssi $x - y$ est rationnel. La relation R est $\mathcal{K}_\sigma$ (i.e. son graphe est un $\mathcal{K}_\sigma$ de $E \times E$), E est métrisable compact, les classes d'équivalence sont dénombrables, et pourtant il n'existe aucune sélection mesurable pour la mesure de Lebesgue.

Supposons E analytique et désignons par $\mathcal{B}(E)/R = \mathcal{B}(E/R)$ la tribu des boréliens de E saturés pour la relation d'équivalence R, tribu que nous identifions avec une tribu sur E/R. Il est claire que $(E/R, \mathcal{B}(E/R))$ est un espace de Blackwell *dès que* $\mathcal{B}(E/R)$ est séparable; mais là est le hic: elle est rarement séparable, et l'ensemble $H = \{(x, y) : x \in y\}$ défini ci-dessus n'est le noyau d'un schéma de Souslin sur $\mathcal{B}(E) \times \mathcal{B}(E/R)$ que si $\mathcal{B}(E/R)$ est séparable, auquel cas il appartient à la tribu produit $\mathcal{B}(E) \times \mathcal{B}(E/R)$ (exercice!). Pourtant, le théorème suivant assure qu'il existe, pour R analytique, assez bien de boréliens saturés pour R.

27 THÉORÈME *Soient E un espace analytique et R une relation d'équivalence analytique sur E. Alors*

(1) *le saturé pour R de toute partie analytique A de E est encore analytique (en particulier, les classes d'équivalences sont analytiques);*
(2) *si A et A' sont deux parties analytiques disjointes et saturées pour R, il existe deux boréliens disjoints et saturés B et B' tels que l'on ait $A \subseteq B$ et $A' \subseteq B'$.*

Démonstration Pour le point (1), il suffit de remarquer que le saturé de A est égal à $\pi_E[R \cap (A \times E)]$. Démontrer (2) demande plus d'astuce (c'est Mokobodzki qui m'a appris le "truc" il y a quelques années, mais il n'est pas le seul à l'avoir découvert). On commence, à l'aide du théorème de séparation ordinaire, par trouver un borélien B_1 contenant A et disjoint de A', puis on sature B_1 et obtient ainsi un nouvel analytique saturé A_1, contenant A et disjoint de A'. Itérant le processus, on obtient une suite croissante (B_n) de boréliens intercalée dans une suite croissante (A_n) d'analytiques saturés contenant A et disjoints de A'. Il ne reste plus qu'à poser $B = \lim_n \uparrow B_n = \lim_n \uparrow A_n$ et $B' = B^c$.

Remarque Le procédé utilisé pour démontrer (2) est plus général qu'il ne parait. Voir Dellacherie (1976*a*), Théorème 23.

Nous donnons à présent une condition nécessaire et suffisante pour que la tribu $\mathscr{B}(E/R)$ soit séparable, et montrons que $\mathscr{B}(E/R)$ ne peut être séparable que si R est une relation borélienne.

28 THÉORÈME *Soient E un espace analytique et R une relation d'équivalence sur E. Alors $\mathscr{B}(E/R)$ est séparable ssi il existe une suite (A_n) de parties analytiques de E saturées pour R vérifiant la condition: si $x, y \in E$ ne sont pas R-équivalents, il existe $p, q \in \mathbb{N}$ tels que l'on ait $x \in A_p$, $y \in A_q$ et $A_p \cap A_q = \varnothing$. De plus, si $\mathscr{B}(E/R)$ est séparable, alors R est une relation borélienne.*

Démonstration Supposons $\mathscr{B}(E/R)$ séparable et soit (G_n) une suite de boréliens saturés l'engendrant. On a xRy ssi $\forall n\ x \in G_n \Leftrightarrow y \in G_n$ et donc R est borélienne. Par ailleurs, quitte à stabiliser (G_n) pour les intersections finies et la complémentation, on peut supposer que les G_n forment une algèbre de Boole: il est alors clair que deux points non équivalents de E peuvent être séparés par deux éléments disjoints de (G_n). Réciproquement, si (A_n) est la suite de l'énoncé, le théorème précédent nous permet de trouver, pour tout couple (p, q) d'entiers tels que $A_p \cap A_q = \varnothing$, un borélien saturé B_q^p contenant A_p et disjoint de A_q. Alors, la tribu $\mathscr{G}$ engendrée par les B_q^p est séparable, $(E/R, \mathscr{G})$ est un espace de Blackwell et ses atomes sont les singletons de E/R. Pour $B \in \mathscr{B}(E/R)$, la tribu $\mathscr{G}^B$ engendrée par $\mathscr{G}$ et B ayant les mêmes propriétés, il résulte de I-36 que $\mathscr{G}$ est égale à $\mathscr{G}^B$, et donc que $\mathscr{G}$ est égale à $\mathscr{B}(E/R)$.

Nous donnons maintenant deux théorèmes classiques de sélection, résultant des théorèmes de section vus aux Sections 1 et 2. Tous deux supposent, au moins implicitement, que $\mathscr{B}(E/R)$ est séparable.

29 THÉORÈME *Soient E un espace analytique et R une relation d'équivalence sur E telle que $\mathscr{B}(E/R)$ soit séparable. Il existe alors une sélection universellement mesurable r pour R.*

Démonstration Nous identifions l'espace de Blackwell séparé $(E/R, \mathscr{B}(E/R))$ à un espace analytique F muni de sa tribu borélienne. L'ensemble $H=\{(x, y)\in E\times F: x\in y\}$, qui n'est autre que le graphe de la surjection canonique de E sur E/R, est borélien et donc analytique dans $E\times F$. D'après le Théorème 7 de Jankov et von Neumann, il existe une application f de F dans E, sectionnant H, et universellement mesurable (cf. 8), laquelle, par composition avec la surjection canonique, fournit notre sélection r.

30 THÉORÈME *Soient E un espace polonais et R une relation d'équivalence sur E vérifiant les conditions suivantes:*

(1) *les classes d'équivalence sont fermées dans E;*
(2) *le saturé de tout ouvert de E est un borélien.*

Il existe alors une sélection borélienne r pour R.

Démonstration Notons $[x]$ la classe d'équivalence de $x\in E$: les conditions (1) et (2) expriment tout simplement le fait que $x\to[x]$ est une application borélienne de E dans $\mathbf{F}(E)$. On obtient alors la sélection voulue en composant cette application avec le "bon" choix fourni par le théorème de Kuratowski et Ryll-Nardzewski 14.

Les remarques qui suivent contiennent en particulier les énoncés de quelques résultats récents (et difficiles).

31 *Remarques* (a) Si, dans 30, on pose $B=\{x\in E: r(x)=r[r(x)]\}$, on voit qu'il existe un borélien B rencontrant chaque classe en un point et un seul. On a un résultat analogue pour 29, avec B universellement mesurable.

(b) Si, dans 30, on suppose seulement E analytique, la remarque 14(b) implique l'existence d'une sélection universellement mesurable, mais on obtient en fait un cas particulier de 29 (exercice!). Il existe cependant un résultat classique, meilleur que 29, dans ce cas, que nous explicitons maintenant comme introduction à un résultat de Burgess. Désignons, pour E analytique, par $\mathscr{T}(E)$ la plus petite tribu sur E contenant $\mathscr{B}(E)$ et stable pour l'opération de Souslin: un résultat classique assure que tout élément de $\mathscr{T}(E)$ est universellement mesurable (on peut aussi déduire cela du théorème de capacitabilité "abstrait" de Choquet, dont nous n'avons pas parlé). On peut montrer que toute relation d'équivalence R sur E analytique, telle que les classes soient fermées et que le saturé de tout ouvert appartienne à $\mathscr{T}(E)$ (cette dernière condition est vérifiée si R est analytique), admet une sélection mesurable relativement à $\mathscr{T}(E)$ (exercice!). Noter que, dans ce cas, les saturés des ouverts engendrent une sous-tribu séparable de $\mathscr{T}(E)$ admettant les classes pour atomes.

(c) Burgess (198?) a démontré tout récemment qu'une relation d'équivalence R sur E analytique admet une sélection $\mathscr{T}(E)$-mesurable lorsque $\mathscr{T}(E)$ contient une sous-tribu séparable admettant les classes pour atomes.

(d) Miller (1978) et Srivastava (1979) ont démontré, tout récemment, que l'on pouvait, dans 30, supposer seulement que les classes sont des $\mathbf{G}_\delta$ (autrement dit, des sous-espaces polonais), auquel cas la tribu $\mathscr{B}(E/R)$ est encore séparable. Burgess en a déduit que son résultat, cité ci-dessus, s'appliquait, pour E polonais, au cas où les classes sont des $\mathbf{G}_\delta$ et le saturé de tout ouvert appartient à $\mathscr{T}(E)$.

Nous terminons ce chapitre en citant un beau résultat de Silver, dont Harrington a fourni une démonstration "simple pour logiciens" mise à la portée de beaucoup de bourses par Louveau (1976/77); on en trouvera une démonstration dans les notes de Kechris de ce volume. Si R est une relation d'équivalence *coanalytique* sur E polonais, alors ou bien R a $\aleph_0$ classes d'équivalence, ou bien R a $2^{\aleph_0}$ classes, auquel cas il existe un parfait non vide rencontrant chaque classe en au plus un point (ce parfait fournit donc une sélection incomplète mais néanmoins "substantielle"). Signalons que Burgess en a déduit, pour une relation R *analytique*, un résultat analogue mais (nécessairement) un peu moins bon: il faut ajouter la possibilité de l'existence de $\aleph_1$ classes d'équivalence, auquel cas on ne peut démontrer l'existence d'une "bonne" sélection partielle "substantielle" en général (c'est ce qui va nous arriver au Chapitre III). Noter que, R étant analytique ou coanalytique, s'il y a $\aleph_0$ classes, alors celles-ci sont boréliennes (I-24(a)), si bien que R est alors une relation borélienne très simple.

III Temps d'arrêt et ensembles analytiques

Soit $\mathbf{X} = (X_s)_{s \in S}$ (la suite vide ϕ étant un élément de S) un schéma de Souslin régulier sur $\mathscr{P}(E)$ et testons si un élément x de E appartient au noyau $\mathbf{N}[\mathbf{X}]$ de ce schéma. On a

(i) $x \in \mathbf{N}[\mathbf{X}]$ ssi $\exists \sigma\ \forall n\ x \in X_{\sigma|n}$,
(ii) $x \notin \mathbf{N}[\mathbf{X}]$ ssi $\forall \sigma\ \exists n\ x \notin X_{\sigma|n}$.

Appelons *arbre* tout sous-ensemble A de S vérifiant

$$t \in A \quad \text{et} \quad s \dashv t \Rightarrow s \in A$$

(faire un dessin pour apprécier le langage imagé utilisé); on voit que $\mathbf{X}$ associe naturellement à x l'arbre $A_x = \{s \in S : x \in X_s\}$ et que (i) exprime que A_x contient une *branche infinie* tandis que (ii) exprime que *toute branche de A_x est finie.* On peut encore voir la même chose d'une manière un peu différente—c'est la vue que nous adopterons ici, la vision arboricole étant développée dans les notes de Martin et Kechris (voir Part 4 de ce volume): pour $\sigma \in \mathbb{N}^{\mathbb{N}}$, posons

$$T_x(\sigma) = \inf\{n \geq 0 : x \notin X_{\sigma|n}\}$$

en convenant que $\sigma|0 = \phi$ et $\inf \varnothing = \infty$. On définit ainsi une application T_x de $\mathbb{N}^{\mathbb{N}}$ dans $\bar{\mathbb{N}}^\circ = \{0, 1, \ldots, \infty\}$ appelée *temps d'arrêt*; il est clair que (i) exprime que T a un *pôle* (i.e. $\exists \sigma\, T_x(\sigma) = \infty$) tandis que (ii) exprime que T est *fini.* Il est par ailleurs très facile de passer de la vision "arbre" à la vision "temps d'arrêt", et réciproquement, à l'aide des formules

$$T_x(\sigma) = \inf\{n \geq 0 : \sigma|n \notin A_x\},$$

$$s \in A_x \text{ ssi } \exists \sigma\, s \dashv \sigma \text{ et } T_x(\sigma) > |s| \text{ ssi } \forall \sigma\, s \dashv \sigma \Rightarrow T_x(\sigma) > |s|,$$

où $|s|$ est la langueur de la suite finie s. Ainsi, le temps d'arrêt (identiquement) 0 est associé à l'arbre vide, tandis que le temps d'arrêt 1 est associé à l'arbre $\{\phi\}$—subtil distinguo!

A la Section 1, nous définissons la notion générale de temps d'arrêt (en temps discret, dirait un probabiliste, les valeurs étant prises dans $\bar{\mathbb{N}}^\circ$) et montrons qu'elle permet une représentation "canonique" des schémas de Souslin (réguliers). Nous y donnons aussi un exemple remarquable d'ensemble analytique non borélien. La Section 2 est consacrée à l'étude de "la difficulté de tester si oui ou non x appartient à $\mathbf{N}[\mathbf{X}]$", soit encore à la complexité d'un temps d'arrêt, par l'intermédiaire de la suite transfinie des *dérivations* d'un temps d'arrêt. Nous donnons à la Section 3 des applications importantes de cette étude: deuxième théorème de séparation, théorème de réduction, *etc.* A la Section 4, on approfondit l'étude de la complexité d'un temps d'arrêt afin de démontrer le théorème de Kondô sur l'uniformisation d'une partie coanalytique d'un espace produit.

1 Temps d'arrêt et ensembles κ-sousliniens

Soit J un ensemble infini, muni de la topologie discrète, que l'on pourra identifier avec son cardinal. lui-même identifié au plus petit ordinal de même cardinalité—exceptionnellement, dans ce paragraphe, nous sortirons un peu de nos espaces ambiants métrisables séparables, ce qui nous permettra une petite incursion du côté des ensembles "$\aleph_n$-sousliniens" dont a parlé Stone (en utilisant une autre terminologie!), mais, très vite, nous prendrons $J = \mathbb{N} = \aleph_0 =$ l'ensemble des entiers naturels[1] avec juste un coup d'oeil, à la Section 4, sur le cas où $J = I = \aleph_1 =$ l'ensemble des ordinaux dénombrables. Soient donc J un ensemble infini muni de la topologie discrète et $\Omega = J^{\mathbb{N}}$ muni de la topologie produit; nous étendons à Ω les notations introduites pour $\Sigma = \mathbb{N}^{\mathbb{N}}$.

1 DÉFINITION Un *temps d'arrêt* (en abrégé: *t.d'a.*)[2] *sur Ω* est une application T de Ω dans $\bar{\mathbb{N}}°$ vérifiant le "*test de Galmarino*", i.e.

$$\forall n \geq 0 \; \forall \omega, w \in \Omega, t(\omega) > n \text{ et } \omega|n = w|n \quad \Rightarrow \quad T(w) > n$$

(où l'on peut remplacer le signe "$>$", *en ses deux occurrences*, par l'un des signes suivants: $\geq$, $<$, $\leq$, $=$).

La terminologie adoptée vient de la pratique probabiliste: considérons un "processus" évoluant au cours du temps (échelle de temps discrète $\mathbb{N}$) et prenant ses "états" dans l'ensemble J. Un point ω de Ω (i.e. une application de $\mathbb{N}$ dans J) est alors une "trajectoire" typique de notre processus. Décidons maintenant d'observer les états successifs de notre processus aux instants $1, 2, \ldots$ jusqu'au moment T où un certain phénomène lié au déroulement du processus (ou de l'horloge! une fonction constante est un t.d'a.) apparait: il est clair que la fonction T est alors un t.d'a. au sens de notre définition 1.

2 Supposons que l'on ait $J = J_1 \times J_2$ où J_1 et J_2 sont deux ensembles infinis (par exemple, $J = \mathbb{N} \times \mathbb{N}$), et identifions $J^{\mathbb{N}} = \Omega$ au produit $J_1^{\mathbb{N}} \times J_2^{\mathbb{N}} = \Omega_1 \times \Omega_2$. Un t.d'a. sur Ω devient alors un t.d'a. "à deux variables" sur $\Omega_1 \times \Omega_2$, i.e. une application T de $\Omega_1 \times \Omega_2$ dans $\bar{\mathbb{N}}°$ vérifiant le test de Galmarino

$$\forall n \geq 0 \; \forall \omega^1, w^1 \in \Omega_1 \; \forall \omega^2, w^2 \in \Omega_2, T(\omega^1, \omega^2) > n \text{ et}$$
$$\omega^1|n = w^1|n \text{ et } \omega^2|n = w^2|n \quad \Rightarrow \quad T(w^1, w^2) > n.$$

On remarquera que, pour $\omega^1 \in \Omega_1$ fixé, l'application $T(\omega^1, .)$ est un t.d'a. sur Ω_2, et de même pour $\omega^2 \in \Omega_2$ fixé.

3 Tout temps d'arrêt est évidemment une fonction continue pour la topologie que nous avons mise sur Ω ($\bar{\mathbb{N}}°$ étant le compactifié d'Alexandrov de $\mathbb{N}°$), et est donc bien déterminé par sa restriction à l'ensemble Ω_s des

[1] Il y a une escroquerie en 0. On prendra aussi $J = \mathbb{N} \times \mathbb{N}$.

[2] En fait, un temps d'arrêt "grossier" car, si J n'est pas dénombrable, un probabiliste demanderait en plus une certaine "mesurabilité" de T.

suites stationnaires[1] qui a même cardinalité que J. Par ailleurs, il est facile de voir que le sup ou l'inf de toute famille de t.d'a. est encore un t.d'a. On en déduit que ce sup ou inf est atteint pour une sous-famille de cardinalité inférieure ou égale à celle de J, et aussi que l'ensemble $\mathcal{T}$ des t.d'a. sur Ω est une lattice complète ayant même cardinalité que $\mathbb{N}^J$. *Nous munissons d'autre part l'ensemble $\mathcal{T}$ de la topologie de la convergence simple*: une famille filtrante (T_α) de t.d'a. converge vers le t.d'a. T ssi $T_\alpha(\omega)$ converge vers $T(\omega)$ pour tout $\omega \in \Omega$. Comme une limite simple de t.d'a. est encore un t.d'a., l'espace $\mathcal{T}$ est compact[2] et, la famille d'applications $T \to T(\omega)$, ω parcourant Ω_s, séparent les points de $\mathcal{T}$, on en déduit que $\mathcal{T}$ a une base de cardinalité égale à celle de J. En particulier, *si J est dénombrable, $\mathcal{T}$ est un espace compact à base dénombrable, et donc métrisable.*

4 Nous dirons que $\omega \in \Omega$ est un *pôle* du t.d'a. T si $T(\omega) = \infty$; un t.d'a. fini est donc un t.d'a. sans pôles. L'ensemble des pôles d'un t.d'a. T est un fermé de Ω; réciproquement, si H est un fermé de Ω, il existe un plus petit t.d'a. T admettant H comme ensemble de ses pôles; ce t.d'a. est défini par

$$T(\omega) = \inf \{n \geq 0 \colon \forall w\ w|n = \omega|n \Rightarrow w \notin H\}.$$

D'autre part, on vérifie aisément que l'application $(T, \omega) \to T(\omega)$ de $\mathcal{T} \times \Omega$ dans $\bar{\mathbb{N}}^\circ$ est continue (du couple) si bien que l'ensemble $\mathcal{P}$ des t.d'a. à pôles, soit $\mathcal{P} = \{T \colon \exists \omega\ T(\omega) = \infty\}$, est la projection du fermé $\{(T, \omega) \colon T(\omega) = \infty\}$ de $\mathcal{T} \times \Omega$ sur $\mathcal{T}$,

5 Posons $\kappa = \text{card } J$; une partie X d'un espace séparé E est dite *κ-souslinienne* si elle est projection sur E d'un fermé de $E \times J^{\mathbb{N}} = E \times \Omega$ (nous verrons un peu plus loin la version "schéma de Souslin" correspondante). Ainsi, $\mathcal{P}$ est une partie κ-souslinienne de $\mathcal{T}$; nous verrons bientôt que l'ensemble $\mathcal{Q}$ des t.d'a. finis, soit $\mathcal{Q} = \mathcal{P}^c = \{T \colon \forall \omega\ T(\omega) < \infty\}$, n'est pas κ-souslinien dans $\mathcal{T}$. Lorsque l'espace E est métrisable séparable, et J dénombrable, la notion de partie $\aleph_0$-souslinienne coincide avec celle de partie E-analytique (cf. II-1 et II-6(2)); ainsi, si J est dénombrable, $\mathcal{P}$ est analytique et $\mathcal{Q}$ est coanalytique, aucun des deux n'étant borélien. Par ailleurs, disons qu'une application $f \colon x \to T_x$ de E dans $\mathcal{T}$ est *semi-continue supérieurement* (en abrégé, *s.c.s.*) si, pour tout $\omega \in \Omega$, la fonction $x \to T_x(\omega)$ est s.c.s., ce qui, vu le test de Galmarino, revient à dire que, pour tout $n \geq 0$, l'ensemble $\{(x, \omega) \colon T_x(\omega) \geq n\}$ est fermé dans $E \times \Omega$. En particulier, toute application continue de E dans $\mathcal{T}$ est s.c.s.; c'est le cas, par exemple, de l'application identique de $\mathcal{T}$ dans lui-même.

[1] (Pour le lecteur de langue anglaise.) Stationnaire = eventually constant. Attention: eventually ≠ éventuellement = possibly!

[2] Et totalement discontinu.

6 THÉORÈME *Soit E un espace séparé. Une partie X de E est κ-souslinienne ssi il existe une application s.c.s. $f: x \to T_x$ de E dans $\mathcal{T}$ telle que $X = f^{-1}(\mathcal{P}) = \{x: \exists\omega\, T_x(\omega) = \infty\}$.*

Démonstration Si f est une application s.c.s. de E dans $\mathcal{T}$, l'ensemble $f^{-1}(\mathcal{P})$ est projection sur E du fermé $\{(x, \omega): T_x(\omega) = \infty\}$ de $E \times \Omega$, et est donc κ-souslinien dans E. Réciproquement, soit X une partie κ-souslinienne de E, projection sur E d'un fermé H de $E \times \Omega$. On a alors $X = f^{-1}(\mathcal{P})$, où $f: x \to T_x$ est définie par

$$T_x(\omega) = \inf\{n \geq 0: \forall w\; w|n = \omega|n \Rightarrow (x, w) \notin H\}$$

(T_x est le t.d'a. associé à la coupe $H(x)$ suivant le procédé du 4), et une application simple du test de Galmarino montre que f est s.c.s.

Remarque Lorsque, par exemple, $J = \mathbb{N}$ et $E = \mathbb{N}^{\mathbb{N}}$, on peut obtenir une meilleure application f que celle introduite dans la démonstration. En effet, soit X une partie analytique de $\mathbb{N}^{\mathbb{N}}$, projection d'un fermé H de $\mathbb{N}^{\mathbb{N}} \times \mathbb{N}^{\mathbb{N}}$, et associons cette fois à H tout entier le t.d'a. sur $(\mathbb{N} \times \mathbb{N})^{\mathbb{N}} = \mathbb{N}^{\mathbb{N}} \times \mathbb{N}^{\mathbb{N}}$ fourni par le no. 4: c'est, d'après le no. 3, un t.d'a. T à deux variables σ, $\tau \in \mathbb{N}^{\mathbb{N}}$. L'application $\sigma \to T_\sigma$ de $\mathbb{N}^{\mathbb{N}}$ dans $\mathcal{T}$ définie par $T_\sigma(\tau) = T(\sigma, \tau)$ est alors continue, et on a

$$\sigma \in X \quad \text{ssi} \quad \exists\tau\; T(\sigma, \tau) = \infty\,.$$

Cette représentation de X jouera un rôle important à la Section 4.

7 THÉORÈME *L'ensemble $\mathcal{Q}$ des t.d'a. finis n'est pas une partie κ-souslinienne de l'espace compact $\mathcal{T}$.*

Démonstration Raisonnons par l'absurde. Si $\mathcal{Q}$ est κ-souslinien, il existe une application s.c.s. $f: x \to T_x$ de $\mathcal{T}$ dans lui-même telle que $\mathcal{Q} = f^{-1}(\mathcal{P})$. Soit alors, pour tout $x \in \mathcal{T}$, $U_x = \inf(x, T_x)$. U_x est un t.d'a. fini pour tout $x \in \mathcal{T}$ et l'application $x \to U_x$ est encore s.c.s. Pour $\omega \in \Omega$ fixé, $x \to U_x(\omega)$ est une application s.c.s. finie de $\mathcal{T}$ dans $\bar{\mathbb{N}}^{\circ}$; par conséquent, $U(\omega) = \sup_{x \in \mathcal{T}} U_x(\omega)$ est fini et donc $U = \sup_x U_x$ est un élément de $\mathcal{Q}$. Soit alors $x_0 = U + 1$. Comme, pour tout $x \in \mathcal{T}$, on a $x(\omega) \wedge T_x(\omega) = U_x(\omega) < x_0(\omega)$ pour tout $\omega \in \Omega$, on a $T_{x_0}(\omega) < x_0(\omega) < \infty$ pour tout $\omega \in \Omega$; ce qui est absurde car, le t.d.'a. x_0 étant fini, T_{x_0} doit avoir au moins un pôle.

Remarque Tout ce qu'on a pu faire jusqu'ici est en fait valable lorsque J est fini non vide (à condition d'adapter les "cardinalités" du no. 3). Seulement, dans ce cas, Ω étant compact, tout t.d'a. fini est borné, et $\mathcal{P}$ est une partie compacte de $\mathcal{T}$. Le marteau-pilon que constitue alors la démonstration précédente montre que la noisette $\mathcal{Q}$, qui est ouverte, n'est pas fermée: ce n'était donc pas la peine de l'écraser . . ..

8 Avant d'en venir aux schémas de Souslin, nous introduisons une opération très simple sur les t.d'a., qui nous sera utile tout au long de ce chapitre (nous laisserons au lecteur le soin de trouver l'analogue de cette opération dans le langage des arbres). Dans les nos. 8 et 9, nous nous permettrons d'écrire S l'ensemble des suites finies d'éléments de J, même pour $J \neq \mathbb{N}$: d'où les notations $s \in S$, $s \dashv t$, etc., auxquelles nous ajouterons les notations $s.t$, $s.\omega$ qui désignerons l'élément de S, Ω obtenu en concaténant t, ω à la droite de s. Ceci dit, pour $T \in \mathcal{T}$ et $s \in S$, on définit un nouveau t.d'a. T^s en posant, pour tout $\omega \in \Omega$,

$$T^s(\omega) = (T(s.\omega) - |s|)^+ ,$$

où $|s|$ est la longueur de s et $a^+ = \sup(a, 0)$. Comme le t.d'a. identiquement nul est le seul t.d'a. à prendre jamais la valeur 0, on a l'avatar suivant du test de Galmarino

$$T^s \neq 0 \quad \text{ssi} \quad T^s \geqslant 1 \text{ ssi } \exists\omega\; s \dashv \omega \text{ et } T(\omega) > |s| \quad \text{ssi} \quad \forall\omega\; s \dashv \omega \Rightarrow T(\omega) > |s| .$$

Par ailleurs, pour s fixé, l'application $T \to T^s$ est continue et, pour s, $t \in S$, on a trivialement $(T^s)^t = T^{s.t}$.

9 Désignant toujours par κ le cardinal de J et par E un espace séparé, nous appellerons *κ-schéma de Souslin dans E* (sous-entendu: régulier, sur les fermés de E—nous n'en verrons point d'autres) une famille $\mathbf{X} = (X_s)_{s \in S}$ de parties fermées de E, indexée par S, telle que $s \dashv t$ implique $X_s \supseteq X_t$; le *noyau* du κ-schéma $\mathbf{X}$ est alors l'ensemble $\mathbf{N}[\mathbf{X}] = \sup_{\omega \in \Omega} \inf_n X_{\omega|n}$. A tout κ-schéma de Souslin $\mathbf{X}$ nous associons l'application $f: x \to T_x$ de E dans $\mathcal{T}$ définie par

$$T_x(\omega) = \inf \{n \geqslant 0 \colon x \notin X_{\omega|n}\} ,$$

qui est évidemment s.c.s., et qui est continue ssi les X_s sont à la fois ouverts et fermés. Réciproquement, à toute application s.c.s. $f: x \to T_x$ de E dans $\mathcal{T}$, nous associons le κ-schéma $\mathbf{X}$ défini par $x \in X_s$ ssi $\forall\omega\; s \dashv \omega \Rightarrow T_x(\omega) > |s|$ (ssi $\exists\omega\; s \dashv \omega$ et $T_x(\omega) > |s|$), ce qui s'écrit tout simplement, en tenant compte du no. 8,

$$x \in X_s \quad \text{ssi} \quad T_x^s \neq 0 .$$

Les applications $\mathbf{X} \to f$ et $f \to \mathbf{X}$ ainsi définies étant inverses l'une de l'autre, on voit qu'il existe une correspondance biunivoque naturelle entre κ-schémas et applications s.c.s. Noter que l'on a

$$x \in \mathbf{N}[\mathbf{X}] \quad \text{ssi} \quad x \in f^{-1}(\mathcal{P}) ,$$

si bien qu'une partie X de E est κ-souslinienne ssi elle est le noyau d'un κ-schéma de Souslin dans E d'après le théorème 6. L'ensemble $\mathcal{P}$ est, quant à lui, noyau du schéma correspondant à l'application identique de $\mathcal{T}$: on a $X_s = \{T: T^s \neq 0\}$.

2 Dérivation et indice d'un temps d'arrêt

Dans tout ce paragraphe, J est un ensemble infini dénombrable que l'on pourra identifier à $\mathbb{N}$. Nous noterons donc S l'ensemble des suites finies d'éléments de J, mais nous continuerons à noter Ω l'ensemble $J^{\mathbb{N}}$ par simple réflexe probabiliste. Les notations $\mathcal{T}$, $\mathcal{P}$, etc., ont été définies à la Section 1—nous n'y revenons pas.

10 DÉFINITION On appelle *dérivé* du t.d'a. T le t.d'a. T' défini par

$$T' = \inf\{U \in \mathcal{T}: U+1 \geq T\},$$

et on appelle *dérivation* dans $\mathcal{T}$ l'application $d: T \to T'$ de $\mathcal{T}$ dans lui-même. Le t.d'a T est dit *parfait* si $d(T) = T$.

Voici quelques propriétés élémentaires importantes de l'opération de dérivation:

11 THÉORÈME (1) *On a l'équivalence, pour tout* $n \geq 0$,

$$T'(\omega) > n \quad \text{ssi} \quad \exists w\ \omega|n = w|n \quad \text{et} \quad T(w) > n+1\,.$$

(2) *Pour tout* $s \in S$, *on a* (cf. 8)

$$(T^s)' = (T')^s \text{ (noté } T'^s)\,.$$

(3) *On a, pour* $U, V \in \mathcal{T}$,

$$d(U) \leq U, \qquad U \leq V \Rightarrow d(U) \leq d(V)\,.$$

(4) *L'application d est borélienne* (*et même de* 1 *ère classe de Baire*).

Démonstration Nous laissons (1) au lecteur comme exercice; regarder aussi l'analogue de la dérivation dans le langage des arbres. (2) est conséquence immédiate de (1) et (3) est trivial. Enfin, d'après (1), l'application $T \to T'(\omega)$ est, pour ω fixé, s.c.i., d'où 4).

12 Soit maintenant I l'ensemble des ordinaux dénombrables et définissons, pour $T \in \mathcal{T}$, la *suite transfinie des dérivés successifs* $(T^i)_{i \in I}$ de T, par récurrence transfinie, comme suit[1]

$$T^0 = T, \qquad T^1 = T',$$

$$T^j = d(\inf_{i<j} T^i) \text{ (et donc } T^j = d(T^{j-1}) \text{ si } j \text{ est de 1-ère espèce)}.$$

Il est clair que $(T^i)_{i\in I}$ est une suite transfinie décroissante; il existe donc (cf. 3) un plus petit ordinal dénombrable $j(T)$ tel que l'on ait $T^j = T^{j(T)}$ pour tout $j \geqslant j(T)$: $T^{j(T)}$ est alors le plus grand t.d'a. parfait majoré par T, et $j(T)$ est parfois appelé l'indice de Hausdorff de T. Mais nous réserverons le nom d'*indice* de T à l'*indice de Lusin et Sierpiński* $i(T)$ défini par

$$i(T) = \inf\{i \in I: T^i = 0\} \; (= j(T) \text{ ssi } T^{j(T)} = 0),$$

$$i(T) = \aleph_1 \text{ si } \{i \in I: T^i = 0\} \text{ est vide}.$$

Il n'est pas difficile de voir que l'on a $i(T) < \aleph_1$ ssi T est fini (exercice!), soit, avec nos notations,

$$i(T) < \aleph_1 \quad \text{ssi} \quad T \in \mathscr{Q},$$

ni, non plus (autre exercice!), que l'application $T \to i(T)$ de $\mathscr{Q}$ dans I est surjective—même si cela surprend le novice—et, qu'en particulier, on a

$$\sup_{T\in\mathscr{Q}} i(T) = \aleph_1$$

(nous identifions $\aleph_1$ avec le premier ordinal non dénombrable). Pour $j \in I$, un ensemble de la forme

$$\mathscr{Q}_j = \{T: i(T) = j\}$$

est appéle un *constituant* de $\mathscr{Q}$. Comme la dérivation est une application borélienne, il est facile de voir, par récurrence transfinie, que les constituants sont boréliens, si bien que l'ensemble coanalytique non borélien $\mathscr{Q}$ s'écrit comme réunion de $\aleph_1$ boréliens disjoints (noter, au passage, que la relation d'équivalence $\mathbf{R}$ sur $\mathscr{T}$ définie par $S\mathbf{R}T$ ssi $i(S) = i(T)$ a exactement $\aleph_1$ classes. Nous verrons plus loin qu'elle est analytique). Par ailleurs, il est possible aussi de décomposer $\mathscr{P}$ en $\aleph_1$ boréliens disjoints. (Exercice! Utiliser l'indice de Hausdorff.)

13 Pour tout t.d'a. *fini* T, nous définissons une application i_T de S dans I, appelée *fonction-indice* de T, comme suit

$$i_T(s) = i(T^s).$$

[1] Il est possible aussi de poser $T^j = \inf_{i<j} d(T^i)$ (cf. II-24), mais alors l'indice $i(T)$ défini plus loin serait toujours un ordinal de 1-ère espèce. Cela aurait pu être évité en excluant de $\mathscr{T}$ le t.d'a. 0—le t.d'a. 1 devenant alors le plus petit élément de $\mathscr{T}$—mais alors l'opération du no. 8 s'écrit moins simplement.

Elle jouera un rôle crucial à la Section 4. Elle nous permet ici de résumer en quelques lignes la construction de l'indice $i(T) = i_T(\phi)$ en écrivant (exercice!) (ne pas confondre T^s, $s \in S$, avec T^i, $i \in I$!)

$$\begin{cases} T^s = 0 & \text{ssi} \quad i_T(s) = 0\,, \\ T^s \neq 0 & \text{ssi} \quad i_T(s) = \sup\limits_{k \in J} [i_T(s\,.\,k) + 1]\,, \end{cases}$$

où l'on a identifié l'élément k de J à la suite de longueur 1 correspondante.

Nous allons étudier maintenant la relation de préordre $i(S) \leqslant i(T)$. Pour ce faire, nous introduisons d'abord un nouveau concept important, venant de la théorie des jeux.

14 DÉFINITION Une application f de S dans S est une *stratégie* si

(1) $\forall s \in S\; |f(s)| = |s|$,
(2) $\forall s, t \in S\; s \dashv t \Rightarrow f(s) \dashv f(t)$.

On peut aussi voir une stratégie f comme une application $\bar{f}$ de Ω dans Ω (poser $\bar{f}(\omega)|n = f(\omega|n)$) vérifiant la condition

$$\forall n \geqslant 0\; \forall \omega, w \in \Omega\; \omega|n = w|n \Rightarrow \bar{f}(\omega)|n = \bar{f}(w)|n\,,$$

et nous nous permettrons de noter aussi f au lieu de $\bar{f}$ cet autre aspect d'une stratégie f. Il est alors clair que, pour tout temps d'arrêt T, l'application ${}^fT = T \circ f$ définie par ${}^fT(\omega) = T[f(\omega)]$ est encore un temps d'arrêt. En fait, cette propriété caractérise les stratégies (exercice!).

15 THÉORÈME *Soient U, V deux t.d'a. et $\mathcal{S}$ l'ensemble des stratégies. On a l'équivalence*

$$i(U) \leqslant i(V) \quad \textit{ssi} \quad \exists f \in \mathcal{S}\; U \leqslant {}^fV.$$

Démonstration Il est clair que l'on a $({}^fV)' \leqslant {}^f(V')$ et donc, par récurrence transfinie, $i({}^fV) \leqslant i(V)$. Ceci nous montre que $U \leqslant {}^fV$ implique $i(U) \leqslant i(V)$. Passons à l'autre sens. Supposons d'abord que $i(V) = \aleph_1$; V a alors un pôle ω_0 et il suffit de poser $f(\omega) = \omega_0$ pour tout $\omega \in \Omega$ pour obtenir une stratégie f telle que $U \leqslant {}^fV$. Supposons maintenant que V (et donc U) soit fini. Nous allons alors construire à l'aide du no. 13, en raisonnant par récurrence sur la longueur d'une suite $s \in S$, une stratégie f de S dans S vérifiant, pour tout $s \in S$, la condition $(\circ)$ suivante

$$i_U(s) \leqslant i_V[f(s)]\,. \tag{$\circ$}$$

On sera alors assuré que $U^s \neq 0$ implique $V^{f(s)} \neq 0$ et donc, puisque $U(\omega) = \inf\{n \geq 0\colon U^{\omega|n} = 0\}$, que l'on a bien $U \leq^f V$. Nous posons donc $f(\phi) = \phi$ (la condition (∘) est vérifiée pour $s = \phi$ par hypothèse) et nous supposons construite f sur l'ensemble S_n des suites de longueur $\leq n$ de sorte que (∘) soit vérifiée pour tout $s \in S_n$. Soient alors p un élément de J et $t = s \,.\, p$ un élément de S_{n+1}. Si on a $i_U(s \,.\, p) = 0$, nous posons $f(s \,.\, p) = f(s) \,.\, q$ où q est un élément quelconque de J. Si on a $i_U(s \,.\, p) \neq 0$, alors on a, d'après 13,

$$0 \neq i_U(s) = \sup_{p \in J} [i_U(s \,.\, p) + 1]$$
$$\leq i_V(f(s)) = \sup_{q \in J} [i_V(f(s) \,.\, q) + 1] \,.$$

Il est alors possible de choisir un $q \in J$ tel que $i_V(f(s) \,.\, q)$ majore $i_U(s \,.\, p)$, et il n'y a plus alors qu'à poser $f(s \,.\, p) = f(s) \,.\, q$.

Munissons l'ensemble S de la topologie discrète, et l'ensemble S^S de la topologie de la convergence simple (si bien que S^S est homéomorphe à $\mathbb{N}^{\mathbb{N}}$): l'ensemble $\mathscr{S}$ des stratégies est alors, trivialement, un fermé de S^S, et donc un espace polonais pour la topologie de la convergence simple. De plus, il est facile de voir que l'application $(f, T) \to {}^fT$ de $\mathscr{S} \times \mathscr{T}$ dans $\mathscr{T}$ est continue. On en déduit que la relation de préordre $i(U) \leq i(V)$ est analytique, et même mieux:

16 THÉORÈME (1) *La relation de "pré-bon-ordre" $i(U) \leq i(V)$ sur $\mathscr{T}$ est analytique.*

(2) *Les relations de préordre partiel sur $\mathscr{T}$ suivantes*

$$U \in \mathscr{Q} \text{ et } i(U) \leq i(V)\,, \qquad U \in \mathscr{Q} \text{ et } i(U) < i(V)\,,$$

sont toutes deux coanalytiques. Suivant les logiciens, on dit que l'application $i(.)$ est une norme coanalytique (*ou une Π^1_1*-norme).

Démonstration Le point (1), et par passage au complémentaire, la deuxième assertion de (2) sont déjà démontrés. Pour démontrer la première assertion de (2), il suffit de savoir qu'il existe une application borélienne $T \to T^*$ de $\mathscr{T}$ dans $\mathscr{T}$ telle que $i(T^*) = i(T) + 1$ pour $T \in \mathscr{Q}$ et $i(T^*) = i(T)$ pour $T \in \mathscr{P}$. On en obtient une, continue, en posant $T^*(\omega) = T(w) + 1$ avec $\omega = \omega|1 \,.\, w$ (noter que $T^{*k} = T$ pour tout $k \in J$).

Remarques (a) Il résulte de (2) que la restriction de la relation $i(U) \leq i(V)$ à $\mathscr{Q}$ est $\mathscr{Q}$-bianalytique. Nous verrons à 20 qu'elle n'est pas borélienne.

(b) Hillard (1978) a montré que la dérivation de Cantor dans un espace métrisable compact (cf. II-24) conduit aussi à une norme coanalytique.

(c) Supposons un instant que, sachant ce qu'est une stratégie, nous ignorions les notions de dérivation, indice, etc. Soit alors R la relation $\exists f \in \mathscr{S}$ $U \leq^f V$ sur $\mathscr{T}$. Cette relation est analytique, et la seule chose qu'il nous faudrait encore savoir sur R pour le Section 3 est que R est totale. Et cela peut se voir simplement en raisonnant par l'absurde (c'est en fait un cas particulier, mais typique, du théorème de Gale et Stewart—voir Part 4 de ce volume, les notes de Martin et Kechris).

Voici enfin le *théorème de la borne* pour notre indice.

17 THÉORÈME *Si A est une partie analytique de $\mathscr{T}$ contenue dans $\mathscr{Q}$, on a*

$$\sup_{T \in A} i(T) < \aleph_1 .$$

Démonstration Nous raisonnons par l'absurde. Si $\sup_{T \in A} i(T)$ vaut $\aleph_1$, on peut écrire

$$U \in \mathscr{Q} \text{ ssi } \exists V \; V \in A \text{ et } i(U) \leq i(V) .$$

Mais, l'ensemble $\{U : \exists V \; V \in A \text{ et } i(U) \leq i(V)\}$ est analytique d'après ce qui précède, si bien que $\mathscr{Q}$ serait analytique. C'est fini.

Remarque Nous avons déduit ici le théorème de la borne du fait que $\mathscr{Q}$ n'est pas analytique. Au Chapitre IV, dans un contexte plus général, nous ferons l'inverse: nous démontrerons le théorème de la borne et déduirons la non-analyticité de $\mathscr{Q}$ de $\sup_{T \in \mathscr{Q}} i(T) = \aleph_1$.

3 Théorèmes de Séparation et applications

Les conventions adoptées au début du Chapitre II sont de nouveau en vigueur. Le lecteur pourra se reporter par ailleurs au préambule de ce Chapitre II pour se remettre en mémoire nos définitions des mots: E-analytique, E-coanalytique, E-bianalytique, etc. Enfin, dans toute cette section, on prend $J = \mathbb{N}$ et donc $\Omega = \Sigma = \mathbb{N}^{\mathbb{N}}$.

A. *Le premier théorème de séparation, revisité*

Nous commençons par résumer, en quelques lignes, l'essentiel de la théorie des constituants.

18 THÉORÈME *Soit $f : x \to T_x$ une application s.c.s. de E dans $\mathscr{T}$ et posons $i(x) = i(T_x)$ pour tout $x \in E$ et $A = f^{-1}(\mathscr{P}) = \{x : i(x) = \aleph_1)$.*

(1) *L'ensemble A est une partie E-analytique de E, et toute partie E-analytique admet une telle représentation.*

(2) *L'ensemble $A^c = E - A$ est la réunion des $\aleph_1$ boréliens $B_j = \{x: i(x) = j\}$, $j \in I$, appelés* constituants *de A^c.*

(3) *Si A' est un sous-espace analytique de E disjoint de A, on a* $\sup_{x \in A'} i(x) < \aleph_1$, *si bien que A' est contenu dans une réunion dénombrable de constituants de A^c.*

(4) *Si l'espace E est analytique, alors A est borélien ssi on a*

$$\sup_{x \notin A} i(x) < \aleph_1 .$$

Démonstration Le point (1) résulte de 6, et (2) de 12 une fois remarqué qu'une application s.c.s. est borélienne. Le point (3) résulte du théorème de la borne: en effet, f est borélienne sur A' et $f(A')$ est une partie analytique de $\mathcal{T}$ contenue dans $\mathcal{Q}$. Enfin, si E est analytique, A^c est un sous-espace analytique si A est borélien, d'où la majoration d'après (3); l'autre sens résulte de (2).

Nous rappelons maintenant un énoncé du premier théorème de séparation (cf. I-22), en nous contentant d'un couple d'ensembles analytiques.

19 THÉORÈME *Soient A_1 et A_2 deux sous-espaces analytiques de E. Si A_1 et A_2 sont disjoints, il existe deux boréliens disjoints B_1, B_2 de E tels que l'on ait $A_1 \subseteq B_1$ et $A_2 \subseteq B_2$.*

Démonstration La démonstration que nous donnons maintenant est très différente de celle du Chapitre I. Soit f une application s.c.s. de E dans T telle que $A_2 = f^{-1}(\mathcal{P})$. D'après 18(3), A_2 est contenu dans la réunion B_1 d'une suite de constituants de A_2 relatifs à f: il n'y a plus qu'à poser $B_2 = B_1^c$. Noter que l'on pouvait se contenter de supposer que A_2 était E-analytique.

Nous remplissons maintenant notre promesse d'exhiber un couple de coanalytiques disjoints non séparables par des boréliens. Pour la démonstration, nous supposons le lecteur un peu au courant de l'existence de la classification de Baire des boréliens.

20 THÉORÈME *Soient C_1 et C_2 les parties disjointes de $\mathcal{T} \times \mathcal{T}$ suivantes*

$$C_1 = \{(U, V): i(U) < i(V)\}, \qquad C_2 = \{(U, V): i(U) > i(V)\} .$$

Elles sont coanalytiques, et tout borélien de $\mathcal{T} \times \mathcal{T}$ contenant C_1 rencontre C_2.

Démonstration La coanalyticité résulte de 16. Nous raisonnons par l'absurde et supposons que H soit un borélien contenant C_1 et disjoint de C_2; nous allons en conclure que la classification de Baire dans Ω est bornée—ce qui est faux. A tout borélien B de Ω faisons correspondre une application continue f_B de Ω dans $\mathcal{T}$ telle que l'on ait $B^c = \Omega - B = \overset{-1}{f_B}(\mathcal{P})$—cf. la remarque de 6; d'après 18, on a $\sup_{\omega \in B} i[f_B(\omega)] < \aleph_1$. Soit alors T_B un t.d.'a. fini dont l'indice majore strictement cet ordinal dénombrable, et soit g_B l'application continue de Ω dans $\mathcal{T} \times \mathcal{T}$ définie par $g_B(\omega) = (f_B(\omega), T_B)$: on a alors $B = \overset{-1}{g_B}(C_1) = \overset{-1}{g_B}(H)$, si bien que la classe de Baire de B dans Ω est majorée par celle de H dans $\mathcal{T} \times \mathcal{T}$. C'est fini.

B. Le deuxième théorème de séparation, première visite

Nous commençons par l'énoncé classique du "deuxième principe" de Lusin, qui porte sur un couple d'ensembles analytiques, et nous l'améliorons ensuite en lui donnant une forme comparable à celle de 19. Mais, ce n'est que sous la rubrique C que nous le donnerons dans toute sa force (i.e. pour une suite finie ou *infinie* d'ensembles analytiques).

21 THÉORÈME *Soient A_1 et A_2 deux sous-espaces analytiques de E. Il existe deux parties E-coanalytiques disjointes C_1, C_2 de E telles que l'on ait $(A_1 - A_2) \subseteq C_1$ et $(A_2 - A_1) \subseteq C_2$.*

Démonstration Soit f_k, pour $k = 1, 2$, une application s.c.s. de E dans $\mathcal{T}$ telle que $A_k = \overset{-1}{f_k}(\mathcal{P})$ et posons $i_k(x) = i[f_k(x)]$ pour tout $x \in E$. Il suffit alors de prendre $C_1 = \{x: i_1(x) > i_2(x)\}$ et $C_2 = \{x: i_2(x) > i_1(x)\}$ d'après 16.

Remarque Comme on ne prend que des images réciproques relatives aux f_k, la démonstration est encore valable si on suppose seulement que A_1 et A_2 sont des parties E-analytiques de E. Mais je profite de l'occasion pour donner un exemple de "renforcement automatique" d'un énoncé, par plongement. Admettons l'énoncé 21, tel qu'il est, en supposant même seulement E métrisable compact (si bien que "E-coanalytique" a une valeur absolue), et soient maintenant A'_k, $k = 1, 2$, des parties F-analytiques d'un espace F. Plongeons F dans un espace métrisable compact E, choisissons des sous-espaces analytiques A_k de E tels que les A'_k soient traces des A_k sur F, puis des sous-espaces coanalytiques C_k de E séparant $(A_1 - A_2)$ et $(A_2 - A_1)$: alors les traces C'_k des C_k sur F sont F-coanalytiques et séparent $(A'_1 - A'_2)$ et $(A'_2 - A'_1)$. On peut même étendre l'énoncé au cas d'espaces topologiques plus généraux, et au cas d'espaces mesurables abstraits (exercice! cf. II-7 $\Rightarrow$ II-8).

Voici maintenant le "bon" énoncé du deuxième théorème de séparation pour un couple. La démonstration que nous en donnons à partir de 21 est due à Mokobodzki (elle n'est pas sans rapport avec celle de II-27):

22 THÉORÈME *Soient A_1 et A_2 deux parties E-analytiques de E. Si A_1 et A_2 sont disjoints, il existe deux parties E-bianalytiques disjointes B_1, B_2 de E telles que l'on ait $A_1 \subseteq B_1$ et $A_2 \subseteq B_2$.*[1]

Démonstration Le théorème précédent, et sa remarque, nous assure l'existence de deux parties E-coanalytiques disjointes C_1, C_2 telles que l'on ait $A_1 \subseteq C_1$ et $A_2 \subseteq C_2$. Regardons la situation de manière dissymétrique en remplaçant A_2 et C_2 par leurs complémentaires (faire un dessin). On voit que, si une partie E-analytique $A' = A_1$ est contenue dans une partie E-coanalytique $C' = A_2^c$, il existe alors une partie E-coanalytique $C_1' = C_1$ et une partie E-analytique $A_1' = C_2^c$ telles que l'on ait la formule "d'interpolation" suivante

$$A' \subseteq C_1' \subseteq A_1' \subseteq C'.$$

On voit alors, en réinterpolant entre A' et C_1', que l'on peut construire par récurrence une suite (A_n') de parties E-analytiques et une suite (C_n') de parties E-coanalytiques telles que l'on ait

$$A' \subseteq \cdots \subseteq C_{n+1}' \subseteq A_{n+1}' \subseteq C_n' \subseteq \cdots \subseteq A_2' \subseteq C_1' \subseteq A_1' \subseteq C'.$$

Il ne reste plus alors qu'a prendre pour B_1 l'ensemble $\inf_n A_n' = \inf_n C_n'$ et à poser $B_2 = B_1^c$.

Remarques (a) L'énoncé s'étend au cas d'un espace mesurable abstrait. Une généralisation plus poussée du type "Soit $\mathscr{E}$ un pavage (cf. II-16) sur E tel que le complémentaire de tout élément de $\mathscr{E}$ soit $\mathscr{E}$-analytique. Alors . . ." est illusoire (exercice!).

(b) L'énoncé 22 a été écrit (intentionnellement, n'en doutons pas) sous une forme voisine de celle de l'énoncé 19. On voit que 22 est plus général dans ses hypothèses, moins bon dans sa conclusion—mais c'est inévitable (cf. l'exemple 20, regardé dans l'espace $E = C_1 \cup C_2$). Noter aussi que, pour affirmer qu'une partie F-bianalytique d'un espace analytique F est borélienne, il faut utiliser le premier théorème de séparation, si bien que 19 n'est pas conséquence immédiate de 22.

(c) L'énoncé 22 est encore équivalent au suivant (exercice!): si H est une partie E-analytique de E, la tribu $\mathscr{BA}(H)$ est la trace sur H de la tribu $\mathscr{BA}(E)$—pour la notation, voir le Chapitre II. Cet énoncé est faux pour H supposé E-coanalytique.

Une autre forme de 22, plus classique: le théorème de réduction de Kuratowski (pour un couple d'ensembles coanalytiques).

[1] Nous nous permettons, dans cette section, de noter B une partie bianalytique.

23 THÉORÈME *Soient C_1 et C_2 deux parties E-coanalytiques de E. Il existe deux parties E-coanalytiques disjointes D_1 et D_2 telles que l'on ait $D_1 \subseteq C_1$, $D_2 \subseteq C_2$ et $D_1 \cup D_2 = C_1 \cup C_2$.*

Démonstration Soit $F = C_1 \cup C_2$. Si $F = E$, alors, le passage au complémentaire refournit l'énoncé 22 avec $A_k = C_k^c$, $k = 1, 2$, et $B_1 = D_2$, $B_2 = B_1^c = D_1$. Dans le cas général, on commence par raisonner dans F: le théorème 22 nous fournit une partition de F en deux ensembles F-bianalytiques B_1 et B_2 tels que $B_1 \supseteq F - C_1$ et $B_2 \supseteq F - C_2$. Comme F est lui-même E-coanalytique, B_1 et B_2 sont aussi des parties E-coanalytiques, et il ne reste plus qu'à prendre $D_1 = B_2$, $D_2 = B_1$.

Remarque L'énoncé 23 peut se déduire facilement du cas particulier où l'on suppose E métrisable compact. Il s'étend aussi au cas d'un espace mesurable abstrait.

Avant de passer aux énoncés généraux du deuxième théorème de séparation et du théorème de réduction, je voudrais essayer d'exprimer ma "philosophie"—confuse—quant à l'utilisation du deuxième théorème de séparation et l'éclairer par un exemple. Supposons que nous soyons dans la situation suivante: $E \times F$ est un espace produit où E est métrisable compact (ou polonais, ou analytique) et on a un énoncé faisant intervenir des parties de E, F et $E \times F$ et la projection π_F de $E \times F$ sur F (par exemple: si G est un borélien de $E \times F$ tel que la coupe $G(y)$ comporte au plus un point pour tout $y \in F$, alors $\pi_F(G)$ est un borélien de F). Si l'énoncé envisagé est vrai (et se démontre à l'aide du premier théorème de séparation) lorsque l'espace F est analytique, alors l'énoncé est encore vrai si on suppose seulement que F est un espace métrisable séparable (ou même un espace mesurable abstrait) à condition de relativiser la notion d'analyticité et de remplacer "borélien" par "bianalytique"; et il faut utiliser le deuxième théorème de séparation pour le démontrer. Attention, on peut beaucoup plus difficilement toucher à E: passer de E compact à E polonais est souvent possible (à condition de remplacer "fermé de E" par "compact de E"), passer de E polonais à E analytique est plus délicat (cf. la remarque (a) de II-19), passer de E analytique à E coanalytique est le plus souvent impossible (dans ce cas intervient principalement le fait que, pour E analytique, l'ensemble $\pi_F(A)$ est F-analytique si A est $(E \times F)$-analytique (exercice!), et que ce n'est pas vrai en général pour E coanalytique).

Démontrons par exemple l'énoncé suivant (le lecteur attentif reconnaitra dans la démonstration des arguments employés aux Chapitres I et II—cf. I-35, I-37 et la remarque suivant II-7—avec utilisation du premier théorème de séparation):

24 THÉORÈME *Soit $E\times F$ un espace produit, où E est analytique, et soit G une partie $(E\times F)$-bianalytique de $E\times F$. Si la coupe $G(y)$ comporte au plus un point pour tout $y\in F$, alors la projection $\pi_F(G)$ de G sur F est F-bianalytique.*

Démonstration Appelons f la restriction de π_F à G et $D=f(G)$ la projection $\pi_F(G)$. D'abord, on montre que f est un isomorphisme de l'espace mesurable $(G, \mathcal{BA}(G))$ sur l'espace mesurable $(D, \mathcal{BA}(D))$: la mesurabilité dans le sens $G\to D$ est claire; pour le sense $D\to G$, on montre que l'image d'un élément A de $\mathcal{BA}(G)$ par f appartient à $\mathcal{BA}(D)$ en appliquant le deuxième théorème de séparation dans D aux ensembles D-analytiques $f(A)$ et $f(G-A)$. Ceci fait, on remarque que l'application g qui à $y\in D$ associe l'unique point de $G(y)$ est mesurable pour les tribus $\mathcal{BA}(D)$ et $\mathcal{B}(E)$ respectivement (elle est la composition de l'inverse de f avec la projection sur E). Maintenant, le deuxième théorème de séparation nous dit (cf. la remarque (c) de 22) que, D étant F-analytique, la tribu $\mathcal{BA}(D)$ est la trace sur D de la tribu $\mathcal{BA}(F)$. Si on plonge alors E dans un espace métrisable compact E', une adaptation simple de l'argument de I-17 montre qu'il existe une application mesurable h de $(F, \mathcal{BA}(F))$ dans $(E', \mathcal{B}(E'))$ telle que sa restriction à D soit égale à g considérée à valeurs dans E'. Le graphe H de h est $(E'\times F)$-bianalytique car la tribu $\mathcal{B}(E')\times\mathcal{BA}(F)$ est contenue dans $\mathcal{BA}(E'\times F)$, et donc l'ensemble $H-G=H\cap((E\times F)-G)$ est $(E'\times F)$-analytique (c'est ici qu'on utilise le fait que G est $(E\times F)$-coanalytique). Par conséquent, l'ensemble $F-D$, égal à la projection de $H-G$ sur F, est F-analytique. D'où, finalement, D est F-bianalytique.

Le corollaire suivant contient, comme cas particulier, le théorème sur "l'ensemble d'unicité" de Lusin.

25 COROLLAIRE *Soit $E\times F$ un espace produit, où E est analytique, et soit H une partie $(E\times F)$-bianalytique de $E\times F$. Alors l'ensemble*

$$D=\{y\in F\colon H(y) \text{ est un singleton}\}$$

est F-coanalytique.

Démonstration Nous montrons d'abord que l'ensemble

$$M=\{y\in F\colon \text{card}\, H(y)\geqslant 2\}$$

est F-analytique. Soit (U_n) une base dénombrable de la topologie de E; l'analyticité de M résulte aisément de l'équivalence

$$y\in M \text{ ssi } \exists m\ \exists n\ U_m\cap U_n=\varnothing \quad \text{et} \quad y\in\pi_F[H\cap(U_m\times F)]\cap\pi_F[H\cap(U_n\times F)].$$

Posons alors $F' = F - \pi_F(M)$ et soit G la trace de H sur $E \times F'$: F' est F-coanalytique, G est $(E \times F')$-bianalytique et D est égal à la projection de G sur F. D'après ce qui précède, D est F'-bianalytique, et donc F-coanalytique.

Remarque Dans l'énoncé de Lusin, F est analytique et H borélien (comme il se doit). On ne sait cependant pas le démontrer sans utiliser le deuxième théorème de séparation, et il est en fait "presque élémentairement équivalent" à l'énoncé 25 (il entraine 24 si on y suppose G borélien): c'est, comme pour le théorème de réduction, un cas où l'énoncé dans le cadre "espaces métrisables compacts" est pratiquement aussi général que l'énoncé dans le cadre "espaces mesurables abstraits".

C. Le deuxième théorème de séparation, deuxième épisode

Nous donnons ici les extensions aux suites d'ensembles analytiques des énoncés 22 (théorème de séparation de Novikov) et 23 (théorème de réduction de Kuratowski). Les démonstrations que nous donnons ne sont pas les plus élégantes (voir les notes de Kechris pour une démonstration élégante de 23, qui s'étend au cas des suites), mais sont à mon avis intéressantes par l'utilisation de l'astuce de Mokobodzki (cf. 21 $\Rightarrow$ 22) et par la mise en évidence de "l'équivalence élémentaire" des énoncés de Novikov et de Kuratowski.

26 THÉORÈME *Si* (A_n) *est une suite de parties E-analytiques de E et si* $\inf_n A_n = \varnothing$, *alors il existe, pour tout n, une partie E-bianalytique* $B_n \supseteq A_n$ *telle que* $\inf_n B_n = \varnothing$.

Démonstration Soit, pour tout n, f_n une application s.c.s. de E dans $\mathcal{T}$ telle que $A_n = f_n^{-1}(\mathcal{P})$ et posons $i_n(x) = i[f_n(x)]$ pour tout $x \in E$. D'après 16, l'ensemble C_n défini, pour tout n, par

$$C_n = \{x \in E : \exists m \; i_m(x) < i_n(x)\}$$

est E-coanalytique; comme $\inf_m A_m = \varnothing$, il contient A_n, et, comme $\inf_m i_m(x)$ est atteint en tout x (nous faisons ici un peu mentir la remarque 16(c)), on est assuré que $\inf_m C_m = \varnothing$. Maintenant, d'après 22 (l'énoncé ou la démonstration), il est possible d'intercaler, pour tout n, un ensemble E-bianalytique B_n entre A_n et C_n; c'est fini.

Remarque Quoique, dans son esprit, la démonstration de cet énoncé soit, à mon avis, moins élémentaire que celle de I-22, elle a été longtemps la seule connue pour établir, conjointement avec le premier théorème de séparation pour un couple, l'énoncé I-22. Voir à ce sujet la note historique. Par ailleurs, les rapports qu'il peut y avoir entre les démonstrations que nous avons données des énoncés I-22 et III-26 restent pour moi un mystère indéchiffrable.

Voyons maintenant ce que nous donne 26, si on ne suppose pas que l'intersection des A_n est vide

27 THÉORÈME *Si (A_n) est une suite de parties E-analytiques de E, il existe une suite (A'_n) de parties E-analytiques vérifiant*

(1) *A'_n contient A_n pour tout n,*
(2) *on a $A'_m \cup A'_n = E$ pour tout couple m, n tel que $m \neq n$,*
(3) *on a $\cap A'_n = \cap A_n$.*

Démonstration Posons $A = \cap A_n$, $F = E - A$ et, pour tout n, $H_n = A_n - A$. Les ensembles H_n forment une suite de parties F-analytiques, d'intersection vide. En appliquant 26 dans F, on obtient une suite (B_n) de parties F-bianalytiques, d'intersection vide, telle que l'on ait $B_n \supseteq H_n$ pour tout n. Par ailleurs, il est facile de voir que, si D est une partie F-analytique de F, alors $A \cup D$ est une partie E-analytique de E. Il est alors clair que les ensembles

$$A'_1 = A \cup B_1, A'_2 = A \cup B_2 \cup (F - B_1), \ldots,$$

$$A'_n = A \cup B_n \cup \left(F - \bigcap_{m<n} B_m\right), \ldots$$

forment une suite de parties E-analytiques ayant les propriétés requises dans l'énoncé.

Par passage au complémentaire, on obtient le théorème de réduction.

28 THÉORÈME *Si (C_n) est une suite de parties E-coanalytiques de E, il existe une suite (D_n) de parties E-coanalytiques vérifiant*

(1) *D_n est contenu dans C_n pour tout n,*
(2) *les D_n sont (deux à deux) disjoints,*
(3) *on a $\cup D_n = \cup C_n$.*

Remarque On vient de voir que l'énoncé 28 résulte aisément de l'énoncé 26. On se gardera de croire (jusqu'à preuve du contraire) que l'on peut déduire l'énoncé 28 de l'énoncé I-22. Un peu de philosophie: premier théorème de séparation ⇔ espaces ambiants métrisables compacts ⇔ théorie des capacités; deuxième théorème de séparation⇔espaces ambiants métrisables séparables⇔théorie de l'indice. L'énoncé 28, écrit dans le cadre d'espaces ambiants métrisables compacts, entraine aisément l'énoncé 28 en toute généralité (i.e. même dans le cadre d'espaces ambiants mesurables abstraits—exercice!), *mais* c'est parce que la *même* démonstration (par la théorie de l'indice) fournit aussi bien l'énoncé particularisé que l'énoncé général. Par ailleurs, un argument de capacitabilité, éventuellement caché sous un argument de compacité, est indispensable pour passer de "bianalytique" à "borélien".

D. Le troisième théorème de séparation

Le nom de "troisième théorème de séparation" a été donné par Rogers à l'énoncé suivant, provenant de Dellacherie (1975), et qui semble n'avoir jamais été dégagé, sous une forme ou une autre, par les "auteurs classiques"—mais j'ai des doutes!

29 THÉORÈME *Si $(A_n)_{n<k}$ est une suite* finie *de parties E-analytiques de E, il existe une suite finie $(A'_n)_{n<k}$ de parties E-analytiques vérifiant*

(1) *A'_n contient A_n pour tout $n<k$,*
(2) *on a $\cup A'_n = E$,*
(3) *on a $A'_m \cap A'_n = A_m \cap A_n$ pour tout couple m, n tel que $m \neq n$.*

Démonstration Nous utiliserons ici le fait que l'indice est une norme coanalytique (cf. 16), mais on peut aussi raisonner par récurrence en partant de l'énoncé 27 pour un couple (exercice!). Soit, pour tout $n<k$, f_n une application s.c.s. de E dans $\mathcal{T}$ telle que $A_n = f_n^{-1}(\mathcal{P})$ et posons $i_n(x) = i[f_n(x)]$ pour tout $x \in E$. Comme l'indice définit une norme coanalytique, les parties de E définies, pour $m, n<k$ fixés, par "$i_m(x) < \aleph_1$ et $i_m(x) \leq i_n(x)$" et par "$i_m(x) < \aleph_1$ et $i_m(x) < i_n(x)$" sont E-coanalytiques. On en déduit que l'ensemble A'_n défini pour tout $n<k$ par

$$x \in A'_n \text{ ssi } \forall m \neq n \; i_m(x) \leq i_n(x) \text{ et } (i_m(x) = i_n(x) < \aleph_1) \Rightarrow m > n$$

est E-analytique. Il est clair que A'_n contient A_n; comme, en tout $x \in E$, $\sup_{m>k} i_m(x)$ est atteint (c'est ici qu'intervient le fait que notre suite est finie), on est assuré que $\sup_{m<k} A'_m = E$; enfin, si x n'appartient pas à $A_n \cap A_n$, avec $m \neq n$, l'un des deux indices $i_m(x)$, $i_n(x)$ est $< \aleph_1$ si bien que, d'après la définition de A'_m et A'_n, x ne peut appartenir à la fois à A'_m et A'_n. D'où la suite $(A'_n)_{n<k}$ a les propriétés requises par l'énoncé.

Par passage au complémentaire, on obtient

30 THÉORÈME *Si* $(C_n)_{n<k}$ *est une suite* finie *de parties E-coanalytiques de E, il existe une suite finie* $(D_n)_{n<k}$ *de parties E-coanalytiques vérifiant*

(1) D_n *est contenu dans* C_n *pour tout* $n<k$,
(2) *l'intersection de tous les* D_n *est vide,*
(3) *on a* $D_m \cup D_n = C_m \cup C_n$ *pour tout couple* m, n *tel que* $m \neq n$.

Remarque Les énoncés 29, 30 sont faux, en général, pour une suite infinie: voir Steel (198?). Par ailleurs, Rogers a montré que 29 est encore vrai pour une suite infinie (A_n) telle que l'ensemble $\limsup_n A_n$ soit E-bianalytique. Nous laissons au lecteur, à titre d'exercice, le soin de démontrer ce dernier résultat à l'aide de l'énoncé 31 qui est, à mon avis, l'extension maximale de 29 (j'entends par là que tout énoncé du genre 29 s'en déduit aisément).

Le résultat suivant, dont la démonstration repose sur le troisième théorème de séparation, permet de simplifier notablement les démonstrations d'un certain nombre de résultats classiques. Nous verrons un exemple de ce fait.

31 THÉORÈME *Soient* $E\times F$ *un espace produit, où* E *est métrisable compact, et* H *une partie* $(E\times F)$*-analytique telle que la coupe* $H(y)$ *soit compacte pour tout* $y\in F$. *Il existe alors une partie* $(E\times F)$*-analytique* L *telle que*

(1) $\forall y \in F\, L(y) \neq \varnothing$,
(2) $\forall y \in F\, H(y) \neq \varnothing \Rightarrow H(y) = L(y)$.

On dira que L *est une* extension analytique complète *de* H *dans* $E\times F$.

Démonstration D'après II-17 (dont la démonstration, très simple, ne fait intervenir en aucune manière un théorème de séparation), l'ensemble H est l'intersection d'une suite décroissante (H_n) où, pour chaque n, H_n est une réunion finie de rectangles de la forme $K_p^n \times A_p^n$ avec K_p^n compact dans E et A_p^n F-analytique dans F. Et, quitte à remplacer A_p^n par sa trace sur la projection $\pi_F(H)$ de H sur F, on peut supposer que $\pi_F(H) = \pi_F(H_n)$ pour tout n. Supposons maintenant que, pour chaque n, nous ayons réussi à construire une extension analytique complète L_n de H_n dans $E\times F$; quitte à remplacer L_n par $\sup_{m\geq n} L_m$, on peut supposer la suite (L_n) décroissante (noter que l'on a, pour tout n, $H_n(y)\neq\varnothing \Leftrightarrow y\in \pi_F(H)$) et, quitte à remplacer L_n par $\bar{L}_n^E$ (cf. II-17 pour la notation), on peut supposer les L_n à coupes compactes. Alors $L=\inf_n L_n$ fournit une extension analytique complète de H dans $E\times F$, à coupes compactes. Nous sommes donc ramenés à démontrer le théorème dans le cas où H s'écrit comme réunion finie de rectangles $K_p \times A_p$ avec K_p compact dans E et A_p F-analytique. Mais alors l'ensemble L, réunion finie des rectangles $K_p \times A'_p$, où les A'_p nous sont fournis par 29 appliqué à F et aux A_p, est une extension analytique complète de H dans $E \times F$.

Remarques (a) Lorsque H est une réunion finie de rectangles $K_p \times A_p$, on n'utilise pas 29 dans toute sa force, mais seulement le fait que la trace de $A'_m \cap A'_n$ sur $\cup A_p$ est, pour $m \neq n$, égale à $A_m \cap A_n$. Steel (198?) a montré que même ce "quatrième" théorème de séparation est faux, en général, pour une suite infinie, si bien qu'une partie $(E \times F)$-analytique H à coupes dénombrables n'admet pas en général d'extension analytique complète dans $E \times F$ (prendre $H = \cup \{p\} \times A_p$).

(b) L'énoncé 28 peut aussi être aisément interprété en termes d'espaces produits. On a, par exemple: si E est analytique, toute partie $(E \times F)$-coanalytique H à coupes ouvertes admet une section par un graphe $(E \times F)$-coanalytique d'application dénombrablement étagée de $\pi_F(H)$ dans E.

Comme application, voici une démonstration très simple d'un théorème dû à Kunugui:

32 THÉORÈME *Soient $E \times F$ un espace produit, où E est polonais, et H une partie $(E \times F)$-bianalytique de $E \times F$. Alors l'ensemble*

$$C = \{y \in H \colon H(y) \text{ est compact, non vide}\}$$

est F-coanalytique.

Démonstration Quitte à plonger E, on peut supposer E compact. Soit alors L une extension analytique complète de $\bar{H}^E$ dans $E \times F$: il est clair que C est le complémentaire dans F de la projection de $L - H$ sur F, d'où la conclusion.

Remarques (a) Si on omet "non vide" dans la définition de C, le résultat est bien plus élémentaire: en considérant la projection de $\bar{H}^E - H$ sur F, on voit aisément que les ensembles $\{y \colon H(y) \text{ est compact}\}$ et $\{y \colon H(y) \text{ est fermé}\}$ sont F-coanalytiques.

(b) Sous les mêmes hypothèses, voici deux autres résultats classiques, dus à Arsenin: les ensembles

$$\{y \in F \colon H(y) \text{ est un } \mathcal{K}_\sigma\},$$

$$\{y \in F \colon H(y) \text{ est un } \mathcal{K}_\sigma \text{ non vide}\},$$

sont tous deux F-coanalytiques. Comme, même avec l'aide de 31, la démonstration classique de ces résultats n'est pas aisée, nous ne la donnerons pas ici (cf. Dellacherie (1975); le cas où l'on omet "non vide" n'est plus élémentaire ici). Nous montrerons cependant au Chapitre IV que $\{y\colon H(y)$ est un $\mathcal{K}_\sigma\}$ et $\{y\colon H(y)$ est un $\mathbf{F}_\sigma\}$ sont F-coanalytiques en nous inspirant de Saint-Raymond (1976). Signalons au passage que Louveau (198?) a aussi montré que ce résultat de coanalyticité est vrai pour toute classe de Baire (dans le cas où l'on omet "non vide", bien entendu).

(c) On déduit immédiatement de 32 l'énoncé "général" correspondant à II-19 (lequel peut aussi se déduire de l'énoncé "général" correspondant à II-18, lui-même démontré à l'aide de III-26). Et, pour E polonais, de cet énoncé à la II-19 résulte immédiatement l'énoncé 24, comme cas particulier.

4 Le théorème de Novikov et Kondô

Avant d'aborder le théorème de Novikov sur le choix "effectif" d'un point dans un ensemble coanalytique et le théorème de Kondô sur l'uniformisation d'une partie coanalytique d'un produit, nous allons regarder les mêmes problèmes dans le cas "analytique": nous commençons ainsi par revenir en arrière, en établissant une forme "canonique" du théorème de Jankov et von Neumann (cf. II-7), non par amour excessif du "canonique", mais pour éclairer la démarche adoptée dans la démonstration du théorème de Novikov et Kondô.

A. Le théorème de Jankov et von Neumann, revisité

Regardons donc le problème du choix "effectif" d'un point dans une partie analytique "effectivement" donnée de $\Omega = \mathbb{N}^{\mathbb{N}}$ (nous entendons ici le mot "effectif", non dans le sens précis des notes de Martin et Kechris (voir part 4 de ce volume), mais dans un sens vague proche de celui donné par les nominalistes français Baire, Borel, Lebesgue, . . . du début du siècle).

D'abord, se donner effectivement une partie analytique A de Ω, c'est, classiquement, se donner une application continue f de Ω dans Ω telle que $A = f(\Omega)$, ou un fermé F de $\Omega \times \Omega$ tel que $A = \pi_1(F)$ (le premier cas se ramenant au second par considération du graphe), ou encore un schéma de Souslin $\mathbf{X}$ sur les ilots de Ω tel que $A = \mathbf{N}[\mathbf{X}]$. Et ces trois descriptions possibles de A reviennent finalement à se donner (exercice!) un temps d'arrêt à deux variables T sur $\Omega \times \Omega$ tel que l'on ait

$$\omega \in A \text{ ssi } \exists w\; T(\omega, w) = \infty ,$$

soit encore, en désignant par T_ω le temps d'arrêt $w \to T(\omega, w)$ à une variable sur Ω,

$$\omega \in A \text{ ssi } T_\omega \text{ a un pôle },$$

le t.d'a. T étant associé au fermé H_T suivant de $\Omega \times \Omega$

$$H_T = \{(\omega, w)\colon T(\omega, w) = \infty\} = \{(\omega, w)\colon w \text{ est un pôle de } T_\omega\}.$$

Munissons $\mathbb{N} \times \mathbb{N}$ du bon ordre $\leqslant$ défini par

$$(m, p) < (n, q) \text{ ssi } p < q \text{ ou } [p = q \text{ et } m < n]$$

puis $\Omega \times \Omega = (\mathbb{N} \times \mathbb{N})^{\mathbb{N}}$ de l'ordre lexicographique associé: toute partie fermée non vide de $\Omega \times \Omega$ a un plus petit élément pour cet ordre. Supposons alors A non vide: H_T est fermé, non vide, et a donc un plus petit élément (ω_0, w_0), qui est le plus petit pôle de T. Le point choisi dans A est alors ω_0, et w_0 est le plus petit pôle de T_{ω_0} pour l'ordre lexicographique sur Ω.

En regardant la dépendance de ω_0 en fonction de T, on obtient finalement la forme canonique suivante du théorème de Jankov et von Neumann

33 THÉORÈME *Soient $\tilde{\mathscr{T}}$ l'espace des temps d'arrêt sur $\Omega \times \Omega$ et $\tilde{\mathscr{P}}$ le sous-espace analytique des temps d'arrêt à pôles. Il existe une application Ψ de $\tilde{\mathscr{P}}$ dans Ω telle que*

(1) *$\Psi(T)$ appartienne à $\{\omega\colon T_\omega$ a un pôle$\}$ pour tout $T \in \tilde{\mathscr{P}}$,*
(2) *Ψ soit mesurable lorsque Ω est muni de sa tribu borélienne et $\tilde{\mathscr{P}}$ de la tribu engendrée par ses parties analytiques.*

Démonstration Posons $\Phi(T) = (\omega_0, w_0)$, où (ω_0, w_0) est, comme ci-dessus le plus petit pôle de $T \in \tilde{\mathscr{P}}$ pour l'ordre lexicographique défini sur $\Omega \times \Omega$, puis $\Psi(T) = \omega_0 = \pi_1[\Phi(T)]$, où π_1 est la première projection. On a

$$\Phi(T) < (\omega. w) \text{ ssi } \exists(\omega', w')\ T(\omega', w') = \infty \text{ et } (\omega', w') < (\omega, w).$$

On en déduit que l'image réciproque par Φ de l'ensemble ouvert

$$J_{\omega,w} = \{(\omega', w')\colon(\omega', w') < (\omega, w)\}$$

est analytique; comme les $J_{\omega,w}$ engendrent la tribu borélienne de $\Omega \times \Omega$ (cf. II-4), on en déduit que Φ est mesurable lorsque $\Omega \times \Omega$ est muni de sa tribu borélienne et $\tilde{\mathscr{P}}$ de la tribu engendrée par ses parties analytiques: d'où la mesurabilité similaire de Ψ.

Et, comme corollaire, nous retrouvons l'énoncé II-7.

34 COROLLAIRE *Soit H une partie $(E \times F)$-analytique d'un produit $E \times F$ où E est un espace analytique. Il existe une application f de F dans E,* *mesurable lorsque E est muni de la tribu $\mathscr{B}(E)$ et F de la tribu engendrée par* $\mathbf{A}(F)$, *et telle que l'on ait $f(y) \in H(y)$ pour tout $y \in F$ tel que $H(y)$ soit non vide.*

Démonstration D'abord, par plongement, on peut supposer que E et F sont polonais. Ensuite, tout espace polonais étant plongeable par un isomorphisme borélien[1] dans Ω (cf. par exemple I-38 et I-36), on peut en fait supposer que $E = F = \Omega$ et donc que H est une partie analytique de $\Omega \times \Omega$. Il existe alors un t.d'a. U, à trois variables notées v, ω, w, sur $\Omega \times \Omega \times \Omega$ tel que l'on ait

$$(v, \omega) \in H \text{ ssi } \exists w \; U(v, \omega, w) = \infty ,$$

et U définit une application continue $v \to U(v, ., .)$ de Ω dans l'espace $\tilde{\tilde{\mathscr{T}}}$ des t.d'a. sur $\Omega \times \Omega$. Pour obtenir notre application f, il suffit alors de composer cette application avec l'application Ψ de l'énoncé 33 (prolongée à $\tilde{\tilde{\mathscr{T}}}$ tout entier par le choix arbitraire d'un point dans Ω).

B. Les théorèmes de Novikov et Kondô

Le théorème de Novikov dont il est question ici montre comment on peut choisir "effectivement" un point dans un ensemble coanalytique tandis que le théorème de Kondô assure qu'une partie coanalytique d'un produit d'espaces polonais a une section par un graphe coanalytique. Le lien entre ces deux théorèmes est clair (cf. le passage de 33 à 34): pour avoir le théorème de Kondô, il doit suffire d'appliquer "correctement" le théorème de Novikov à chaque coupe. Pourtant, l'article de Novikov (et Lusin) date de 1935, celui de Kondô de 1938 et, jusque vers 1960, le théorème de Kondô passait auprès des spécialistes (par exemple, Sierpiński et Mostowski) pour un des théorèmes les plus difficiles de la théorie descriptive des ensembles. Or, nous allons donner ici une démonstration courte et lumineuse de ces deux théorèmes, issue des travaux des logiciens (mais bien dans le style classique).

35 Nous rappelons d'abord la définition de la *fonction-indice* i_T associée à un t.d.'a. fini T sur $\Omega = \mathbb{N}^{\mathbb{N}}$. C'est l'application de S dans I définie par

$$i_T(s) = i(T^s)$$

où l'on a $T^s(\omega) = [T(s\,.\,\omega) - |s|]^+$ (cf. 8 et 13). Elle est uniquement déterminée par les propriétés suivantes:

$$\begin{cases} T^s = 0 & \text{ssi} \quad i_T(s) = 0 , \\ T^s \neq 0 & \text{ssi} \quad i_T(s) = \sup\limits_{k \in \mathbb{N}} [i_T(s\,.\,k) + 1] . \end{cases}$$

[1] En fait, un théorème de Kuratowski assure que tout borélien non dénombrable d'un espace polonais est Borel-isomorphe à Ω.

Définissons maintenant la notion, plus générale, de *fonction de rang r* associée à un t.d'a. T sur Ω comme étant une application de S dans I vérifiant

$$\begin{cases} T^s = 0 & \text{ssi} \quad r(s) = 0\,, \\ T^s \neq 0 & \text{ssi} \quad r(s) \geq \sup_{k\in\mathbb{N}} [r(s\,.\,k)+1]\,. \end{cases}$$

On vérifie aisément que le temps d'arrêt T admet (au moins) une fonction de rang ssi il est fini et que, si T est fini, alors sa fonction-indice est sa plus petite fonction de rang pour l'ordre naturel sur I^S. Nous munissons maintenant I de la topologie discrète et I^S de la topologie de la convergence simple. Voici alors le lemme fondamental pour notre propos, dont la démonstration, à peu près immédiate, est laissée au lecteur.

36 LEMME *L'ensemble*

$$\{(T, r)\colon r \text{ est une fonction de rang pour } T\}$$

est une partie fermée de l'espace produit $\mathcal{T} \times I^S$.

Remarque L'ensemble (analytique) $\mathcal{P}$ est projection sur $\mathcal{T}$ du fermé $\{(T, \omega)\colon \omega$ est un pôle de $T\}$ de $\mathcal{T} \times \mathbb{N}^{\mathbb{N}}$ et est donc $\aleph_0$-souslinien (cf. 5). On trouve ici que l'ensemble (coanalytique) $\mathcal{Q}$ est projection sur $\mathcal{T}$ du fermé $\{(T, r)\colon r$ est une fonction de rang de $T\}$ de $\mathcal{T} \times I^S$. Si on identifie S à $\mathbb{N}$ par une bijection, on voit que $\mathcal{Q}$ est $\aleph_1$-souslinien. Cela n'est en fait pas étonnant: on sait déjà par la théorie des constituants que $\mathcal{Q}$ est la réunion de $\aleph_1$-boréliens, d'où l'on déduit sans peine son caractère $\aleph_1$-soulinien. Le fait remarquable est la représentation canonique de $\mathcal{Q}$ que nous fournit 36, beaucoup plus riche que toute autre représentation $\aleph_1$-souslinienne de $\mathcal{Q}$.

Nous passons maintenant au théorème de Novikov et Kondô, en paraphrasant ce qui a été dit à propos du théorème de Jankov et von Neumann.

37 Se donner "effectivement" une partie coanalytique C de Ω, c'est finalement se donner un t.d'a. à deux variables T sur $\Omega \times \Omega$ tel que

$$\omega \in C \quad \text{ssi} \quad \forall w\; T(\omega, w) < \infty\,,$$

soit encore, en introduisant l'application continue $\omega \to T_\omega$ de Ω dans $\mathcal{T}$ induite par T,

$$\omega \in C \quad \text{ssi} \quad T_\omega \textit{ est fini} \quad \text{ssi} \quad T_\omega \text{ a une fonction de rang}$$

et, d'après le lemme précédent, l'ensemble

$$H_T = \{(\omega, r)\colon r \text{ est une fonction de rang pour } T_\omega\}$$

est fermé dans $\Omega \times I^S$. Identifions S et $\mathbb{N}$ par une bijection, ce qui permet d'identifier $\Omega \times I^S$ et $(\mathbb{N} \times I)^{\mathbb{N}}$, y compris pour les topologies. Puis, munissons $\mathbb{N} \times I$ du bon ordre $\leqslant$ défini par

$$(m, i) < (n, j) \quad \text{ssi} \quad i < j \text{ ou } [i = j \text{ et } m < n],$$

et $\Omega \times I^S = (\mathbb{N} \times I)^{\mathbb{N}}$ de l'ordre lexicographique associé. Si C n'est pas vide, H_T est une partie fermée non vide de $(\mathbb{N} \times I)^{\mathbb{N}}$ et a donc un plus petit élément (ω_0, r_0). Le point choisi dans C est alors ω_0, et r_0 est la fonction-indice du temps d'arrêt T_{ω_0}. En regardant la dépendance de ω_0 en fonction de T, on obtient la forme canonique suivante du théorème de Novikov et Kondô.

38 THÉORÈME *Soient $\tilde{\mathcal{T}}$ l'espace des temps d'arrêt sur $\Omega \times \Omega$ et $\tilde{\mathcal{R}}$ le sous-ensemble $\{T \in \tilde{\mathcal{T}} : \exists \omega \; \forall w \; T(\omega, w) < \infty\}$ de $\tilde{\mathcal{T}}$. Il existe une application Ψ de $\tilde{\mathcal{R}}$ dans Ω telle que*

(1) *$\Psi(T)$ appartienne à $\{\omega : T_\omega$ est fini$\}$ pour tout $T \in \tilde{\mathcal{R}}$,*
(2) *le graphe de Ψ soit coanalytique dans $\tilde{\mathcal{T}} \times \Omega$.*

Démonstration Après avoir identifié S à $\mathbb{N}$ par une bijection $k \to s_k$, telle que $s_1 = \varnothing$, on pose comme ci-dessus

$$\Psi(T) = \text{la première composante du plus petit élément de } H_T \text{ pour l'ordre lexicographique sur } \Omega \times I^S.$$

Pour T et ω fixés, la trace de H_T sur $\{\omega\} \times I^S$ est fermée dans $\Omega \times I^S$; elle a donc, si elle n'est pas vide (i.e. si T_ω est fini), un plus petit élément qui n'est autre que la fonction-indice de T_ω. Par conséquent, on a $\Psi(T) = \omega$ ssi on a

$$T_\omega \in Q \text{ et } \forall w \; [w \neq \omega \Rightarrow i(T_\omega) < i(T_w) \text{ ou } (T_w \in \mathcal{Q} \text{ et} \ldots)],$$

où "..." est le prédicat

$$\exists m \; \forall k < m \;\; \omega(k) = w(k) \text{ et } i_{T_\omega}(s_k) = i_{T_w}(s_k) \text{ et}$$
$$[i_{T_\omega}(s_m) < i_{T_w}(s_m) \text{ ou } (i_{T_\omega}(s_m) = i_{T_w}(s_m) \text{ et } \omega(m) < w(m))].$$

Le point (2) du théorème résulte alors aisément du Théorème 16, l'application $(T, \omega) \to T_\omega$ de $\tilde{\mathcal{T}} \times \Omega$ dans $\mathcal{T}$ étant continue.

En ramenant la situation "générale" à la situation canonique comme dans la démonstration de 34, on obtient l'énoncé classique du théorème de Kondô.

39 COROLLAIRE *Soit H une partie $(E \times F)$-coanalytique d'un produit $E \times F$ où E est un espace polonais. Il existe une partie $(E \times F)$-coanalytique G,* *contenue dans H, telle que la coupe $G(y)$ soit un singleton pour tout y appartenant à la projection de H sur F.*

Remarques (a) On déduit aisément de 39 que toute partie PCA d'un produit d'espaces polonais admet une section par un graphe PCA. (Exercice! Une partie d'un espace polonais X est dite PCA si c'est la projection sur X d'une partie coanalytique d'un espace produit $X \times Y$, où Y est un espace polonais auxiliaire.)

(b) Contrairement à 34, on ne peut affirmer que l'on obtient grâce à 39 une section par le graphe d'une application universellement mesurable: les axiomes habituels de la théorie des ensembles sont trop faibles pour trancher une telle question. On sait qu'il existe des parties PCA de $[0, 1]$ non Lebesgue-mesurables si on ajoute aux axiomes habituels (axiome de choix compris) l'axiome de constructibilité "$V = L$" de Gödel (dans ce cas, il existe un bon ordre PCA sur $[0, 1]$); mais, on sait aussi que toute partie PCA d'un espace polonais est universellement mesurable si, niant l'hypothèse du continu, on ajoute aux axiomes habituels l'axiome de Martin.

IV Dérivations et applications

Nous avons vu, au cours des Chapitres II et III, deux exemples de dérivation: la dérivation de Cantor, et la dérivation d'un temps d'arrêt. Nous allons étudier systématiquement cette notion à la Section 1, sans chercher la plus grande généralité (notre règle de conduite ayant été "avoir des énoncés simples, applicables à de nombreux exemples"). Le résultat fondamental est un *théorème de la borne* (cf. II-26 et III-17), qui aura un énoncé du genre:

$$\text{si } A \text{ est un sous-espace analytique contenu dans } Q\text{, alors on a } \sup_{x \in A} i(x) < \aleph_1 .$$

Nous illustrons l'utilisation de ce théorème à la Section 3 en montrant que certains espaces fonctionnels "naturels" ne sont pas analytiques (on utilise là le fait que, pour l'espace Q en question, on a $\sup_{x \in Q} i(x) = \aleph_1$), et à la Section 4 en démontrant le théorème de Saint-Raymond sur les ensembles boréliens à coupes $\mathbf{F}_\sigma$ (on utilise là le fait que, pour un certain ensemble analytique A, $\sup_{x \in A} i(x)$ est un ordinal dénombrable). La Section 2 est consacré au développement de la machinerie pour démontrer ce théorème de la borne, à savoir l'étude des relations bien-fondées analytiques. En fait, nous irons au delà de nos stricts besoins puisque nous y démontrerons le théorème de Kunen et Martin, qui affirme que toute relation bien-fondée et $\aleph_n$-souslinienne sur un espace polonais a une longueur $< \aleph_{n+1}$. Cette section pouvant apparaitre de lecture difficile à un lecteur novice en la matière, nous avons résumé à la Section 0 (eh oui!) la partie de la Section 2 indispensable pour démontrer (mais non pour comprendre ou savoir utiliser) les énoncés de la Section 1. Et, pour éviter d'avoir à introduire à la Section 0 la notion de relation bien-fondée (familière aux logiciens—cf. l'axiome de fondation en théorie des ensembles—mais suffisamment sophistiquée pour avoir échappé aux "auteurs classiques", voir les commentaires), nous avons écrit cette Section 0 dans le langage, maintenant familier (?), des temps d'arrêt.

0 Temps d'arrêt analytiques

1 Nous dirons qu'une fonction f de E dans $\bar{\mathbb{R}}_+$ est *E-analytique* (resp. *E-coanalytique*) si son "sous-graphe ouvert" $\{(x, t): f(x) > t\}$ (ou, cela revient au même, son "sous-graphe fermé" $\{(x, t): f(x) \geq t\}$) est $E \times \bar{\mathbb{R}}_+$-analytique (resp. $E \times \bar{\mathbb{R}}_+$-coanalytique); ainsi, une partie de E est E-analytique (resp. E-coanalytique) ssi sa fonction indicatrice est E-analytique (resp. E-coanalytique). Il est facile de voir (exercice!) que f est E-analytique (resp. E-coanalytique) ssi, pour tout $t \in \mathbb{R}_+$, l'ensemble $\{x: f(x) > t\}$ (ou $\{x: f(x) \geq t\}$) est E-analytique (resp. E-coanalytique).

2 *Exemples* (Exercices!) (a) Soit $E \times F$ un espace produit, où E est polonais, et soit f une fonction borélienne positive sur $E \times F$. Alors, la fonction $g(y) = \sup_{x \in E} f(x, y)$ est F-analytique.

(b) Soit B une partie borélienne de $E \times \mathbb{R}_+$ et définissons le *début* D_B de B (très utilisé en probabilités) par

$$D_B(x) = \inf \{t \geq 0: (x, t) \in B\} \qquad (\text{avec } \inf \varnothing = \infty).$$

Alors D_B est une fonction E-coanalytique.

3 Soit maintenant E un espace analytique, et munissons $E^{\mathbb{N}}$ de sa topologie "naturelle", la topologie produit, qui en fait aussi un espace analytique. Soit par ailleurs T un temps d'arrêt sur $E^{\mathbb{N}}$ (bien entendu, la topologie naturelle de $E^{\mathbb{N}}$ n'intervient pas dans la définition d'un temps d'arrêt—cf. III-1): si T est continu pour $E^{\mathbb{N}}$ muni de la topologie produit associée à la topologie discrète sur E, T est par contre, en général, bien loin d'être une fonction continue et même borélienne pour la topologie naturelle de $E^{\mathbb{N}}$ lorsque E n'est pas dénombrable (par exemple, si H est une partie "quelconque" de E, le t.d'a. T défini par $T(\omega) = 1$ si $\omega|1$ appartient à H et $T(\omega) = \infty$ sinon est évidemment très "quelconque"). Cela n'empêche que l'on peut, comme en III-10, définir le dérivé T' d'un t.d'a. T sur $E^{\mathbb{N}}$, puis la suite transfinie des dérivés successifs $(T^i)_{i \in I}$ de T, indexée par les ordinaux dénombrables, et finalement l'*indice* de T par $i(T) = \inf\{i \in I : T^i = 0\}$, avec $\inf \emptyset = \aleph_1$. Il est ici encore vrai (et trivial) que, si on a $i(T) < \aleph_1$, alors le t.d'a. T est fini. Mais la réciproque, vraie si E est dénombrable, est fausse pour E non dénombrable (il est vrai que, pour T fini, si on a défini de manière évidente le α-ième dérivé T^α de T pour tout ordinal α, alors on a $T^\alpha = 0$ pour α suffisamment grand, mais en général bien plus grand que $\aleph_1$, premier ordinal non dénombrable).

Voici alors, dans le langage des temps d'arrêt, la traduction du théorème sur la longueur d'une relation analytique bien fondée (que nous démontrerons à la Section 2).

4 THÉORÈME *Soient E un espace analytique et T un t.d'a.* analytique *sur $E^{\mathbb{N}}$ (muni de sa topologie naturelle). On a alors*

$$T \text{ est fini ssi } i(T) < \aleph_1 .$$

Remarque Bourgain et Saint-Raymond savent démontrer directement ce théorème, mais leur "virtuosité" est hors de ma portée . . ..

Nous illustrons tout de suite l'emploi de ce théorème en redémontrant le théorème de la borne III-17

5 COROLLAIRE *Soit $\mathcal{T}$ l'espace des t.d'a. sur $\mathbb{N}^{\mathbb{N}}$ et soit A une partie analytique de $\mathcal{T}$ constituée de t.d'a. finis. Alors on a*

$$\sup_{T \in A} i(T) < \aleph_1 .$$

Démonstration Nous prenons pour espace analytique E la somme topologique de $\mathcal{T}$ et de $\mathbb{N}$, et nous définissons un t.d'a. τ sur $E^{\mathbb{N}}$ en posant, pour $\omega = \omega|1 . w$ élément de $E^{\mathbb{N}}$,

$$\tau(\omega) = \inf\{n \geq 1 : \omega|n \in A \times \mathbb{N}^{n-1} \Rightarrow (\omega|1)^{w|n-1} = 0\}$$

(remarquer que, si $\omega|n$ est un élément de $A \times \mathbb{N}^{n-1}$, alors $\omega|1$ est un élément de $\mathcal{T}$ et $w|n-1$ un élément de $\mathbb{N}^{n-1}$, si bien que la notation $(\omega|1)^{w|n-1}$ a un sens, celle de III-8; noter aussi que, pour $\omega|1 = T$ élément de A et w élément de $\mathbb{N}^{\mathbb{N}}$, on a $\tau(\omega) = 1 + T(w)$). Une fois compris de quoi il s'agit, on vérifie aisément que τ est un t.d'a. fini analytique et que l'on a $i(\tau) \geq \sup_{T \in A} i(T)$; d'où la conclusion.

Remarque La démonstration précédente est un peu obscurcie par le fait qu'on est obligé de considérer tout $E^{\mathbb{N}}$ alors qu'on n'est vraiment intéressé que par les $\omega = \omega|1 . w$ avec $\omega|1$ élément de $\mathcal{T}$ et w élément de $\mathbb{N}^{\mathbb{N}}$. On aurait pu éviter cela en présentant la définition d'un temps d'arrêt sur un produit "inhomogène" $\prod_{n\in\mathbb{N}} E_n$, puis en prenant ci-dessus $E_1 = \mathcal{T}$ et $E_n = \mathbb{N}$ pour $n > 1$: on pouvait alors poser tout simplement $\tau(\omega) = 1$ si $\omega|1$ n'appartient pas à A et, sinon, $\tau(\omega) = 1 + (\omega|1)(w)$. Mais, agissant ainsi, on aurait compliqué au Chapitre III la définition de la notion de temps d'arrêt. Bien entendu, il n'y a plus maintenant d'objection à étendre un peu le concept de t.d'a.

1 Dérivations analytiques

Pour bien comprendre les définitions générales qui vont suivre, le lecteur est prié de garder en mémoire les exemples des Chapitres II et III, auxquels nous ferons allusion de temps en temps.

6 On se donne un espace (métrisable séparable) E muni d'une relation d'ordre partiel, notée $\leqslant$, vérifiant les conditions suivantes:

(a) il existe un plus petit élément pour $\leqslant$, noté ϕ;
(b) toute famille (non vide) filtrante décroissante d'éléments de E admet une borne inférieure;
(c) il existe un espace analytique E^* et une relation analytique $\dashv$ sur E^* de sorte que E soit un sous-espace de E^* et que l'on ait, pour tout $x, y \in E^*$,

$$x \dashv y \text{ et } y \in E \quad \text{ssi} \quad x \in E \text{ et } x \leqslant y \text{ et } y \in E$$

(cela implique que $\dashv$ est une extension de $\leqslant$ et que $\leqslant$ est un ordre $E \times E$-analytique sur E).

Nous dirons, pour abréger, que $\leqslant$ est un ordre *adéquat*. Lorsque E est analytique, tout ordre analytique $\leqslant$ sur E vérifiant (a) et (b) est adéquat (prendre $E^* = E$ et $\dashv = \leqslant$). Au Chapitre II, E est l'espace des parties compactes d'un espace métrisable compact, et $\leqslant$ est l'inclusion; au Chapitre III, E est l'espace des temps d'arrêt sur $\mathbb{N}^{\mathbb{N}}$ et $\leqslant$ est l'ordre naturel.

7 DÉFINITION Une *dérivation analytique sur* $(E, \leqslant)$ est une application d de E dans E vérifiant les conditions

(1) $\forall x\, d(x) \leqslant x$,
(2) $\forall x, y\;\; x \leqslant y \Rightarrow d(x) \leqslant d(y)$,
(3) la relation $x \leqslant d(y)$ est $E \times E$-analytique (c'est le cas si d est une application borélienne).

On dit que $x \in E$ est *parfait* si $d(x) = x$.

8 Nous nous donnons maintenant une dérivation analytique d sur $(E, \leq)$ et, à tout $x \in E$, nous associons la *suite transfinie* $(x^i)_{i\in I}$ *de ses dérivés successifs*, indexée par l'ensemble I des ordinaux dénombrables; on a par définition

$$x^0 = x\,, \qquad x^1 = d(x)$$

$$x^j = d(\inf_{i<j} x^i) \text{ (et donc } x^j = d(x^{j-1}) \text{ si } j \text{ est de 1-ère espèce)}$$

Puis, nous définissons l'*indice* $i(x)$ de $x \in E$ par

$$i(x) = \inf\{i \in I: x^i = \phi\}, \text{ avec } \inf \varnothing = \aleph_1\,.$$

Au Chapitre II, nous avions "$i(x) < \aleph_1$ ssi x est dénombrable" et, au Chapitre III, "$i(x) < \aleph_1$ ssi x est fini"; on n'a pas ici de caractérisation "naturelle" des ensembles

$$P = \{x \in E: i(x) = \aleph_1\}\,, \qquad Q = \{x \in E: i(x) < \aleph_1\}\,,$$

ne faisant pas appel à la dérivation d. On en a cependant la caractérisation suivante, "à la Cantor".

9 THÉORÈME *On a $x \in P$ ssi il existe $y \neq \phi$ parfait tel que $y \leq x$.*

Démonstration Nous supposons, pour simplifier, que E est analytique; nous verrons, à la fin de la démonstration, comment faire pour traiter le cas général. La clé pour l'étude de notre dérivation d réside en la considération du t.d'a. T_x sur $W = E^{\mathbb{N}}$ défini, pour tout $x \in E$, comme suit: pour $w = (y_n)_{n\in\mathbb{N}}$ élément de W, on a

$$T_x(w) = \inf\{n \in \mathbb{N}: y_{n-1} \leq d(y_n) \text{ et } y_n \leq x \Rightarrow y_1 = \phi\}\,,$$

où l'on convient que $y_0 = \phi$ et $\inf \varnothing = \infty$, et où l'on interprète, bien entendu, "$A \Rightarrow B$" comme synonyme de "non A ou B". Afin de ne pas alourdir notre texte, nous laissons au lecteur le soin de vérifier les propriétés suivantes, élémentaires mais un peu longuettes à établir: pour $x \in E$ fixé, T_x est un t.d'a. analytique sur l'espace analytique W; pour toute famille filtrante décroissante (x^α) d'éléments de E, on a $\inf_\alpha T_{x^\alpha} = T_{\inf_\alpha x^\alpha}$; enfin, on a $T_\phi = 1$ et, pour $x \neq \phi$, $T_{d(x)} = d(T_x)$, $d(T_x)$ dénotant le dérivé du t.d'a. T_x. Il est alors clair que l'on a, si $i(T_x)$ dénote l'indice du t.d'a. T_x,

$$i(x) \leq i(T_x) \leq i(x) + 1$$

(on paie ici le fait que "1" ne soit pas le plus petit t.d'a.). Supposons maintenant que $x \in E$ majore un parfait $y \neq \phi$. Alors, on a évidemment $i(x) = \aleph_1$, et T_x admet $w = y, y, \ldots, y, \ldots$ pour pôle. Réciproquement, si $i(x) = \aleph_1$, alors on a aussi $i(T_x) = \aleph_1$ et donc, d'après le théorème 4, T_x

admet un pôle $w=(y_n)_{n\in\mathbb{N}}$; vu la définition de T_x, le pôle w vérifie, pour tout n,

$$\phi \neq y_1 \leq d(y_2) \leq y_2 \leq \cdots \leq d(y_n) \leq y_n \leq x\,.$$

Si on définit maintenant, pour tout ordinal α, le α-ième dérivé x^α de x par $x^\alpha = d(\inf_{\beta<\alpha} x^\beta)$, on voit aisément, par récurrence transfinie, que l'on a aussi pour tout n et tout α

$$\phi \neq y_1 \leq d(y_2) \leq y_2 \leq \cdots \leq d(y_n) \leq y_n \leq x^\alpha\,.$$

Par ailleurs, un raisonnement élémentaire de cardinalité montre que l'on a $x^\alpha = x^{\alpha+1}$ pour α suffisamment grand. On en déduit finalement que x majore l'élément parfait $y=\inf_\alpha x^\alpha \neq \phi$. Pour terminer, si E n'est pas analytique, on le plonge d'abord dans l'espace analytique E^* de 6, et on construit, pour $x\in E$, le t.d'a. T_x sur $W=E^{*\mathbb{N}}$ en remplaçant, dans la définition ci-dessus de T_x, la relation "$u \leq v$" sur E par son extension "$u \mathrel{=\!\!|} v$" sur E^* (cf. 6(c)), et la relation $E\times E$-analytique "$u\leq d(v)$" sur E par une extension analytique "$u \dashv v$" sur E^* satisfaisant $u \dashv v \Rightarrow u \mathrel{=\!\!|} v$ (prolonger "$u\leq d(v)$" en une relation analytique "uRv" sur E^* et poser $u\dashv v$ ssi $u \mathrel{=\!\!|} v$ et uRv). Notez que, d'après 6(c), si on a $z \dashv y$ et $y \mathrel{=\!\!|} x$ pour x élément de E, alors z et y appartiennent aussi à E si bien que "$z\dashv y$ et $y \mathrel{=\!\!|} x$" équivaut alors à "$z\leq d(y)$ et $y\leq x$".

10 COROLLAIRE *L'ensemble P est une partie E-analytique et l'ensemble Q une partie E-coanalytique de E.*

Démonstration Conservant les notations introduites ci-dessus, on a

$$x\in P \quad \text{ssi} \quad \exists y \;\; y\neq\phi \text{ et } y\dashv y \text{ et } y \mathrel{=\!\!|} x \text{ et } x\in E\,,$$

d'où P est E-analytique, et Q est E-coanalytique.

Nous passons maintenant au théorème de la borne.

11 THÉORÈME *Si A est un* sous-espace analytique *de Q, on a*

$$\sup_{x\in A} i(x) < \aleph_1\,.$$

Démonstration Nous reprenons les notations de la démonstration du théorème 9, et nous définissons un nouveau t.d'a. τ sur W en posant, pour $\omega=\omega|1\,.w$ élément de W,

$$\tau(\omega)=1 \quad \text{si } \omega|1 \text{ n'appartient pas à } A\,,$$

$$\tau(\omega)=1+T_{\omega|1}(w) \quad \text{si } \omega|1 \text{ appartient à } A\,,$$

ce qui peut encore s'écrire (cf. la démonstration de 5)

$$\tau(\omega)=\inf\{n\geq 1: \omega|1\in A \Rightarrow T_{\omega|1}^{w|n-1}=0\}.$$

On vérifie sans peine que τ est un t.d'a. analytique sur W; par ailleurs, A étant inclus dans Q, τ est un t.d'a. fini. Comme on a $i(\tau)\geq \sup_{x\in A} i(T_x)\geq \sup_{x\in A} i(x)$, il résulte alors du théorème 4 que l'on a $\sup_{x\in A} i(x)<\aleph_1$.

12 Corollaire *Si on a* $\sup_{x\in Q} i(x)=\aleph_1$, *alors* Q *n'est pas un espace analytique.*

Remarque Prenons pour E l'espace polonais des fermés de $\mathbb{N}^{\mathbb{N}}$ (cf. II-3), pour $\leq$ l'inclusion (c'est un ordre fermé, adéquat) et pour d la dérivation de Cantor (d n'est pas une application borélienne, mais $x\leq d(y)$ est une relation analytique—cf. la remarque (b) de II-25). Alors, d'après ce qui précède, l'indice $i(\cdot)$ associé à d vérifie le théorème de la borne, et l'ensemble Q des fermés dénombrables est coanalytique, non borélien. Mais, l'ensemble des fermés discrets étant lui-aussi coanalytique non borélien (cf. la remarque (b) de II-25), il n'est pas vrai que $i(\cdot)$ définit une norme coanalytique (cf. III-16). Je conjecture que, pour E polonais et $\leq$ fermé, l'indice $i(\cdot)$ définit une norme coanalytique ssi d est une application borélienne.

2 Le théorème de Kunen et Martin

Il existe trois caractérisations importantes de la notion de relation bien-fondée. Nous avons choisi comme définition la caractérisation "la plus parlante" (donc aussi la plus verbeuse . . .), et rejeté les deux autres dans un théorème.[1]

13 Soit R une relation sur E. Nous dirons que, pour $x, y\in E$, x est un *ancêtre* de y si l'on a xRy, et que y est *primitif* s'il n'a pas d'ancêtre. La relation R est alors dite *bien-fondée* si elle vérifie le "*principe du raisonnement par induction*", c'est à dire "si une proposition P est vérifiée par tout élément primitif et si elle est vérifiée par $y\in E$ dès qu'elle l'est par tout ancêtre x de y, alors elle est vérifiée par tout $y\in E$" (dans l'énoncé de ce principe, on sous-entend souvent "Si $P(y)$ est vraie pour tout y primitif" car, si on formalise la seconde condition en

$$(\forall x\; xRy \Rightarrow P(x))\Rightarrow P(y),$$

[1] Comme au Chapitre III, Section 1, nous sortons ici exceptionnellement de notre cadre habituel. En particulier, E peut désigner un ensemble sans topologie.

la première condition en résulte à l'aide de l'interprétation usuelle de l'implication).

14 THÉORÈME *La relation R sur E est bien-fondée ssi elle vérifie l'une des conditions suivantes*:

(1) *Pour toute partie non vide H de E, la restriction de R à H admet un élément primitif.*
(2) *Il n'existe pas de suite infinie $(x_n)_{n\in\mathbb{N}}$ telle que x_{n+1} soit un ancêtre de x_n pour tout n.*

Démonstration Si R est bien-fondée, alors le principe du raisonnement par induction entraine que la proposition "il n'existe pas de suite infinie $(x_n)_{n\in\mathbb{N}}$ telle que x_{n+1} soit un ancêtre de x_n pour tout n et que $x_1 = y$" est vraie pour tout $y \in E$, d'où (2). Si (2) est vrai, alors (1) est vrai: en effet, si la restriction de R à H non vide n'admet pas d'élément primitif, on construit aisément par récurrence une suite $(x_n)_{n\in\mathbb{N}}$ de points de H tels que $x_{n+1}Rx_n$ pour tout n. Enfin, si (1) est vrai, alors R est bien-fondée: en effet, si P est une proposition qui n'est pas vérifiée par tout $z \in E$, alors l'ensemble $H = \{z : P(z) \text{ est faux}\}$ est non vide, et si y est un élément primitif de la restriction de R à H, il n'est pas possible d'avoir $(\forall x\; xRy \Rightarrow P(x)) \Rightarrow P(y)$.

15 *Quelques exemples* (a) La relation $\in$, en théorie des ensembles, est bien-fondée: c'est un axiome, dit "de fondation" ou "de régularité".

(b) L'ordre strict associé à une relation de bon-ordre est bien-fondé.

(c) Soit T un t.d'a. sur $\Omega = E^{\mathbb{N}}$ et soit $\mathbf{R}$ la relation sur l'ensemble S des suites finies d'éléments de E définie par

$$s\mathbf{R}t \quad \text{ssi} \quad T^t \neq 0 \text{ et } t \dashv s .$$

Alors $\mathbf{R}$ est bien-fondée ssi T est fini.

(d) Sous les hypothèses de la Section 2, soit, pour $x \in E$, la relation R_x sur E définie par

$$yR_xz \quad \text{ssi} \quad \phi \neq z \text{ et } z \leqslant d(y) \text{ et } y \leqslant x .$$

Alors R_x est bien-fondée ssi x appartient à Q.

16 Soit R une relation sur E. On lui associe son *arbre généalogique* A_R de la manière suivante: une suite finie $x_1, \ldots, x_n$ d'éléments de E appartient à A_R ssi on a x_nRx_{n-1} et ... et x_2Rx_1 (on convient que toute suite finie de longueur $\leqslant 1$ se trouve dans A_R). Il est alors clair, d'après 14(2), que R est bien-fondée ssi toute branche de A_R est finie. On peut aussi traduire cela en termes de temps d'arrêt; à R est associé le t.d'a. T_R sur $\Omega = E^{\mathbb{N}}$ défini comme suit (cf. la démonstration de 9): pour $\omega = (x_n)_{n\in\mathbb{N}}$ élément de Ω,

$$T_R(\omega) = \inf\{n \geqslant 2 \colon x_n \text{ n'est pas un ancêtre de } x_{n-1}\},$$

et R est bien-fondée ssi T_R est un t.d'a. fini.

17 Nous nous donnons maintenant une relation bien-fondée R sur E. On définit alors la *relation dérivée* $R' = d(R)$ de R par

$$xR'y \quad \text{ssi} \quad xRy \quad \text{et} \quad x \text{ a un } R\text{-ancêtre},$$

et la famille (R^α) des dérivés successifs de R, indexée par les ordinaux, par la récurrence transfinie habituelle

$$R^0 = R, \qquad R^1 = R', \qquad R^\beta = d(\inf_{\alpha<\beta} R^\alpha).$$

Comme, pour $\alpha < \beta$, on a $R^\alpha \supseteq R^\beta$ et $R^\alpha \neq R^\beta$ si $R^\alpha \neq \varnothing$, un raisonnement élémentaire de cardinalité montre qu'il existe un ordinal α, ayant même cardinalité que $E \times E$, tel que $R^\alpha = \varnothing$. Définissons alors une application $x \to \rho(x)$ de E dans les ordinaux par

$$\rho(x) = \inf\{\alpha \colon x \text{ est } R^\alpha\text{-primitif}\}.$$

L'application $\rho(\cdot)$ est la *fonction de rang* de R et, pour $R \neq \varnothing$, l'ordinal $|R| = \sup_{x\in E}[\rho(x)+1]$ la *longueur* de R. On peut évidemment voir les choses en termes d'arbres ou de t.d'a. (exercice!). La seule chose qui nous importe est de savoir que, si T_R^α est le α-ième dérivé du t.d'a. généalogique T_R, alors $|R|$ a même cardinalité que $|T_R| = \inf\{\alpha \colon T_R^\alpha = 0\}$ (en fait, on a $|R| \leqslant |T_R| \leqslant |R|+2$). Noter au passage que, si $\mathbf{R}$ est la relation associée en 15(c) à un t.d'a. fini T de *longueur* $|T| = \inf\{\alpha \colon T^\alpha = 0\}$, alors on a $|\mathbf{R}| = |T|+1$.

18 Nous nous donnons maintenant un ensemble infini J, muni de la topologie discrète, désignons par κ le cardinal de J, et supposons que E est un *espace* κ*-analytique*. Nous entendons par là que E est un espace topologique séparé et qu'il existe une application continue de $\Omega = J^{\mathbb{N}}$ *sur* E (par conséquent, si $\kappa = \aleph_0$, E est souslinien au sens de Bourbaki, soit mieux qu'analytique au sens du Chapitre I si E n'est pas métrisable séparable). Il est facile de voir que tout espace λ-analytique, où λ est un cardinal $< \kappa$, est *a fortiori* κ-analytique (exercice!); on peut aussi montrer qu'un sous-espace d'un espace κ-analytique est κ-analytique ssi c'est une partie κ-souslinienne, mais nous n'utiliserons pas ce fait.

Voici alors un énoncé du théorème de Kunen et Martin, un peu plus général que l'énoncé habituel.

19 THÉORÈME *Soit R une relation bien-fondée sur un espace κ-analytique E. Si R est une partie κ-souslinienne de $E \times E$, alors on a*

$$|R| < \kappa^+,$$

où κ^+ désigne le cardinal successeur de κ.

Démonstration Dans un premier temps, nous montrons, rapidement, comment on se ramène aisément au cas où R est une relation bien-fondée fermée sur $\Omega = J^{\mathbb{N}}$. D'abord, E étant κ-analytique, il existe une application continue f de Ω sur E; définissons alors une relation $\mathbf{R}$ sur Ω en posant $\omega \mathbf{R} w$ ssi $f(\omega) R f(w)$: on vérifie sans peine que $\mathbf{R}$ est une relation bien-fondée, κ-souslinienne, ayant même longueur que R. On peut donc supposer que $E = \Omega$. Ceci fait, comme R est une partie κ-souslinienne de $\Omega \times \Omega$, il existe un fermé H de $\Omega \times \Omega \times \Omega$, tel que l'on ait $\omega R w$ ssi $\exists \omega' \, (\omega, w, \omega') \in H$; définissons alors une relation $\mathbf{R}$ sur $\Omega \times \Omega$ par

$$(\omega, \omega')\mathbf{R}(w, w') \quad \text{ssi} \quad \omega' = w' \quad \text{et} \quad (\omega, w, \omega') \in H:$$

il est clair que $\mathbf{R}$ est une relation fermée sur $\Omega \times \Omega$ et, ici encore, on vérifie aisément que $\mathbf{R}$ est bien-fondée et a même longueur que R. Enfin, $\Omega \times \Omega$ étant homéomorphe à Ω, on peut supposer que R est une relation bien-fondée et fermée sur Ω. Voici maintenant ce que nous allons faire, en termes d'arbres—nous le ferons en termes de t.d'a.! Disons qu'une application f de l'ensemble $\bar{S}$ des suites finies d'éléments de Ω dans l'ensemble S des suites finies d'éléments de J est une stratégie si elle vérfie les deux conditions suivantes (un peu moins restrictives que celles de III-14):

(a) il existe une application croissante $\Phi : n \to \bar{n}$ de $\mathbb{N}^\circ$ dans $\mathbb{N}^\circ$ telle que $\Phi(0) = 0$ et que, si $\bar{s} \in \bar{S}$ a pour longueur $|\bar{s}| = n$, alors $s = f(\bar{s})$ a pour longueur $|s| = \bar{n}$;

(b) pour tout $\bar{s}$, $\bar{t} \in \bar{S}$, on a $\bar{s} \dashv \bar{t} \Rightarrow f(\bar{s}) \dashv f(\bar{t})$ (si bien que Φ est strictement croissante).

Voici un exemple de stratégie, que l'on pourra utiliser tout au long de notre démonstration: pour $\bar{s} = (\omega_1, \omega_2, \ldots, \omega_n)$, poser

$$\mathbf{f}(\bar{s}) = \omega_1|1.\omega_1|2.\omega_2|2. \ldots .\omega_1|n.\omega_2|n. \ldots .\omega_n|n$$

si bien que $\bar{n} = 1^2 + 2^2 + \cdots + n^2$. Ayant choisi une stratégie "convenable" f de $\bar{S}$ dans S (notre exemple $\mathbf{f}$ sera "convenable"), nous associerons à l'arbre généalogique A de R dans $\bar{S}$ un arbre B dans S, sans branche infinie (là interviendra le fait que R est fermée) tel que $f(A) \subseteq B$: comme B est dans S, qui a même cardinalité que J, on aura $|B| < \kappa^+$ et la condition $f(A) \subseteq B$ entrainera que $|A| \leq |B|$. Au travail, maintenant. Nous nous donnons une stratégie f de $\bar{S}$ dans S, vérifiant la condition suivante:

(c) si $\bar{s}=(\omega_1,\ldots,\omega_n)$ et $\bar{t}=(w_1,\ldots,w_n)$ sont deux éléments de $\bar{S}$ de longueur n, alors on a

$$f(\bar{s})=f(\bar{t}) \quad\Rightarrow\quad \omega_1|n=w_1|n \text{ et } \ldots \text{ et } \omega_n|n=w_n|n\,.$$

Voici une autre manière de voir (c). Pour $\bar{\omega}=(\omega_1,\ldots,\omega_n,\ldots)$ élément de $\bar{\Omega}=\Omega^{\mathbb{N}}$, désignons par $f(\bar{\omega})$ l'élément ω de Ω tel que l'on ait $\omega|\bar{n}=f(\bar{\omega}|n)$ pour tout n—on prolonge ainsi f en une application, encore notée f, de $\bar{\Omega}$ dans Ω; alors (c) exprime que la connaissance des $\bar{n}$ premiers termes de $f(\bar{\omega})$ implique celle de l'élément $\bar{\omega}\wr n=(\omega_1|n,\ldots,\omega_n|n)$ de S, soit un "fragment substantiel" de $\bar{\omega}|n$. En particulier, f est un homéomorphisme de $\bar{\Omega}=\Omega^{\mathbb{N}}$ sur une partie fermée de Ω (qu'un tel homéomorphisme existe n'a rien d'extraordinaire: c'est l'un des "moteurs" de la théorie des ensembles analytiques), et f fait tout son possible pour être aussi un homéomorphisme de $\bar{S}_n=\{\bar{s}\colon|\bar{s}|=n\}$ sur une partie de $S_n=\{S\colon|s|=\bar{n}\}$, sans y arriver naturellement—cardinalité oblige. Donc, nous nous donnons une stratégie "convenable" f, et nous définissons une application $T\to\bar{T}$ de l'ensemble des t.d'a. sur $\bar{\Omega}$ dans celui des t.d'a. sur Ω en posant, pour $\omega\in\Omega$,

$$\bar{T}(\omega)=\inf\{\bar{n}\colon \forall\bar{\omega}\in\bar{\Omega}\ \omega|\bar{n}=f(\bar{\omega}|n)\Rightarrow T(\bar{\omega})\leq n\}\,.$$

On peut montrer (mais nous ne nous en servirons pas) que $\bar{T}$ est le plus petit t.d'a. majorant la fonction T^* sur Ω définie par $T^*(\omega)=\bar{n}$ si $\omega=f(\bar{\omega})$ et $T(\bar{\omega})=n$, et $T^*(\omega)=0$ si $\omega\notin f(\bar{\Omega})$. Quelques calculs élémentaires permettent de vérifier, d'abord que $\bar{T}$ est bien un t.d'a., ensuite que, si U est l'inf. d'une famille (V_α) de t.d'a. sur $\bar{\Omega}$, alors $\bar{U}$ est l'inf. de la famille $(\bar{V}_\alpha)$, et enfin que que l'on a $d(T)\leq d(\bar{T})$. D'où l'on déduit que, si $\bar{T}$ est fini, alors on a $|T|\leq|\bar{T}|<\kappa^+$. Il n'est pas vrai, en général, que $\bar{T}$ est fini si T l'est, mais c'est le cas si T est une application s.c.s. En effet, sinon, $\bar{T}$ aurait un pôle w et il existerait donc une suite $(\bar{\omega}^n)$ d'éléments de $\bar{\Omega}$ telle que $w|\bar{n}=f(\bar{\omega}^n|n)$ et $T(\bar{\omega}^n)>n$ pour tout n. Mais, la connaissance de $w|\bar{n}$ entrainant celle de $\bar{\omega}^n\wr n$ (cf. plus haut), on en déduirait l'existence d'un unique élément $\bar{w}$ de $\bar{\Omega}$ tel que $\bar{w}\wr n=\bar{\omega}^n\wr n$ pour tout n: en particulier, $\bar{w}$ serait la limite de la suite $(\bar{\omega}^n)$ dans $\bar{\Omega}$ et on aurait alors $T(\bar{w})\geq\limsup T(\bar{\omega}^n)=\infty$. Pour achever la démonstration du théorème, il ne reste plus alors qu'à vérifier que le t.d'a. généalogique d'une relation est fini ssi la relation est bien-fondée et s.c.s. ssi elle est fermée. Le premier point a déjà été signalé; nous laissons le soin de vérifier le second au lecteur.

Comme une partie E-analytique (resp. E-coanalytique) d'un espace séparé est $\aleph_0$-souslinienne (resp. $\aleph_1$-souslinienne; cf. la remarque de III-36), on a

20 COROLLAIRE *Soit R une relation bien-fondée sur un espace polonais. Alors on a*

$$R \text{ analytique} \Rightarrow |R| < \aleph_1,$$

$$R \text{ coanalytique} \Rightarrow |R| < \aleph_2.$$

Remarques (a) En particulier, il n'existe pas de bon-ordre analytique sur un espace polonais non dénombrable; ce résultat était connu des auteurs "classiques".

(b) Le résultat concernant une relation bien-fondée coanalytique n'est évidemment intéressant que si l'on ne suppose pas l'hypothèse du continu satisfaite.

Nous démontrons maintenant notre énoncé 4, qui n'est en fait qu'une variante du premier point du corollaire précédent.

21 COROLLAIRE *Soit E un espace (métrisable séparable) analytique et soit T un t.d'a. analytique sur $E^{\mathbb{N}}$. On a alors*

$$T \textit{ est fini} \quad \textit{ssi} \quad i(T) < \aleph_1.$$

Démonstration Il est clair que T est fini si $i(T) < \aleph_1$. Réciproquement, supposons T fini. On peut alors montrer que l'on a $i(T) < \aleph_1$ en appliquant 19 à la relation **R** associée à T en 15(c). Mais nous préférons présenter une variante, que s'écrit bien dans le langage des t.d'a. Nous définissons une relation R_T sur $\Omega = E^{\mathbb{N}}$ en posant

$$\begin{aligned} \omega R_T w \quad &\text{ssi} \quad \forall n\, [T(w) > n \text{ et } \omega|n = w|n \quad \Rightarrow \quad T(\omega) > n+1] \\ &\text{ssi} \quad \forall n\, [\omega|n = w|n \text{ et } T(w) \leq n+1 \quad \Rightarrow \quad T(\omega) > n+1]. \end{aligned}$$

La deuxième écriture de R_T a l'avantage de montrer clairement que R_T est analytique, et aussi bien-fondée. Par ailleurs, on vérifie aisément que $R_T = \varnothing$ ssi T est constant, et que, si le α-ième dérivé R_T^α est $\neq \varnothing$, alors $R_T^\alpha = R_{T^\alpha}$. D'où $i(T) \leq |R_T| + \aleph_0 < \aleph_1$.

3 Quelques espaces fonctionnels

Les quelques espaces fonctionnels que nous allons envisager dans cette section sont bien connus des probabilistes: ce sont des espaces de trajectoires en théorie des processus stochastiques en temps continu. Nous commençons par fixer les hypothèses et notations qui aurons cours tout au long de cette section.

22 On se donne un espace polonais E (par exemple, $E=\mathbb{R}$), et on désigne par W l'ensemble des applications de $\mathbb{R}_+$ dans E: intuitivement un élément w de W est une trajectoire possible, paramétrée par l'échelle de temps $\mathbb{R}_+$, d'un processus prenant ses états dans E. On définit sur W la famille $(X_t)_{t\in\mathbb{R}_+}$ des *applications coordonnées* (i.e. on a $X_t(w)=w(t)$ pour $w\in W$) et la famille $(\theta_t)_{t\in\mathbb{R}_+}$ des *opérateurs de translation*: on a pour tout $w\in W$ et tout $s, t\in\mathbb{R}_+$

$$X_t[\theta_s(w)]=X_{t+s}(w)\,.$$

Noter que (θ_t) est une semi-groupe pour la composition: on a $\theta_{t+s}=\theta_t\circ\theta_s$ pour tout s, t. La lettre Ω, agrémentée d'indices divers, désignera un espace (métrisable, séparable) vérifiant les conditions suivantes:

(a) en tant qu'ensemble, Ω est une partie de W, stable pour les opérateurs de translation (les restrictions de X_t, θ_t à Ω seront encore notées X_t, θ_t);

(b) en tant qu'espace, Ω assure que les applications

$$(t,\omega)\to X_t(\omega)\,,\qquad (t,\omega)\to\theta_t(\omega)\,,$$

sont boréliennes, $\mathbb{R}_+\times\Omega$ étant muni de la topologie produit (en fait, seule la structure mesurable de Ω entre en jeu dans tout ce que nous ferons, mais nous n'insisterons pas là-dessus. Cela explique cependant pourquoi nous ne chercherons pas toujours à mettre la "plus belle" topologie sur notre Ω).

On réservera cependant la notation Ω^* à divers usages non nécessairement compatibles avec les propriétés (a) et (b) ci-dessus.

23 *Exemples* (a) D'abord, deux exemples chers aux probabilistes: l'ensemble $\Omega=\Omega_{\text{càd}}$ des applications continues à droite (en abrégé, application càd) et l'ensemble $\Omega=\Omega_{\text{càd}}^{\text{làg}}$ des applications continues à droite et pourvues de limites à gauche[(1)] (en abrégé, application càdlàg), que l'on munit de la tribu $\mathscr{F}^\circ$ engendrée par les X_t. On plonge Ω dans l'espace polonais produit $E^{\mathbb{Q}_+}$ en identifiant $\omega\in\Omega$ à sa restriction aux rationnels $\geqslant 0$: Ω se trouve alors équipé de la topologie de la convergence simple sur les rationnels, et il est facile de voir que $\mathscr{F}^\circ$ est égale à la tribu borélienne $\mathscr{B}(\Omega)$. Par ailleurs, l'application $(t,\omega)\to X_t(\omega)$ est borélienne comme étant séparément borélienne en ω et càd en t (résultat classique en théorie des processus), et $(t,\omega)\to\theta_t(\omega)$ est également borélienne car, pour s fixé, $(t,\omega)\to X_s[\theta_t(\omega)]=X_{s+t}(\omega)$ est borélienne, et $\mathscr{B}(\Omega)$ est la tribu engendrée par les X_s.

(1) Sous-entendu: pour $t>0$.

(b) Un exemple plus commun: l'ensemble $\Omega = \Omega_{\text{con}}$ des applications continues, qui est un espace polonais pour la topologie de la convergence uniforme sur tout compact de $\mathbb{R}_+$. Cette topologie est bien plus fine que celle de la convergence simple sur les rationnels, mais définit cependant la même tribu borélienne d'après le marteau-pilon I-36 (il y a plus élémentaire!).

24 L'espace Ω étant donné, nous dirons, pour abréger, qu'une application T de Ω dans $\bar{\mathbb{R}}_+$ est un *temps adéquat* (sous-entendu, pour la théorie des dérivations—nous verrons cela plus loin) si

(a) T est une fonction Ω-coanalytique (cf. no. 1);
(b) T vérifie, pour tout $t \geq 0$ et tout $\omega \in \Omega$,

$$T(\omega) \leq t + T[\theta_t(\omega)] .$$

Donnons tout de suite un exemple pour fixer les idées: B étant un borélien de E (par exemple, $B = \{0\}$ pour $E = \mathbb{R}$), on pose

$$T_B(\omega) = \inf \{t > 0 \colon X_t(\omega) \in B\} \qquad (\text{avec } \inf \varnothing = \infty) .$$

On dit que T_B est le *temps d'entrée* dans B même si, en général, on n'a pas $X_{T(\omega)}(\omega) \in B$ pour $T(\omega) < \infty$; c'est par ailleurs un exemple de temps d'arrêt en temps continu, mais nous ne chercherons pas ici à définir exactement cette notion, nous contentant de vérifier que T_B possède les propriétés (a) et (b) (non partagées par tous les t.d'a.!): d'abord, on a $T_B(\omega) < t$ ssi $\exists s\ s < t$ et $X_s(\omega) \in B$, si bien que T_B est Ω-coanalytique; ensuite, on a $T(\omega) = t + T[\theta_t(\omega)]$ pour tout $t < T(\omega)$, si bien que l'on a $T(\omega) \leq t + T[\theta_t(\omega)]$ pour tout t. Le temps d'entrée T_B est aussi, en un certain sens, le premier temps de passage T^1 dans B. Il est naturel de considérer le second temps de passage T^2 dans B, égal à $T^1 + T^1 \circ \theta_{T^1}$ (on entend par là que, pour $T^1(\omega) < \infty$, on a $T^2(\omega) = T^1(\omega) + T^1[\theta_{T^1(\omega)}(\omega)]$, et que $T^2(\omega) = \infty$ si $T^1(\omega) = \infty$), puis le troisième temps de passage $T^3 = T^2 + T^1 \circ \theta_{T^2}$, etc.; remarquer que, si on a $T^1(\omega) < T^2(\omega) < T^3(\omega)$, alors $T^1(\omega)$ est effectivement le plus petit élément de $\{t : X_t(\omega) \in B\}$ et $T^2(\omega)$ celui de $\{t > T_1(\omega) : X_t(\omega) \in B\}$. Nous allons définir maintenant, de manière générale, les itérés successifs d'un temps adéquat.

25 On associe à un temps adéquat T sur Ω la *suite transfinie de ses itérés successifs* $(T^i)_{i \in I}$, indexée par les ordinaux dénombrables, définie comme suit (ne pas confondre avec (T^i) de III-12!)

$$T^0 = 0 , \qquad T^1 = T ,$$

$$T^j = U + T \circ \theta_U , \quad \text{où} \quad U = \sup\nolimits_{i<j} T^i (= T^{j-1} \text{ si } j-1 \text{ existe}) ,$$

où $T \circ \theta_U(\omega) = T[\theta_{U(\omega)}(\omega)]$ pour $U(\omega) < \infty$ et $T \circ \theta_U(\omega) = \infty$ pour $U(\omega) = \infty$. Pour tout $\omega \in \Omega$, il existe un plus petit ordinal dénombrable $\mathbf{j} = j_T(\omega)$ tel que $T^{\mathbf{j}}(\omega) = T^{\mathbf{j}+1}(\omega) = \ldots$: si $T^{\mathbf{j}}(\omega)$ est fini, alors on a $T \circ \theta_{T^{\mathbf{j}}}(\omega) = 0$; réciproquement, il est facile de voir qu'on a $T^{\mathbf{j}}(\omega) = \infty$ ssi on a $T \circ \theta_t(\omega) > 0$ pour tout $t \in \mathbb{R}_+$. On associe enfin au temps adéquat T sa *fonction-indice* i_T de Ω dans $I \cup \{\aleph_1\}$ définie par $i_T(\omega) = \inf\{i \colon T^i(\omega) = \infty\}$, avec $\inf \varnothing = \aleph_1$: on a alors $i_T(\omega) = j_T(\omega)$ si l'on a $\forall t\, T \circ \theta_t(\omega) > 0$, et $i_T(\omega) = \aleph_1$ sinon. Noter au passage que le sous-espace $Q_T = \{\omega : i_T(\omega) < \aleph_1\}$ de Ω est stable pour les opérateurs de translation, et peut donc être considéré comme un nouvel espace Ω. Revenons une fois encore à l'exemple T_B du temps d'entrée dans le borélien B: si $\mathbf{i} = i_{T_B}(\omega)$ est $< \aleph_1$, alors la suite transfinie $(T^j(\omega))_{j<\mathbf{i}}$ est strictement croissante, et dénote successivement les temps de passage "réels" de ω dans B—à ceci près qu'il peut y avoir des omissions dans ce décompte au niveau des ordinaux limites.

Voici maintenant le résultat fondamental de cette section.

26 THÉORÈME *Soit T un temps adéquat sur l'espace Ω.*

(1) *L'ensemble Q_T défini par*

$$\omega \in Q_T \quad \text{ssi} \quad i_T(\omega) < \aleph_1 \quad \text{ssi} \quad \exists i\, T^i(\omega) = \infty \quad \text{ssi} \quad \forall t\, T \circ \theta_t(\omega) > 0$$

est une partie Ω-coanalytique de Ω.

(2) *Si A est un sous-espace analytique de Q_T, on a alors*

$$\sup_{\omega \in A} i_T(\omega) < \aleph_1 \,;$$

soit encore, il existe un ordinal dénombrable $\mathbf{i}$ tel que l'on ait $T^{\mathbf{i}}(\omega) = \infty$ pour tout $\omega \in A$. En particulier, si $\sup_{\omega \in Q_T} i_T(\omega)$ vaut $\aleph_1$, alors Q_T n'est pas un espace analytique.

Démonstration Le point (1) est élémentaire, en écrivant que l'on a $\omega \in Q_T$ ssi on a $\forall t\, T[\theta_t(\omega)] > 0$. Pour démontrer le point (2), nous allons introduire une dérivation notée Θ_T, non pas sur Ω mais plutôt sur $\mathbb{R}_+ \times \Omega$: on aura $\Theta_T(t, \omega) = (t + T(\omega), \theta_{T(\omega)}(\omega))$ pour $T(\omega) < \infty$. Définissons cependant, d'abord, un ordre adéquat. Afin d'assurer, entre autres, que nous aurons un plus petit élément pour notre ordre, nous augmentons l'espace produit $\mathbb{R}_+ \times \Omega$ d'un point isolé, noté $(\infty, [\delta])$: on peut considérer que ∞ est le point de compactification habituel de $\mathbb{R}_+$, tandis que $[\delta]$ est un point isolé rajouté à Ω, trajectoire constante égale à un point isolé δ ajouté à E. On prolonge alors le semi-groupe des translations en posant, pour $\omega \in \Omega$ ou pour $\omega = [\delta]$, $\theta_\infty(\omega) = [\delta]$, $\theta_t([\delta]) = [\delta]$, et la famille des coordonnées en posant $X_\infty(\omega) = \delta$, $X_t([\delta]) = \delta$. Ceci fait, nous munissons l'espace $F = (\mathbb{R}_+ \times \Omega) \cup \{(\infty, [\delta])$ de l'ordre $\leqslant$ suivant

$$(t, \omega) \leqslant (s, w) \quad \text{ssi} \quad t \geqslant s \quad \text{et} \quad \omega = \theta_{t-s}(w)$$

en convenant que $t-s=\infty$ si $t=\infty$. Cet ordre a une interprétation intuitive: si l'on regarde un point (t, ω) de $\mathbb{R}_+\times\Omega$ comme dénotant que la trajectoire de notre processus, regardée à partir de l'instant t, est égale à ω (i.e. sa valeur en $t+h$, pour $t, h\geqslant 0$, est égale à $\omega(h)=X_h(\omega)$), il est naturel de considérer que (t, ω) vient après (s, w) si $t\geqslant s$ et $\omega=\theta_{t-s}(w)$. Vérifions que notre ordre $\leqslant$ est adéquat, au sens de 6:

(a) on a $\phi=(\infty, [\delta])$ comme plus petit élément;
(b) une famille filtrante décroissante (t_j, ω_j) admet pour borne inférieure l'élément (y, ω) où $t=\sup_j t_j$ et où $\omega=\theta_{t-t_j}(\omega_j)$—qui ne dépend pas de j;
(c) plongeons $\Omega\cup\{[\delta]\}$ dans un espace compact métrisable Ω^*.

Alors $(t, \omega)\to\theta_t(\omega)$ est une application borélienne de $\bar{\mathbb{R}}_+\times(\Omega\cup\{[\delta]\})$ dans Ω^*, que l'on peut donc prolonger, d'après I-17(2), en une application borélienne $(t, \omega)\to\Phi_t(\omega)$ de $\bar{\mathbb{R}}_+\times\Omega^*$ dans Ω^*. Définissons alors une relation $\dashv$ sur l'espace produit $\bar{\mathbb{R}}_+\times\Omega^*$ par

$$(t, \omega)\dashv(s, w) \quad \text{ssi} \quad t\geqslant s \text{ et } \omega=\Phi_{t-s}(w).$$

La relation $\dashv$ est borélienne, et l'on a clairement

$$(t, \omega)\dashv(s, w) \text{ et } (s, w)\in F \quad \text{ssi} \quad (s, w)\in F \text{ et } (t, \omega)\in F \text{ et } (t, \omega)\leqslant(s, w).$$

Notre ordre $\leqslant$ étant adéquat, nous associons à notre temps adéquat T une dérivation analytique sur $(F, \leqslant)$, notée Θ_T, définie par

$$\Theta_T(t, \omega)=(t+T(\omega), \theta_{T(\omega)}(\omega)).$$

Vérifions que θ_T satisfait aux conditions de 7. D'abord, il est clair que l'on a $\Theta_T(t, \omega)\leqslant(t, \omega)$. Ensuite, la relation définie par $(t, \omega)\leqslant\Theta_T(s, w)$ est F-analytique puisqu'on a

$$(t, \omega)\leqslant\Theta_T(s, w) \quad \text{ssi} \quad s+T(w)\leqslant t \quad \text{et} \quad \omega=\theta_{t-s}(w).$$

Enfin—nous l'avons gardé pour la bonne bouche—on a

$$(t, \omega)\leqslant(s, w)\Rightarrow\Theta_T(t, \omega)\leqslant\Theta_T(s, w);$$

il s'agit là d'un calcul élémentaire (classique en théorie des processus de Markov), mais dans lequel on se mêle facilement les pieds. Faisons le, pour édifier le lecteur. On suppose que l'on a $t\geqslant s$ et $\omega=\theta_{t-s}(w)$; on a alors, d'après 24(b),

$$t+T(\omega)=s+t-s+T[\theta_{t-s}(w)]\geqslant s+T(w),$$

et d'autre part

$$\theta_{t+T(\omega)-s-T(w)}(w) = \theta_{T(\omega)+t-s}(w) = \theta_{T(\omega)}(\omega) .$$

Ceci fait, il est clair que le i-ème dérivé de (t, ω) pour la dérivation Θ_T est égal à $(t+T^i(\omega), \theta_{T^i(\omega)}(\omega))$, où T^i est le i-ème itéré de T: on a donc (t, ω) parfait $\neq \phi$ ssi $t<\infty$ et $T(\omega)=0$. Par ailleurs, l'indice $i(t, \omega)$ associé à $(t, \omega) \in \mathbb{R}_+ \times \Omega$ par Θ_T ne dépend pas de t et est égal à l'indice $i_T(\omega)$ défini en 25. Le point (2) de notre énoncé résulte alors immédiatement du théorème 11.

Avant de donner quelques applications de ce théorème, nous rappelons qu'un espace (métrisable, séparable) est dit coanalytique si, plongé dans un espace polonais, il s'y trouve être une partie coanalytique (la propriété d'être une partie coanalytique ne dépendant pas du plongement considéré—cf. I-24(b)).

Voici une première application, simple, de notre théorème.

27 THÉORÈME *Soit $\Omega = \Omega_{\text{con}}$ l'espace des fonctions continues de $\mathbb{R}_+$ dans $\mathbb{R}$. Le sous-espace Ω^0 constitué des ω vérifiant*

$$X_t(\omega)=0 \quad \Rightarrow \quad \exists \varepsilon>0 \; \forall s \in \,]t, t+\varepsilon[\;\; X_s(\omega) \neq 0$$

pour tout $t \geqslant 0$, est coanalytique mais n'est pas analytique.

Démonstration Désignons par T le temps d'entrée dans $\{0\}$: on a $T(\omega) = \inf\{t>0 \colon X_t(\omega)=0\}$. C'est un temps adéquat, et on a $\Omega^0 = Q_T$. Montrons que $\sup_{\omega \in Q_T} i_T(\omega)$ vaut $\aleph_1$. La démonstration que nous donnons de ce fait s'adapte aisément à de nombreuses situations. Soit w un élément fixé de Q_T tel que $T(w)<\infty$ et que l'on ait $X_0(w)=0$ et $\sup_t |X_t(\omega)|<\infty$. Fixons nous un ordinal dénombrable k et partitionnons $\mathbb{R}_+$ en intervalles (non vides) $[t_i, t_{i+1}[$ pour $i \leqslant k$ avec $t_0=0$ et $t_{k+1}=\infty$ (qu'une telle partition existe se montre aisément par récurrence transfinie). Désignons par π_i un homéomorphisme de $[t_i, t_{i+1}[$ sur $[0, \infty[$ et définissons un "analogue" w_i de w sur $[t_i, t_{i+1}[$ en posant, pour $t \in [t_i, t_{i+1}[= [t_i, t_i+h_i[$,

$$w_i(t) = \mathrm{e}^{-1/h_i} w[\pi_i(t)]$$

puis recollons les w_i en un élément ω^k de Ω en posant $\omega^k(t)=w_i(t)$ pour $t \in [t_i, t_{i+1}[$. Il est alors clair que l'on a $i_T(\omega^k) \geqslant k$ (et en fait $=k$ si w est bien choisie).

Remarques (a) L'espace Ω^0 de l'énoncé est constitué des ω tels que l'ensemble $\{t : \omega(t)=0\}$ soit discret pour la topologie droite sur $\mathbb{R}_+$ (i.e. la topologie dont les ouverts fondamentaux sont les intervalles de la forme

$[u, v[$; cette topologie, qui est séparable mais non métrisable, intervient souvent en théorie des processus). On peut montrer que, par contre, l'espace des ω tels que $\{t : \omega(t) = 0\}$ soit discret pour la topologie ordinaire est un borélien de Ω_{con} et donc un espace lusinien au sens de Bourbaki (exercice!).

(b) On peut montrer que, dans $\Omega = \Omega_{\text{càd}}$ (ou $\Omega = \Omega_{\text{càd}}^{\text{làg}}$), l'ensemble Ω^0 des ω tels que $\{t : \omega(t) = 0\}$ soit discret pour la topologie droite est une partie Ω-coanalytique qui n'est pas Ω-analytique. (Exercice! Attention, comme l'espace $\Omega_{\text{càd}}$ n'est pas analytique—nous allons le voir ci-dessous—une application directe de 26 fournirait un résultat moins fort, à savoir que l'espace Ω^0 n'est pas analytique.)

(c) Pour $\Omega = \Omega_{\text{con}}$, le temps d'entrée T dans $\{0\}$ est une fonction s.c.i., et donc une fonction borélienne. Nous verrons que ce n'est pas le cas pour $\Omega = \Omega_{\text{càd}}$ ou $\Omega = \Omega_{\text{càd}}^{\text{làg}}$.

28 THÉORÈME *Supposons que l'espace polonais E a au moins deux éléments. Alors l'espace $\Omega_{\text{càd}}$ des applications continues à droite de $\mathbb{R}_+$ dans E est coanalytique, mais n'est pas analytique.*

Démonstration Plongeons E dans $F = [0, 1]^{\mathbb{N}}$, ce qui induit un plongement de $\Omega_{\text{càd}}$ dans l'espace métrisable compact $\Omega^* = F^{\mathbb{Q}_+}$. Définissons, pour tout $t \in \mathbb{R}_+$, deux applications $\bar{X}_t$, $\underline{X}_t$ de Ω^* dans F en posant, pour tout $w \in \Omega^*$,

$$\bar{X}_t(w) = \limsup_{s > t,\, s \to t} w(s),$$

$$\underline{X}_t(w) = \liminf_{s > t,\, s \to t} w(s)\,,$$

où, bien entendu, s est astreint à être rationnel, et où l'on prend les lim sup, lim inf pour chaque coordonnée sur $F = [0, 1]^{\mathbb{N}}$. Comme $\mathbb{Q}_+$ est dénombrable, on vérifie sans peine que les applications $(t, w) \to \bar{X}_t(w)$ et $(t, w) \to \underline{X}_t(w)$ sont boréliennes. Soit alors B la partie borélienne de $\mathbb{R}_+ \times \Omega^*$ définie comme suit: on a $(t, w) \in B$ ssi on a $\bar{X}_t(w) \notin E$ ou $\bar{X}_t(w) \neq \underline{X}_t(w)$ ou ($t \in \mathbb{Q}_+$ et $\bar{X}_t(w) \neq w(t)$). Et soit D le "début" de B, défini par $D(w) = \inf\{t \geq 0 \colon (t, w) \in B\}$: D est une fonction coanalytique sur Ω^*, et donc $\Omega_{\text{càd}} = \{w \colon D(w) = \infty\}$ est une partie coanalytique de Ω^*. Nous démontrons maintenant que $\Omega_{\text{càd}}$ n'est pas un espace analytique. Comme, pour toute partie fermée H de E, l'ensemble $\{\omega \in \Omega_{\text{càd}} \colon \forall t\, \omega(t) \in H\}$ est un borélien de $\Omega_{\text{càd}}$, il suffit de traiter le cas où $E = \{0, 1\}$. Nous définissons alors une fonction T de Ω dans $\bar{\mathbb{R}}_+$ par

$$T(\omega) = \inf\left\{t > 0;\ \limsup_{s < t,\, s \to t} \omega(s) = 1,\ \liminf_{s < t,\, s \to t} \omega(s) = 0\right\}.$$

Comme ω est càd, on peut astreindre s à parcourir les rationnels. Il est alors facile de vérifier que T est un temps adéquat, partout > 0 sur $\Omega_{\text{càd}}$. Nous laissons au lecteur le soin de vérifier que l'on a $\sup_{\omega\in\Omega_{\text{càd}}} i_T(\omega)=\aleph_1$ (cf. la fin de la démonstration de 27).

L'on a, par contre,

29 THÉORÈME *L'espace $\Omega_{\text{càd}}^{\text{làg}}$ des applications continues à droite et pourvues de limites à gauche de $\mathbb{R}_+$ dans l'espace polonais E est un espace lusinien.*

Démonstration Nous allons montrer que $\Omega=\Omega_{\text{càd}}^{\text{làg}}$ est une partie borélienne de $\Omega^*=E^{\mathbb{Q}_+}$. Nous munissons E d'une distance d qui en fait un espace métrique complet; nous augmentons d'autre part E d'un point isolé δ et posons $d(x,\delta)=\infty$ pour tout $x\in E$. Pour tout $\varepsilon>0$, nous définissons maintenant, par récurrence pour tout $n\in\mathbb{N}^\circ$, une fonction T_ε^n de Ω^* dans $\bar{\mathbb{R}}_+$ et une application Y_ε^n de Ω^* dans $E\cup\{\delta\}$ comme suit: s étant astreint à être rationnel, nous posons, pour tout $w\in\Omega^*$, en écrivant "$s\downarrow t$" pour "$s\geqslant t$, $s\to t$",

$$T_\varepsilon^0(w)=0\,,$$

$$Y_\varepsilon^0(w)=\lim_{s\downarrow 0} w(s) \text{ si cette limite existe, et} =\delta \text{ sinon, puis,}$$

$$T_\varepsilon^{n+1}(w)=\inf\{s: s>T_\varepsilon^n(w) \text{ et } d(w(s),Y_\varepsilon^n(w))>\varepsilon\}\quad(\inf\varnothing=\infty)\,,$$

$$Y_\varepsilon^{n+1}(w)=\lim_{s\downarrow T_\varepsilon^{n+1}(w)} w(s) \text{ si } T_\varepsilon^{n+1}(w)<\infty \text{ et si cette limite existe, et } =\delta$$

sinon.

On vérifie sans peine (autre que dactylographique) que les T_ε^n, Y_ε^n sont des applications boréliennes. Par ailleurs, il n'est pas non plus difficile de voir que l'on a

$$w\in\Omega_{\text{càd}}^{\text{làg}} \qquad \text{ssi} \qquad \forall m\in\mathbb{N}\ \lim_{n\to\infty} T_{1/m}^n(w)=\infty\,,$$

si bien que notre Ω est une partie borélienne de $E^{\mathbb{Q}_+}$.

Nous nous bornerons à citer un dernier exemple d'application intéressant du théorème 26.

30 THÉORÈME *Soit F un fermé non ouvert de l'espace polonais E et soit $\Omega_{\text{càd}}^F$ le sous-espace de $\Omega_{\text{càd}}$ constitué des ω évitant F (i.e. $\omega\in\Omega_{\text{càd}}^F$ ssi $\forall t$ $\omega(t)\notin F$). Alors l'espace $\Omega_{\text{càd}}^F$ est coanalytique, mais n'est pas une partie $\Omega_{\text{càd}}$-analytique, et $\Omega_{\text{càd}}^F\cap\Omega_{\text{càd}}^{\text{làg}}$ n'est pas un espace analytique.*

Démonstration Comme $\Omega_{\text{càd}}$ est coanalytique, il est clair que $\Omega_{\text{càd}}^F$ est coanalytique. D'autre part, comme $\Omega_{\text{càd}}^{\text{làg}}$ est un borélien de $\Omega_{\text{càd}}$, le fait que

$\Omega^F_{\text{càd}}$ ne soit pas $\Omega_{\text{càd}}$-analytique résulte immédiatement du fait que l'espace $\Omega^{\text{làg}}_{\text{càd}} \cap \Omega^F_{\text{càd}}$ n'est pas analytique. Démontrons ce dernier point. Pour tout élément ω de $\Omega^{\text{làg}}_{\text{càd}}$, nous noterons $X_{t-}(\omega) = \omega(t-)$ la limite à gauche en $t > 0$, en poserons $X_{0-}(\omega) = X_0(\omega)$: il est clair que l'application $(t, \omega) \to X_{t-}(\omega)$ ainsi définie est borélienne. Nous définissons maintenant une fonction T de $\Omega^{\text{làg}}_{\text{càd}}$ dans $\bar{\mathbb{R}}_+$ par

$$T(\omega) = \inf\{t \geq 0: X_{t-}(\omega) \text{ ou } X_t(\omega) \text{ appartient à } F\}.$$

On vérifie aisément que T est un temps adéquat. D'autre part, si $(T^i)_{i \in I}$ est la suite transfinie des itérés successifs de T et si on pose $U = \sup_{i \in I} T^i$ (on a $U(\omega) = T^{i_T(\omega)}(\omega)$), il n'est pas difficile de voir que l'on a

$$U(\omega) = \inf\{t \geq 0: X_t(\omega) \in F\},$$

si bien que $\Omega^{\text{làg}}_{\text{càd}} \cap \Omega^F_{\text{càd}}$ est égal à $Q_T = \{\omega: i_T(\omega) < \aleph_1\}$. Il nous reste donc à vérifier que $\sup_{\omega \in Q_T} i_T(\omega)$ vaut $\aleph_1$. Nous adaptons pour cela la fin de la démonstration de 27 comme suit. Nous choisissons une suite (x_n) d'éléments de $E - F$ convergeant vers un élément x de F et nous désignons par w l'élément de $\Omega^{\text{làg}}_{\text{càd}}$ qui vaut x_n pour $t \in [n-1, n[$. Soit k un ordinal dénombrable et partitionnons $\mathbb{R}_+$ en intervalles $[t_i, t_{i+1}[$ pour $i \leq k$. Notant d une distance sur E compatible avec sa topologie, posons, pour tout $\varepsilon > 0$,

$$N(\varepsilon) = \inf\{n: d(x_n, x) < \varepsilon\},$$

en convenant que $N(\infty) = 1$, et soit, pour tout $i \leq k$, un homéomorphisme π_i de $[t_i, t_{i+1}[= [t_i, t_i + \varepsilon_i[$ sur $[N(\varepsilon_i) - 1, \infty[$: nous prenons pour analogue w_i de w sur $[t_i, t_{i+1}[$ l'application $w \circ \pi_i$. On peut alors recoller les w_i en un élément ω^k de $\Omega^{\text{làg}}_{\text{càd}}$, et on voit facilement qu'on a $i_T(\omega^k) \geq k$, et même $= k$.

Remarques (a) En prenant $E = \bar{\mathbb{R}}$ et $F = \{-\infty, +\infty\}$, on voit que l'espace des fonctions continues à droite dans $\mathbb{R}$ et pourvues de limites à gauche dans $\bar{\mathbb{R}}$ n'est pas analytique!

(b) En prenant $E = \mathbb{R}$ et $F = \{0\}$, on voit aisément que le temps d'entrée T dans $\{0\}$ n'est pas une fonction borélienne sur $\Omega_{\text{càd}}$ ou $\Omega^{\text{làg}}_{\text{càd}}$. On peut même montrer que $\{\omega: T(\omega) \geq t\}$ n'est pas un borélien, pour tout $t \in \bar{\mathbb{R}}_+$ différent de 0 (exercice!).

Nous terminons cette Section 3 par une digression, qui nous ramènera au chapitre I. Du point de vue de la théorie de la mesure, un espace coanalytique, ce n'est pas mal car

(a) toute partie coanalytique d'un espace polonais est universellement mesurable,
(b) tout espace (métrisable, séparable) Borel-isomorphe à un espace coanalytique est coanalytique.

Mais, en ce qui concerne ces deux propriétés, un espace analytique, c'est mieux car

(a) toute partie analytique d'un espace polonais est universellement capacitable,
(b) tout espace (métrisable, séparable) image borélienne d'un espace analytique est analytique.

Il existe cependant des cas où il vaut mieux avoir affaire à un espace coanalytique plutôt qu'à un espace analytique. Nous établissons d'abord une espèce de théorème de séparation, qui eût pu trouver sa place au Chapitre I comme exemple d'application du théorème I-21, puis donnons un exemple de ce fait relatif à nos espaces fonctionnels Ω.

31 THÉORÈME *Soient F un espace polonais, $\mathcal{M}^1(F)$ l'espace polonais des mesures (positives) de masse ≤ 1 sur F* (cf. I-3(e)), *et A' une partie analytique de $\mathcal{M}^1(F)$. Si A est une partie analytique de F telle que l'on ait $m(A)=0$ pour tout $m \in A'$, alors il existe un borélien B de F contenant A tel que l'on ait $m(B)=0$ pour tout $m \in A'$.*

Démonstration Considérons la fonction J de $\mathcal{P}(F)\times\mathcal{P}[\mathcal{M}^1(F)]$ dans $\mathbb{R}_+$ définie par

$$J(H, H') = \sup_{m \in H'} m(H).$$

Nous avons noté en I-6(a) que J est une bicapacité sur $F\times\mathcal{M}^1(F)$. Comme on a $J(A, A')=0$ par hypothèse, il résulte de I-21 qu'il existe des boréliens B, B' de F, $\mathcal{M}^1(F)$ respectivement tels que $A \subseteq B$, $A' \subseteq B'$ et que $J(B, B')=0$. D'où la conclusion.

32 THÉORÈME *Soit $(m_x)_{x\in E}$ une famille de mesures de masse ≤ 1 sur Ω indexée par l'espace polonais E, et supposons (m_x) borélienne (i.e. la fonction $x \to m_x(B)$ est borélienne sur E pour tout $B \in \mathcal{B}(\Omega)$). Si l'espace Ω est coanalytique, alors il existe un sous-espace lusinien Ω_0 de Ω, stable pour les opérateurs de translation, portant toutes les mesures m_x (i.e. $m_x(\Omega-\Omega_0)=0$ pour tout $x \in E$).*

Démonstration Nous plongeons Ω dans un espace polonais F: comme Ω est universellement mesurable dans F, on peut supposer que m_x est une mesure sur F portée par Ω, et donc que $(m_x)_{x\in E}$ est une famille borélienne d'éléments de $\mathcal{M}^1(F)$. Il est alors facile de voir, par un argument de classes monotones, que $x \to m_x$ est une application borélienne de E dans $\mathcal{M}^1(F)$ si bien que $A'=\{m_x, x\in E\}$ est une partie analytique de $\mathcal{M}^1(F)$. Posons d'autre part $A = F - \Omega$, qui est une partie analytique de F. On a alors $m(A) = 0$ pour tout $m \in A'$, si bien qu'il existe, d'après 31, un borélien H de F contenu dans Ω (donc un sous-espace lusinien de Ω) portant toutes les mesures $m \in A'$. Il reste à assurer que H est stable pour les opérateurs de translations. Nous allons définer, par récurrence, une suite croissante (H_n) de parties de F contenues dans Ω telle que

(a) pour n impair, H_n soit un borélien de F portant les m_x,
(b) pour n pair, H_n, soit un analytique de F, stable pour les opérateurs de translation.

Il n'y aura plus alors qu'à poser $\Omega_0 = \bigcup_n H_n$ pour achever la démonstration du théorème. Nous posons $H_1 = H$, puis définissons H_2 par

$$\omega \in H_2 \qquad \text{ssi} \qquad \exists t \; \exists w \in H_1 \;\; \omega = \theta_t(w) \, .$$

Comme H_1 est un espace lusinien et que $(t, w) \to \theta_t(w)$ est une application borélienne de $\mathbb{R}_+ \times H_1$ dans Ω et donc dans F, il est clair que H_2 est une partie analytique de F, contenue dans Ω et contenant H_1. Comme Ω est coanalytique, il existe, d'après le théorème classique de séparation, un borélien H_3 de F contenant H_2 et contenu dans Ω. On recommence alors avec H_3, etc. C'est fini.

Remarques (a) Revenons à l'énoncé 31. Si l'on décompose l'ensemble coanalytique A^c en ses constituants (cf. III-18), il résulte du théorème de la borne que B^c est contenu dans une réunion dénombrable de constituants si bien que l'on peut toujours prendre pour B le complémentaire d'une réunion dénombrable de constituants. Cette remarque s'étend bien entendu à l'énoncé 32, mais il est ici plus naturel de rechercher des "constituants adaptés à notre situation fonctionnelle". Or, supposons que T soit un temps adéquat sur Ω tel que $i_T(\omega) < \aleph_1$ pour tout $\omega \in \Omega$, et posons, pour tout ordinal dénombrable i, $\Omega^i = \{\omega \in \Omega : i_T(\omega) \leq i\}$: Ω^i est stable pour les opérateurs de translation, et, comme Ω est la réunion des Ω^i, il existe, d'après 32 et le théorème de la borne, un ordinal dénombrable j tel que Ω^j porte toutes les mesures m_x, $x \in E$. Cependant, en général, les Ω^i, qui sont coanalytiques, ne sont pas lusiniens, et même pas boréliens dans Ω.

(b) L'énoncé 32 est, en général, faux pour Ω analytique. Prenons $E = \mathbb{R}$, désignons par H un borélien de $\mathbb{R}$ homéomorphe à $\mathbb{N}^{\mathbb{N}}$ et par f une application continue de H sur Ω supposé analytique, non lusinien. Prenons alors pour m_x la mesure de Dirac en $f(x)$ pour $x \in H$ et la mesure nulle pour $x \notin H$: il est clair qu'aucun sous-espace lusinien de Ω ne pourra porter toutes les mesures m_x.

4 Une forme générale du théorème de Saint-Raymond

Nous allons démontrer ici une généralisation intéressante du théorème de Saint-Raymond sur les boréliens à coupes $\mathbf{F}_\sigma$, due indépendamment à Burgess (1979) et Hillard (1979). Nous verrons plus loin que s'introduit comme outil essentiel pour démontrer ce théorème une dérivation analytique sur l'espace des fermés de $\mathbb{N}^{\mathbb{N}}$. Aussi commencerons nous par étudier de manière approfondie les dérivations analytiques sur l'espace des fermés d'un espace analytique; nous obtiendrons ainsi des résultats sur les ensembles analytiques à coupes fermées ayant leur intérêt propre, desquels nous déduirons assez facilement la généralisation annoncée.

A. Dérivations sur l'espace des fermés d'un espace analytique

Nous désignons par E un espace *analytique* et par $\mathbf{F}(E)$ l'espace des parties fermées de E muni de sa structure d'Effros (cf. II-9); on munit $\mathbf{F}(E)$ de l'ordre de l'inclusion (noté indifféremment $\leqslant$ ou $\subseteq$). Comme on a, (U_n) étant une base dénombrable de la topologie de E,

$$X \leqslant Y \qquad \text{ssi} \qquad \forall n\,[U_n \cap X \neq \varnothing \Rightarrow U_n \cap Y \neq \varnothing],$$

l'ordre $\leqslant$ est analytique, et donc adéquat puisque $\mathbf{F}(E)$ est analytique (ici, quelques abus de langage sans importance).

Nous nous donnons une dérivation analytique d sur $\mathbf{F}(E)$. Nous verrons un peu plus loin un procédé général pour construire de telles dérivations, mais le lecteur désireux d'avoir tout de suite un exemple concret à se mettre sous la dent pourra songer à la dérivation de Cantor (cf. II-23). Nous commençons par établir un résultat sur les "constituants" de $Q = \{X \in \mathbf{F}(E) : i(X) < \aleph_1\}$, où $i(X)$ est l'indice de X pour notre dérivation d—résultat en fait vrai dans le cadre général de la Section 1.

33 THÉORÈME *L'ensemble* $Q = \{X \in \mathbf{F}(E) : i(X) < \aleph_1\}$ *est une partie* $\mathbf{F}(E)$*-coanalytique de* $\mathbf{F}(E)$ *ainsi que, pour tout* $j \in I$, *l'ensemble* $Q_j = \{X \in \mathbf{F}(E) : i(X) < j\}$.

Démonstration On sait déjà que Q est $\mathbf{F}(E)$-coanalytique (10). Pour Q_j, cela résulte aisément, par récurrence transfinie, de l'équivalence, pour $X, Y \in \mathbf{F}(E)$

$$i(X) \geqslant j \qquad \text{ssi} \qquad \forall k < j\ \exists Y\, i(Y) \geqslant k \text{ et } Y \leqslant d(X)\,.$$

34 Nous nous donnons, enfin, un autre espace *analytique* F. Si H est une partie de $E \times F$ telle que la coupe $H(y)$ soit fermée dans E pour tout $y \in F$, nous noterons, pour tout ordinal dénombrable j, ou pour $j=\aleph_1$, H^j la partie de $E\times F$ définie par ses coupes comme suit:

$$(x, y)\in H^j \qquad \text{ssi} \qquad x\in H^j(y)$$

où $H^j(y)$ est le j-ième dérivé de $H(y)$ pour j dénombrable et où $H^j(y)=\inf_{i<j} H^i(y)$ pour $j=\aleph_1$. Noter que, comme E est à base dénombrable, il existe, pour toute suite transfinie décroissante $(X_i)_{i\in I}$ de fermés de E, un ordinal dénombrable j tel que l'on ait $X_j=X_{j+1}$; par conséquent, pour tout $y\in F$, il existe un ordinal dénombrable $j(y)$ tel que $H^{j(y)}(y)=H^{j(y)+1}(y)=\cdots=H^{\aleph_1}(y)$, qui est alors le plus grand fermé "d-parfait" inclus dans $H(y)$.

34 Théorème *Si H est une partie analytique de $E\times F$ à coupes fermées dans E, alors H^j est, pour tout $j\leqslant\aleph_1$, une partie analytique de $E\times F$ à coupes fermées.*

Cet énoncé résulte immédiatement de 33 et de la première partie du lemme suivant.

35 Lemme *Soit H une partie analytique de $E\times F$ à coupes fermées dans E.*

(1) *Si A est une partie analytique de $\mathbf{F}(E)$, l'ensemble*

$$H_A=\{(x, y)\colon \exists X\in A\ x\in X \text{ et } X\leqslant H(y)\}$$

est une partie analytique de $E\times F$.

(2) *Inversement, l'ensemble*

$$A_H=\{X\in\mathbf{F}(E)\colon \exists y\in F\quad X\leqslant H(y)\}$$

est une partie analytique de $\mathbf{F}(E)$.

Démonstration Nous désignons par (U_n) une base dénombrable de la topologie de E. Remarquons, pour (1), que l'on a $(x, y)\in H_A$ ssi

$$(x, y)\in H \text{ et } \exists X\, X\in A \text{ et } x\in X \text{ et } \forall n\ [U_n\cap X\neq\varnothing \Rightarrow U_n\cap H(y)\neq\varnothing]$$

et, pour (2), que l'on a $X\in A$ ssi

$$\exists y\ \forall n\ [U_n\cap X\neq\varnothing\Rightarrow U_n\cap H(y)\neq\varnothing].$$

Il est alors aisé de conclure (nous laissons les détails au lecteur).

Nous avons vu plus haut que, pour tout y, $H^j(y)$ est stationnaire à partir d'un certain ordinal dénombrable $j(y)$. Il n'est pas vrai, en général, que $j(y)$ soit uniformément borné par un ordinal dénombrable (prendre $F=\mathbf{F}(E)$ et $H=\{(x, y)\colon x\in y\}$). On a cependant le résultat fondamental suivant:

36 THÉORÈME *Soit H une partie analytique de $E \times F$ à coupes fermées dans E. Si $H^{\aleph_1}$ est vide, il existe un ordinal dénombrable i tel que H^i soit vide.*

Démonstration D'après 35, $A = \{X \in \mathbf{F}(E) : \exists y \in F\, X \leqslant H(y)\}$ est une partie analytique de $\mathbf{F}(E)$. Et alors, d'après le théorème de la borne (cf. 11), $\sup_{y \in F} i[H(y)]$ est un ordinal dénombrable. Si i désigne cet ordinal, on a évidemment $H^i = \varnothing$.

Nous énonçons maintenant un résultat intéressant sur les analytiques à coupes fermées, qui a sa place légitime ici, et que nous utiliserons d'ailleurs dans la démonstration de notre version du théorème de Saint-Raymond, mais dont la démonstration sera en partie rejetée en appendice: il s'agit en effet d'un corollaire d'un théorème très général de séparation établi, dans l'appendice, à l'aide de la théorie des multicapacités.

37 THÉORÈME *Soit $\mathscr{C}$ une partie $\mathbf{F}(E)$-coanalytique de $\mathbf{F}(E)$, héréditaire (i.e. $X \in \mathscr{C}$ et $Y \leqslant X \Rightarrow Y \in \mathscr{C}$) et soient H, H^* deux parties analytiques disjointes de $E \times F$, où H est à coupes fermées appartenant à $\mathscr{C}$. Alors H est contenu dans un borélien L de $E \times F$, disjoint de H^*, et à coupes fermées appartenant à $\mathscr{C}$.*

Démonstration Il résulte du théorème 12 de l'appendice que H est contenu dans un borélien M de $E \times F$ disjoint de H^* tel que $\overline{M(y)}$ appartienne à $\mathscr{C}$ pour tout $y \in F$. Maintenant, d'après II-17, on peut écrire $H^c = \sup_n (U_n \times C_n)$ où (U_n) est une base dénombrable de la topologie de E et (C_n) une suite de parties F-coanalytiques de F. La fin de la démonstration est alors analogue à celle de II-18. Les ensembles $(U_n \times C_n) \cup M$ sont $(E \times F)$-coanalytiques et recouvrent $E \times F$. D'après le théorème de séparation I-22 de Novikov, il existe alors une suite (M_n) de boréliens de $E \times F$ telle que $M^n - M$ soit inclus dans $(U_n \times C_n)$ pour tout n et que $M^c = \sup_n (M^n - M)$. Une nouvelle application du théorème de séparation (dans F cette fois) permet de trouver, pour tout n, un borélien B_n de F tel que l'on ait

$$(M^n - M) \subseteq (U_n \times B_n) \subseteq (U_n \times C_n)\,.$$

On prend alors pour L le complémentaire de la réunion des $(U_n \times B_n)$.

Etant donné 33, on a alors (en prenant $H^* = \varnothing$)

38 COROLLAIRE *Une partie analytique à coupes fermées H de $E \times F$ telle que H^j soit vide pour un ordinal $j \leqslant \aleph_1$ est contenu dans un borélien L de $E \times F$ à coupes fermées tel que L^j soit vide.*

Remarque Les cas les plus intéressants sont les extrêmes: $j=1$ et $j=\aleph_1$, encore que, d'après 36, le cas $j=\aleph_1$ se ramène à un cas "intermédiaire".

Maintenant, si $X \in \mathbf{F}(E)$ vérifie $i(X) < \aleph_1$, on a

$$X = \sup_{j<i(X)} (X^j - X^{j+1}) ;$$

par ailleurs, on a $d(Y) = \varnothing$ pour tout $Y \in \mathbf{F}(E)$ contenu dans un des $(X^j - X^{j+1})$. On en déduit aisément que tout fermé X ayant son indice $i(X) < \aleph_1$ est la réunion d'une suite de fermés (X_n) de dérivé $d(X_n)$ vide (la réciproque n'est pas vraie en général, mais le sera, si E est polonais, pour les dérivations locales que nous définirons plus loin).

D'après 38, une partie analytique H de $E \times F$ à coupes fermées telle que $H^{\aleph_1}$ soit vide est contenu dans un borélien ayant les mêmes propriétés. Voici alors la "quinte-essence" du théorème de Saint-Raymond.

39 THÉORÈME *Soit H un borélien de $E \times F$ à coupes fermées dans E tel que $H^{\aleph_1}$ soit vide. Alors H est la réunion d'une suite (H_n) de boréliens à coupes fermées de dérivé vide.*

Démonstration Disons, pour abréger, qu'un ordinal dénombrable j vérifie la propriété $\mathscr{P}$ si, pour tout borélien L contenant H^j, le borélien $H - L$ est contenu dans une réunion dénombrable de boréliens à coupes fermées de dérivé vide. Nous alons montrer, par récurrence transfinie, que tout $j \in I$ vérifie $\mathscr{P}$. Nous obtiendrons alors l'énoncé en prenant $j = \sup_{y \in F} i[H(y)]$, dénombrable d'après 36, et, tout bêtement, $L = \varnothing$ (tout ce travail est rendu nécessaire par le fait que les H^j ne veulent pas être boréliens en général). Montrons d'abord que $j=1$ vérifie $\mathscr{P}$. Soit L un borélien contenant H^1; d'après II-18, le complémentaire de L peut s'écrire

$$L^c = \sup_n (U_n \times B_n) ,$$

où (U_n) est une base dénombrable de la topologie de E et (B_n) une suite de boréliens de F. Ecrivons, pour n fixé, l'ouvert U_n comme réunion d'une suite (X_p^n) de fermés de E: on voit alors que $H - L$ est la réunion, pour n, p parcourant $\mathbb{N}$, des ensembles $H \cap (X_p^n \times B_n)$, et ces derniers sont des boréliens à coupes fermées contenues dans celles de $H - H_1$ et donc de dérivé vide. Donc, j vérifie $\mathscr{P}$ si $j=1$. Supposons maintenant que $\mathscr{P}$ soit vérifiée pour tout $j < i$ et voyons que i vérifie encore $\mathscr{P}$. Soit L un borélien contenant H^i et souvenons nous que l'on a

$$H^i = A^1 \qquad \text{où} \qquad A = \inf_{j<i} H^j .$$

Procédant comme ci-dessus, avec A à la place de H, on voit que $A-L$ est réunion, pour n, p parcourant $\mathbb{N}$, d'ensembles $A\cap(X_p^n\times B_n)$ à coupes fermées de dérivé vide; mais on peut seulement affirmer que ces ensembles sont analytiques, d'après 34. Cependant 38 nous assure que chacun d'eux est inclus dans un borélien à coupes fermées de dérivé vide si bien que $A-L$ est finalement inclus dans la réunion M d'une suite (M_n) de boréliens à coupes fermées de dérivé vide. Comme on a $A=\inf_{j<i}H^i$, on en déduit en particulier, grâce au théorème de séparation de Novikov appliqué aux H^j-M, qu'il existe, pour tout $j<i$, un borélien L_j de $E\times F$ contenant H^i tel que $\inf_{j<i}L_j$ soit encore contenu dans M. Par hypothèse, $H-L_j$ est contenu, pour tout $j<i$, dans la réunion d'une suite de boréliens à coupes fermées de dérivé vide et on a vu que $A-L$ possède la même propriété, M dénotant la réunion d'une telle suite de boréliens pour $A-L$. Compte tenu du fait que l'on a

$$A=\inf_{j<i}H^j\subseteq\inf_{j<i}L_j\subseteq M,$$

on voit, en recollant les morceaux, que $H-L$ a aussi cette propriété, et donc que i vérifie $\mathcal{P}$. C'est fini.

Nous allons enfin montrer comment construire certaines dérivations sur $\mathbf{F}(E)$; nous avons attendu jusqu'à maintenant pour ce faire car c'est ce type de dérivation que nous allons rencontrer dans la partie B.

40 Nous dirons que la dérivation d sur $\mathbf{F}(E)$ est *locale* si, comme la dérivation de Cantor, elle vérifie

$$d(X)=\{x\in X:\forall U\ x\in U\Rightarrow d(\bar{U}\cap X)\neq\varnothing\},$$

où U parcourt les ouverts de E ou, cela revient au même, les éléments d'une base dénombrable de la topologie de E. Une telle dérivation est bien déterminée par l'ensemble $\{X\in\mathbf{F}(E)\colon d(X)=\varnothing\}$, qui est une partie héréditaire et $\mathbf{F}(E)$-coanalytique de $\mathbf{F}(E)$ (cf. 33). Réciproquement, on a

41 THÉORÈME *Soit $\mathscr{C}$ une partie héréditaire et $\mathbf{F}(E)$-coanalytique de $\mathbf{F}(E)$. L'application d de $\mathbf{F}(E)$ dans $\mathbf{F}(E)$ définie par*

$$d(X)=\{x\in X\colon\ \forall U\ x\in U\Rightarrow\bar{U}\cap X\notin\mathscr{C}\},$$

où U parcourt les ouverts de E, est une dérivation analytique locale sur $\mathbf{F}(E)$. Nous dirons que d est engendrée *par $\mathscr{C}$.*

Démonstration Il est clair que d est une dérivation, et facile de voir qu'elle est locale (quoique $\{X\colon d(X)=\varnothing\}$, qui contient $\mathscr{C}$, ne lui soit pas égal en général). D'autre part, si (U_n) est une base dénombrable de la topologie de E, on a, avec $X, Y, Z \in \mathbf{F}(E)$,

$$Y \leqslant d(X) \qquad \text{ssi} \qquad \forall n\ [U_n \cap Y \neq \emptyset \Rightarrow \exists Z\ Z \subseteq \bar{U}_n \cap X \text{ et } Z \notin \mathscr{C}],$$

d'où l'on déduit aisément que d est analytique. La présence d'un "$\exists Z$" explique pourquoi, si E n'est pas métrisable compact, la relation $Y \leqslant d(X)$ n'est pas borélienne même si $\mathscr{C}$ est une partie fermée de $\mathbf{F}(E)$.

Nous terminons cette partie A en démontrant, comme promis

42 THÉORÈME *Supposons que d soit engendrée par une partie héréditaire et* $\mathbf{F}(E)$*-coanalytique* $\mathscr{C}$ *de* $\mathbf{F}(E)$. *Alors tout* $X \in \mathbf{F}(E)$ *tel que* $i(X) < \aleph_1$ *est réunion dénombrable d'éléments de* $\mathscr{C}$. *Réciproquement, si E est polonais, on a* $i(X) < \aleph_1$ *pour tout* $X \in \mathbf{F}(E)$ *réunion dénombrable d'éléments de* $\mathscr{C}$.

Démonstration D'abord, si on a $i(X) < \aleph_1$, on a vu que X est réunion dénombrable de fermés de dérive vide, et, E étant à base dénombrable, il est facile de voir que tout fermé de dérivé vide est réunion dénombrable d'éléments de $\mathscr{C}$. Réciproquement, supposons E polonais et soit X fermé non vide, réunion d'une suite (X_n) d'éléments de $\mathscr{C}$. Alors, comme X est un espace de Baire, un des X_n a un intérieur relatif à X non vide; il est alors clair que l'on a $d(X) < X$. Comme $\mathscr{C}$ est héréditaire, $d(X)$ est lui-même réunion dénombrable d'éléments de $\mathscr{C}$, d'où finalement on a $i(X) < \aleph_1$.

B. Une généralisation du théorème de Saint-Raymond

43 On désigne par E, F des espaces analytiques fixés. On dira qu'une classe $\mathscr{C}$ de parties de E est une *classe de parties modestes de* E si

(1) $\mathscr{C}$ est héréditaire,
(2) l'ensemble $\mathscr{C} \cap \mathscr{F}(E)$ est une partie $\mathbf{F}(E)$-coanalytique de $\mathbf{F}(E)$.

Nous laissons au lecteur le soin de vérifier que les classes suivantes sont des classes de parties modestes:

(a) la classe des parties de E ayant au plus un élément;
(b) la classe des parties de E ayant un intérieur vide;
(c) pour E lusinien (i.e. borélien d'un espace polonais), la classes des parties relativement compactes de E.

On se donne dans toute la suite une classe $\mathscr{C}$ de parties modestes de E. Vu l'exemple (b) ci-dessus, nous dirons, de manière générale, qu'une partie X de E est $\mathscr{C}$*-rare* si son adhérence $\bar{X}$ appartient à $\mathscr{C}$ et est $\mathscr{C}$*-maigre* si elle est réunion dénombrable de parties $\mathscr{C}$-rares. Il est clair qu'une partie $\mathbf{F}_\sigma$ de E, $\mathscr{C}$-maigre, est réunion dénombrable de fermés $\mathscr{C}$-rares.

Voici alors la généralisation annoncée

44 THÉORÈME *Soit H une partie analytique (resp. borélienne) de $E\times F$ à coupes $\mathscr{C}$-maigres dans E. Alors H est la réunion d'une suite (H_n) de parties analytiques (resp. boréliennes) de $E\times F$ à coupes $\mathscr{C}$-rares. Si, de plus, H est borélien à coupes $\mathbf{F}_\sigma$, on peut prendre les H_n boréliens à coupes fermées.*

Démonstration Soient π_E, π_F les projections de $E\times F$ sur E, F respectivement et soit f une application continue de $\mathbb{N}^\mathbb{N}$ sur H. Nous désignons par $\mathscr{C}^f$ la classe des parties X de $\mathbb{N}^\mathbb{N}$ telles que $\pi_E[f(X)]$ soit $\mathscr{C}$-rare (et que, de plus, pour la seconde partie de l'énoncé, l'adhérence de $f(X)$ dans $E\times F$ soit contenue dans H): $\mathscr{C}^f$ est une classe de parties modestes de $\mathbb{N}^\mathbb{N}$. L'hérédité est claire; le fait que $\mathscr{C}^f\cap\mathbf{F}(\mathbb{N}^\mathbb{N})$ soit coanalytique résulte de ce que les applications associant respectivement à $X\in\mathbf{F}(\mathbb{N}^\mathbb{N})$ l'adhérence de $\pi_E[f(X)]$ dans E et celle de $f(X)$ dans $E\times F$ sont boréliennes (cf. II-10) tandis que, si H est borélien, l'ensemble des fermés de $E\times F$ contenus dans H est $(E\times F)$-coanalytique (cf. la remarque (c) de II-13). Soit alors H^f la partie fermée de $\mathbb{N}^\mathbb{N}\times F$ définie par

$$(\sigma, y)\in H^f \qquad \text{ssi} \qquad \pi_F[f(\sigma)]=y.$$

Toutes les coupes dans $\mathbb{N}^\mathbb{N}$ de H^f sont clairement $\mathscr{C}^f$-maigres. Par conséquent, si on a su décomposer H^f en la réunion d'une suite (H^f_n) d'ensembles analytiques à coupes $\mathscr{C}^f$-rares, les ensembles analytiques $H_n = f(H^f_n)$ seront à coupes $\mathscr{C}$-rares (et à adhérence contenue dans la coupe correspondante de H), et de réunion égale à H; et une application de 37 aux "adhérences coupe par coupe" des H_n—qui sont analytiques d'après II-17—permet, si nécessaire, de grossir les H_n en boréliens idoines. Finalement, quitte à remplacer E par $\mathbb{N}^\mathbb{N}$, $\mathscr{C}$ par $\mathscr{C}^f$ et H par H^f, on s'est ramené à démontrer l'énoncé dans le cas où E est polonais et H une partie fermée de $E\times F$. Considérons alors la dérivation d sur $\mathbf{F}(E)$ engendrée par $\mathscr{C}$. Comme E est polonais, il résulte de 42 que $X\in\mathbf{F}(E)$ est $\mathscr{C}$-maigre ssi on a $i(X)<\aleph_1$ et, comme H est fermé dans $E\times F$, une application de 39 permet d'écrire H comme réunion dénombrable de boréliens à coupes fermées de dérivé vide. Ce n'est pas tout à fait fini, car nous voulons des coupes appartenant à $\mathscr{C}$, ce qui est un peu mieux (on aurait pu regarder cela à propos des dérivations locales, mais il faut bien qu'il nous reste quelquechose à faire ici!). Or, soit L un borélien de $E\times F$ à coupes fermées de dérivé vide et posons, pour tout n,

$$C_n=\{y\in F\colon \bar{U}_n\cap L(y) \text{ appartient à } \mathscr{C}\},$$

où (U_n) est une base dénombrable de la topologie de E; C_n est F-coanalytique d'après 34, $(\bar{U}_n \times C_n) \cap L$ a ses coupes $\mathscr{C}$-rares, et les ensembles F-coanalytiques $(\bar{U}_n \times C_n)$ recouvrent L. Deux applications successives du théorème de séparation, comme dans la démonstration de 37, permettent alors de trouver des boréliens B_n de F tels que l'on ait $B_n \subseteq C_n$ pour tout n et que les $(\bar{U}_n \times B_n)$ recouvrent encore L, si bien que L est réunion dénombrable de boréliens à coupes fermées appartenant à $\mathscr{C}$. C'est fini.

Remarques (a) Une application de 37 (effectuée d'ailleurs dans la démonstration) montre que H, analytique à coupes $\mathscr{C}$-maigres, est contenu dans la réunion L d'une suite de boréliens à coupes fermées et $\mathscr{C}$-rares. Si, de plus, H est disjoint d'un autre analytique H^*, il n'est pas difficile de voir que l'on peut supposer de plus L disjoint de H^* (exercice!).

(b) Il arrive souvent que, pour tout sous-espace analytique A de E, la classe trace $\mathscr{C}_A$ de $\mathscr{C}$ sur A soit une classe de parties modestes de A. On peut alors "relativiser" l'énoncé du théorème comme suit. Disons qu'une partie X de E est *relativement $\mathscr{C}$-rare* (resp. *relativement $\mathscr{C}$-maigre*) si elle est $\mathscr{C}_X$-rare (resp. $\mathscr{C}_X$-maigre), où $\mathscr{C}_X$ est la trace de $\mathscr{C}$ sur X: cela revient à dire que X appartient à $\mathscr{C}$ (resp. est réunion dénombrable d'éléments de $\mathscr{C}$ relativement fermés dans X). Il est clair que toute partie $\mathscr{C}$-rare (resp. $\mathscr{C}$-maigre) de E l'est aussi "relativement", et qu'une partie fermée (resp. $\mathbf{F}_\sigma$) de E relativement $\mathscr{C}$-rare (resp. relativement $\mathscr{C}$-maigre) est en fait $\mathscr{C}$-rare (resp. $\mathscr{C}$-maigre). Ceci dit, si, pour tout sous-espace analytique A de E, la classe trace $\mathscr{C}_A$ est une classe de parties modestes de A, alors on peut remplacer dans l'énoncé 44 l'expression "à coupes $\mathscr{C}$-maigres" par "à coupes relativement $\mathscr{C}$-maigres" à condition de remplacer l'expression "à coupes $\mathscr{C}$-rares" par "à coupes appartenant à $\mathscr{C}$" (exercice!).

Nous énonçons maintenant quelques corollaires attendus, en prenant la peine de les écrire entiérement.

D'abord, le théorème de Lusin et Novikov, dans son intégralité

45 COROLLAIRE *Soient E, F des espaces analytiques et H une partie analytique* (*resp. borélienne*) *de $E \times F$ à coupes dénombrables dans E. Alors H est la réunion d'une suite* (H_n) *de parties analytiques* (*resp. boréliennes*) *de $E \times F$ dont les coupes comportent au plus un point* (*autrement dit, des graphes*).

Démonstration Prendre pour $\mathscr{C}$ la classe des parties de E comportant au plus un point.

Puis un résultat relatif aux ensembles maigres ou rares, au sens classique, noté semble-t-il pour la première fois par Burgess et Hillard

46 COROLLAIRE *Soient E, F des espaces analytiques et H une partie analytique (resp. borélienne) de* $E \times F$ *à coupes maigres. Alors H est la réunion d'une suite* (H_n) *de parties analytiques (resp. boréliennes) à coupes rares. Si, de plus, H est borélien à coupes* $\mathbf{F}_\sigma$, *on peut prendre les* H_n *boréliens à coupes fermées.*

Démonstration Prendre pour $\mathscr{C}$ la classe des parties de E d'intérieur vide. Noter que l'énoncé peut être "relativisé".

Le théorème de Saint-Raymond (1976) comporte, dans son esprit, deux parties: l'une concernant les ensembles à coupes $\mathscr{K}_\sigma$, l'autre les ensembles à coupes $\mathbf{F}_\sigma$. Cela donne, dans notre cadre, deux énoncés distincts (on prendra garde aux hypothèses).

47 COROLLAIRE *Soient E, F des espaces analytiques et H une partie borélienne de* $E \times F$ *à coupes* $\mathbf{F}_\sigma$. *Alors H est la réunion d'une suite* (H_n) *de boréliens à coupes fermées.*

Démonstration Prendre tout simplement pour $\mathscr{C}$ la classe de toutes les parties de E (auquel cas seule la seconde partie de 44 est intéressante!).

Remarque Cela ne s'étend pas aux analytiques à coupes $\mathbf{F}_\sigma$. Steel a un contre-exemple où E et F sont des espaces compacts métrisables.

48 COROLLAIRE *Soient E un espace* lusinien, *F un espace analytique et H une partie analytique (resp. borélienne) de* $E \times F$ *à coupes* $\mathscr{K}_\sigma$*-bornées (i.e. chaque coupe est contenue dans un* $\mathscr{K}_\sigma$ *de E). Alors H est la réunion d'une suite* (H_n) *de parties analytiques (resp. boréliennes) de* $E \times F$ *à coupes relativement compactes. Si, de plus, H est borélien à coupes* $\mathscr{K}_\sigma$, *on peut prendre les* H_n *boréliens à coupes compactes.*

Démonstration Prendre pour $\mathscr{C}$ la classe des parties relativement compactes de E.

49 *Remarque* Nous citons ici un corollaire de Hillard sous la forme d'une remarque parce qu'il fait intervenir une notion sans doute inconnue de beaucoup de lecteurs. Supposons, pour simplifier, que E soit un espace euclidien $\mathbb{R}^d$ et soit m une mesure de Hausdorff sur E (la mesure de Lebesgue est une telle mesure, mais, en général, une mesure de Hausdorff *n'est pas σ-finie*: la plus simple de ce type est la mesure de comptage des points, où $m(X) = \text{card}\, X$ si X est fini et $m(X) = \infty$ sinon). Comme m est la limite d'une suite croissante de capacités (cf. Dellacherie, 1972*b*), la classe $\mathscr{C}$ des parties de E de mesure finie est une classe de parties modestes, qui peut se "relativiser". On obtient alors l'énoncé suivant, qui constitue en quelque sorte une généralisation du théorème de Lusin et Novikov: "Soient F un espace analytique et H une partie analytique (resp. borélienne) de $\mathbb{R}^d \times F$ telle que, pour tout $y \in F$, $H(y)$ soit réunion dénombrable d'ensembles de mesure finie relativement fermés dans $H(y)$. Alors H est la réunion d'une suite (H_n) de parties analytiques (resp. boréliennes) de $\mathbb{R}^d \times F$ telle que, pour tout $y \in F$ et tout n, $H_n(y)$ soit de mesure finie. Si, de plus, H est borélien à coupes $\mathbf{F}_\sigma$, on peut prendre les H_n boréliens à coupes fermées". Il serait très intéressant de pouvoir, au moins dans ce cadre, éliminer la condition "relativement fermés dans $H(y)$". Je doute que cela soit possible, dans le cadre général de l'énoncé 44, même si $\mathscr{C}$ est une classe qui se relativise.

Nous revenons à notre cadre général, et terminons ce chapitre en étudiant quelques "cribles généralisés", selon l'expression de Sierpiński. D'abord, un résultat élémentaire (corollaire par ailleurs d'un théorème très général établi, dans l'appendice, à l'aide de la théorie des multicapacités)

50 THÉORÈME *Soit H une partie analytique de $E \times F$. L'ensemble*

$$\{y \in F\colon H(y) \text{ n'est pas } \mathscr{C}\text{-rare}\}$$

est analytique. Si H est borélien, il en est de même pour

$$\{y \in F\colon H(y) \text{ n'est pas un fermé } \mathscr{C}\text{-rare}\}\,.$$

Démonstration Désignons par L la partie de $E \times F$ telle que l'on ait $L(y) = \overline{H(y)}$ pour tout $y \in F$: L est analytique d'après II-17. La première partie résulte alors de 35(1), et la seconde du fait que $\{y \in F\colon H(y)$ n'est pas fermé$\}$ est la projection de $L - H$ sur F.

Nous passons à l'énoncé analogue, pour les coupes $\mathscr{C}$-maigres, qui est nettement plus difficile (et néanmoins très utile!).

51 THÉORÈME *Soit H une partie analytique de $E \times F$. L'ensemble*

$$\{y \in F\colon H(y) \text{ n'est pas } \mathscr{C}\text{-maigre}\}$$

est analytique. Si H est borélien, il en est de même pour

$$\{y \in F\colon H(y) \text{ n'est pas un } \mathbf{F}_\sigma\ \mathscr{C}\text{-maigre}\}\,.$$

Démonstration La démonstration va être courte, parce que nous disposons de toute la machinerie nécessaire. En procédant comme au début de la démonstration de 44, on se ramène au cas où E est un espace polonais et H est une partie fermée de $E \times F$ (la seconde partie de l'énoncé se "dissolvant" dans la première par le choix fait, dans ce cas, de la classe $\mathscr{C}$). Comme E est polonais, les fermés $\mathscr{C}$-maigres coincident alors avec les fermés d'indice $< \aleph_1$ pour la dérivation engendrée par $\mathscr{C}$, et l'énoncé résulte alors de 33 et 35(1) (ou, d'un coup, de 36).

Nous énonçons maintenant les corollaires attendus sous forme de remarques.

52 *Remarques* (a) Si $\mathscr{C}$ est l'ensemble des parties de E ayant au plus un point, la première partie de l'énoncé est un résultat classique de Mazurkiewicz et Sierpiński.

(b) Si $\mathscr{C}$ est la classe des parties de E d'intérieur vide, la première partie de l'énoncé est un résultat classique de Novikov, remis en vogue par Vaught (1974).

(c) Si $\mathscr{C}$ est l'ensemble des parties de E, la seconde partie de l'énoncé est un cas particulier d'un théorème très général de Louveau (cf. la remarque (b) de III-32).

(d) Si E est lusinien, et si $\mathscr{C}$ est la classe des parties relativement compactes de E, la seconde partie de l'énoncé fournit la "moitié" du théorème d'Arsenin (cf. la remarque (b) de III-32).

(e) Nous revenons ici à 49 mais, au lieu de parler de ce que donne 51 dans ce cas, je préfère signaler qu'on peut montrer que

$$\{y \in F : H(y) \text{ n'est pas } \sigma\text{-finie pour } m\}$$

est aussi analytique (cf. Dellacherie, 1972*b*). D'où le (petit) espoir exprimé à la fin de 49.

Signalons, enfin, que les énoncés 50 et 51 s'étendent, sans difficulté, au cas où F est "quelconque", et aussi au cas F est un espace mesurable abstrait (exercice!).

Appendice: Théorie générale des capacités

Nous présentons ici deux manières distinctes (qui peuvent se combiner) d'étendre les notions de multicapacité et d'opération capacitaire, préservant les résultats essentiels de la théorie: stabilité par composition (cf. I-12), invariance de l'analyticité (cf. I-15), approximation par l'intérieur (cf. I-19) et approximation par l'extérieur (cf. I-20). Personnellement, ces extensions "ne me coûtent pas cher", sans être gratuites (elles ont réellement utiles, comme nous le verrons). Mais, il n'en est sans doute pas de même pour le lecteur. C'est pourquoi nous avons attendu cet appendice pour les présenter et, par souci pédagogique, nous mettrons un peu la charrue avant les boeufs: l'extension que nous présentons en premier devrait, en toute logique, venir après la seconde.

Dans la première extension, on se permet de fixer ou d'identifier des variables dans une multicapacité ou une opération capacitaire (cf. la remarque (a) de I-12). Cela permet d'étendre considérablement le champ d'application de la théorie. On peut, par exemple, montrer que, pour E analytique,

(1) la fonction d'ensemble I obtenue en posant, pour $H \in \mathcal{P}(E)$, $I(H) = 0$ si H est dénombrable et $I(H) = 1$ sinon est une "capacité généralisée". Noter que cette fonction n'est pas continue à droite sur les compacts.
(2) l'opération qui à $H \in \mathcal{P}(E)$ associe son adhérence $\bar{H}$ est une "opération capacitaire généralisée". Noter que cette opération n'est pas montante (ce qui oblige de modifier un peu I-19; nous verrons cela plus loin).

Dans la deuxième extension, on étend le domaine de définition des capacités et opérations en les faisant aussi opérer sur les fonctions (réelles) positives. Cela permet d'une part d'unifier les notions de multicapacités (généralisées ou non) et d'opérations capacitaires (généralisées ou non), et d'autre part de faire entrer dans la théorie, par exemple, pour E analytique,

(1) la fonction d'ensemble $I = \sup_\sigma \inf_n I_{\sigma|n}$ où, pour tout $s \in S$, I_s est une capacité sur E;
(2) l'opération qui à $f \geq 0$ associe sa transformée Vf par un noyau borélien V au sens de la théorie de la mesure sur E.

Par habitude, et aussi pour ne pas avoir à surcharger les exemples d'hypothèses, nous adoptons dans cet appendice les conventions du Chapitre II: tous les espaces topologiques sont métrisables séparables, etc.

1 Multicalibres et opérations calibrées

Nous dénotons, comme au Chapitre I, par $\mathcal{I}$ ou $\mathcal{J}$ (avec ou sans indices), un ensemble dénombrable (fini ou non), que nous identifions souvent avec un segment initial de $\mathbb{N}$.

Nous commençons par définir les notions généralisées de capacité et d'opération capacitaire à un argument: c'est à la fois plus simple pour le lecteur et le dactylographe . . . .

1 Soit I une multicapacité (resp. une opération capacitaire) sur $\prod_{i \in \mathscr{I}} E_i$ (resp. et à valeurs dans F), et supposons distinguée une partie non vide $\mathscr{J}$ de $\mathscr{I}$ telle que $E_i = E$ ne dépende pas de $i \in \mathscr{J}$. Donnons nous, pour tout $i \notin \mathscr{J}$ (il n'est pas exclu que $\mathscr{I} - \mathscr{J} = \varnothing$), un argument *analytique* A_i pour I et contemplons la fonction d'ensembles J sur $\mathscr{P}(E)$ (resp. à valeurs dans F) "définie" comme suit: pour $X \in \mathscr{P}(E)$ *analytique*, on a

$$J(X) = I(\ldots, X_i, \ldots) \quad \text{où} \quad X_i = A_i \quad \text{pour} \quad i \notin \mathscr{J} \quad \text{et} \quad X_i = X \quad \text{pour} \quad i \in \mathscr{J}\ ;$$

pour $X \in \mathscr{P}(E)$, non analytique, $J(X)$ égale n'importe quoi assurant que J est croissante sur $\mathscr{P}(E)$ (seuls les arguments analytiques nous intéressent, et la convention que nous venons de faire sera commode pour la présentation de certains exemples). Nous dirons que $(I, \mathscr{J}, (A_i)_{i \notin \mathscr{J}})$ est une *représentation analytique* de J. Noter que J n'est pas, en général, continue à droite sur les compacts si les "paramètres" A_i ne sont pas compacts, et n'est pas, en général, montante si l'ensemble $\mathscr{J}$ est infini.

2 DÉFINITION Un *calibre* (resp. une *opération calibrée*) J sur E est une application de $\mathscr{P}(E)$ dans $\bar{\mathbb{R}}_+$ (resp. dans $\mathscr{P}(F)$) admettant une représentation analytique.

Remarque En remontant à la définition I-14 d'une ensemble analytique, on voit sans peine que J admet une représentation analytique où les paramètres A_i sont ou compacts ou égaux à $\mathbb{N}$, et même tous égaux a $\mathbb{N}$ (exercice!).

Bien entendu, toute capacité (resp. opération capacitaire) à un argument est un calibre (resp. une opération calibrée). Nous allons voir des exemples plus originaux, dont ceux de l'introduction. Voici d'abord quelques exemples de calibres.

3 *Exemples* (a) Si (I_n) est une suite de capacités (ou plus généralement, de calibres) sur E, alors $\sup_n I_n$ et $\inf_n I_n$ (le sup et l'inf s'entendant pour chaque argument) sont des calibres. (Exercice! Utiliser un argument semblable à celui de I-13(f).) On en déduit aisément que, si I est une capacité (ou, plus généralement un calibre), alors, pour tout $t \in \bar{\mathbb{R}}_+$, les indicatrices des ensembles $\{X : I(X) > t\}$ et $\{X : I(X) \geqslant t\}$ sont des calibres (exercice!).

(b) Nous montrons ici, brièvement, que l'indicatrice J de l'ensemble $\{X \in \mathscr{P}(E)$: X n'est pas dénombrable$\}$ est un calibre. Plongeons l'espace E dans un espace métrisable compact E^* et considérons la bicapacité I sur $E \times \mathscr{M}^1(E^*)$ définie par (cf. I-6(a))

$$I(X, Y) = \sup_{m \in Y} m(X)\,.$$

On vérifie aisément que la partie D de $\mathcal{M}^1(E^*)$ constituée par les mesures diffuses (i.e. ne chargeant pas les points) est un $\mathbf{G}_\delta$ de l'espace métrisable compact $\mathcal{M}^1(E^*)$. D'autre part, on sait que tout compact non dénombrable porte un élément non nul de D et que tout analytique non dénombrable contient un compact non dénombrable (résultat classique de Souslin, qu'il est un peu dommage de ne pas déduire d'un théorème de capacitabilité). Il est alors clair que J admet $(I, \{1\}, D)$ comme représentation analytique.

(c) Nous indiquons ici une généralisation de l'exemple précédent. Soit I une capacité sur E telle que la réunion de deux ensembles de capacité nulle soit encore de capacité nulle (dans l'exemple précédent, on prend la capacité qui vaut 0 si son argument est vide et 1 sinon) et définissons J sur $\mathcal{P}(E)$ en posant $J(X)=1$ si X contient les éléments d'une famille non dénombrable (K_α) de compacts disjoints tels que $I(K_\alpha)>0$ pour tout indice α, et $J(X)=0$ sinon. On peut alors montrer que J est un calibre (la démonstration de Dellacherie (1972*b*) est insuffisante, car la notion de calibre introduite dans ce travail est différente de la "bonne" notion introduite ici). Cet exemple a des applications en théorie du potentiel (étude des ensembles semi-polaires) et en théorie des mesures de Hausdorff (étude des ensembles σ-finis).

Maintenant, quelques exemples d'opérations calibrées.

4 *Exemples* (a) Soit R une relation d'équivalence analytique sur E. L'application qui à $X \in \mathcal{P}(E)$ associe son saturé pour R est une opération calibrée (exercice!).

(b) Soit E un espace analytique: nous allons montrer que $X \to \bar{X}$ est une opération calibrée. Soit I l'opération capacitaire définie pour $X, Y \in \mathcal{P}(E)$ et (X_n) suite de parties de $E \times E$ par

$$I[X, Y, X_1, .\,, X_n, \ldots] = \inf_n \pi[X \times Y) \cap X_n],$$

où π est la seconde projection de $E \times E$ sur E. Pour obtenir l'adhérence, il suffit de prendre pour paramètres analytiques $Y=E$ et (X_n) une base d'entourages ouverts de la diagonale de $E \times E$. On peut montrer que, par contre, pour $E=[0, 1]$, l'opération qui à X associe son intérieur n'est pas une opération calibrée. Elle est cependant duale (en un sens clair) de l'opération d'adhérence. Nous ne tenterons pas d'étudier ici les opérations duales des opérations calibrées.

(c) Soient E, F deux espaces et I un calibre sur E tel que $I(\emptyset)=0$. On peut montrer que, pour tout $t \in \mathbb{R}_+$, on définit une opération calibrée J^t de $E \times F$ dans F en posant, pour $H \in \mathcal{P}(E \times F)$,

$$J^t(H) = \{y \in F : I[H(y)] > t\}.$$

(Exercice! Qui ne me semble pas facile au stade où nous en sommes; nous verrons une démonstration "standard" à la fin de la Section 2.)

Avant d'énumérer les principales propriétés de notre nouvelle notion, nous nous paierons le luxe d'écrire la définition des multicalibres et des opérations calibrées à plusieurs arguments.

5 Soit I une multicapacité (resp. une opération capacitaire) sur $\prod_{i\in\mathcal{I}} E_i$ (resp. et à valeurs dans F). Supposons distinguée une suite (non vide) $(\mathcal{I}_j)$ de parties non vides, disjointes, de $\mathcal{I}$ telle que $E_i = E^j$ ne dépende pas de $i \in \mathcal{I}_j$ pour j fixé, et donnons nous pour tout $i \notin \bigcup_j \mathcal{I}_j$ (s'il en existe!), un argument *analytique* A_i pour I. On dit alors que $(I, (\mathcal{I}_j), (A_i)_{i\notin\bigcup_j\mathcal{I}_j})$ est une *représentation analytique* d'une application J, globalement croissante, de $\prod_j \mathcal{P}(E^j)$ dans $\bar{\mathbb{R}}_+$ (resp. $\mathcal{P}(F)$) si, pour toute suite (A^j) où A^j est, pour chaque j, une partie *analytique* de E^j, on a

$$J(\ldots, A^j, \ldots) = I(\ldots, X_i, \ldots),$$

où $X_i = A_i$ pour $i \notin \cup \mathcal{I}_j$ et $X_i = A^j$ pour $i \in \mathcal{I}_j$. Enfin, l'on dit qu'une application globalement croissante J de $\prod_j \mathcal{P}(E^j)$ dans $\bar{\mathbb{R}}_+$ (resp. $\mathcal{P}(F)$) est un *multicalibre* (resp. une *opération calibrée*) sur $\prod_j E^j$ si elle admet une représentation analytique.

Les démonstrations des théorèmes suivants sont laissées au lecteur, à titre d'exercice: une fois compris de quoi il s'agit, on a à appliquer sans grande malice les théorèmes correspondants du Chapitre I.

6 THÉORÈME (d'analyticité) *Si I est une opération calibrée et si $A_1, ., A_i, \ldots$ sont des arguments analytiques pour I, alors $I(A_1, ., A_i, \ldots)$ est analytique.*

7 THÉORÈME (de composition) *Soit I un multicalibre (resp. une opération calibrée) sur $\prod_i E_i$ et supposons donné, pour tout i, une opération calibrée J_i sur $\prod_j E^i_j$, à valeurs dans E_i. Alors l'application J obtenue en composant I avec les J_i est encore un multicalibre (resp. une opération calibrée).*

8 THÉORÈME (d'approximation) *Soit I un multicalibre (resp. une opération calibrée) sur $\prod_i E_i$, et soit, pour tout i, A_i un argument analytique pour I. On a alors*

$$I(A_1, ., A_i, \ldots) = \sup I(L_1, ., L_i, \ldots) = \inf I(B_1, ., B_i, \ldots)$$

où L_i parcourt les $\mathcal{K}_\sigma$ contenus dans A_i et B_i parcourt les boréliens contenant A_i. De plus, si I est un multicalibre, l'inf et le sup sont atteints.

Le lecteur aura remarqué le seul changement notable par rapport aux énoncés du Chapitre I: la présence dans l'énoncé précédent de "$\mathcal{K}_\sigma$" au lieu des habituels "compacts". On n'a pas mieux, en général, si I n'est pas au moins séparément montante.

Nous intéresse particulièrement ici le corollaire suivant du théorème d'approximation:

9 THÉORÈME (de séparation) *Soit I un multicalibre, ou une opération calibrée, et soient $A_1, ., A_i, \ldots$ des arguments analytiques pour I. Si l'on a*

$$I(A_1, ., A_i, \ldots) = 0 \text{ (ou } \varnothing) ,$$

alors il existe, pour toute i, un borélien $B_i \supseteq A_i$ tel que

$$I(B_1, ., B_i, \ldots) = 0 \text{ (ou } \varnothing) .$$

Voici une application très intéressante du théorème de séparation que nous illustrerons en démontrant, très facilement, un résultat de Cenzer et Mauldin (1979) sur les "classes Π^1_1-monotones".

10 COROLLAIRE *Soient E, F des espaces, I un calibre sur E et H, H^* des parties analytiques disjointes de $E \times F$. Si, pour tout $y \in F$, la coupe $H(y)$ est de calibre nul, alors il existe un borélien L de $E \times F$ contenant H et disjoint de H^* tel que $L(y)$ soit de calibre nul pour tout $y \in F$.*

Démonstration D'après le théorème de séparation 9 appliqué à l'exemple 4(c) pour $t = 0$, il existe un borélien B contenant H tel que $B(y)$ soit de calibre nul pour tout $y \in F$. Par ailleurs, d'après le théorème de séparation "ordinaire", il existe un borélien B' contenant H et disjoint de H^*. Il n'y a plus qu'à poser $L = B \cap B'$.

11 Disons, suivant Cenzer et Mauldin, qu'une classe $\mathscr{C}$ de parties de E est *Π^1_1-monotone* s'il existe une partie analytique T de $E^{\mathbb{N}}$, appelée *test*, telle que l'on ait, pour $X \in \mathscr{P}(E)$,

$$X \in \mathscr{C} \quad \text{ssi} \quad X^{\mathbb{N}} \cap T = \varnothing .$$

Par exemple, si E est analytique et si $\mathscr{C}_0$ est une classe de parties modestes de E (cf. IV-43), alors la classe $\mathscr{C}$ des parties de E $\mathscr{C}_0$-rares est Π^1_1-monotone (prendre pour test T l'ensemble des suites (x_n) telles que l'adhérence de $\{x_1, ., x_n, \ldots\}$ dans E n'appartienne pas à $\mathscr{C}_0$). Revenons au cas général. Si $\mathscr{C}$ est une classe Π^1_1-monotone de parties de E, alors l'indicatrice J du complémentaire de $\mathscr{C}$ dans $\mathscr{P}(E)$ est un calibre. En effet, si T est un test pour

$\mathscr{C}$ et si I est la capacité sur $E^{\mathbb{N}}$ valant 0 si son argument est vide et 1 sinon, on a pour $X \in \mathscr{P}(E)$

$$J(X) = I(X^{\mathbb{N}} \cap T),$$

d'où J est une calibre d'après le théorème de composition. D'où le théorème de Cenzer et Mauldin, que nous prenons la peine d'écrire in extenso (cf. IV-37).

12 Théorème *Soient E, F des espaces, $\mathscr{C}$ une classe Π_1^1-monotone de parties de E et H, H^* des parties analytiques disjointes de $E \times F$. Si, pour tout $y \in F$, $H(y)$ appartient à $\mathscr{C}$, alors il existe un borélien L de $E \times F$ contenant H et disjoint de H^* tel que $L(y)$ appartienne à $\mathscr{C}$ pour tout $y \in F$.*

13 *Remarques* (a) La notion de classe d'ensembles de calibre nul est plus générale que celle de classe Π_1^1-monotone: ainsi, la classe des parties de $\mathbb{R}$ de mesure lebesguienne nulle est un exemple tout à fait élémentaire de classe d'ensembles de calibre nul, qui n'est pas Π_1^1-monotone. Noter, par ailleurs, que, d'après le théorème de composition, la notion de classe d'ensembles de calibre nul est stable pour de nombreuses opérations.

(b) Si I est un calibre sur un espace analytique E, on peut montrer que la classe $\mathscr{C}_I$ des ensembles de calibre nul est une classe de parties modestes de E (exercice! se ramener au cas où E est métrisable compact), même "relativisable", à laquelle on peut donc appliquer les résultats de la Section 4 du Chapitre IV. Et, d'après 11, la classe des parties $\mathscr{C}_I$-rares est aussi une classe d'ensembles de calibre nul, pour un calibre plus "grossier". Par contre, en général, la classe des parties $\mathscr{C}_I$-maigres n'est pas une classe d'ensembles de calibre nul (exercice!)—et donc, *a fortiori*, n'est pas Π_1^1-monotone—quoiqu'elle satisfasse à un énoncé analogue à 12 d'après la remarque (a) de IV-44. Noter cependant que la démarche adoptée au début de la démonstration de IV-44 revient finalement à ramener l'étude des ensembles $\mathscr{C}_I$-maigres à celle d'un calibre (celui associé à la classe Π_1^1-monotone engendrée par les fermés $\mathscr{C}_I^f$-maigres); il est possible de développer cette idée, mais le jeu ne semble pas en valoir la chandelle.

2 Multinoyaux capacitaires et calibrés

14 Dans toute la suite, nous entendrons par "fonction" une application à valeurs dans $\bar{\mathbb{R}}_+$. Pour tout espace E, nous désignons par $\mathfrak{P}(E)$ l'ensemble des fonctions sur E: une partie de E étant identifiée à son indicatrice, $\mathscr{P}(E)$ est un sous-ensemble de $\mathfrak{P}(E)$. Nous noterons $\mathfrak{B}(E)$ l'ensemble des fonctions boréliennes, $\mathfrak{U}(E)$ l'ensembles des fonctions s.c.i. et $\mathfrak{K}(E)$

l'ensemble des fonctions s.c.s. finies à support compact (et donc bornées; en abrégé, nous dirons, sans en être très fier, fonction s.c.s.s.c.). Etant données f s.c.i. et g s.c.s.s.c., nous dirons que f *majore largement* g si l'on a $f(x) > g(x)$ pour tout x appartenant au support de g; cela revient à dire qu'il existe un $\varepsilon > 0$ tel que l'on ait $f(x) > g(x) + \varepsilon$ pour tout x appartenant au support de g.

15 Les multinoyaux capacitaires seront en tout point analogues aux opérations capacitaires, sauf que leurs arguments et valeurs peuvent être des fonctions qui ne sont pas nécessairement des indicatrices. Afin d'obtenir une réelle extension des notions de multicapacités et d'opérations capacitaires, il nous est donc nécessaire d'indiquer comment on peut prolonger une application I de $\prod_i \mathscr{P}(E_i)$ dans $\mathfrak{P}(F)$ (les éléments de $\bar{\mathbb{R}}_+$ étant identifiés avec les fonctions sur F où F est un singleton) en une application I^+ de $\prod_i \mathfrak{P}(E_i)$ dans $\mathfrak{P}(F)$. Notant $I_y[X_1, ., X_i, \ldots]$ la valeur de la fonction $I[X_1, ., X_i, \ldots]$ au point $y \in F$, nous conviendrons (sauf exception) de prolonger I en I^+ comme suit: pour $f_i \in \mathfrak{P}(E_i)$ pour tout i, et pour tout $y \in F$, on pose que $I_y^+[f_1, ., f_i, \ldots]$ vaut

$$-\log\left[\int_0^1 . \int_0^1 \ldots \exp(-I_y[\{f_1 \geqslant t_1\}, ., \{f_i \geqslant t_i\}, \ldots])\, dt_1 \,.\, dt_i \ldots\right]$$

où $\{f_i \geqslant t_i\}$ désigne évidemment l'ensemble $\{x \in E_i : f_i(x) \geqslant t_i\}$; la fonction "exp" ne sert qu'à rendre "bien convergente" notre intégrale, et la fonction "$-\log$" à assurer que I^+ est bien un prolongement de I. Il existe d'autres prolongements possibles, suffisamment réguliers pour préserver plus loin les propriétés qui nous intéressent (croissance globale, montée séparée, continuité à droite sur les fonctions s.c.s.s.c.). Par exemple, si I est une capacité sur E, il est plus joli de poser, pour $f \in \mathfrak{P}(E)$,

$$I^*(f) = \int_{\mathbb{R}_+} I[\{f > t\}]\, dt \;;$$

noter que, si I est une mesure (au sens de I-1), alors $I^*(f)$ est l'intégrale (supérieure) de f par rapport à I et que, si I est la capacité sur E valant 0 si son argument est vide et 1 sinon, on obtient $I^*(f) = \sup_{x \in E} f(x)$. Nous laissons au lecteur la joie de découvrir les prolongements "naturels" de certaines des opérations capacitaires considérées en I-13.

Voici finalement notre définition des "multinoyaux capacitaires", qui comprendra comme cas particuliers les définitions I-1 des multicapacités ($\bar{\mathbb{R}}_+$ étant identifié à $\mathfrak{P}(F)$ où F est un singleton) et I-11 des opérations capacitaires, à condition d'avoir prolongé convenablement multicapacités et opérations capacitaires aux fonctions. Il y a une petite subtilité de langage dans la définition, expliquée dans la remarque qui lui succède:

16 DÉFINITION Un *multinoyau capacitaire I de F dans* $\prod_{i \in \mathscr{I}} E_i$ est une application de $\prod_i \mathfrak{P}(E_i)$ dans $\mathfrak{P}(F)$ vérifiant les propriétés suivantes:

(1) elle est *globalement croissante*:

$$f_1 \leq g_1, \ldots, f_i \leq g_i, \ldots \Rightarrow I[f_1, \ldots, f_i, \ldots] \leq I[g_1, \ldots, g_i, \ldots] ;$$

(2) elle est *séparément montante*: si tous les arguments f_i sont fixés, sauf pour $i = j$, on a

$$f_j^n \uparrow f_i \Rightarrow I[f_1, \ldots, f_{j-1}, f_j^n, f_{j+1}, \ldots] \uparrow I[f_1, \ldots, f_{j-1}, f_j, f_{j+1}, \ldots] ;$$

(3) elle est *globalement continue à droite sur les fonctions s.c.s.s.c.*: si les arguments g_i sont s.c.s.s.c., $I[g_1, \ldots, g_i, \ldots]$ est aussi s.c.s.s.c. et, pour toute fonction s.c.i. f majorant *largement* $I[g_1, \ldots, g_i, \ldots]$, il existe, pour tout i, une fonction s.c.i. f_i majorant *largement* g_i telle que f_i soit constante sauf pour un nombre fini d'indices, et que f majore (non nécessairement largement!) la fonction $I[f_1, \ldots, f_i, \ldots]$. (Ouf!)

Remarques (a) Le lecteur (s'il en reste!) a peut-être été surpris par le fait que l'on dise "multinoyau de F dans $\prod E_i$ "et non" de $\prod E_i$ dans F". Cette manière de parler vient du langage utilisé généralement (mais pas toujours!) en théorie de la mesure lorsqu'on y définit les noyaux. Il se justifie comme suit: si, pour $y \in F$ fixé, on considère l'application

$$(f_1, \ldots, f_i, \ldots) \to I_y[f_1, \ldots, f_i, \ldots],$$

il est facile de voir que l'on définit ainsi une application $y \to I_y$ de F dans l'ensemble des multicapacités sur $\prod E_i$ (application astreinte par ailleurs à une condition de régularité incluse dans le point (3) de la définition).

(b) Nous laissons au lecteur le soin patient de vérifier la légitimité des quelques lignes précédant la définition. Remarquer que, dans la superbe intégrale de 15, on peut remplacer une quantité finie d'accolades $\{f_i \geq t_i\}$ par les accolades $\{f_i > t_i\}$ correspondantes sans changer la valeur de l'intégrale.

(c) Si les espaces ambiants sont métrisables compacts, la seconde partie de la condition (3) est équivalente à la descente globale sur les fonctions s.c.s., plus agréable à écrire.

Nous nous contentons de donner quelques exemples de noyaux capacitaires (i.e. de multinoyaux à un argument), le Chapitre I ayant dû persuader le lecteur de l'intérêt à considérer des multinoyaux ayant même une infinité dénombrable d'arguments.

17 *Exemples* (a) L'application qui à $f \in \mathfrak{P}(E \times F)$ associe la fonction πf sur F définie par $\pi f(y) = \sup_{x \in E} f(x, y)$ est un noyau capacitaire de F dans $E \times F$, prolongeant la projection π de $E \times F$ sur F.

(b) Supposons E, F compacts métrisables pour simplifier. On dit en théorie de la mesure qu'une application $V\colon y \to m_y$ de F dans $\mathcal{M}^1(E)$ est un *noyau sous-markovien*; un tel noyau induit une application $f \to Vf$ de $\mathfrak{P}(E)$ dans $\mathfrak{P}(F)$ grâce à l'intégrale (supérieure): on a $Vf(y) = \int_E f(x)\,\mathrm{d}m_y(x)$. Le noyau V est dit *fellerien* si Vf est continue lorsque f est continue. On peut montrer que tout noyau sous-markovien et fellerien est un noyau capacitaire (exercice!). Plus généralement, soit $y \to I_y$ une application de F dans l'ensemble des capacités sur E, que nous identifions à l'application I de $\mathfrak{P}(E)$ dans $\mathfrak{P}(F)$ qu'elle induit. Si, pour tout $f \in \mathfrak{K}(E)$, $I(f)$ appartient à $\mathfrak{K}(F)$, alors I est un noyau capacitaire.

(c) Un dernier exemple, moins farfelu qu'il n'y parait au premier abord. Soit J une bicapacité sur $E \times F$ et définissons une application I de $\mathcal{P}(E)$ dans $\mathfrak{P}(F)$ en posant, pour tout $y \in F$ et $X \in \mathcal{P}(E)$,

$$I_y(X) = J(X, \{y\}) .$$

On peut montrer que, si $I(K)$ est une fonction à support compact pour tout $K \in \mathcal{K}(E)$ (ce qui est certainement le cas si F est compact), alors I est un noyau capacitaire (exercice!). Réciproquement, soit I un noyau capacitaire de F dans E et définissons une fonction J sur $\mathcal{P}(E) \times \mathcal{P}(F)$ en posant, pour $X \in \mathcal{P}(E)$ et $Y \in \mathcal{P}(Y)$,

$$J(X, Y) = \sup_{y \in Y} I_y(X) .$$

On peut montrer que J est une bicapacité (exercice!) si bien que tout noyau capacitaire est du type envisagé précédemment. Supposons E métrisable compact pour simplifier, et prenons $F = \mathcal{M}^1(E)$. Si on applique ce qui précède au noyau sous-markovian et fellerien $m \to m$ de $\mathcal{M}^1(E)$ dans E, on retrouve l'exemple de bicapacité de I-6(a).

La théorie se déroule alors comme prévu, c'est à dire dans les sillons de celle du Chapitre I, les lettres "gothiques" (par exemple $\mathfrak{K}$) remplaçant les lettres "rondes" (par exemple $\mathcal{K}$). Nous n'allons pas tout réécrire, et nous nous contenterons de nommer les principaux jalons, agrémentés de quelques remarques.

18 On a successivement

(a) *Le théorème de composition* (I-12).

On n'a plus ici à distinguer deux sortes de notion.

(b) *La définition des fonctions analytiques* (I-14).

Le rôle des ensembles $\mathscr{K}_\sigma$ est joué ici par les fonctions qui sont enveloppe supérieure d'une suite de fonctions s.c.s.s.c. Il n'est pas tout à fait évident que l'on n'obtient pas plus d'ensembles analytiques qu'avec I-14; en fait, on peut montrer, grâce au théorème de capacitabilité ci-dessous (ou plutôt sa démonstration, cf. I-27), que les fonctions analytiques définies ici sont, pour E analytique, celles que nous avions définies en IV-1.

(c) *Le théorème de l'invariance de l'analyticité* (I-15).

accompagné d'une kyrielle de corollaires.

(d) *Le théorème de capacitabilité* (I-19).

Ici, un petit ennui: la démonstration de I-10 ne s'étend pas bien au cas d'une infinité dénombrable d'arguments fonctions, à cause de la présence de "large majoration" dans 16(3). Cependant il est facile de voir, en appliquant judicieusement le théorème de composition à la définition d'une fonction analytique, que toute fonction analytique g a une représentation de la forme

$$g = I[f_1, \ldots, f_i, \ldots]$$

où I est un multinoyau capacitaire et où les arguments f_i sont ou des fonctions s.c.s.s.c. ou égaux à l'indicatrice de $\mathbb{N}$. On peut alors ramener la démonstration du théorème de capacitabilité à un cas particulier "sans problèmes".

(e) *Le théorème d'approximation par l'extérieur* (I-20)

et ses divers corollaires, regroupés sous le nom de *théorème de séparation* (I-21, I-22).

Voici un exemple de noyau capacitaire, qui montre comment, tout naturellement, on sort du cadre du Chapitre I pour entrer dans notre cadre actuel.

19 THÉORÈME *Soient E, F des espaces et C une capacité sur E telle que $C(\emptyset) = 0$. On définit un noyau capacitaire I de F dans $E \times F$ en posant, pour tout $H \in \mathscr{P}(E \times F)$*

$$I_y(H) = C[H(y)],$$

où $H(y)$ est la coupe de H suivant $y \in F$.

Démonstration Plongeons E dans un espace compact métrisable E' (ceci n'est pas vraiment nécessaire, mais nous simplifie un peu la vie) et posons, pour $H \in \mathscr{P}(E \times F)$ et $L \in \mathscr{P}(F)$,

$$J(H, L) = C[\pi_E(H \cap (E' \times L))],$$

où π_E est la projection sur E. D'après le théorème de composition on définit ainsi une bicapacité J et comme on a $I_y(H)=J(H,\{y\})$, on voit que I est un noyau capacitaire d'après 17(c).

D'où des corollaires intéressants (pour H analytique, la fonction $y \to C[H(y)]$ est analytique, etc.); nous ne développons pas: nous allons voir mieux un peu plus loin (on pourra prendre pour C un calibre).

Nous arrivons maintenant à la notion "nec plus ultra" de multinoyau calibré. Il reste maintenant si peu de lecteurs que je peux me payer le luxe de donner tout de suite la définition générale. Nous allons sagement recopier 5.

20 Soit I un multinoyau capacitaire de F dans $\prod_{i\in\mathscr{I}} E_i$. Supposons distinguée une suite (finie ou infinie, mais non vide) $(\mathscr{I}_j)$ de parties non vides, disjointes, de $\mathscr{I}$ telle que $E_i=E^j$ ne depende pas de $i\in\mathscr{I}_j$ pour j fixé, et donnons nous, pour tout $i\notin\bigcup_j\mathscr{I}_j$ (il peut ne pas y avoir de tel i), un argument *analytique* f_i pour I. On dit alors que $(I, (\mathscr{I}_j), (f_i)_{i\notin\bigcup_j I_j})$ est une *représentation analytique* d'une application J, globalement croissante, de $\prod_j \mathfrak{P}(E^j)$ dans $\mathfrak{P}(F)$ si, pour toute suite (f^j) où f^j est pour chaque j une fonction *analytique* sur E^j, on a

$$J(\ldots, f^j, \ldots)=I(\ldots, g_i, \ldots),$$

où $g_i=f_i$ pour $i\notin\bigcup_j\mathscr{I}_j$ et $g_i=f^j$ pour $i\in\mathscr{I}_j$. Enfin, l'on dit qu'une application globalement croissante J de $\prod_j \mathfrak{P}(E^j)$ dans $\mathfrak{P}(F)$ est un *multinoyau calibré de F dans $\prod_j E_j$* si elle admet une représentation analytique.

Remarques (a) On peut montrer qu'on multinoyau calibré admet une représentation analytique où les "paramètres" f_i sont égaux à l'indicatrice de $\mathbb{N}$.

(b) On peut montrer qu'un multicalibre ou une opération multicalibrée au sens de 5 admet un prolongement en un multinoyau calibré, en prolongeant d'abord la multicapacité ou l'opération capacitaire d'une représentation analytique au sens de 5 grâce au procédé de 15.

(c) On peut montrer ..., mieux vaut s'arrêter à temps!

21 *Exemples* (a) Posons, pour tout $f\in\mathfrak{P}(E)$ et tout $t\in\bar{\mathbb{R}}_+$,

$$I^t(f)=\{x: f(x)\geq t\}, \qquad J^t(f)=\{x: f(x)>t\}.$$

On définit ainsi deux noyaux calibrés I, J de $\bar{\mathbb{R}}_+\times E$ dans E, l'un fournissant le sous-graphe fermé de f et l'autre le sous-graphe ouvert de f (cf. IV-1). (Exercice! Montrer, en utilisant une intégrale, que I est la limite d'une suite décroissante et J la limite d'une suite croissante de noyaux capacitaires.)

(b) Posons, pour toute suite (f_n) d'éléments de $\mathfrak{P}(E)$,

$$I[f_1, \ldots, f_n, \ldots] = \liminf f_n\,, \qquad J[f_1, \ldots, f_n, \ldots] = \limsup f_n\,.$$

On peut montrer que I, J sont des multinoyaux calibrés (exercice!).

(c) Supposons que E et F sont analytiques, et soit V un noyau sous-markovien et borélien de F dans E, au sens de la théorie de la mesure: V est une application $y \to m_y$ de F dans $\mathcal{M}^1(E)$ telle que Vf soit une fonction borélienne sur F lorsque f est borélienne sur E (rappelons que $Vf(y) = \int_E f(x)\,\mathrm{d}m_y(x)$). On peut montrer que V est un noyau calibré de F dans E (cf. Dellacherie, 1976*a*).

(d) Soit J un bicalibre sur $E \times F$, où F est analytique, et définissons une application I de $\mathcal{P}(E)$ dans $\mathfrak{P}(F)$ en posant, pour $y \in F$ et $X \in \mathcal{P}(E)$,

$$I_y(X) = J(X, \{y\})\,.$$

On peut montrer que I est un noyau calibré de F dans E. (Exercice! Se ramener au cas où F est compact, puis considérer une représentation analytique de J et appliquer à la multicapacité de cette représentation une extension de 17(c).)

22 On a maintenant, dans ce cadre élargi (les lettres "gothiques" remplaçant les lettres "rondes"), les mêmes résultats qu'aux numéros 6, 7, 8 et 9: théorèmes d'*analyticité*, de *composition*, d'*approximation* (intérieure et extérieure) et de *séparation*. Nous nous contenterons d'en avoir cité les noms. Il résulte, par exemple, du théorème de composition que le noyau d'un schéma de Souslin sur l'ensemble des capacités sur E est un calibre sur E (il y a une analogie entre "capacités" et "compacts" d'une part—quoique l'inf d'une suite de capacités ne soit pas une capacité—et entre "calibres" et "ensembles analytiques" d'autre part; je ne sais si cela va très loin et si, en particulier, tout calibre sur E métrisable compact est, pour ses arguments analytiques, noyau d'un schéma de Souslin sur les capacités sur E).

Nous terminons par une étude de certains "cribles généralisés". Dans toute la suite, *on désigne par E, F des espaces et par C un calibre sur E tel que* $C(\varnothing) = 0$.

23 THÉORÈME *Posons, pour $H \in \mathcal{P}(E \times F)$, $y \in F$ et $t \in \bar{\mathbb{R}}_+$,*

$$I_y(H) = C[H(y)], \qquad J^t(H) = \{y : I_y(H) > t\}\,.$$

On définit ainsi un noyau calibré I de F dans $E \times F$ et un noyau calibré J de $\bar{\mathbb{R}}_+ \times F$ dans $E \times F$.

Démonstration Par plongement, on se ramène au cas où F est métrisable compact. On démontre alors que I est un noyau calibré comme en 19, en utilisant 21(d), et que J est un noyau calibré en composant I avec le noyau calibré de 21(a).

Remarque L'énoncé s'étend au cas où, plus généralement, C est un noyau calibré de G dans E tel que $C(\varnothing)=0$, I étant alors un noyau de $G\times F$ dans $E\times F$, etc.

Le théorème d'analyticité nous donne alors

24 THÉORÈME *Supposons que H soit une partie analytique de $E\times F$. Alors la fonction $y\to C[H(y)]$ est analytique et, en particulier, l'ensemble $\{y: C[H(y)]>t\}$ est analytique pour tout $t\in\mathbb{R}_+$.*

Le théorème d'approximation, joint au théorème de composition, nous donne aisément

25 THÉORÈME *Supposons que H soit une partie analytique de $E\times F$ et soit Q un calibre sur F. Il existe alors $L\in\mathcal{K}_\sigma(E\times F)$ et $B\in\mathcal{B}(E\times F)$ tels que H contienne L et soit contenu dans B, et que l'on ait*

$$Q[J^t(L)]=Q[J^t(H)]=Q[J^t(B)]$$

pour tout $t\in\bar{\mathbb{R}}_+$, avec $J^t(X)=\{y\in F: C[X(y)]>t\}$ pour $X\in\mathcal{P}(E\times F)$.

De ce théorème résulte en particulier le théorème de séparation de 10, que nous recopions avec nos notations actuelles:

26 THÉORÈME *Supposons que H soit une partie analytique de $E\times F$, disjointe d'une autre partie analytique H' de $E\times F$. Si, pour tout $y\in F$, on a $C[H(y)]=0$, alors il existe un borélien M de $E\times F$, contenant H et disjoint de H', tel que l'on ait $C[M(y)]=0$ pour tout $y\in F$.*

Commentaires

Chapitre I

La théorie des capacités trouve sa source dans le mémoire fondamental de Choquet (1955) (le nom de "capacité" provenant de la théorie du potentiel: l'étude la capacité newtonienne est à l'origine du travail de Choquet). De ce mémoire ne subsiste ici que la notion de capacité (et sa possible extension aux fonctions, cf. l'appendice) et l'idée fondamentale que capacités et ensembles analytiques vont de pair, par l'intermédiaire du théorème de capacitabilité. Nous n'avons malheureusement pas abordé d'autres points fondamentaux du mémoire comme la construction de capacités alternées d'ordre deux, et l'étude des capacités alternées d'ordre infini. Outre le mémoire de Choquet, dont la richesse n'est sans doute pas encore épuisée, on pourra consulter Dellacherie (1972*b*) et Dellacherie et Meyer (1975*b*) pour la construction de capacités; Mokobodzki (1978) et Talagrand (1978) ont récemment approfondi, de manière très différente, l'étude des capacités alternées d'ordre infini. Comme il a été dit dans le texte, c'est Sion (1963) qui a trouvé la bonne notion de continuité à droite dans le cas topologique, et c'est lui aussi qui, sans doute le premier, a remarqué que le premier théorème de séparation (pour deux ensembles) est lié au théorème de capacitabilité. L'introduction d'une infinité dénombrable possible d'arguments a une histoire qui me semble valoir le coup d'être racontée. J'avais tôt remarqué que la notion de multicapacité à un nombre fini d'arguments était intéressante, mais, malheureusement (?), j'avais aussi remarqué que, comme pour une bimesure, la notion de bicapacité, par exemple, se ramenait aisément, par prolongement, à celle de capacité sur un espace produit (cf. Dellacherie, 1971, 1976*a*). Dans mon article de survol (1975) de la théorie classique des ensembles analytiques, j'avais noté qu'il serait très intéressant de savoir démontrer le "premier" théorème de séparation de Novikov (i.e. l'énoncé I-22, non l'énoncé III-26) à l'aide de la théorie des capacités; voulant à tout prix ramener une multicapacité à un nombre infini d'arguments à une capacité sur un espace produit infini, je n'avais personnellement réussi à trouver que des démonstrations fausses.... C'est Mokobodzki (1976), et, indépendamment, Saint-Raymond (1976), qui, les premiers, ont montré comment déduire cette version du théorème de Novikov à partir d'arguments du type "capacitabilité". La démonstration de Mokobodzki se prêtait bien à une généralisation à n'importe quelle multicapacité, sous la forme d'un théorème d'approximation extérieure, que j'ai écrite (Dellacherie, 1976*b*), sans me rendre compte de son intérêt. Pendant ce temps Davies, Jayne, Ostaszewski et Rogers (1977) d'une part, et, indépendamment, Saint-Pierre (1977) d'autre part, se sont rendu compte que l'on pouvait, dans le même esprit que Mokobodzki et Saint-Raymond, démontrer le "premier" théorème de séparation de Liapunov (cf. I-24). Et c'est seulement alors que je me suis aperçu que "ma" version (Dellacherie, 1976*b*) du résultat de Mokobodzki donnait aisément cette version du théorème de Liapunov, à condition de remarquer que, en considérant encore une fois une infinité dénombrable d'arguments, l'opération de Souslin pouvait aussi se traduire dans le langage des capacités à l'aide de la notion d'opération capacitaire. Dans cette direction, le premier pas important avait aussi été franchi par Mokobodzki (1966/67), dans un article assez peu connu, où il introduisait la notion de capacité fonctionnelle (qui correspond à notre notion de noyau capacitaire—à un argument—de l'appendice): il y montrait, de manière frappante, comment analyticité et capacitabilité étaient intimement liées, et c'est en grande partie ce résultat de Mokobodzki (1978) qui m'avait poussé, il y a sept ans, à écrire, Dellacherie (1972*b*). Pour obtenir notre présentation des ensembles analytiques du Chapitre I, il n'y avait plus alors qu'à mettre ensemble ces deux idées: suite finie ou *infinie* d'arguments, et capacité fonctionnelle; cette conjonction a été faite dans Dellacherie (1978*a*), où l'on trouvera aussi, dans le même esprit, une présentation de la théorie dans le cas "abstrait" (le lecteur attentif aura remarqué que, si, dans ce Chapitre I, on remplace la continuité à droite sur les compacts par la notion, plus faible, de descente sur les compacts, de nombreux résultats sont encore valables, sans jamais utiliser sérieusement la compacité; l'endroit où la compacité intervient de manière essentielle est le théorème de séparation, où l'on utilise le fait que l'indicatrice de la classe des parties non vide est une capacité, d'où la notion de pavage semi-compact en théorie abstraite). Il reste cependant encore, pour moi, un grand mystère dans la théorie: l'élucidation des rapports, s'il y en a, entre capacités et *deuxième* théorème de séparation (cf. Chapitre III). Une première étape dans cette voie pourrait être l'extension du "deuxième" théorème de séparation de Novikov (cf. III-26) pour obtenir un énoncé du type I-21, avec "E-analytique" (cf. Chapitre II) à la place de "analytique" et "E-bianalytique" à la place de "boréliens": une étude minutieuse du "deuxième" théorème de séparation de Liapunov (que je n'ai jamais eu le courage de faire) pourrait s'avérer utile pour avancer dans cette voie.

Avant de quitter les Sections 1 et 2 de ce chapitre, je tiens à signaler que, de son côté, Hoffmann–Jørgensen est arrivé à un point de vue analogue au nôtre, qui sera sans doute exposé dans la "bible danoise" de théorie de la mesure, mise en chantier depuis plusieurs années en collaboration avec Christensen et Topsøe.

La Section 3 de ce chapitre, ainsi que les Sections 1, 2, 5 du Chapitre II, sont là pour achever une revue à peu près complète des résultats de la théorie des ensembles analytiques les plus utiles pour l'analyste ou le probabiliste. Signalons que Blackwell avait appelé "ses" espaces, espaces de Lusin, ce qui prêtait à confusion avec la terminologie bourbachique, et que d'autres auteurs entendent par espaces de Blackwell des espaces plus généraux que ceux en notre sens (le théorème I-35 de Blackwell leur sert de définition). Les quelques grains de sel peut-être originaux dans les démonstrations de cette Section 3 proviennent de Dellacherie et Meyer (1975*b*).

Chapitre II

Le théorème de Jankov (à "l'est") et von Neumann (à "l'ouest") est maintenant bien connu, et très utilisé dans divers domaines des mathématiques, ce qui lui a valu diverses extensions intéressantes; l'extension II-8, qui provient de Dellacherie et Meyer (1975*b*), devrait convaincre le lecteur que la plupart de ces extensions sont des corollaires très simples de l'énoncé primitif. Exposer les théorèmes de Jankov et von Neumann et de Kuratowski et Ryll–Nardzewski m'a donné l'occasion d'essayer de populariser la notion de tribu d'Effros. Utiliser cette notion allonge quelque peu les démonstrations, mais permet, à mon avis, de serrer de plus près le fond du problème. Les Sections 3 et 4 ont été rédigées essentiellement pour fournir une introduction au Chapitre IV, et aussi pour avoir l'occasion de parler des résultats récents de Louveau (198?); les théorèmes de section borélienne qui y sont énoncés sont, bien entendu, très intéressants, mais nous n'avons pas voulu aborder systématiquement ce sujet ici. Quant à la Section 5, elle est là en quelque sorte pour boucler la boucle; je n'ai pas la prétention d'y avoir étudié très sérieusement les relations d'équivalence.

Chapitre III

La notion de temps d'arrêt est une des notions fondamentales de la théorie des processus stochastiques. L'idée d'écrire dans le langage des t.d'a. tout ce qui tourne autour du deuxième théorème de séparation nous est venue, à Meyer et moi (cf. Dellacherie et Meyer 1975*a*), après lecture du bel article de Rogers (1973) (avant cet article, nous n'avions jamais rien compris à ce théorème . . .). Et c'est seulement après avoir écrit notre version que nous nous sommes aperçus que, dans le langage des arbres, tout cela était bien connu des logiciens (encore qu'il ne me semble pas avoir vu apparaitre chez eux une topologie naturelle sur les arbres qui permettent de considérer l'espace "canonique" des arbres, sans passer par le truchement d'un codage). A mon avis, il est bon de voir *à la fois* les choses en termes de t.d'a. et en termes d'arbres; c'est par simple coquetterie que nous avons systématiquement utilisé le langage des t.d'a., moins employé jusqu'ici que le langage des arbres. La démonstration que nous avons donnée du théorème de Kondô est fortement inspirée de celle que Kechris a écrite dans un cours (non publié, mais aux copies "circulantes") fait il y a quelques années au M.I.T.; elle me semble devoir bannir à tout jamais la crainte que ce théorème inspirait auparavant.

Chapitre IV

La notion de dérivation analytique, et le théorème de la borne qui s'y rattache, proviennent de Dellacherie (1977), fortement motivé par l'article de Saint-Raymond (1976) et la thèse de 3-ème cycle de Hillard (1978). Il est probable que ces résultats sont bien connus des logiciens, mais à peine moins probable qu'ils n'aient rien publié "officiellement" à ce sujet (le "folklore" californien doit être essentiellement de tradition orale . . .). A ce sujet, il est heureux que Kunen et Martin aient trouvé un théorème général sur la longueur des relations bien-fondées, sinon seuls les logiciens sauraient encore à l'heure présente que toute relation bien-fondée analytique a une longueur $<\aleph_1$ (c'est pour eux un résultat "classique"; il est vrai que les auteurs classiques "russo-polonais" n'étaient pas bien loin de cet énoncé, mais il me semble que, ignorant la notion de relation bien-fondée, ils n'ont jamais écrit leurs résultats en des termes à la fois si généraux et si simples). Tempérons cependant notre humeur: quand paraitront ces lignes, Moschovakis (1980) aura sans doute publié son merveilleux livre sur la théorie descriptive des ensembles, où il a pris la peine de mettre à la portée de tous les principaux outils forgés, depuis une douzaine d'années, par les logiciens en théorie effective.

Les applications à l'étude d'espaces fonctionnels proviennent de Dellacherie (1978*b*); je ne suis pas arrivé à faire entrer dans ce cadre l'exemple classique de Mazurkiewicz (l'espace des fonctions partout dérivables de $\mathbb{R}_+$ dans $\mathbb{R}$), ni l'exemple récent de Mauldin (1979) (l'espace des fonctions nulle part dérivables!). Les théorèmes IV-31 et 32 proviennent de Dellacherie (1976*a*), qui contient en outre d'autres applications du Chapitre I et de l'appendice que nous n'avons pas exposées ici (on n'y considère cependant que des multicapacités à un nombre fini d'arguments).

Comme il a été dit dans le texte, notre traitement du théorème de Saint-Raymond a été fortement inspiré par celui de Burgess (1979) et surtout de Hillard (19). Cela a été aussi pour moi l'occasion de découvrir que, également pour les calibres (que nous avions introduits dans Dellacherie (1976*a*); pour la définition, voir l'appendice), il était intéressant de considérer aussi la possibilité d'une infinité dénombrable d'arguments: d'où la démonstration que nous avons donnée, dans l'appendice, du théorème de séparation de Cenzer et Maudlin (198?).

Bibliographie

Bourgain, J.
198?*a* $\mathbf{F}_{\sigma\delta}$-sections of Borel sets, *Fund. Math.*
198?*b* Borels sets with $\mathbf{F}_{\sigma\delta}$-sections, *Fund. Math.*
Burgess, J. P.
1979 A reflection phenomenon in descriptive set theory, *Fund. Math.*, **104**, 127–139.

198? Measurable selectors for countably generated partitions, *Fund. Math.*
Cenzer, D. et Mauldin, R.
198? Faithful extensions of analytic sets to Borel sets, *Houston J. Math.*
Choquet, G.
1955 Theory of capacities, *Ann. Inst. Fourier, Grenoble*, **5**, 131–295.
Christensen, J. P. R.
1974 *Borel Structures*, Notes in Mathematics no. 10, North Holland, Amsterdam.
Davies, R. O., Jayne, J. E., Ostaszewski, A. J. et Rogers, C. A.
1977 Theorems of Novikov type, *Mathematika*, **24**, 97–114.
Dellacherie, C.
1971 Une démonstration du théorème de séparation des ensembles analytiques, dans *Séminaire de Probabilités V*, Lecture Notes in Mathematics no. 191, Springer-Verlag, Berlin, Heidelberg et New York, pp. 82–85.
1972*a* *Capacités et Processus Stochastiques*, Ergebnisse no. 67, Springer Verlag, Berlin, Heidelberg et New York.
1972*b* *Ensembles Analytiques, Capacités, Mesures de Hausdorff*, Lecture Notes in Mathematics no 295, Springer-Verlag, Berlin, Heidelberg et New York.
1975 Ensembles analytiques: théorèmes de séparation et applications, dans *Séminaire de Probabilités IX*, Lecture Notes in Mathematics no. 465, Springer-Verlag, Berlin, Heidelberg et New York, pp. 336–372.
1976*a* Sur la construction de noyaux boréliens, dans *Séminaire de Probabilités X*, Lecture Notes in Mathematics no. 511, Springer-Verlag, Berlin, Heidelberg et New York, pp. 545–577.
1976*b* Compléments aux exposés sur les ensembles analytiques, dans *Séminaire de Probabilités X*, Lecture Notes in Mathematics no. 511, Springer-Verlag, Berlin, Heidelberg et New York, pp. 579–593.
1977 Les dérivations en théorie descriptive des ensembles et le théorème de la borne, dans *Séminaire de Probabilités XI*, Lecture Notes in Mathematics no. 581, Springer-Verlag, Berlin, Heidelberg et New York, pp. 34–46. (Voir aussi: Erratum et addendum, dans *Séminaire de Probabilités XII*, p. 523.)
1978*a* Théorie unifiée des capacités et des ensembles analytiques, dans *Séminaire de Probabilités XII*, Lecture Notes in Mathematics no. 649, Springer-Verlag, Berlin, Heidelberg et New York, pp. 707–738.
1978*b* Quelques exemples familiers, en probabilités, d'ensembles analytiques non boréliens, dans *Séminaire de Probabilités XII*, Lecture Notes in Mathematics no. 649, Springer-Verlag, Berlin, Heidelberg et New York, pp. 746–756.
Dellacherie, C. et Meyer, P. A.
1975*a* Ensembles analytiques et temps d'arrêt, dans *Séminaire de Probabilités IX*, Lecture Notes in Mathematics no. 465, Springer-Verlag, Berlin, Heidelberg et New York, pp. 373–389.
1975*b* *Probabilités et Potentiel*, Hermann, Paris, Chaps. I–IV.
Hillard, G.
1978 Exemples de normes en théorie descriptive des ensembles, dans *Séminaire de Probabilités XII*, Lecture Notes in Mathematics no. 649, Springer-Verlag, Berlin, Heidelberg et New York, pp. 524–563.
1979 Une généralisation du théorème de Sainte-Raymond sur les boréliens à coupes $\mathcal{K}_\sigma$, *C.R. Acad. Sci., Paris*, **288**, 749–751.

Jayne, J. E.
1976 Structure of analytic Hausdorff spaces, *Mathematika*, **23**, 208–211.

Louveau, A.
1976/77 Relations d'équivalence coanalytiques, d'apres Harrington, Séminaire Choquet d'Initiation à l'analyse, no. 19, Institut H. Poincaré, Paris.
198? A separation theorem for Σ_1^1 sets with applications to Borel hierarchies in product spaces, *Trans. Amer. Math. Soc.*

Mauldin, R. P.
1979 The set of continuous nowhere differentiable functions, *Pacific J. Math.*, **83**, 199–205.

Miller, D. E.
1978 A selector for equivalence relations with $\mathbf{F}_\delta$ orbits, *Proc. Amer. Math. Soc.*, **72**, 365–369.

Mokobodzki, G.
1966/67 Capacités fonctionelles, Séminaire Choquet d'Initiation à l'analyse no. 9, Institut H. Poincaré, Paris.
1976 Démonstration élémentaire d'un théorème de Novikov, dans *Séminaire de Probabilités X*, Lecture Notes in Mathematics no. 511, Springer-Verlag, Berlin, Heidelberg et New York, pp. 539–543.
1978 Ensembles à coupes dénombrables et capacités dominées par une mesure, *Séminare de Probabilités XII*, Lecture Notes in Mathematics no. 649, Springer-Verlag, Berlin, Heidelberg et New York, pp. 491–508. 508.

Moschovakis, Y. N.
1980 *Descriptive Set Theory*, North Holland, Amsterdam.

Rogers, C. A.
1973 Lusin's second theorem of separation, *J. London Math. Soc.*, **6**, 491–503.

Saint-Pierre, J.
1977 Séparation simultanée d'un ensemble analytique de suites d'ensembles **F**-sousliniens, *C. R. Acad. Sci., Paris*, **285**, 933–936.

Saint-Raymond, J.
1971/73 Caractérisations d'espaces polonais, Séminaire Choquet d'Initiation à l'analyse no. 14–15, Institut H. Poincaré, Paris.
1976 Boréliens à coupes $\mathcal{K}_\sigma$, *Bull. Soc. Math. France*, **104**, 389–406.
1977 Sélections mesurables pour la structure borélienne d'Effros, *Bull. Sci. Math.* **101**, 353–355.
1978 La structure borélienne d'Effros est-elle standard? *Fund. Math.*, **100**, 201–210.

Sion, M.
1963 On capacitability and measurability, *Ann. Inst. Fourier, Grenoble*, **13**, 88–99.

Srivastava, S. M.
1979 Selection theorems for $\mathbf{G}_\delta$-valued multifunctions, *Trans. Amer. Math. Soc.*, **254**, 283–293.

Steel, J. R.
198? A note on analytic sets, *Cabal Seminar*, 78–79.

Talagrand, M.
1978 Capacités invariantes extremales, *Ann. Inst. Fourier, Grenoble*, **28**, 79–146.

Vaught, R.
1974 Invariant sets in topology and logic, *Fund. Math.*, **82**, 269–294.

Notations

Chapitre I

Dans ce chapitre, I, J désignent des (multi)capacités ou des opérations capacitaires, tandis que $\mathscr{I}$, $\mathscr{J}$ désignent des ensembles (finis ou) dénombrables d'indices.

On se reportera au début du chapitre en ce qui concerne l'utilisation des notations suivantes:

$$E, F, A, B, K, U\ ;$$

$$\mathscr{A}(E), \mathscr{B}(E), \mathscr{K}(E), \mathscr{U}(E) \quad \text{et} \quad \mathscr{P}(E)\,.$$

Réunion et intersection sont souvent notées sup et inf.

$\uparrow, \downarrow$	montée et descente de suites, I-1
$\mathscr{M}^1(E)$	espace des mesures $\geqslant 0$ de masse $\leqslant 1$, I-3
Σ, σ, τ	espace $\mathbb{N}^{\mathbb{N}}$ et éléments génériques, I-25
S, s, t	ensemble $\bigcup_n \mathbb{N}^n$ et éléments génériques, I-25
$s \dashv t, \sigma$	suites se prolongeant strictement, I-25
$\sigma\vert n$	n premiers termes de $\sigma \in \mathbb{N}^{\mathbb{N}}$, I-25
I_s	ilot (ouvert-fermé fondamental) de $\mathbb{N}^{\mathbb{N}}$, I-25
$\mathbf{X}, \mathbf{N}(\mathbf{X})$	schéma de Souslin et son noyau, I-25
$\Psi_{\mathbf{X}}$	application s.c.s. associée à un schéma, I-25
Ψ_A	application s.c.s. représant A, I-28
$(\Omega, \mathscr{F})$	espace mesurable, I-31
$(\bar{\Omega}, \bar{\mathscr{F}})$	séparé associé, I-31
$\Vert s\Vert$	somme des termes de $s \in S$, I-38

Chapitre II

Dans ce chapitre, I désigne l'ensemble des ordinaux dénombrables. On se reportera au début du chapitre en ce qui concerne les conventions sur les espaces topologiques, ainsi que l'emploi des notations suivantes:

$$\mathbf{F}, \mathbf{G}, \mathbf{F}_\sigma, \mathbf{G}_\delta, \text{etc.}\ ;$$

$$\mathbf{A}(E), \mathbf{CA}(E), \mathbf{BA}(E)\ ;$$

et du préfixe $E-$, $F-$, etc.

$\mathbf{F}(\Sigma)$	espace des fermés de $\mathbb{N}^{\mathbb{N}}$, II-3
$\mathbf{F}(E)$	espace (mesurable) des fermés de E, II-9

$\mathscr{E}$, etc.	notations pour pavages, II-16
X'	dérivé d'un fermé X, II-24
(X^i)	dérivés successifs, indexés par I, II-24
$\mathscr{B}(E/R)$	tribu des boréliens saturés pour R, II-26/27

Chapitre III

Dans ce chapitre, I désigne l'ensemble des ordinaux dénombrables (identifié à $\aleph_1$) tandis que J désigne un ensemble infini (égal le plus souvent à $\mathbb{N}$).

Ω, ω, w	espace $J^{\mathbb{N}}$ et éléments génériques, III-1
$\overline{\mathbb{N}}^{\circ}$	ensemble $\{0, 1, 2, \ldots, n, \ldots, \infty\}$, III-1
$\mathscr{T}$, T	espace des t.d'a. et élément générique, III-3
$\mathscr{P}$, $\mathscr{Q}$	espaces des t.d'a. à pôles et finis, III-4/5
T_x, $x \in E$	application s.c.s. de E dans $\mathscr{T}$, III-5
S, s, t	ensemble $\bigcup_n J^n$ et éléments génériques, III-8
$\lvert s \rvert$	longeur de $s \in S$, III-8
$s.t$, ω	concaténation de suites, III-8
$s \dashv t$, ω	comme au chapitre I
T^s, $s \in S$	on a $T^s(\omega) = (T(\omega) - \lvert s \rvert)^+$, III-8
$\mathbf{X}$, $\mathbf{N}(\mathbf{X})$	κ-schéma de Souslin et son noyau, III-9
T', $d(T)$	dérivé d'un t.d'a., III-10
T^i, $i \in I$	dérivés successifs, indexés par I, III-12
$i(T)$	indice d'un t.d'a., III-12
$i_T(\cdot)$	fonction indice d'un t.d'a., III-12
Q_j, $j \in I$	constituant, III-12
${}^fT$	composition d'un t.d'a. avec une stratégie, III-14

Chapitre IV

Dans ce chapitre, I désigne toujours l'ensemble des ordinaux dénombrables (identifié à $\aleph_1$).

$i(T)$	indice d'un t.d'a., IV-3
T^α	α-ième dérivé de T pour un ordinal α, IV-3
ϕ, $\dashv$	plus petit élément et prolongement d'un ordre adéquat, IV-6
d, $d(x)$	dérivation et dérivé, IV-7
x^i, $i \in I$	dérivés successifs indexés par I, IV-8
$i(x)$	indice de x, IV-8
P, Q	on a $x \in P(\in Q)$ ssi $i(x) = \aleph_1$ $(< \aleph_1)$, IV-8

R', R^{α}	dérivé et α-ième dérivé de R bien-fondée, IV-17
$\|R\|$	longeur de R bien-fondée, IV-17
$\|T\|$	longeur d'un t.d'a. T, IV-17
W, w	ensemble $E^{\mathbb{R}_+}$ et élément générique, IV-22
X_t, θ_t	coordonnée et translation sur W pour $t \in \mathbb{R}_+$, IV-22
$\Omega, \Omega_{\text{etc.}}$	espaces fonctionnels, IV-22/23
$T^i, i \in I$	itérés successifs de T adéquat, indexés par I, IV-25
$i_T(\cdot)$	indice associé à T adéquat, IV-25
Q_T	espace Q associé à T adéquat, IV-25
Q_j	variété de constituant, IV-33
H', H^j	dérivé et j-ième dérivé, coupe par coupe, IV-34
$\mathscr{C}$	classe de parties modestes, IV-43

Appendice

Les notations sont essentiellement celles du chapitre I: s'y reporter. S'y ajoutent les notations $\mathfrak{P}(E)$, $\mathfrak{K}(E)$, etc.: voir A-14.

Index terminologique

Part 3

Analytic spaces and their application

F. TOPSØE AND J. HOFFMANN-JØRGENSEN

1 Souslin schemes and standard models of separable metrizable spaces

1.1 Introduction

The word "standard" will be used here with its common meaning. We shall study three standard models. *The rationals, the irrationals* and *the Cantor set*, for which we use the notation $\mathbb{Q}$, $\mathbb{N}^{\mathbb{N}}$ and $2^{\mathbb{N}}$ respectively. Of course, the reason why we may use $\mathbb{N}^{\mathbb{N}}$ for the irrationals is that we are only interested in topological properties and that the set of irrational numbers is topologically isomorphic, i.e. homeomorphic, to $\mathbb{N}^{\mathbb{N}}$. For similar reasons, $2^{\mathbb{N}} = \{0, 1\}^{\mathbb{N}}$ may be used for the Cantor set.

Basically, we are interested in four kinds of results.

The authors were supported by the Danish Natural Science Research Council.

I. Characterization of standard models
- $\mathbb{Q}$ Countable metrizable space without isolated points
- $\mathbb{N}^{\mathbb{N}}$ Zero-dimensional Polish space such that every compact subset is nowhere dense
- $2^{\mathbb{N}}$ Zero-dimensional compact metrizable space without isolated points

II. Standard models as universal spaces
- $\mathbb{Q}$ Closed subsets give all countable metrizable spaces
- $\mathbb{N}^{\mathbb{N}}$ (a) Subsets give all zero-dimensional separable metrizable spaces
  (b) Closed subsets give all zero-dimensional Polish spaces
- $2^{\mathbb{N}}$ (a) Subsets give all zero-dimensional separable metrizable spaces
  (b) Closed (= compact) subsets give all zero-dimensional compact metrizable spaces

III. Continuous images of standard models
- $\mathbb{N}^{\mathbb{N}}$ (a) Every Polish space can be obtained as the continuous image of $\mathbb{N}^{\mathbb{N}}$
  (b) Every Polish space can be obtained as the continuous injective image of a closed subset of $\mathbb{N}^{\mathbb{N}}$
- $2^{\mathbb{N}}$ Every compact metrizable space can be obtained as the continuous image of $2^{\mathbb{N}}$

IV. Standard models embedded in other spaces
- $\mathbb{Q}$ Every coanalytic separable metrizable space which is non-Polish contains a copy of $\mathbb{Q}$ as a closed subset
- $\mathbb{N}^{\mathbb{N}}$ (a) Every uncountable Polish space contains a copy of $\mathbb{N}^{\mathbb{N}}$ as a $\mathcal{G}_\delta$-set
  (b) Every analytic separable metrizable space that is not σ-compact contains a copy of $\mathbb{N}^{\mathbb{N}}$ as a closed subset
- $2^{\mathbb{N}}$ Every uncountable Polish space contains a copy of $2^{\mathbb{N}}$ (necessarily as a closed set).

The bibliographical indications given in the text are very sketchy. As sources for further information we recommend Kuratowski's classical treatise on topology, Hoffmann-Jørgensen's notes on analytic spaces and Engelking's *General Topology* (see Kuratowski, 1966; Hoffmann-Jørgensen, 1970; Engelking, 1977).

1.2 Characterization of the rationals and the irrationals

First, let us prove the characterization theorem for $\mathbb{Q}$. We begin with the following order-theoretic result of Cantor:

THEOREM 1.2.1 (Cantor, 1895) *Two countable, totally and densely ordered sets without smallest and largest elements are order-isomorphic.*

(Densely ordered means: $\forall a<b, \exists c: a<c<b$.)

Brief proof Let X and Y be the sets, say,

$$X=\{x_1, x_2, \ldots\}, \qquad Y=\{y_1, y_2, \ldots\}.$$

To each $x \in X$, say $x=x_n$, define $x^- \in X \cup \{-\infty\}$ and $x^+ \in X \cup \{\infty\}$ by

x^- = largest among $-\infty, x_1, \ldots, x_{n-1}$ which is less than x,
x^+ = smallest among $+\infty, x_1, \ldots, x_{n-1}$ which is larger than x.

Similarly, for $y \in Y$ define y^- and y^+.
Construct $f: X \to Y$ such that

(a) $f(x_1)=y_1$ $(f(-\infty)=-\infty, f(\infty)=\infty)$,
(b) $f(x_n)$ = first element y among $y_1, y_2, \ldots$ such that $y \notin \{f(x_k): k<n\}$ and such that $f(x_n^-)<y<f(x_n^+)$.

Clearly, $f: X \to f(X)$ is an order isomorphism.

Assume $f(X) \neq Y$ and let $n=\min\{j \geq 1: y_j \notin f(X)\}$. Choose k and l such that $f(x_k)=y_n^-$, $f(x_l)=y_n^+$ (conventions regarding $\pm\infty$ clear). Consider that index α among $1, 2, \ldots, k_l$ for which $f(x_\alpha)$ is closest to y_n, but smaller than y_n. Consider $\beta \in \{1, \ldots, k_l\}$ with $f(x_\beta)$ closest to y_n, but larger than y_n. Let m be the first among $1, 2, \ldots$ for which $x_\alpha<x_m<x_\beta$. Then $x_m^-=x_\alpha$ and $x_m^+=x_\beta$. Clearly, $f(x_m)=y_n$. This contradiction completes the proof.

COROLLARY 1.2.2 *Let I be an open interval in $\mathbb{R}$. If X is a countable dense subset of I, then X is order-isomorphic and homeomorphic to $\mathbb{Q}$.*

Proof This follows upon noting that, under the assumptions of the corollary, X is densely ordered with no maximal or minimal elements, thus X is order-isomorphic to $\mathbb{Q}$. Moreover, in the situation observed, the induced topology on X is generated by the "intervals": $]a, b[\cap X$ for $a, b \in X$, and so X is homeomorphic to $\mathbb{Q}$.

Most of our remaining results will rely on the notion of a Souslin scheme. Recall that a *Souslin scheme* is a map, $A(\cdot)$, from the set of *multi-indices*

$$\mathbb{N}^{(\mathbb{N})}=\bigcup_{n=0}^{\infty} \mathbb{N}^n \qquad (\mathbb{N}^0=\{\varnothing\})$$

into the *power set* 2^X of some given set X. If $A(\cdot)$ is a Souslin scheme, we may perform the *Souslin operation* on A as follows:

$$S(A)=\bigcup_{\sigma \in \mathbb{N}^{\mathbb{N}}} \bigcap_{n=1}^{\infty} A(\sigma|n),$$

where $\sigma|n = (\sigma_1, \ldots, \sigma_n)$ whenever $\sigma = (\sigma_j) \in \mathbb{N}^{\mathbb{N}}$. If the set X is a topological space we shall say that a Souslin scheme $A(\cdot)$ is *open* (*closed*) if all its values, $A(\sigma|n)$, $\sigma|n \in \mathbb{N}^{(\mathbb{N})}$, are open (closed).

Here we shall use the Souslin scheme to construct continuous functions from $\mathbb{N}^{\mathbb{N}}$ (or subsets of $\mathbb{N}^{\mathbb{N}}$) to a given metric space. This construction proceeds as follows.

Let (X, d) be a metric space and $A(\cdot)$ a Souslin scheme with values in 2^X. We say that $A(\cdot)$ satisfies *the diameter condition* if

$$\lim_{n\to\infty} \operatorname{diam} A(\sigma|n) = 0, \qquad \forall \sigma \in \mathbb{N}^{\mathbb{N}}, \tag{1}$$

with the convention: diam $\varnothing = 0$. If (1) is satisfied then $\bigcap_n A(\sigma|n)$ contains at most one point, thus we may define a function φ from a subset Z of $\mathbb{N}^{\mathbb{N}}$ to X by

$$Z = \{\sigma \in \mathbb{N}^{\mathbb{N}} : \bigcap_n A(\sigma|n) \neq \varnothing\}, \tag{2}$$

$$\{\varphi(\sigma)\} = \bigcap_n A(\sigma|n), \qquad \forall \sigma \in Z. \tag{3}$$

Thus Z is the domain of definition for φ. This map is referred to as *the map associated to* $A(\cdot)$.

If $A(\cdot)$ is a Souslin scheme in a metric space (X, d) satisfying the diameter condition (1), and φ is the associated map, then we have the following simple propositions:

$$\varphi \text{ is continuous;} \tag{4}$$

$$\varphi(Z) = S(A) = \bigcup_\sigma \bigcap_n A(\sigma|n); \tag{5}$$

if (X, d) is complete and A is closed, then Z is a closed subset of $\mathbb{N}^{\mathbb{N}}$; (6)

if $\bigcap_{n=1}^{k} A(\sigma|n) \neq \varnothing$ and $A(\sigma|k) \subseteq \bigcup_{q=1}^{\infty} A(\sigma|k, q)$ for all $\sigma|k \in \mathbb{N}^{\mathbb{N}}$, then Z is a sense subset of $\mathbb{N}^{\mathbb{N}}$; (7)

if A is open and $A(\sigma|n) \subseteq \bigcup_{q=1}^{\infty} A(\sigma|n, q)$ for all $\sigma|n \in \mathbb{N}^{(\mathbb{N})}$, then φ is an open map: $Z \to \varphi(Z)$; (8)

if $\bigcap_n A(\sigma|n) \cap \bigcap_n A(\tau|n) = \varnothing$ for all $\sigma \neq \tau$, then φ is injective. (9)

$$\left.\begin{array}{l}\text{if } A \text{ is open and } \{A(\sigma|n, q): q\in\mathbb{N}\} \text{ are disjoint subsets}\\ \text{of } A(\sigma|n) \text{ for all } \sigma|n\in\mathbb{N}^{(\mathbb{N})}, \text{ then } \varphi \text{ is a homeomorphism}\\ \text{of } Z \text{ onto } \varphi(Z).\end{array}\right\} \tag{10}$$

Here we have used the convention $(\sigma|n, q) = (\sigma_1, \ldots, \sigma_n, q)$ if $\sigma|n = (\sigma_1, \ldots, \sigma_n) \in \mathbb{N}^n$ and $q \in \mathbb{N}$.

We shall only indicate the proofs of (7) and (8) and leave the rest (which are all extremely simple) to the reader. Let us put

$$\mathbb{N}(\sigma|n) = \{\tau \in \mathbb{N}^{\mathbb{N}}: \tau|n = \sigma|n\}.$$

Then $\{\mathbb{N}(\sigma|n): \sigma|n \in \mathbb{N}^{(\mathbb{N})}\}$ is a base for the topology on $\mathbb{N}^{\mathbb{N}}$. And if A satisfies

$$A(\sigma|n) \subseteq \bigcup_{q=1}^{\infty} A(\sigma|n, q),$$

then one easily verifies that

$$\varphi(Z \cap \mathbb{N}(\sigma|n)) = \bigcap_{j=1}^{n} A(\sigma|j), \qquad \forall \sigma|n \in \mathbb{N}^{(\mathbb{N})},$$

from which (7) and (8) easily follow

We may also achieve a Lipschitz condition for the associated map φ, but then we need to specify a metric on $\mathbb{N}^{\mathbb{N}}$. This is done as follows:

$$\rho(\sigma, \tau) = 2^{-n}, \quad \text{where} \quad n = \min\{j: \sigma_j \neq \tau_j\},$$

with the convention $\min \varnothing = \infty$ and $2^{-\infty} = 0$. Then the reader easily verifies:

$$\left.\begin{array}{l}\text{if } \operatorname{diam} A(\sigma|n) \leqslant 2^{-n-1} \text{ for all } \sigma|n \in \mathbb{N}^{(\mathbb{N})}, \text{ then we have}\\ d(\varphi(\sigma), \varphi(\tau)) \leqslant \rho(\sigma, \tau) \quad \text{if} \quad \rho(\sigma, \tau) < \tfrac{1}{2}.\end{array}\right\} \tag{11}$$

We can now turn to the structure theorems for separable metrizable spaces. First we consider *zero-dimensional spaces*, i.e. spaces admitting a basis of clopen sets.

So let X be a separable, zero-dimensional, metrizable space and choose a metric d for X. We can then construct a Souslin scheme $A(\cdot)$ on X with the following properties:

(i) $A(\varnothing) = X$;
(ii) $A(\sigma|n)$ is clopen, $\forall \sigma|n \in \mathbb{N}^{(\mathbb{N})}$;
(iii) $\{A(\sigma|n, q): q \in \mathbb{N}\}$ is a partition of $A(\sigma|n)$, $\forall \sigma|n \in \mathbb{N}^{(\mathbb{N})}$;
(iv) $\operatorname{diam} A(\sigma|n) \leqslant 2^{-n-1}$, $\forall \sigma|n \in \mathbb{N}^{(\mathbb{N})}$.

Hence the associated map φ is a homeomorphism of Z onto X (note that $X = S(A)$) by (5) and (10) and φ is also Lipschitzian in the sense of (11). Thus we have proved

THEOREM 1.2.3 *If (X, d) is a zero-dimensional separable metric space, then X is homeomorphic to a subset $Z \subseteq \mathbb{N}^{\mathbb{N}}$. Furthermore, the homeomorphism may be taken to be Lipschitzian from (Z, ρ) to (X, d).*

If X is Polish and d is chosen to be complete, then by (6) we obtain

THEOREM 1.2.4 *If (X, d) is a zero-dimensional, complete separable metric space, then X is homeomorphic to a closed subset Z of $\mathbb{N}^{\mathbb{N}}$. And the homeomorphism may be taken to be a Lipschitzian map from (Z, ρ) to (X, d).*

If $A(\sigma|n)$ may be chosen non-empty for all $\sigma|n \in \mathbb{N}^{(\mathbb{N})}$, then Z will be dense in $\mathbb{N}^{\mathbb{N}}$. This happens in several cases.

Suppose that the metric d is complete and that int $K = \varnothing$ for all compact subsets K of X. Since relative compactness equals total boundedness (=precompactness) in a complete metric space, we find that no open non-empty set is totally bounded. Thus if $A(\sigma|n)$ is open and non-empty and

$$A(\sigma|n) \subseteq \bigcup_{q=1}^{\infty} B_q,$$

with diam $B_q < \varepsilon$, then for ε sufficiently small we must have that $B_q \neq \varnothing$ for infinitely many values of q. But this shows that in our construction of $A(\sigma|n)$ we can ensure that $A(\sigma|n) \neq \varnothing$ for all $\sigma|n \in \mathbb{N}^{(\mathbb{N})}$. Hence Z is dense by (7) and closed by (6) so $Z = \mathbb{N}^{\mathbb{N}}$. Thus we have proved

THEOREM 1.2.5 (Alexandrov and Urysohn, 1928) *If (X, d) is a zero-dimensional, complete separable metric space satisfying*

(a) int $K = \varnothing$, $\forall K$ *that are compact subsets of X,*

then X is homeomorphic to $\mathbb{N}^{\mathbb{N}}$. And the homeomorphism may be chosen to be Lipschitzian from $(\mathbb{N}^{\mathbb{N}}, \rho)$ to (X, d).

If X is countable and has no isolated points, then it is fairly easy to check that we may choose $A(\sigma|n) \neq \varnothing$ for all $\sigma|n$. To see this, let $A \neq \varnothing$ be a clopen subset of X and $\varepsilon > 0$. Let $A = \{x_1, x_2, \ldots\}$ be an enumeration of A (note that A is infinite, since X has no isolated points). Then we choose a clopen neighbourhood A_1 of x_1 such that $A_1 \subseteq A$, diam $A_1 \leqslant \varepsilon$ and $A_1 \neq A$. Now we put

$$n = \min \{j \in \mathbb{N} \colon x_j \notin A_1\},$$

and pick a clopen neighbourhood A_2 of x_n, such that $A_2 \subseteq A \backslash A_1$, diam $A_2 \leq \varepsilon$ and $A_2 \neq A \backslash A_1$.

Continuing this construction gives us a sequence of disjoint clopen non-empty sets A_j, with

$$A = \bigcup_{j=1}^{\infty} A_j \quad \text{and} \quad \operatorname{diam} A_j \leq \varepsilon, \qquad \forall j.$$

Hence it is possible to obtain a Souslin scheme satisfying (i)–(iv), such that $A(\sigma|n) \neq \emptyset$. But then X is homeomorphic to a dense countable subset of $\mathbb{N}^{\mathbb{N}}$, and, since $\mathbb{N}^{\mathbb{N}}$ is homeomorphic to the irrational numbers, we have that X is homeomorphic to a countable dense subset of $\mathbb{R}$. Thus from Corollary 1.2.2 we conclude

THEOREM 1.2.6 (Sierpiński, 1920*a*) *If X is a countable metric space without isolated points, then X is homeomorphic to $\mathbb{Q}$.*

Finally, we remark that by choosing a suitable metric d for the topology on X, we may obtain $A(\sigma|n) \neq \emptyset$ under the condition $\operatorname{int}(K) = \emptyset$ for all K compact, without completeness.

To see this we begin by embedding X as a dense subset of a compact metric space $(\bar{X}, \bar{d})$ (cf. appendix to Section 1). The condition $\operatorname{int} K = \emptyset$ then implies that $\bar{X} \backslash X$ is dense in $\bar{X}$; hence we can find a countable set $T = \{a_n\} \subseteq \bar{X} \backslash X$, such that T is dense in $\bar{X}$. Then we put

$$d(x, y) = \bar{d}(x, y) + \sum_{n=1}^{\infty} 2^{-n} \arctan|\bar{d}(x, a_n)^{-1} - \bar{d}(y, a_n)^{-1}|,$$

which evidently is a metric on X generating the original topology on X. If C is totally bounded with respect to d, then its closure, $\bar{C}$, in $\bar{X}$ is contained in $\bar{X} \backslash T$. However, if U is an open subset of $\bar{X}$, then $U \subseteq \overline{U \cap X}$, and so no open non-empty subset of X is totally d-bounded. Hence an argument similar to the proof of Theorem 1.2.5 gives

THEOREM 1.2.7 *If X is a zero-dimensional, separable, metrizable space, satisfying*

(a) $\operatorname{int}(K) = \emptyset$, $\forall K$ *that are compact subsets of* X,

then X is homeomorphic to a dense subset of $\mathbb{N}^{\mathbb{N}}$.

Let us elaborate a little further on Sierpiński's theorem. Suppose that (X, d) is a countable metric space which may have isolated points. We still know that φ is a homeomorphism of Z onto X where φ is the map associated with the Souslin scheme satisfying (i)–(iv). If $\sigma|n \in \mathbb{N}^{(\mathbb{N})}$ and $A(\sigma|n) = \varnothing$, then we pick a point $\xi(\sigma|n)$ in $\mathbb{N}(\sigma|n)$, and we may then consider the set

$$Y = Z \cup \{\xi(\sigma|n) \colon \sigma|n \in \mathbb{N}^{(\mathbb{N})}, A(\sigma|n) = \varnothing\}.$$

Then $\mathbb{N}(\sigma|n) \cap Y \neq \varnothing$ for all $\sigma|n \in \mathbb{N}^{(\mathbb{N})}$; thus Y is a countable dense subset of $\mathbb{N}^{\mathbb{N}}$, and moreover

$$Z = Y \backslash \cup \{\mathbb{N}(\sigma|n) \colon \sigma|n \in \mathbb{N}^{(\mathbb{N})}, \quad A(\sigma|n) = \varnothing\},$$

since $Z \cap \mathbb{N}(\sigma|n) = \varnothing$ if $A(\sigma|n) = \varnothing$. But this shows that Z is closed relatively in Y, and since Y is homeomorphic to $\mathbb{Q}$ (cf. Theorem 1.2.6) we have proved

THEOREM 1.2.8 (Fréchet, 1910) *Every countable metrizable space is homeomorphic to a closed subset of* $\mathbb{Q}$.

To complete our study of countable metrizable spaces, we shall mention the following theorem which gives a complete structure theorem for countable metrizable spaces:

THEOREM 1.2.9 *Any countable Polish space is homeomorphic to a countable subspace of the ordinal space* $[0, \omega_1[$.

Any countable metrizable space X *admits a unique partition* $X = P \cup Q$, *where* P *is open and Polish and* $Q = X \backslash P$ *is either empty or homeomorphic to* $\mathbb{Q}$.

We shall not prove the first part here. In the second part we take Q equal to the perfect kernel of X, then $Q = \varnothing$ or $Q \approx \mathbb{Q}$ by Theorem 1.2.6. Moreover, $P = X \backslash Q$ is open without perfect subset from which we can conclude that P is Polish (cf. Corollary 1.4.6).

The structure theorems above have a lot of immediate consequences and we shall list a few below.

COROLLARY 1.2.10 $\mathbb{Q}, \mathbb{Q}^2, \mathbb{Q}^3, \ldots$ *are all homeomorphic.*

COROLLARY 1.2.11 (Baire, 1909) *All the spaces*

$$(\mathbb{R} \backslash \mathbb{Q}), (\mathbb{R} \backslash \mathbb{Q})^2, \ldots, (\mathbb{R} \backslash \mathbb{Q})^{\mathbb{N}}, \mathbb{N}^{\mathbb{N}}, \mathbb{N}^{\mathbb{N} \times \mathbb{N}}, \textit{etc.},$$

are homeomorphic.

COROLLARY 1.2.12 (Mazurkiewicz, 1917–18) *Let* X *be a zero-dimensional Polish space contained in a Hausdorff space* X^*. *If* X *and* $X^* \backslash X$ *both are dense in* X^*, *then* X *is homeomorphic to* $\mathbb{N}^{\mathbb{N}}$.

Proof Let K be a compact subset of X. If G is open in X^* and $G \cap X \subseteq K$, then by denseness of X we have $G \subseteq K \subseteq X$; hence by denseness of $X^* \backslash X$ we have $G = \varnothing$. So the corollary follows from Theorem 1.2.5.

If $X^* = \mathbb{R}$, then zero-dimensionality is redundant, since $\{X \cap]a, b[: a, b \in \mathbb{R}\backslash X\}$ is a clopen base for X if $\mathbb{R}\backslash X$ is dense in $\mathbb{R}$. Thus we have

COROLLARY 1.2.13 *If X is a Polish subset of $\mathbb{R}$ with both X and $\mathbb{R}\backslash X$ dense in $\mathbb{R}$, then X is homeomorphic to $\mathbb{N}^{\mathbb{N}}$.*

Note that this result is no longer true if $\mathbb{R}$ is replaced by $\mathbb{R}^2$ (consider $X = \mathbb{R} \times (\mathbb{R}\backslash\mathbb{Q})$).

The results above exhaust the possibilities of the Souslin scheme $A(\cdot)$ introduced at the beginning of this section; the method is completely restricted to zero-dimensional spaces. We shall now see what may be done for general separable metric spaces.

So let (X, d) be a separable metric space. It is then straightforward to construct a Souslin scheme $B(\cdot)$ satisfying

(i)* $B(\varnothing) = X$;
(ii)* $B(\sigma|n)$ is non-empty and open for all $\sigma|n \in \mathbb{N}^{(\mathbb{N})}$;
(iii)* $B(\sigma|n) = \bigcup_{q=1}^{\infty} B(\sigma|n, q) = \bigcup_{q=1}^{\infty} \overline{B(\sigma|n, q)}, \forall \sigma|n \in \mathbb{N}^{(\mathbb{N})}$;
(iv)* $\operatorname{diam} B(\sigma|n) \leq 2^{-n-1}, \forall \sigma|n \in \mathbb{N}^{(\mathbb{N})}$.

Then the map φ associated with B coincides with the one associated with $\overline{B(\cdot)}$ by (iii)*. Hence, by (4), (5) and (8), we have that φ is continuous and open from Z onto X. In general Z is a dense subset of $\mathbb{N}^{\mathbb{N}}$ by (7), and if (X, d) is complete then $Z = \mathbb{N}^{\mathbb{N}}$ by (6) and (7). Hence we find

THEOREM 1.2.14 *Let (X, d) be a separable metric space, then X is a continuous, open, Lipschitzian image of a dense subset Z of $\mathbb{N}^{\mathbb{N}}$.*

Moreover, if (X, d) is complete and separable, then we may take $Z = \mathbb{N}^{\mathbb{N}}$. That is, every Polish space is a continuous, open image of $\mathbb{N}^{\mathbb{N}}$.

A topological Hausdorff space X which is a continuous image of $\mathbb{N}^{\mathbb{N}}$ is called analytic. The theorem above thus states that every Polish space is analytic. Much more is true: every Borel subset of an analytic space is analytic. This important result is treated earlier in this volume (see Theorem 2.5.3 of Part I). However, in general one cannot obtain all analytic spaces as an open continuous image of $\mathbb{N}^{\mathbb{N}}$; in fact every open continuous image of a Polish space is itself Polish provided it is regular and Hausdorff.

In general a metric space is neither a homeomorphic image of $\mathbb{N}^{\mathbb{N}}$ nor of its subsets, but we can seek injective continuous maps. (9) tells what is needed, and with a little more care we can construct a Souslin scheme $C(\cdot)$ satisfying

(i)** $C(\varnothing)=X$;

(ii)** $C(\sigma|n)=\bigcup_{q=1}^{\infty} C(\sigma|n, q)=\bigcup_{q=1}^{\infty} \overline{C(\sigma|n, q)}, \forall \sigma|n \in \mathbb{N}^{(\mathbb{N})}$;

(iii)** $\operatorname{diam} C(\sigma|n) \leq 2^{-n-1}, \forall \sigma|n \in \mathbb{N}^{(\mathbb{N})}$;

(iv)** $C(\sigma|n) \cap C(\tau|n)=\varnothing, \forall \sigma|n, \tau|n \in \mathbb{N}^{(\mathbb{N})}$ with $\sigma|n \neq \tau|n$.

By (ii)** the sets $C(\sigma|n)$ are all $\mathscr{F}_\sigma$-sets, and the observation needed to carry through the construction is the following easy fact:

$$\left.\begin{array}{l}\text{if } C \text{ is an } \mathscr{F}_\sigma\text{-set and } \varepsilon>0, \text{ then there exist disjoint} \\ \mathscr{F}_\sigma\text{-sets } C_k, \text{ such that } \bar{C}_k \subseteq C, \operatorname{diam} C_k<\varepsilon \text{ and } C=\bigcup_{k=1}^{\infty} C_k.\end{array}\right\} \tag{12}$$

To see this we first choose closed sets F_n with $F_n \uparrow C$. Then we note that $F_1, F_2 \backslash F_1, F_3 \backslash F_2, \ldots$ all may be written as a disjoint union of $\mathscr{F}_\sigma$-sets of diameter $<\varepsilon$, and then we note that $\overline{F_{n+1} \backslash F_n} \subseteq F_{n+1} \subseteq C$.

Again the map φ associated with $C(\cdot)$ coincides with the one associated with $\overline{C(\cdot)}$. Now by (4) and (9) we have that φ is continuous and injective from Z onto X, and, if (X, d) is complete, we have that Z is closed. Thus we have

THEOREM 1.2.15 *Every separable metric space is a continuous, injective, Lipschitzian image of a subset of* $\mathbb{N}^{\mathbb{N}}$.

Every complete separable metric space is a continuous, injective, Lipschitzian image of a closed subset of $\mathbb{N}^{\mathbb{N}}$.

We remark that Jayne and Rogers (1980) give necessary and sufficient conditions on a Polish space Z (expressed in terms of the quasi-components or in terms of the retraction properties of Z) for every Polish space to be representable as a continuous image of Z.

1.3 Characterization of the Cantor set

The Souslin scheme introduced in Section 1.2 is particularly appropriate for the treatment of the structure of $\mathbb{N}^{\mathbb{N}}$. When we turn to $2^{\mathbb{N}}$ we need a similar device, namely the Hausdorff scheme.

Let $2^{(\mathbb{N})}$ be the set of finite sequences of zeros and ones:

$$2^{(\mathbb{N})} = \bigcup_{k=0}^{\infty} \{0, 1\}^k \qquad (\{0, 1\}^0 = \{\emptyset\}).$$

Then a *Hausdorff scheme* is a map $A: 2^{(\mathbb{N})} \to 2^X$.

If (X, d) is a metric space and $A(\cdot)$ is a Hausdorff scheme in X, then we say that $A(\cdot)$ satisfies *the diameter condition* if we have

$$\lim_{n\to\infty} \operatorname{diam} A(\sigma|n) = 0, \qquad \forall \sigma \in 2^{\mathbb{N}}. \tag{13}$$

If this condition is satisfied we can, as in Section 1.2, define *the map φ associated with $A(\cdot)$ by*

$$Z = \{\sigma \in 2^{\mathbb{N}}: \bigcap_n A(\sigma|n) \neq \emptyset\}, \tag{14}$$

$$\{\varphi(\sigma)\} = \bigcap_n A(\sigma|n), \qquad \forall \sigma \in Z. \tag{15}$$

One can then formulate and prove statements in complete analogy with (4)–(10). We shall, however, leave the details of this to the reader.

Using Hausdorff schemes in much the same way as we used Souslin schemes in the proof of Theorem 1.2.5, we can now prove

THEOREM 1.3.1 (Brouwer, 1910) *A zero-dimensional compact metrizable space without isolated points is homeomorphic to* $2^{\mathbb{N}}$.

Proof Let X be zero-dimensional compact metrizable without isolated points. Choose a metric d on X. First observe that: *for every $\varepsilon > 0$ there exists a decomposition of X into at least* 2, *but only finitely many non-empty clopen subsets of diameter* $< \varepsilon$.

By a repeated application of this observation, we can construct a Hausdorff scheme $A(\cdot)$ such that

> $A(\cdot)$ is a decomposition scheme (i.e. (i) and (iii) of Section 1.2 hold) satisfying the diameter condition, and each $A(\sigma|n)$, $\sigma|n \in 2^{(\mathbb{N})}$ is a non-empty clopen subset of X. (16)

The first step is to write X in the form $X = \bigcup_{i=1}^{\infty} X_i$ with the X_i being non-empty clopen subsets of X of diameter < 1 and $n \geq 2$ (see Figure 1). We

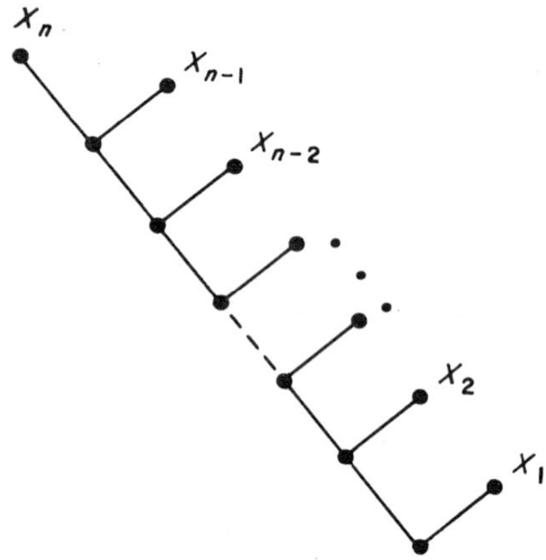

Figure 1

use the X_i's to define the following sets:

$$A(1) = X_1,$$
$$A(01) = X_2,$$
$$\vdots$$
$$A(00\cdots 01) = X_{n-1},$$
$$A(00\cdots 00) = X_n.$$

With (16) in mind, we define the sets

$$A(00\cdots 0) = X_n \cup X_{n-1} \cup \cdots \cup X_{i+1} \qquad (i\ 0\text{'s}\ 1 \leq i \leq n-2).$$

Then we apply the same procedure to each of the sets $X_1, X_2, \ldots, X_n$, this time making sure that the resulting sets have diameter $< \frac{1}{2}$. Continuing, we construct $A(\cdot)$ such that (16) holds.

Clearly, the map associated with the Hausdorff scheme $A(\cdot)$ is a homeomorphism of $2^{\mathbb{N}}$ onto X.

As a corollary, the usual Cantor set $\mathbb{C}$ is homeomorphic to $2^{\mathbb{N}}$. In this specific case, it is clear that we can explicitly find a Hausdorff scheme $A(\cdot)$ having the properties mentioned in the above proof. Indeed, referring to the standard construction of $\mathbb{C}$ outlined in Figure 2, we may take

$$A(\sigma|n) = \mathbb{C} \cap F(\sigma|n), \qquad \sigma|n \in 2^{(\mathbb{N})}.$$

Also, referring to the notation from Figure 2,

$$\mathbb{C} = [0, 1] \Big\backslash \bigcup_{\sigma|n \in 2^{(\mathbb{N})}} G(\sigma|n) = \bigcup_{\sigma \in 2^{\mathbb{N}}} \bigcap_{n \in \mathbb{N}} F(\sigma|n).$$

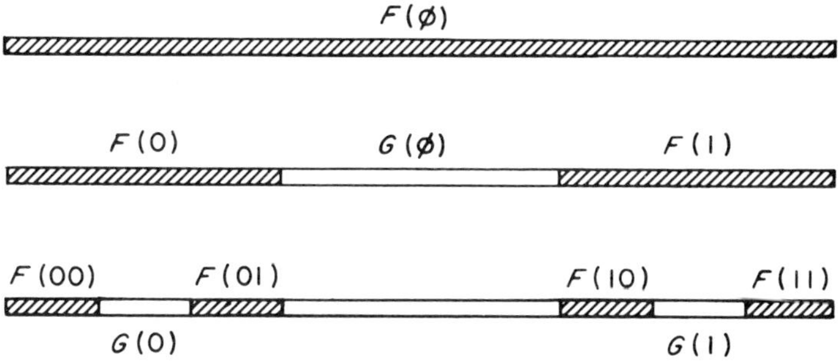

Figure 2

As another corollary to the characterization theorem, we see that the spaces $\mathbb{C}, \mathbb{C}\times\mathbb{C}, \ldots$ and $\mathbb{C}^{\mathbb{N}}$ are all homeomorphic. Also, it may be noted that for any finite set X (with discrete topology) containing at least two points, $X^{\mathbb{N}}$ is homeomorphic to $\mathbb{C}$.

THEOREM 1.3.2 (Alexandrov, 1927; Hausdorff, 1927) *Every compact metrizable space is a continuous image of* $2^{\mathbb{N}}$.

This follows from some lemmas:

LEMMA 1.3.3 *There exists a continuous surjective map* $\theta: 2^{\mathbb{N}} \to [0, 1]^{\mathbb{N}}$.

Proof As $2^{\mathbb{N}}$ is homeomorphic to $(2^{\mathbb{N}})^{\mathbb{N}}$, we need only prove that there exists a continuous surjective map $2^{\mathbb{N}} \to [0, 1]$. To see this consider the map

$$(\sigma_1, \sigma_2, \ldots) \to \sum_{i=1}^{\infty} \sigma_i 2^{-i}.$$

LEMMA 1.3.4 *Every compact metrizable space X can be obtained as the continuous image of a closed subset of* $2^{\mathbb{N}}$.

Proof As is well known, X is homeomorphic to a closed subset K of $[0, 1]^{\mathbb{N}}$. The restriction of θ from Lemma 1.3.3 to $\theta^{-1}(K)$ is a continuous surjection: $\theta^{-1}(K) \to K$.

Note that this lemma could equally well have been obtained by constructing a suitable Hausdorff scheme.

The next lemma is of interest in itself (Sierpiński 1928*b*).

LEMMA 1.3.5 *For every closed subset $F \subseteq 2^{\mathbb{N}}$, there exists a continuous map $f: 2^{\mathbb{N}} \to F$ extending the identity on F.*

Proof The metric

$$d(\sigma,\tau)=\sum_{n=1}^{\infty}|\sigma_n-\tau_n|\times 10^{-n},\qquad \sigma,\tau\in 2^{\mathbb{N}},$$

induces the product topology on $2^{\mathbb{N}}$. We put

$$\|\sigma\|=\sum_{n=1}^{\infty}|\sigma_n|\times 10^{-n}$$

for every $\sigma\in\{-1,0,1\}^{\mathbb{N}}$. Then $d(\sigma,\tau)=\|\sigma-\tau\|$. Furthermore, for a sequence (σ^n) in $\{-1,0,1\}^{\mathbb{N}}$,

$$\|\sigma^n\|\to\|\sigma\|\Rightarrow|\sigma^n|\to|\sigma|,$$

where $|\sigma|=(|\sigma_1|,|\sigma_2|,\ldots)$. In particular for $\tau,\nu\in 2^{\mathbb{N}}$ we have

$$d(\sigma,\tau)=d(\sigma,\nu)\Rightarrow\tau=\nu.$$

It follows that for every $\sigma\in 2^{\mathbb{N}}$ there exists a unique $\tau\in F$ such that $d(\sigma,\tau)$ is minimal $(=d(\sigma,F))$. Let f be the map $\sigma\to\tau$ thus defined. Then $f(\sigma)=\sigma$ for $\sigma\in F$.

To prove that f is continuous, assume that $\sigma^n\to\sigma$. Put $\tau^n=f(\sigma^n)$, $\tau=f(\sigma)$. Then

$$d(\sigma^n,\tau^n)=d(\sigma^n,F)\to d(\sigma,F)=d(\sigma,\tau);$$

i.e.

$$\|\sigma^n-\tau^n\|\to\|\sigma-\tau\|.$$

Then $|\sigma^n-\tau^n|\to|\sigma-\tau|$ and as $\sigma^n\to\sigma$, $\tau^n\to\tau$ follows.

THEOREM 1.3.6 (Young, 1903) *Let X be a Polish space, Y be an uncountable Hausdorff space and let $f\colon X\to Y$ be a continuous surjection. Then there exists a subset $A\subseteq X$ homeomorphic to the Cantor space such that $f|A$ is a homeomorphism of A onto $f(A)$.*

Proof The facts that f need not be injective and that X may contain isolated points complicate this proof somewhat. To handle these complications, we first choose an uncountable subset X_0 of X such that $f|X_0$ is injective. In X_0 we consider the subset $X_1\subseteq X_0$ of *condensation points*, i.e. points for which all neighbourhoods (in X_0) are uncountable. By a well-known argument (Cantor–Bendixson's lemma), X_1 is uncountable and the topological space X_1 has no isolated points (indeed, $X_0\backslash X_1$ is countable).

Choose a complete metric on X.

Construct a Hausdorff scheme $A(\cdot)$ on X such that

(i) $A(\sigma|n)$ is open, $\forall\sigma|n \in 2^{(\mathbb{N})}$;
(ii) $A(\sigma|n) \cap X_1 \neq \emptyset$, $\forall\sigma|n \in 2^{(\mathbb{N})}$;
(iii) diam $A(\sigma|n) \leq 1/n$;
(iv) $\overline{f(A(\sigma|n))} \cap f(A(\tau|n)) = \emptyset$, for $\sigma|n \neq \tau|n$;
(v) $\overline{A(\sigma|n+1)} \subseteq A(\sigma|n)$.

Let $\varphi: Z \to X$ be the associated map. Then it is easy to see that φ is a homeomorphism of $2^{\mathbb{N}}$ onto $\varphi(2^{\mathbb{N}})$ and that the restriction of f to $\varphi(2^{\mathbb{N}})$ is also a homeomorphism.

An immediate corollary is the following:

THEOREM 1.3.7 *Any uncountable analytic space contains a copy of the Cantor space.*

In particular, the continuum hypothesis holds for analytic spaces in the sense that an infinite analytic space either has cardinality $\aleph_0$ or $2^{\aleph_0}$.

As the Cantor space contains a copy of the irrationals, every uncountable analytic space X contains a subset $Y \subseteq X$ homeomorphic to $\mathbb{N}^{\mathbb{N}}$. By a well-known result (cf. the appendix to Section 1) Y is then a $\mathcal{G}_\delta$ in $\bar{Y}$. If the analytic space X is regular, it can be shown that it is perfectly normal. Then Y is a $\mathcal{G}_\delta$ in X too. To sum up:

THEOREM 1.3.8 *Any uncountable regular analytic space contains a copy of $\mathbb{N}^{\mathbb{N}}$ as a $\mathcal{G}_\delta$-subset.*

We remark that Jayne and Rogers (1980) show that, in Theorem 1.3.6, if Y is not σ-compact, then the subset A on which f is a homeomorphism can be taken to be a closed subset of X that is homeomorphic to $\mathbb{N}^{\mathbb{N}}$ rather than to $2^{\mathbb{N}}$.

1.4 The rationals and irrationals embedded in other spaces

We shall construct copies of $\mathbb{Q}$ by constructing special Souslin schemes of one-point sets.

For the following considerations, $X = (X, d)$ denotes a complete metric space. We define a map $r: \mathbb{N}^{(\mathbb{N})} \to \mathbb{R}_+$ by

$$r(\sigma_1, \ldots, \sigma_k) = 2^{-(\sigma_1 + \cdots + \sigma_k)}.$$

By a *condensation scheme* in X, we shall understand a map: $\sigma|n \to x(\sigma|n)$ of $\mathbb{N}^{(\mathbb{N})}$ into X satisfying the following conditions:

$$0 < d(x(\sigma|n), x(\sigma|n+1)) \leq r(\sigma|n+1), \qquad \forall \sigma|n+1 \in \mathbb{N}^{(\mathbb{N})}. \tag{17}$$

Somewhat more general schemes were discussed by Hurewicz in 1928 under the name of "Häufungssysteme".

It follows from (17) that

$$x(\sigma_1, \ldots, \sigma_{n-1}, \sigma_n) \to x(\sigma_1, \ldots, \sigma_{n-1}) \quad \text{for} \quad \sigma_n \to \infty,$$

and also that, for each $\sigma \in \mathbb{N}^{\mathbb{N}}$, the sequence $x(\sigma|n)$, $n \geq 1$, converges. We put

$$\varphi(\sigma) = \lim_{n \to \infty} x(\sigma|n), \qquad \sigma \in \mathbb{N}^{\mathbb{N}}.$$

We refer to the mapping $\sigma \to \varphi(\sigma)$ of $\mathbb{N}^{\mathbb{N}}$ into X as the mapping *associated* with the condensation scheme. Usually, we denote this map by φ.

With the given condensation scheme we also associate the following sets:

$$D_n = \{x(\sigma|n) \colon \sigma \in \mathbb{N}^{\mathbb{N}}\},$$

$$D = \{x(\sigma|n) \colon \sigma|n \in \mathbb{N}^{(\mathbb{N})}\} = \bigcup_{n=0}^{\infty} D_n.$$

Then we evidently have (cf. Theorem 1.2.6):

$$\varphi \text{ is continuous}; \tag{18}$$

$$\bigcup_{k=0}^{n} D_k \text{ is compact}, \quad \forall n \geq 0; \tag{19}$$

$$D \text{ is homeomorphic to } \mathbb{Q}; \tag{20}$$

$$\bar{D} \backslash D \subseteq \varphi(\mathbb{N}^{\mathbb{N}}). \tag{21}$$

Note that a metric space admits a condensation scheme if, and only if, the space contains a non-empty perfect set. Thus we have

THEOREM 1.4.1 *A metrizable space X contains a copy of $\mathbb{Q}$, if, and only if, X contains a non-empty perfect subset.*

The subset homeomorphic to $\mathbb{Q}$ in Theorem 1.4.1 will of course be an $\mathscr{F}_\sigma$-set in X. Similarly in Theorem 1.3.8 we saw that $\mathbb{N}^{\mathbb{N}}$ sits as a $\mathscr{G}_\sigma$-subset in any uncountable analytic metrizable space. We shall now show that it is possible to embed $\mathbb{Q}$ and $\mathbb{N}^{\mathbb{N}}$ simultaneously as relatively closed subsets. As a matter of fact we shall now prove the following theorem:

THEOREM 1.4.2 (Hurewicz, 1928) *Let X be a Polish space and Y an analytic subset of X, which is not an $\mathscr{F}_\sigma$-set. Then there is a closed set $F \subseteq X$, such that*

(a) $P = F \cap Y$ *is homeomorphic to* $\mathbb{N}^{\mathbb{N}}$,
(b) $Q = F \backslash Y$ *is homeomorphic to* $\mathbb{Q}$,
(c) $\bar{Q} = F$.

The proof, which is a mixture of Hurewicz' original proof and an argument from Saint-Raymond (1978) will be based on two lemmas. The first connects the associated map for condensation schemes with the associated map for Souslin schemes.

LEMMA 1.4.3 *Let $x(\cdot)$ be a condensation scheme and $H(\cdot)$ an open Souslin scheme on a complete metric space (X, d). Suppose that we have*

(a) $x(\sigma|n) \in H(\sigma|n) \Big\backslash \bigcup_{q=1}^{\infty} \overline{H(\sigma|n, q)}, \ \forall \sigma|n \in \mathbb{N}^{(\mathbb{N})}$;
(b) $H(\sigma|n) \cap H(\tau|n) = \varnothing$, whenever $\sigma|n \neq \tau|n$;
(c) $\overline{H(\sigma|n+1)} \subseteq H(\sigma|n), \ \forall \sigma|n \in \mathbb{N}^{(\mathbb{N})}$;
(d) $\operatorname{diam} H(\sigma|n) \leq r(\sigma|n), \ \forall \sigma|n \in \mathbb{N}^{(\mathbb{N})}$.

Then the associated maps for $x(\cdot)$, $H(\cdot)$ and $\overline{H(\cdot)}$ coincide; call it φ. And we have that φ is a homeomorphism of $\mathbb{N}^{\mathbb{N}}$ onto $\bar{D} \backslash D$.

Proof It follows immediately from (a) and (c) that the associated maps to $x(\cdot)$, $H(\cdot)$ and $\overline{H(\cdot)}$ coincide. Moreover, by (10) we have that φ is a homeomorphism of $\mathbb{N}^{\mathbb{N}}$ onto $\varphi(\mathbb{N}^{\mathbb{N}})$. The condition (a) clearly implies that

$$\varphi(\mathbb{N}^{\mathbb{N}}) \subseteq \bar{D} \backslash D,$$

and so the lemma follows from (21).

LEMMA 1.4.4 *Let X be a Polish space and Y an analytic subset of X, which is not an $\mathscr{F}_\sigma$-set. Then there exists a non-empty Polish space T and a continuous map $\theta: T \to Y$, such that*

(a) $\overline{\theta(G)} \backslash Y \neq \varnothing, \qquad \forall G$ *open non-empty* $\subseteq T$.

Proof Let S be Polish and θ_0 a continuous surjection of S onto Y. And consider the set

$$G_0 = \left\{ \begin{array}{l} x \in S\colon \exists U \text{ open in } T, \exists A \text{ an } \mathscr{F}_\sigma\text{-set in } X \\ \qquad \text{such that } x \in U, \ \theta_0(U) \subseteq A \subseteq Y \end{array} \right\}.$$

Then G_0 is open and by the Lindelöf property of S there exists an $\mathscr{F}_\sigma$-set A_0 in X, such that

(i) $$\theta_0(G_0) \subseteq A_0 \subseteq Y.$$

Now let $T = S \setminus \theta_0^{-1}(A_0)$ and put $\theta = \theta_0|T$. Then T is $\mathscr{G}_\delta$ in S, since A_0 is $\mathscr{F}_\sigma$. Hence T is Polish, and $T \neq \varnothing$ since otherwise we would have $Y = A_0$, contrary to our assumption on Y. Also θ clearly maps T into Y.

Now take an open set G in S, then we have

$$\theta_0(G) \subseteq \theta_0(G \cap \theta_0^{-1}A_0) \cup \theta_0(G \setminus \theta_0^{-1}A_0) \subseteq A_0 \cup \overline{\theta(G \cap T)}.$$

The last set is an $\mathscr{F}_\sigma$-set which cannot be contained in Y unless $G \subseteq G_0$. Hence we have

(ii) $$G \cap T \neq \varnothing \Rightarrow \overline{\theta(G \cap T)} \setminus Y \neq \varnothing,$$

since $G \subseteq G_0$ implies $G \cap T = \varnothing$. As $G \cap T$ is a general open set in T we have proved the lemma.

Proof of Theorem 1.4.2 Let T and θ be chosen as in Lemma 1.4.4. Then we construct an open Souslin scheme $G(\cdot)$ in T, an open Souslin scheme $H(\cdot)$ in X and a condensation scheme $x(\cdot)$ on X such that

(i) $x(\sigma|n) \in (X \setminus Y) \cap \overline{\theta(G(\sigma|n))} \Big\backslash \bigcup_{q=1}^{\infty} \overline{H(\sigma|n, q)}, \forall \sigma|n \in \mathbb{N}^{(\mathbb{N})}$;
(ii) $0 < d(x(\sigma|n), x(\sigma|n+1) < r(\sigma|n+1), \forall \sigma|n \in \mathbb{N}^{(\mathbb{N})}$;
(iii) $\overline{H(\sigma|n)} \cap \overline{H(\tau|n)} = \varnothing$, whenever $\sigma|n \neq \tau|n$;
(iv) $\overline{H(\sigma|n+1)} \subseteq H(\sigma|n), \forall \sigma|n \in \mathbb{N}^{(\mathbb{N})}$;
(v) $\operatorname{diam} H(\sigma|n) \leq r(\sigma|n), \forall \sigma|n \in \mathbb{N}^{(\mathbb{N})}$;
(vi) $\overline{G(\sigma|n+1)} \subseteq G(\sigma|n), \forall \sigma|n \in \mathbb{N}^{(\mathbb{N})}$;
(vii) $\operatorname{diam} \overline{G(\sigma|n)} \leq 2^{-n}, \forall \sigma|n \in \mathbb{N}^{(\mathbb{N})}, \forall n \geq 1$;
(viii) $\overline{\theta(G(\sigma|n))} \subseteq H(\sigma|n), \forall \sigma|n \in \mathbb{N}^{(\mathbb{N})}$.

We shall leave the details of the straightforward construction to the reader. Now put

$$D = \{x(\sigma|n) \colon \sigma|n \in \mathbb{N}^{(\mathbb{N})}\}.$$

Then $\bar{D}$ is closed and we have, by (20) and Lemma 1.4.3,

(ix) D is homeomorphic to $\mathbb{Q}$, $D \subseteq X \setminus Y$;
(x) $\bar{D} \setminus D$ is homeomorphic to $\mathbb{N}^{\mathbb{N}}$.

We claim that

(xi) $\bar{D} \setminus D \subseteq Y$.

So let $x \in \bar{D} \backslash D$, then by (21) there exists an $\sigma \in \mathbb{N}^{\mathbb{N}}$ with

$$x = \lim_{n \to \infty} x(\sigma|n).$$

So by (i) we may choose $t_n \in G(\sigma|n)$ so that

$$d(\theta(t_n), x(\sigma|n)) \leqslant 2^{-n}, \qquad \forall n \geqslant 1.$$

Then $\theta(t_n) \to x$, and from (vi) and (vii) we infer that $\{t_n\}$ converges (T is Polish!). That is $t_n \to t$ and so $x = \theta(t) \in Y$, by Lemma 1.4.4. Hence (xi) is proved.

From (ix) and (xi) it follows that:

$$D = [D \cap (X \backslash Y)] \cup [(\bar{D} \backslash D) \cap (X \backslash Y)] = \bar{D} \backslash Y,$$

$$\bar{D} \backslash D = [(\bar{D} \backslash D) \cap Y] \cup [D \cap Y] = \bar{D} \cap Y,$$

and so the theorem is proved.

If Y is an analytic metric space that is not a $\mathscr{K}_\sigma$-set, then we may embed Y in a compact metric space in which Y is not an $\mathscr{F}_\sigma$.

COROLLARY 1.4.5 *If Y is an analytic metrizable space that is not a $\mathscr{K}_\sigma$, then Y contains $\mathbb{N}^{\mathbb{N}}$ as a closed subset.*

Similarly, if Z is a coanalytic non-Polish space, then we may embed Z in some Polish space X. And we have that $Y = X \backslash Z$ is analytic and not $\mathscr{F}_\sigma$, thus we find

COROLLARY 1.4.6 *If Y is a coanalytic metrizable space that is not Polish, then Y contains $\mathbb{Q}$ as a closed subset.*

In particular, every non-Polish countable metric space contains a closed copy of the rationals (cf. Theorem 1.2.9).

Appendix: Basic notions from topology

A topological space is *topologically complete* if it is metrizable with a complete metric. A *Polish space* is a separable, topologically complete space.

If the spaces X_n, $n \geqslant 1$, are metrizable, topologically complete or Polish, then so is $\prod_{n=1}^{\infty} X_n$.

$[0, 1]^{\mathbb{N}}$ is a universal separable metrizable space in the sense that every separable metrizable space is homeomorphic to a subspace of $[0, 1]^{\mathbb{N}}$. To see this, let $X=(X,d)$ be a separable metric space with $d\leqslant 1$; choose (x_n) dense in X and consider the map $x\to(d(x,x_n))_{n\geqslant 1}$ of X into $[0,1]^{\mathbb{N}}$.

It follows that a separable metrizable space has a countable basis. In the converse direction we have Urysohn's metrization theorem:

$$X \text{ regular, } X \text{ has a countable base} \Rightarrow X \text{ metrizable.}$$

Indication of proof From a general result it follows that X is normal. Let $(G_n)_{n\geqslant 1}$ be a base. Consider all pairs $\alpha=(G_n, G_m)$ with $\overline{G_n}\subseteq G_m$ and denote by f_α a continuous function $X\to[0,1]$ such that $f_\alpha=0$ on $\bar{G}_n$ and $f_\alpha=1$ on $X\backslash G_m$. The map $x\to(f_\alpha(x))_\alpha$ is a homeomorphism.

The following are equivalent: X is separable and metrizable, X is regular and has a countable base, X is homeomorphic with a subspace of a compact metrizable space, X is homeomorphic with a subspace of $[0,1]^{\mathbb{N}}$.

A compact space is metrizable, if, and only if, it has a countable base. There exists a compact, separable, perfectly normal (hence first-countable) space which is not metrizable (cf. Bourbaki, 1966, Chap. 9, § 2, Ex. 13). Recall that a topological space is perfectly normal if it is normal and if every closed set is a $\mathscr{G}_\delta$ (i.e. a countable intersection of open sets).

A well-known result states that the metric space (X,d) is complete if, and only if, $\bigcap F_n\neq\varnothing$ for every decreasing sequence (F_n) of closed non-empty sets for which diam $(F_n)\to 0$. The result may be used to prove that every topologically complete space is a Baire space. Another consequence is that a compact metrizable space is complete in any metric compatible with the topology.

The following results, due to Alexandrov, Hausdorff, and to Sierpiński show an important connection between completeness and $\mathscr{G}_\delta$-properties:

(i) X topologically complete, Y a $\mathscr{G}_\delta$-subspace of $X\Rightarrow Y$ topologically complete;

(ii) X Hausdorff, Y topologically complete subspace of $X\Rightarrow Y$ is a $\mathscr{G}_\delta$-subset of $\bar{Y}$.

Indication of proof (i) $Y=\bigcap_n G_n$. Put $F_n=X\backslash G_n$. The map

$$y\to(y, d(y,F_1)^{-1}, d(y,F_2)^{-1},\ldots)$$

is a homeomorphism and the image is closed in $X\times\mathbb{R}^{\mathbb{N}}$.

(ii) Put

$$G_n=\bar{Y}\cap\bigcup\{H\colon H \text{ open in } X, \operatorname{diam}(H\cap Y)\leqslant 1/n\}.$$

Then $Y=\bigcap G_n$.

In (ii), Y will of course be a $\mathscr{G}_\delta$ in X if X is, say metrizable.

The space of irrationals $\mathbb{R}\backslash\mathbb{Q}$ is topologically complete by (i). The space of rationals $\mathbb{Q}$ is not topologically complete, it is not even a Baire space.

Every Polish space X is homeomorphic to a closed subset of $\mathbb{R}^{\mathbb{N}}$. To see this, first notice that X is homeomorphic to a $\mathscr{G}_\delta$-set in $[0, 1]^{\mathbb{N}}$, say $X = \bigcap G_n$ (neglecting the homeomorphism). For each $n \in \mathbb{N}$ define a map $f_n: X \to \mathbb{R}$ as follows: f_{2n+1} is taken as the nth projection and f_{2n} is taken as the map $x \to (d(x, X\backslash G_n))^{-1}$. The map $(f_n): X \to \mathbb{R}^{\mathbb{N}}$ is a homeomorphism of X onto the closed subset $(f_n)(X)$.

As a corollary, we see that for every Polish space X there exists a family $(\pi_n)_{n \geq 1}$ of real-valued maps on X such that a subset $K \subseteq X$ is relatively compact if, and only if, $\pi_n(K)$ is bounded for every $n \geq 1$. (Consider X as a closed subset of $\mathbb{R}^{\mathbb{N}}$ and let π_n be the nth coordinate map $X \to \mathbb{R}$.) Conversely, every metrizable space admitting a family $(\pi_n)_{n \geq 1}$ of continuous real-valued functions with the property above is Polish (Kerstan, 1972).

2 Automatic continuity

2.1 Introduction

In 1821 Cauchy stated the problem of finding all functions $f: \mathbb{R} \to \mathbb{R}$ satisfying

$$f(x+y) = f(x) + f(y), \qquad \forall x, y \in \mathbb{R} \tag{22}$$

(*the Cauchy equation*). Cauchy showed that the only continuous solutions to (22) are of the form

$$f(x) = cx, \tag{23}$$

where $c \in \mathbb{R}$.

For a long time it was an open question whether (23) exhausted all solutions to the Cauchy equation. And it was the attempts to solve this, and similar problems, that gave birth to the investigations into the automatic continuity theorems, by which we shall mean a result stating that every not too pathological function satisfying certain functional equations is continuous.

To return to the history of equation (22), we can mention that Darboux (1880) showed that solutions to (22) which do not change sign in a small interval to the right of 0 are of the form (23). Hamel (1905) showed that there exist non-continuous solutions to (22). Sierpiński (1920*b*) showed that measurable solutions to (22) are given by (23), and this was extended by Ostrowski (1929) to functions which are bounded on some set of positive Lebesgue measure.

The results of Sierpiński and Ostrowski will be typical for many of the results treated in Section 2. And many of the theorems to come will contain Sierpiński's and/or Ostrowski's result as a special case. As in Section 1 the bibliography will be very sketchy and incomplete. The references will only be given by name and year.

The results fall in three groups. In the first group we assume that the function restricted to certain sets is continuous (separate continuity or has the Baire property). In the second group f is a homomorphism and f is measurable with respect to certain σ-algebras. And in the third group f is defined on a subset of a locally compact abelian group and f satisfies a certain integral equation. To guide the reader we have sketched the results in Table 1. The hypothesis in Table 1 is only vaguely indicated, so the reader is referred to the statement in the text for the precise formulation.

In the table we use BP as an abbreviation for the "Baire property", and C, respectively C_s, denotes the set of continuity points, respectively sequential continuity points, of f.

Table 1

Dom f	Range f	Hypothesis	Conclusion	See theorem no.
$X \times Y$, Y metric	Metric	Separate continuity	C is co-meagre	2.2.4
Hausdorff	Metric	First Baire class	C is co-meagre	2.2.5
Hausdorff	Hausdorff	BP, commutes with an action	$C_s = \text{Dom } f$	2.2.12
Baire group	Group	BP, homomorphism	$C_s = \text{Dom } f$	2.3.1
Baire group	Baire group	Group multiplication	$C = \text{Dom } f$	2.3.2–2.3.4
$2^{\mathbb{N}}$	Group	BP, finitely additive	$C = \text{Dom } f$	2.3.13
σ-algebra	Group	Finitely additive, "measurable"	Countably additive	2.3.15
Semigroup	Semigroup	"Thick" homomorphism	$1 \in C$	2.4.1
Semigroup	Semigroup	"Bounded" homomorphism	$1 \in C$	2.4.9
Semigroup	Semigroup	"Measurable" homomorphism	$1 \in C$	2.4.7–2.4.8
$\subseteq$ group	Banach space	$f(x+y) = \int_S g(s,x)h(s,y)\,\mu(ds)$	$C_s = \text{Dom } f$	2.5.2

Let us finally remark that there are large areas of automatic continuity theorems which we do not even touch; in particular, there exists a highly developed theory for algebra homomorphisms and derivations in Banach algebras. We refer the reader to a recent survey article by H. G. Dales (1978) for this subject. However, the results presented here are of such a general nature that they also apply to certain questions in Banach algebras.

2.2 The Baire property and automatic continuity

The Baire property and the concepts attached to the Baire property (e.g. nowhere dense, meagre and non-meagre) plays an important role in our first two versions of the automatic continuity theorem. The idea, which goes back to Banach and Kuratowski, is that a function on a topological space X is either wildly pathological or has some sort of continuity properties on fairly "large" subsets of X.

We shall now remind the reader about some of the basic notions around the Baire property (Kuratowski (1966) serves as an excellent reference for the subject). A condensed account of some of the results is given in Part 1 of the present volume, see Section 2.9. Let X be a topological space (*all topological spaces will be assumed to be Hausdorff*), and let A be a subset of X, then we say A is:

nowhere dense, if $\operatorname{int}(\bar{A}) = \emptyset$;
meagre (*first category*), if A is a countable union of nowhere dense sets;
non-meagre (*second category*), if A is not meagre;
co-meagre, if $X \backslash A$ is meagre.

If X and Y are topological spaces, then a function $f: X \to Y$ is said to have the *Baire property* if there exists a co-meagre set A in X such that $f|A$ is continuous. A set $A \subseteq X$ is said to have the *Baire property* if its indicator function $1_A: X \to \{0, 1\}$ has the Baire property.

All the notions above are relative to the space X. And one can easily give examples of sets A which are meagre in X and contained in X_0, but which do not even have the Baire property with respect to X_0. To see this, let us consider the Cantor set $\mathbb{C} \subseteq [0, 1] = I$, and let A be a subset of $\mathbb{C}$ which does not have the Baire property relative to $\mathbb{C}$ (which is possible by the axiom of choice). Then A is nowhere dense in I. The following proposition clarifies this point.

PROPOSITION 2.2.1 *Let X and Y be Hausdorff spaces and φ a continuous map*: $X \to Y$.

If φ is a homeomorphism of X onto $\varphi(X)$, then we have

(a) *A meagre in $X \Rightarrow \varphi(A)$ meagre in Y;*
(b) *$B \subseteq \varphi(X)$, B non-meagre in $Y \Rightarrow \varphi^{-1}(B)$ non-meagre in X.*

If φ is open relative to $\varphi(X)$ and $\varphi(X)$ is dense in Y, then we have

(c) *A non-meagre in $X \Rightarrow \varphi(A)$ non-meagre in Y;*
(d) *B meagre in $Y \Rightarrow \varphi^{-1}(B)$ meagre in X;*
(e) *B co-meagre in $Y \Rightarrow \varphi^{-1}(B)$ co-meagre in X.*

The proof of this is straightforward. Note that we have

$$\left.\begin{array}{l}\text{if } g \text{ is continuous and } f \text{ has the Baire property}\\ \text{then } g \circ f \text{ has the Baire property}\end{array}\right\} \tag{24}$$

(assuming that $g \circ f$ is defined). However, the example above shows that $f \circ h$ may fail to have the Baire property, even when h is a homeomorphism onto its image and f has the Baire property. But the proposition shows that we have

$$\left.\begin{array}{l}\text{if } h \text{ is open continuous and surjective, and } f \text{ has}\\ \text{the Baire property, then so has } f \circ h.\end{array}\right\} \tag{25}$$

Now let us list some of the most immediate consequences of the definitions.

$$\left.\begin{array}{l}\text{Every subset of a meagre set and every countable}\\ \text{union of meagre sets is itself meagre.}\end{array}\right\} \tag{26}$$

$$A \text{ is co-meagre} \Leftrightarrow A \supseteq \bigcap_{1}^{\infty} U_n \text{ with } U_n \text{ open and dense.} \tag{27}$$

$$\left.\begin{array}{l}X \text{ is non-meagre if, and only if, every co-meagre set}\\ \text{in } X \text{ is non-empty.}\end{array}\right\} \tag{28}$$

The property (28) gives birth to the definition of a Baire space. A topological space X is said to be a *Baire space* if every co-meagre subset is dense in X, or equivalently (see (27)), if every countable intersection of open dense sets is itself dense in X. Clearly we have

$$X \text{ a Baire space} \Rightarrow X \text{ is non-meagre in itself.} \tag{29}$$

The converse of this even fails for subsets of $\mathbb{R}$.

We shall in a moment give sufficient conditions for the converse implication, but first we must look at the local structure of meagre and non-meagre sets.

Let X be a topological space and A a subset of X, then we define

$$D(A)=\{x: U\cap A \text{ is non-meagre}, \forall U \text{ a neighbourhood of } x\},$$
$$M(A)=\{x: \exists U \text{ a neighbourhood of } x, \text{ so that } U\cap A \text{ is meagre}\};$$

that is $M(A)=D(A)^c$, and $D(A)$ is the locally non-meagre part of A and $M(A)$ the locally meagre part of A. The main fact about $D(A)$ and $M(A)$ is given by the following proposition:

$$A\backslash \text{int}\,(D(A))=A\cap\overline{M(A)} \quad \text{is meagre in } X, \qquad \forall A\subseteq X. \tag{30}$$

The statement is non-trivial and we refer the reader to Kuratowski (1966, p. 84), or see Part 1 of the present volume, Corollary 2.9.5. From (30) we deduce immediately that we have

$$A \text{ meagre} \Leftrightarrow D(A)=\varnothing \Leftrightarrow A\backslash\overline{M(A)} \text{ is meagre.} \tag{31}$$

In particular, we have that X is non-meagre in itself if, and only if, $D(X)\neq\varnothing$. For Baire spaces we have a similar characterization:

PROPOSITION 2.2.2 *Let X be a Hausdorff space, then the following three statements are equivalent*:

(a) *X is a Baire space*;
(*b*) *every open non-empty set in X is non-meagre*;
(c) $D(X)=X$.

The equivalence of (b) and (c) is evident from the definition of $D(X)$. The equivalence of (a) and (b) is a straightforward consequence of the definition of a Baire space.

We are now ready for the converse of (29):

PROPOSITION 2.2.3 *Let X be a Hausdorff space such that the set of all open, continuous surjections*: $X\to X$ *is almost transitive, that is*

(a) $\forall U$, *V open and non-empty*, $\exists\varphi$ *a continuous, open surjection*: $X\to X$, *such that* $\varphi(V)\cap U\neq\varnothing$.

If X is non-meagre in itself, then X is a Baire space.

Proof Let $V=\text{int}\,D(X)$, then V is open and non-empty by (30), hence if $U\neq\varnothing$ and open, we have that $\varphi(V)\cap U\neq\varnothing$ for some continuous open surjective φ, so there exists an $x_0\in V$ with $\varphi(x_0)\in U$. Now $\varphi^{-1}(U)$ is a neighbourhood of $x_0\in D(X)$, so $\varphi^{-1}(U)$ is non-meagre. Hence, by Proposition 2.2.1(c), we have that $U=\varphi(\varphi^{-1}(U))$ is non-meagre, and so X is a Baire space by Proposition 2.2.2.

Note that a set Φ of functions: $X \to X$ is almost transitive in the sense of Proposition 2.2.3 if, for instance, the trajectory $\Phi(x) = \{\varphi(x): \varphi \in \Phi\}$ is dense in X for all x in a dense subset of X (e.g. if Φ is transitive).

We shall now turn to our first result on "automatic" continuity. The result states that a separately continuous function $f: X \times Y \to Z$ is continuous "almost everywhere". The idea is due to Kuratowski (1923), but we use it here in a different setting.

THEOREM 2.2.4 *Let $f: X \times Y \to Z$ be a map, where X is Hausdorff and Y and Z are metrizable. Consider the sets of partial continuity*

$$C_1 = \{(x, y): f(\cdot, y) \text{ is continuous at } x\},$$

$$C_2 = \{(x, y): f(x, \cdot) \text{ is continuous at } y\}.$$

Suppose that we have

(a) $\forall (x, y) \in X \times Y, \exists \{y_j\} \subseteq C_1(x)$ *with* $y_j \to y$ *and* $f(x, y_j) \to f(x, y)$,

where $C_0(x) = \{y \in Y : (x, y) \in C_1\}$. Then there exists a co-meagre set in A in $X \times Y$, so that all its sections:

$$A(y) = \{x \in X: (x, y) \in A\}, \qquad y \in Y,$$

are co-meagre in X, and with f continuous at every point $(x, y) \in A \cap C_2$.

Proof Let d be a metric for Y and ρ a metric for Z, and let $S(y, r)$ denote the open d-ball in Y with centre y and radius r. Then we put

$$F_{nk} = \{(x, y): \rho(f(x, w), f(x, v)) \leq 2^{-n}, \qquad \forall w, v \in S(y, 2^{-k})\}.$$

Then clearly we have

$$C_2 = \bigcap_{n=1}^{\infty} \bigcup_{k=1}^{\infty} F_{nk}. \tag{32}$$

Now suppose that $(x_\alpha, y_\alpha) \in F_{nk}$ and $(x_\alpha, y_\alpha) \to (x, y)$. If $v, w \in S(y, 2^{-k}) \cap C_1(x)$, then for some α_0 we have

$$d(y_\alpha, v) < 2^{-k} \quad \text{and} \quad d(y_\alpha, w) < 2^{-k}, \qquad \forall \alpha \geq \alpha_0.$$

And since $(x_\alpha, y_\alpha) \in F_{nk}$ we find

$$\rho(f(x_\alpha, v), f(x_\alpha, w)) \leq 2^{-n}, \qquad \forall \alpha \geq \alpha_0.$$

The functions $f(\cdot, v)$ and $f(\cdot, w)$ are continuous at x, since $v, w \in C_1(x)$, so passing to the limit we find

$$\rho(f(x, v), f(x, w)) \leq 2^{-n}, \qquad \forall v, w \in S(y, 2^{-k}) \cap C_1(x).$$

Now let $v, w \in S(y, 2^{-k})$, then by (a) we can find sequences $\{v_j\}$ and $\{w_j\}$ in $S(y, 2^{-k}) \cap C_1(x)$, so that

$$\lim_{j\to\infty} f(x, v_j) = f(x, v), \qquad \lim_{j\to\infty} f(x, w_j) = f(x, w).$$

Hence we find

$$\rho(f(x, v), f(x, w)) \leqslant 2^{-n}, \qquad \forall v, w \in S(y, 2^{-k}).$$

That is $(x, y) \in F_{nk}$ and so we have proved

$$F_{nk} \text{ is closed}, \qquad \forall n, k \geqslant 1. \tag{33}$$

We can now turn to the definition of A. If $y \in Y$ and n, k are integers we define

$$F_{nk}(y) = \{x : (x, y) \in F_{nk}\},$$
$$D_{nk} = \{(x, y) : x \in F_{nk}(y) \backslash \operatorname{int}(F_{nk}(y))\},$$
$$D = \bigcup_{n=1}^{\infty} \bigcup_{k=1}^{\infty} D_{nk}.$$

It is then easy to verify that we have

$$D_{nk}(y) = \{x \in X : (x, y) \in D_{nk}\} = F_{nk}(y) \backslash \operatorname{int} F_{nk}(y);$$
$$D_{nk} \subseteq F_{nk} \backslash \operatorname{int} F_{nk}.$$

Hence $A = (X \times Y) \backslash D$ is co-meagre in $X \times Y$ and $A(y)$ is co-meagre in X for all $y \in Y$, since F_{nk} is closed.

We shall now show that f is continuous at every point in $A \cap C_2$. So let $(x, y) \in A \cap C_2$, and let $\varepsilon > 0$ be given, then we choose $n \geqslant 1$ so that $2^{-n} \leqslant \varepsilon$. Now from (32) we know that there exist $k \geqslant 1$, so that $(x, y) \in F_{nk}$. That is

$$x \in F_{nk}(y) \backslash D(y) \subseteq F_{nk}(y) \backslash (F_{nk}(y) \backslash \operatorname{int} F_{nk}(y))$$
$$= \operatorname{int} F_{nk}(y).$$

Now by (a) there exist $v \in C_1(x) \cap S(y, 2^{-k})$, and a neighbourhood U of x, such that $U \subseteq F_{nk}(y)$ and

$$\rho(f(x, y), f(x, v)) \leqslant \tfrac{1}{2}\varepsilon,$$
$$\rho(f(x, v), f(s, v)) \leqslant \tfrac{1}{2}\varepsilon, \qquad \forall s \in U.$$

Now let $(s, t) \in U \times S(y, 2^{-k})$, then $(s, y) \in F_{nk}$ and v and t both belong to $S(y, 2^{-k})$; hence we have

$$\begin{aligned}\rho(f(x, y), f(s, t)) &\leq \rho(f(x, y), f(x, v)) + \rho(f(x, v), f(s, v)) \\ &\quad + \rho(f(s, v), f(s, t)) \\ &\leq \tfrac{1}{2}\varepsilon + \tfrac{1}{2}\varepsilon + 2^{-n} \leq 2\varepsilon.\end{aligned}$$

Hence f is continuous at (x, y), and the theorem is proved.

The condition (a) may seem a little obscure. It clearly implies that $C_1(x)$ is dense in Y for all x. What is said is that for each $x \in X$ and each $y \in Y$ we have that $f(x, y)$ is a limit point of $f(x, z)$ as $z \to y$ through $C_1(x)$. It is clearly satisfied in each of the following three cases:

(a)′ $C_1(x)$ is dense in Y and $Y = C_1(x) \cup C_2(x)$, $\forall x \in X$.
(a)″ $f(\cdot, y)$ is continuous, $\forall y \in Y$, i.e. $C_1 = X \times Y$.
(a)‴ $f(x, \cdot)$ is continuous and $C_1(x)$ is dense in Y, $\forall x \in X$.

Theorem 2.2.4 is normally stated for functions f which are *separately continuous*, i.e. $f(\cdot, y)$ and $f(x, \cdot)$ are continuous for all $y \in Y$ and all $x \in X$, or in our terminology $C_1 = C_2 = X \times Y$.

Let us give an application showing that our formulation is useful. Let X be a Hausdorff space and Z a metric space and $\{f_n\}$ a sequence of continuous functions: $X \to Z$, such that

$$f(x) = \lim_{n\to\infty} f_n(x)$$

exists for all $n \geq 1$. Let

$$Y = \{0, 1, \tfrac{1}{2}, \tfrac{1}{3}, \ldots, 1/n, \ldots\}$$

with its usual topology and put

$$f(x, y) = \begin{cases} f_n(x), & \text{if } y = 1/n, \\ f(x), & \text{if } y = 0. \end{cases}$$

Then we have

$$C_1 \supseteq X \times (Y \backslash \{0\}), \qquad C_2 = X \times Y.$$

So condition (a)‴ holds and hence (a) is satisfied. In particular, we conclude that $f(x) = f(x, 0)$ is continuous at every x in the co-meagre set $A(0) \subseteq X$. Thus we have proved (Baire, 1909; Kuratowski, 1923)

THEOREM 2.2.5 *Let X be a Hausdorff space and Z metrizable. If $\{f_n\}$ is a sequence of functions, $f_n: X \to Z$, such that*

$$f(x) = \lim_{n \to \infty} f_n(x)$$

exists for all $x \in X$, then we have

(a) *if f_n is continuous for all $n \geq 1$, then the set of continuity points of f is co-meagre in X;*
(b) *if f_n has the Baire property for all $n \geq 1$, then so has f.*

Let us remark that Namioka (1974) has shown that, if X is Čech complete, Y is locally compact and σ-compact and Z is metrizable, then every separately continuous function $f: X \times Y \to Z$ is continuous at every $(x, y) \in A \times Y$, for some co-meagre set A in X.

Noting that 1_F is continuous at all points in $X \backslash \mathrm{bd}(F)$, we find that 1_F has the Baire property if F is closed. So by Theorem 2.2.5 we find the following result due to Lebesgue (1905):

THEOREM 2.2.6 *$\mathscr{B}_p(X)$ is a σ-algebra on X containing the Borel sets $\mathscr{B}(X)$.*

Here we have introduced the pavings

$$\mathscr{B}(X) = \text{the Borel subsets of } X;$$

$$\mathscr{B}_p(X) = \text{the subsets of } X \text{ with the Baire property.}$$

We shall need a few more pavings in the sequel, so let us use the opportunity to introduce them at this moment:

$$\mathscr{S}(X) = \text{the smallest Souslin closed } \sigma\text{-algebra containing } \mathscr{B}(X).$$

$$\mathscr{B}_u(X) = \text{the universally measurable subsets of } X.$$

$$\mathscr{A}(X) = \text{the analytic subsets of } X.$$

$$\mathscr{A}_K(X) = \text{the } K\text{-analytic subsets of } X.$$

Here "*universally measurable*" means measurable with respect to every finite (or equivalently every σ-finite) measure on the Borel sets.

Let us continue our study of the Baire property, and first let us give a sample of equivalent statements.

PROPOSITION 2.2.7 *Let X be a Hausdorff space, and let A be a subset of X. Then the following statements are equivalent:*

(a) *A has the Baire property (i.e.* $A \in \mathscr{B}_p(X)$);
(b) $\exists U$ *open so that* $A \triangle U$ *is meagre*;
(c) $\exists F$ *closed so that* $A \triangle F$ *is meagre*;
(d) $\exists G \in \mathscr{G}_\delta$, $\exists P$ *meagre so that* $A = G \cup P$;
(e) $\exists H \in \mathscr{F}_\sigma$, $\exists P$ *meagre so that* $A = H \backslash P$;
(f) $D(A) \backslash A$ *is meagre.*

The proof is fairly straightforward and may be found in several elementary topology books, e.g. Kuratowski (1966).

Note that for any set A we have that $A \backslash D(A)$ is meagre, so (f) in the proposition above shows that the closed set in (c) may be taken to be $D(A)$. Now since $A \backslash D(A)$ is meagre, it is contained in some meagre $\mathscr{F}_\sigma$-set H_0. Hence

$$A \subseteq H = D(A) \cup H_0,$$

and H is an $\mathscr{F}_\sigma$-set. It is now easily checked that H satisfies the following (Szpilrajn-Marczewski, 1929; 1933):

PROPOSITION 2.2.8 *Let* A *be a subset of a Hausdorff space* X, *then there exists an* $\mathscr{F}_\sigma$-*set* H, *so that* $A \subseteq H$, *and*

$$A \subseteq B \in \mathscr{B}_p(X) \Rightarrow H \backslash B \text{ is meagre.}$$

The set H in the proposition above is, apart from a meagre set, the smallest Baire property-set containing A. Now the "abstract Souslin closure theorem", due to Szpilrajn-Marczewski (see Part 1 in the present volume, Theorem 2.9.1), then shows that we have the following theorem:

THEOREM 2.2.9 *Let* X *be a Hausdorff space, then the* σ-*algebras* $\mathscr{B}_u(X)$ *and* $\mathscr{B}_p(X)$ *are Souslin closed and both contain* $\mathscr{S}(X)$.

The theorem also shows that we have the following set of inclusions:

$$\mathscr{B}(X) \subseteq \mathscr{S}(X) \subseteq \mathscr{B}_u(X) \cap \mathscr{B}_p(X), \tag{34}$$

$$\mathscr{A}(X) \subseteq \mathscr{A}_K(X) \subseteq \mathscr{S}(X). \tag{35}$$

We have already noted that the composition of two functions with the Baire property may fail to have the Baire property; however, the following result is helpful in this respect.

THEOREM 2.2.10 *Let* $f: X \to Y$ *be a map. If* f *has the Baire property, then* $f: (X, \mathscr{B}_p(X)) \to (Y, \mathscr{B}(Y))$ *is measurable.*

Conversely, if Y has a countable base for the topology (e.g. if Y is separable and metrizable), then every measurable function $f: (X, \mathscr{B}_p(X)) \to (Y, \mathscr{B}(Y))$ has the Baire property.

Proof The first part is trivial. For the second we choose open sets $V_n \subseteq X$ such that $P_n = V_n \triangle f^{-1}(U_n)$ is meagre for all $n \geq 1$, where $\{U_n: n \geq 1\}$ is the countable base for the topology in Y. Let

$$A = X \Big\backslash \bigcup_1^\infty P_n$$

then

$$A \cap f^{-1}(U_n) = (A \cap V_n) \triangle (A \cap P_n) = A \cap V_n,$$

since $f^{-1}(U_n) = V_n \triangle P_n$ and $A \cap P_n = \varnothing$. Hence $f|A$ is continuous and A is co-meagre.

Suppose that $f: X \to Y$ is a function, then since f^{-1} commutes with countable unions, complementation and the Souslin operation, we have

$$\left.\begin{array}{l}\text{if } f: (X, \mathscr{B}_p(X)) \to (Y, \mathscr{B}(Y)) \text{ is measurable, then } f \text{ is} \\ \text{measurable: } (X, \mathscr{B}_p(X)) \to (Y, \mathscr{S}(Y)).\end{array}\right\} \quad (36)$$

A $(\mathscr{B}_p(X), \mathscr{B}(Y))$-measurable map will simply be called *$\mathscr{B}_p$-measurable.* Similarly a $(\mathscr{S}(X), \mathscr{B}(Y))$-measurable map will be called *$\mathscr{S}$-measurable*, and as above we have

$$\text{if } f \text{ is } \mathscr{S}\text{-measurable, then } f: (X, \mathscr{S}(X)) \to (Y, \mathscr{S}(Y)) \text{ is measurable.} \quad (37)$$

A $(\mathscr{B}_u(X), \mathscr{B}(Y))$-measurable map will be called *universally measurable.* Now since $\mu \circ f^{-1}$ defines a finite Borel measure on Y, whenever f is μ-measurable and μ is a finite Borel measure on X, we have

$$\left.\begin{array}{l}\text{if } f \text{ is universally measurable, then } f \text{ is measurable:} \\ (X, \mathscr{B}_u(X)) \to (Y, \mathscr{B}_u(Y)).\end{array}\right\} \quad (38)$$

Hence we find the following rules for measurability of composite functions:

$$\left.\begin{array}{l}\text{if } f \text{ is } \mathscr{B}_p\text{-measurable, and } g \text{ is } \mathscr{S}\text{-measurable, then} \\ g \circ f \text{ is } \mathscr{B}_p\text{-measurable;}\end{array}\right\} \quad (39)$$

$$\text{if } f \text{ and } g \text{ are } \mathscr{S}\text{-measurable, then so is } g \circ f; \quad (40)$$

$$\text{if } f \text{ and } g \text{ are universally measurable, then so is } g \circ f. \quad (41)$$

Let us in passing note the following "measurable graph theorem", which is an immediate consequence of the separation theorem for K-analytic sets:

$$\left.\begin{array}{l}\text{if } f \text{ has a } K\text{-analytic graph, then } f \text{ is Borel measurable and}\\ f(A) \text{ is } K\text{-analytic for all } A \in \mathscr{A}_K(X).\end{array}\right\} \tag{42}$$

The proposition is intimately connected to "the measurable selection theorem" which follows.

THEOREM 2.2.11 *Let A be an analytic subset of $X \times Y$, such that $X = p(A)$ where $p: X \times Y \to X$ is the projection. Then there exists an $\mathscr{S}$-measurable map $f: X \to Y$, so that*

$$\text{Graph}(f) = \{(x, y): y = f(x)\} \subseteq A.$$

Proof Let φ be a continuous surjection: $\mathbb{N}^{\mathbb{N}} \to A$, and let $\mathbb{N}^{\mathbb{N}}$ be ordered lexicographic. Then every closed non-empty set in $\mathbb{N}^{\mathbb{N}}$ admits a minimum (belonging to the given set!). Let $\varphi(\sigma) = (\varphi_1(\sigma), \varphi_2(\sigma))$, then we may define

$$f(x) = \varphi_2(\min \overset{-1}{\varphi_1}(x)): X \to Y,$$

and the reader will find no difficulties in checking that f has the required properties.

Now suppose that Y is an analytic space, then $Y = \varphi(\mathbb{N}^{\mathbb{N}})$ for some continuous map $\varphi: \mathbb{N}^{\mathbb{N}} \to Y$. Hence, considering the reflected graph:

$$A = \{(y, \sigma): y = \varphi(\sigma)\}.$$

Theorem 2.2.11 shows that φ has an $\mathscr{S}$-measurable right inverse ψ (i.e. $\varphi(\psi(y)) = y \ \forall y \in Y$). Hence, from (39), we deduce the following extension of Theorem 2.2.10:

$$\left.\begin{array}{l}\text{if } f: X \to Y \text{ is } \mathscr{B}_p\text{-measurable and } Y \text{ is analytic,}\\ \text{then } f \text{ has the Baire property.}\end{array}\right\} \tag{43}$$

This could of course be generalized a little further, namely to $Y = \varphi(Z)$, where Z has a countable base, φ is continuous and φ has an $\mathscr{S}$-measurable inverse.

We have already given one kind of automatic continuity theorem (Theorem 2.2.4), but there we already assumed some continuity properties (partial continuity). We shall now see how we can get continuity just from the Baire property and certain functional equations. The proof is essentially due to Banach (1932) and it is remarkable in its simplicity and generality. Recall that, if T and X are sets, we say that *T acts on X*, if to each $t \in T$ is assigned a map denoted: $x \mapsto tx$ from X into X. Then we have

THEOREM 2.2.12 *Let X, Y and T be Hausdorff spaces such that T acts on both X and Y. Suppose that the action satisfies*

(a) $t \mapsto tx$ *is open, continuous and surjective*: $T \to X$, $\forall x \in X$.
(b) $x \mapsto tx$ *is sequentially continuous*: $X \to X$, $\forall t \in T$.
(c) $y \mapsto ty$ *is injective*: $Y \to Y$ *and open relatively to* tY *for all* $t \in T$.

If T is non-meagre in itself and $f: X \to Y$ *is a function with the Baire property, satisfying*

(d) $f(tx) = tf(x)$, $\forall t \in T$, $\forall x \in X$.

Then f is sequentially continuous (i.e.

$$\lim_{n\to\infty} f(x_n) = f(x_0) \quad \textit{whenever} \quad x_0 = \lim_{n\to\infty} x_n).$$

Proof Let A be a co-meagre set in X, so that $f|A$ is continuous, and suppose that $x_n \to x_0$. Then by (a) it follows that (cf. Proposition 2.2.1(e))

$$B = \bigcap_{n=0}^{\infty} \{t \in T: tx_n \in A\}$$

is co-meagre in T. Hence $B \neq \varnothing$ since T is non-meagre, so let $t \in B$. Then $tx_n \in A$ for all $n \geqslant 0$, and by (b) we have $tx_n \to tx_0$. Hence from (d) we find that

$$f(tx_n) = tf(x_n) \xrightarrow[n\to\infty]{} f(tx_0) = tf(x_0).$$

Hence (c) shows that

$$f(x_n) \xrightarrow[n\to\infty]{} f(x_0).$$

Sequential continuity implies ordinary continuity in the spaces that are called s-spaces. If X is a Hausdorff space and F is a subset of X, then F is said to be *sequentially closed* if for every convergent sequence $\{x_n\} \subseteq F$ we have $\lim_{n\to\infty} x_n \in F$. Clearly, every closed set is sequentially closed; conversely, if every sequentially closed set is closed, then X is called an *s-space*. Evidently we have

every Hausdorff space satisfying the first-countability axiom (e.g. every metric space) is an s-space. (44)

The reader easily verifies that

if X is an s-space and $f: X \to Y$ is sequentially continuous, then f is continuous. (45)

Note that in an arbitrary Hausdorff space, (X, τ), the *sequentially open* sets (i.e. complements of sequentially closed sets) form a topology, which we denote τ_s, and call the *s-extension*. It is straightforward to check that τ_s is an s-topology, which is stronger than τ, and that convergence of sequences coincides for τ and τ_s. Moreover, we have

$$\left.\begin{array}{l} f: X \to Y \text{ is sequentially continuous if, and only if, } f \text{ is} \\ \text{continuous in the } s\text{-extension of the topology on } X. \end{array}\right\} \quad (46)$$

A Hausdorff space X is called an s^*-*space* if any real-valued sequentially continuous function is continuous. Clearly every s-space is an s^*-space, but the converse is not true in general.

We can weaken the s-condition in several other ways. A Hausdorff space X is called a *k-space* if every set $F \subseteq X$ is closed if, and only if, $F \cap K$ is compact for all compact sets $K \subseteq X$. And X is called a k^*-*space* if a function $f: X \to \mathbb{R}$ is continuous if, and only if, the restriction of f to any compact subset of X is continuous. Clearly we have

$$\left.\begin{array}{l} \text{every } s^*\text{-space is a } k^*\text{-space, every } s\text{-space is a } k\text{-space and} \\ \text{every } k\text{-space is a } k^*\text{-space.} \end{array}\right\} \quad (47)$$

2.3 Topological groups and semigroups

The automatic continuity theorems have found many of their applications in the theory of topological groups and various kinds of topological semigroups, typically the multiplicative semigroup associated to a Banach algebra. To give the flavour of this sort of application we start with a simple corollary to Theorem 2.2.12.

THEOREM 2.3.1 *Let G and H be groups endowed with some Hausdorff topologies, satisfying the conditions that G is non-meagre,*

(a) *$x \mapsto ax$ and $x \mapsto xa$ are continuous: $G \to G$, $\forall a \in G$, and*
(b) *$y \mapsto by$ is continuous: $H \to H$, $\forall b \in H$.*

Then every homomorphism $f: G \to H$ having the Baire property is sequentially continuous.

Proof G acts on itself by left multiplication, and let G act on H by $gh = f(g)h$. Then putting $T = G$, $X = G$ and $Y = H$ in Theorem 2.2.12, we find that the maps

$$t \mapsto tg_0, \qquad g \mapsto t_0 g, \qquad h \mapsto t_0 h$$

are all surjective homeomorphisms, and f satisfies

$$f(tx) = f(t)f(x) = tf(x),$$

so the theorem follows from Theorem 2.2.12.

We shall now briefly describe part of the theory of topological groups. We shall assume that the reader is familiar with the definition and elementary theory of topological groups. Recall that *a topological group* is a group $(G, \cdot)$ endowed with a Hausdorff topology satisfying

$$(x, y) \mapsto xy \text{ and } x \mapsto x^{-1} \text{ are continuous;} \tag{48}$$

or equivalently

$$(x, y) \mapsto xy^{-1} \text{ is continuous: } G \times G \to G. \tag{49}$$

Often it is much easier to verify separate continuity of the multiplication map than the simultaneous continuity. In such cases the following automatic continuity theorem due to R. Ellis (1957) may be helpful.

THEOREM 2.3.2 *Let $(G, \cdot)$ be a group with a Hausdorff topology and suppose that G acts on the Hausdorff space X, such that*

(a) $1x = x, \forall x \in X$ *and* $(gh)x = g(hx), \forall g, h \in G, \forall x \in X$;
(b) $(g, x) \mapsto gx$ *is separately continuous*: $G \times X \to X$;
(c) $(g, h) \mapsto gh$ *is separately continuous*: $G \times G \to G$.

Then the action $(g, x) \mapsto gx$ is jointly continuous in either of the following two cases:

(d) *X is metrizable and G is non-meagre*; or
(e) *X and G are locally compact.*

Proof (d) Let $(g_0, x_0) \in G \times X$, then by Theorem 2.2.4 there exist $h_0 \in G$, so that m is continuous at (h_0, x_0), where $m(g, x) = gx$. Moreover, we have

$$m(g, x) = g_0 h_0^{-1} m(h_0 g_0^{-1} g, x).$$

Thus m is continuous at (g_0, x_0).

(e) This case is much more involved and consists of a long and cumbersome reduction to case (d). We shall not go into the details here but refer the reader to R. Ellis (1957).

If we put $G = X$ in the theorem above and let G act on itself by left translation, we find that separate continuity of the multiplication in a group G implies joint continuity of the multiplication, if either G is non-meagre and metrizable, or G is locally compact.

Now, suppose that G is a group with a locally compact topology making the multiplication separately continuous. Let $\hat{G} = G \cup \{\omega\}$ be the one-point compactification of G, and let G act on $\hat{G}$ by the left translation on G and $g\omega = \omega$ for all $g \in G$. Then (a)–(c) in Theorem 2.3.2 are satisfied, and so $(g, h) \mapsto gh$ is jointly continuous on $G \times \hat{G}$. Thus

$$F = \{(g, h) \in G \times \hat{G} : gh = 1\}$$

is closed in $G \times \hat{G}$, and since F is the graph of the map $g \mapsto g^{-1}$ considered as a map from G in $\hat{G}$ we have that this map is continuous. (A function taking values in a *compact space*, is continuous whenever it has a closed graph.) Hence we conclude (R. Ellis, 1957):

THEOREM 2.3.3 *Let G be a group with a locally compact topology making the multiplication map: $(g, h) \mapsto gh$ separately continuous. Then G is a topological group.*

Note that if G is a Hausdorff space with a jointly continuous group multiplication, then the graph of $g \mapsto g^{-1}$ is given by

$$F = \{(g, h) \in G \times G : gh = 1\}.$$

Thus the graph is closed, and $g \mapsto g^{-1}$ is a homomorphism $G \to G^*$, where $G^* = G$ has the group structure $\circ$ defined by $g \circ h = hg$. Thus by (42), Theorem 2.2.10 and Theorem 2.3.1 we have

THEOREM 2.3.4 *Let G be a group endowed with a Hausdorff topology satisfying*

(a) *$(g, h) \mapsto gh$ is separately continuous: $G \times G \to G$;*
(b) *G is non-meagre, metrizable and analytic.*

Then G is a topological group.

In the theorems above we have considered non-meagre groups. Now, since the translations form a transitive family of homeomorphisms it follows from Proposition 2.2.3 that we have

$$\text{any non-meagre topological group is a Baire space.} \tag{50}$$

And we shall now see that any analytic Baire group is Polish. But first we need some more information about metrizable groups.

THEOREM 2.3.5 *Let G be a topological group. Then G is metrizable if, and only if, G has a countable neighbourhood base at 1, and if so the topology on G is generated by a left invariant metric d, i.e. a metric satisfying*

$$d(x, y) = d(xz, yz), \qquad \forall x, y, z \in G.$$

Moreover, let d be a left invariant metric defining the topology on C, and consider the metric

$$\rho(x, y) = d(x, y) + d(x^{-1}, y^{-1}),$$

which clearly generates the topology on G. If $(\bar{G}, \bar{\rho})$ *denotes the* ρ*-completion of G, then the group multiplication extends uniquely to* $\bar{G}$*, in such a way that* $(\bar{G}, \bar{\rho})$ *becomes a topological group.*

In other words, any metrizable group can be embedded as a dense subgroup of a complete metric group.

Proof The first part is elementary and can be found in any book on the subject. The second part is more subtle and the proof presented here is taken from Wilansky (1970).

Let $p(x) = d(x, 1) = \frac{1}{2}\rho(x, 1)$, then $p: G \to \mathbb{R}$ is uniformly ρ-continuous and so p extends uniquely to $\bar{G}$ by continuity, and p and its extension $\bar{p}$ clearly satisfies

$$p(1) = 0, \qquad p(x) = p(x^{-1}), \qquad \forall x \in G; \tag{51}$$

$$p(xy) \leq p(x) + p(y), \qquad \forall x, y \in G; \tag{52}$$

$$\rho(x, y) = p(y^{-1}x) + p(yx^{-1}), \qquad \forall x, y \in G; \tag{53}$$

$$\bar{p}(\bar{x}) = \tfrac{1}{2}\bar{\rho}(\bar{x}, 1), \qquad \forall \bar{x} \in \bar{G}. \tag{54}$$

Now let $\bar{x}_0$ and $\bar{y}_0 \in \bar{G}$ and let $\varepsilon > 0$ be given. We can then find $x_0, y_0 \in G$, so that

$$\bar{\rho}(x_0, \bar{x}_0) < \varepsilon/10, \qquad \bar{\rho}(y_0, \bar{y}_0) < \varepsilon/10.$$

Since the maps: $x \mapsto p(x_0 x x_0^{-1})$ and $x \mapsto p(y_0^{-1} x y_0)$ are continuous and are equal to 0 for $x = 1$, we can find $\delta > 0$ so that

$$p(x_0 x x_0^{-1}) < \varepsilon/10, \qquad p(y_0^{-1} x y_0) < \varepsilon/10, \tag{55}$$

whenever $p(x) < 2\delta$; and we may of course assume that $\delta < \varepsilon/10$. Now let $B(\bar{x}_0, \delta)$ and $B(\bar{y}_0, \delta)$ denote the open $\bar{\rho}$-balls in $\bar{G}$, with radius δ and centre $\bar{x}_0$ (resp. $\bar{y}_0$).

If $x, u \in G \cap B(\bar{x}_0, \delta)$ and $y, v \in G \cap B(\bar{y}_0, \delta)$, then we have

$$\left.\begin{array}{lll} \rho(x, x_0) < \varepsilon/5, & \rho(u, x_0) < \varepsilon/5, & \rho(u, x) < 2\delta, \\ \rho(y, y_0) < \varepsilon/5, & \rho(v, y_0) < \varepsilon/5, & \rho(v, y) < 2\delta. \end{array}\right\} \tag{56}$$

In particular we see that $p(u^{-1}x) < 2\delta$ and $p(vy^{-1}) < 2\delta$. Hence, by (55), we have

$$p(y_0^{-1}(u^{-1}x)y_0) < \varepsilon/10,$$

$$p(x_0(vy^{-1})x_0^{-1}) < \varepsilon/10.$$

These inequalities together with (56) allow us to deduce that

$$\begin{aligned}\rho(xy, uv) &= p(v^{-1}u^{-1}xy) + p(uvy^{-1}x^{-1})\\ &\leqslant p(v^{-1}y_0) + p(y_0^{-1}u^{-1}xy_0) + p(y_0^{-1}y) + p(ux_0^{-1})\\ &\quad + p(x_0vy^{-1}x_0^{-1}) + p(x_0x^{-1})\\ &\leqslant \varepsilon/5 + 4\varepsilon/5 = \varepsilon.\end{aligned}$$

That is, we have shown

$$\left.\begin{array}{l}\forall \bar{x}_0, \bar{y}_0 \in \bar{G}, \quad \forall \varepsilon > 0, \quad \exists \delta > 0, \quad \text{so that} \quad \rho(xy, uv) < \varepsilon,\\ \quad \text{for all } x, u \in B(\bar{x}_0, \delta) \cap G \quad \text{and all } y, v \in B(\bar{y}_0, \delta) \cap G\end{array}\right\} \tag{57}$$

Hence the completeness of $(\bar{G}, \bar{\rho})$ implies that the multiplication map: $(x, y) \mapsto xy$ from $G \times G$ into $\bar{G}$ extends uniquely by continuity to $\bar{G} \times \bar{G}$. This extension clearly defines an associative multiplication denoted $\bar{x}\bar{y}$, having the unit element of G as a unit.

Now let $x, y \in G$, then we have

$$\rho(x^{-1}, y^{-1}) = p(yx^{-1}) + p(y^{-1}x) = \rho(x, y),$$

so $x \mapsto x^{-1}$ is a ρ-isometry $G \to G$. Hence this map extends uniquely by continuity to a $\bar{\rho}$-isometry $\bar{G} \to \bar{G}$, which we denote $\bar{x} \mapsto \bar{x}^{-1}$. Since $xx^{-1} = x^{-1}x = 1$, we find by continuity that

$$\bar{x}\bar{x}^{-1} = \bar{x}^{-1}\bar{x} = 1,$$

and so $\bar{G}$ is a group under the extended group multiplication. Since multiplication and the inverse are both continuous, we have shown that $\bar{G}$ is a topological group under the metric $\bar{\rho}$.

Remark If G has a two-sided invariant metric, then

$$\begin{aligned}d(xy, uv) &\leqslant d(xy, uy) + d(uy, uv)\\ &\leqslant d(x, u) + d(y, v),\end{aligned}$$

so multiplication is uniformly continuous on $G \times G$ and then trivially extends by continuity to $\bar{G}$.

However, there exist metrizable groups without a two-sided invariant metric. One such example (Dieudonné?) is the set of homeomorphisms of the unit interval $[0, 1]$ onto itself, with the sup-norm metric

$$d(f, g) = \sup_{x \in I} |f(x) - g(x)|$$

and composition as multiplication. Note that the d is right-invariant but certainly not left-invariant.

With this theorem at our disposal we can prove that analytic Baire groups are Polish (R. J. Loy, 1976*a*, *b*; J. P. R. Christensen, 1974).

THEOREM 2.3.6 *Let G be a topological Baire group, and suppose that G is a continuous image of a regular, K-analytic space P satisfying the first countability axiom (e.g. if G is analytic). Then G is Polish.*

Proof For the proof of this we will refer to a result, Pettis's theorem, which we shall prove later (Theorem 2.4.2).

Let φ be the continuous surjection $P \to G$, where P is K-analytic, regular and first-countable. Let us consider the set

$$P_0 = \{x \in P: \exists U \text{ neighbourhood of } x \text{ so that } \varphi(U) \text{ is meagre}\}.$$

Since a K-analytic space is Lindelöf, we see that, if $P = P_0$, then

$$P = \bigcup_{n=1}^{\infty} U_n$$

where $\varphi(U_n)$ is meagre. But this is impossible since

$$\varphi(P) = G = \bigcup_{n=1}^{\infty} \varphi(U_n),$$

which is non-meagre; that is $P_0 \neq P$, and so there exists $x_0 \in P$ with

$$\varphi(U) \text{ is non-meagre in } G, \forall U \text{ a neighbourhood of } x_0. \tag{58}$$

Since P is regular and first-countable, we can find a sequence of closed subsets $\{V_k\}$, which form a neighbourhood base at x_0. We shall now show that

$$\{\varphi(V_k)\varphi(V_k)^{-1}: k \geq 1\} \text{ is a neighbourhood base at } 1. \tag{59}$$

First we note that since V_k is closed, we have that $\varphi(V_k)$ is K-analytic, and so $\varphi(V_k)$ has the Baire property (cf. (34) and (35)). By (58) we know that $\varphi(V_k)$ is non-meagre so by Pettis's theorem (see Theorem 2.4.2), we have that $\varphi(V_k)\varphi(V_k)^{-1}$ is a neighbourhood of 1, for all $k \geq 1$.

Conversely, if U is a neighbourhood of 1 in G, then we can find a neighbourhood W of 1 in G, such that $WW^{-1} \subseteq U$. By continuity of φ there exists a $k \geq 1$ with $\varphi(V_k) \subseteq W\varphi(x_0)$, and so

$$\varphi(V_k)\varphi(V_k)^{-1} \subseteq W\varphi(x_0)\varphi(x_0)^{-1}W^{-1} = WW^{-1} \subseteq U,$$

which proves (59).

From Theorem 2.3.5, we then deduce that G is metrizable and, since G is Lindelöf, G is embedded as a dense subset of some Polish group $\bar{G}$. The K-analyticity of G implies that G has the Baire property in $\bar{G}$ and, by

Proposition 2.2.1(c), we have that G is non-meagre in $\bar{G}$. So applying Pettis's theorem to G (as a subset of $\bar{G}$), we have that $G = GG^{-1}$ contains a neighbourhood of 1 in $\bar{G}$. Now if $\bar{x} \in \bar{G} \backslash G$, then $(\bar{x}G) \cap G = \varnothing$, since G is a subgroup, hence G is closed and dense in $\bar{G}$. In other words $G = \bar{G}$, and the theorem is proved.

Theorem 2.3.6 shows that the Baire property is a much more severe restriction on a group than it seems at the first glance.

This impression is supported by our next theorem, which may be considered as a substitute for a zero–one law for Baire groups (Oxtoby, 1961):

THEOREM 2.3.7 *Let* $(G, \cdot)$ *be a topological Baire group, and let* T *be a countable dense subset of* G. *Then for every co-meagre set* $A \subseteq G$ *we have that the set*

$$\{x \in A: (xT) \cap A \textit{ is dense in } G\}$$

is co-meagre in G, *and so is non-empty.*

Moreover, if $f: G \to Y$ *has the Baire property and* f *is right* T*-invariant (i.e.* $f(x) = f(xt), \forall x \in G, \forall t \in T$), *then there exists a co-meagre set* A *in* G *such that* f *is constant on* A.

Proof Since all the sets At^{-1} are co-meagre (cf. Proposition 2.2.1(e)), we have that

$$B = A \cap \bigcap_{t \in T} At^{-1}$$

is co-meagre, since T is countable. Now let $x \in B$, and let V be an open non-empty set. Since T^{-1} is dense in G we have that $VT^{-1} = G$, and so $x = vt^{-1}$ for some $v \in V$ and some $t \in T$. Hence

$$v = xt \in (xT) \cap A \cap V,$$

since $xt \in A$ whenever $x \in B$ and $t \in T$; that is $(xT) \cap A$ is dense in G for $x \in B$, and so the first part of Theorem 2.3.5 is proved.

For the second part we consider a co-meagre set A so that $f|A$ is continuous and choose $x \in A$ so that $(xT) \cap A$ is dense in G. Since f is constant on xT and $f|A$ is continuous on A, we find that f is constant on A.

So far we have only considered groups, but we shall now turn to semigroups. A *semigroup* is simply a set G with an associative operation: $G \times G \to G$. We shall usually use multiplicative notation, i.e. xy, for the "product" of x and y, but if G is abelian we shall allow ourselves, whenever it is convenient, to use additive notation $x + y$ for the semigroup operation.

We shall always assume that our semigroup has a unit element 1 (i.e. $1x = x1 = x, \forall x \in G$).

In spite of the lack of inverses we can still define AB^{-1} and $A^{-1}B$ in a semigroup. So let A and B be subsets of a semigroup G. Then we define

$$AB^{-1} = \{x \in G: (xB) \cap A \neq \emptyset\},$$
$$A^{-1}B = \{x \in G: (Ax) \cap B \neq \emptyset\},$$
$$AB = \{x \in G: x = ab \text{ for some } a \in A, b \in B\},$$

which coincides with the usual definition in the case when G is a group.

If G and H are semigroups and $f: G \to H$ is a function, then f is called *multiplicative* (or *additive* if we use additive notation in both G and H) if f satisfies

$$f(xy) = f(x)f(y).$$

Note that if f is multiplicative, then $f(1)^2 = f(1)$. If f is multiplicative and $f(1) = 1$ we say that f is an *homomorphism.*

Let us note the following immediate consequences of the definitions above:

$$A \cap B \neq \emptyset \Leftrightarrow 1 \in AB^{-1} \Leftrightarrow 1 \in A^{-1}B; \tag{60}$$

$$1 \in B \Rightarrow A \subseteq AB^{-1} \quad \text{and} \quad A \subseteq B^{-1}A, \tag{61}$$

$$f(AB^{-1}) \subseteq f(A)f(B)^{-1} \quad \text{if } f \text{ is multiplicative;} \tag{62}$$

$$f(A^{-1}B) \subseteq f(A)^{-1}f(B) \quad \text{if } f \text{ is multiplicative.} \tag{63}$$

There does not seem to be any generally accepted definition of a topological semigroup, and, rather than give one here, we shall now list a number of possible axioms for a topologized semigroup, and refer to those axioms which are necessary.

AXIOM I *The map* $(x, y) \mapsto xy$ *is continuous.*

If we only require continuity at (1, 1) we shall speak about axiom I_0.

AXIOM II *The left translations* $\lambda_a(x) = ax$ *and the right translations* $\rho_a(x) = xa$ *are continuous,* $\forall a \in G$.

This axiom is the conjunction of a left version, denoted II^{l}, and a right version denoted II^{r}. And again the subscript 0 indicates that we only require continuity at 1.

Axiom III $(xV)V^{-1}$ *and* $V^{-1}(Vx)$ *are neighbourhoods of* x, *for all* $x \in G$ *and all neighbourhoods* V *of* 1.

With the same conventions as above, we shall use the notation $\mathrm{III}^{\mathrm{l}}$, $\mathrm{III}^{\mathrm{r}}$, III, etc. Note, however, that III_0 is trivially satisfied for any topologized semigroup. Our fourth and last axiom is the converse of III:

Axiom IV $\forall x \in G$, $\forall U \in \mathcal{N}(x)$, $\exists V \in \mathcal{N}(1)$: *so that* $(xV)V^{-1} \subseteq U$ *and* $V^{-1}(Vx) \subseteq U$.

Here $\mathcal{N}(x)$ denotes the set of neighbourhoods of x. Again, we may ramify this axiom into IV^{r}, $\mathrm{IV}_0^{\mathrm{l}}$, IV_0, etc.

Just to see how these axioms work, we shall leave the following as an exercise to the reader.

$$\left.\begin{array}{l}\text{Let } f\colon G \to H \text{ be multiplicative, and suppose that } G \text{ satisfies } \mathrm{III}^{\mathrm{l}}\\ \text{and } H \text{ satisfies } \mathrm{IV}^{\mathrm{l}}. \text{ Then } f \text{ is everywhere continuous if}\\ f \text{ is continuous at } 1.\end{array}\right\} \qquad (64)$$

A topological group G of course satisfies all four axioms, and actually, for a group G, axioms I and $\mathrm{IV}_0^{\mathrm{l}}$ imply that G is a topological group. To illustrate the axioms we shall consider some examples.

Example 2.3.1 $\mathbb{R}_+ = [0, \infty[$ with addition satisfies Axioms I–IV. Note that if $V = [0, \varepsilon[$ then

$$\begin{aligned}(x + V) - V &= \{y \geq 0\colon \exists v_1, v_2 \in V,\ y + v_1 = x + v_2\}\\ &= \mathbb{R}_+ \cap\,]x - \varepsilon, x + \varepsilon[.\end{aligned}$$

Note also that $x + V = [x, x + \varepsilon[$ is *not* a neighbourhood of x.

More generally, if G is a topological group, then every subsemigroup G_0 containing 1 satisfies Axioms I, II and IV. However, Axiom III need not be satisfied; this can be seen by considering $G = \mathbb{R}$, with addition and

$$G_0 = \{0\} \cup [1, \infty[.$$

Example 2.3.2 Let $(G, +)$ be an abelian locally compact group and let M denote the set of non-negative finite Radon measures. Then M with convolution is a semigroup, and we may consider the vague topology on M, i.e. $\mu_\alpha \to \mu$, if, and only if,

$$\int_G f\, \mathrm{d}\mu_\alpha \to \int_G f\, \mathrm{d}\mu, \qquad \forall f \in C_0(G),$$

where $C_0(G)$ is the set of continuous functions vanishing at ∞.

Then $(M, *)$ satisfies Axiom II, but if G is not compact, then $(M, *)$ does not satisfy Axiom I_0.

Example 2.3.3 Let E be a Banach space and let G be the set of linear bounded operators $T: E \to E$ with $\|T\| \leq 1$. Then G with composition is a semigroup. On G we may consider the operator norm $\|\cdot\|$ or we may consider the strong operator topology τ_s:

$$T_\alpha \to T \quad \text{in} \quad \tau_s \Leftrightarrow T_\alpha x \to Tx, \qquad \forall x \in E.$$

Then $(G, \|\cdot\|)$ and (G, τ_s) both satisfy Axioms I and II. $(G, \|\cdot\|)$ satisfies Axiom III_0, but not Axiom III, if $\dim E \geq 1$, and $(G, \|\cdot\|)$ satisfies Axiom IV. (G, τ_s) satisfies Axiom IV^1.

These examples justify the fact that there is no natural definition of a topological semigroup, and they show that for various applications we need various axioms.

Let G be a topologized semigroup and $\{x_j\}$ a sequence in G, then we put

$$p_n = \prod_{j=1}^{n} x_j = x_1 x_2 \ldots x_n,$$

and we say that the infinite product

$$\prod_{j=1}^{\infty} x_j \tag{65}$$

converges, if $\lim_{n\to\infty} p_n$ exists. If S is a finite subset of $\mathbb{N}$, then we define

$$\prod_{j\in S} x_j = \prod_{i=1}^{k} x_{\sigma(i)},$$

where $\sigma(1) < \cdots < \sigma(k)$ are the elements of S in their natural order. If S is an infinite subset of $\mathbb{N}$ and $\sigma(1) < \sigma(2) < \ldots$ are the elements of S again in their natural order we put

$$\prod_{j\in S} x_j = \prod_{i=1}^{\infty} x_{\sigma(i)} = \lim_{n\to\infty} \prod_{i=1}^{n} x_{\sigma(i)},$$

whenever the limit on the right-hand side exists.

The infinite product (65) is said to be *unconditionally convergent* if all the products

$$\mu(S) = \prod_{j\in S} x_j$$

exist for all $S \subseteq \mathbb{N}$ (with the convention $\mu(\varnothing) = 1$) and the map $\mu: 2^{\mathbb{N}} \to G$ is continuous. Here $2^{\mathbb{N}}$ is the power set of $\mathbb{N}$ which we shall identify with the

Cantor group $\{0, 1\}^{\mathbb{N}}$, by the natural identification: $T \leftrightarrow 1_T$, and shall frequently use the symbol $T(n)$ for $1_T(n)$. Then $2^{\mathbb{N}}$ becomes a compact Polish abelian group with the group operation

$$T+S = T\triangle S = (T\backslash S)\cup(S\backslash T),$$

or, in terms of the coordinates,

$$(T+S)(n) \equiv T(n)+S(n) \qquad (\text{mod } 2).$$

Note that if $T_0 \in 2^{\mathbb{N}}$, then

$$\begin{aligned} N_k(T_0) &= \{T\in 2^{\mathbb{N}}\colon T\cap\{1,\ldots,k\} = T_0\cap\{1,\ldots,k\}\} \\ &= \{T\in 2^{\mathbb{N}}\colon T(n) = T_0(n), \forall 1\leq n\leq k\} \end{aligned}$$

is a neighbourhood base at T_0, consisting of clopen sets. Let $2^{(\mathbb{N})}$ denote the set of finite subsets of $\mathbb{N}$; that is

$$\begin{aligned} 2^{(\mathbb{N})} &= \{T\in 2^{\mathbb{N}}\colon T \text{ is a finite subset of } \mathbb{N}\} \\ &= \{T\in 2^{\mathbb{N}}\colon T(n) = 0 \text{ eventually}\}. \end{aligned}$$

Putting $x^0 = 1$ and $x^1 = x$ we have

$$\prod_{j\in T} x_j = \prod_{j=1}^{\infty} (x_j)^{T(j)}, \tag{66}$$

in the sense that, if one of the products exists, then so does the other and they are equal.

If G is abelian and additive notation is used, then we write $\sum$ instead of $\prod$, and the reader easily verifies that in this case we have

PROPOSITION 2.3.8 *Let $\sum_{j=1}^{\infty} x_j$ be an unconditionally convergent series in a topologized abelian semigroup $(G, +)$, and let*

$$\mu(T) = \sum_{j\in T} x_j, \qquad \forall T\subseteq \mathbb{N}.$$

Then μ is a continuous map $2^{\mathbb{N}}\to G$, satisfying

(a) $\mu(T\cup S)+\mu(T\cap S) = \mu(T)+\mu(S), \qquad \forall T, S\in 2^{\mathbb{N}}.$

And if G satisfies Axiom II, *then*

(b) $\mu\left(\bigcup_{n=1}^{\infty} T_n\right) = \sum_1^{\infty} \mu(T_n)$, *whenever* $T_n\cap T_k = \varnothing, \forall n\neq k.$

This leads naturally to the following definition. Let $\mathscr{A}$ be an algebra of subsets of a set X and G a topologized semigroup. Then a map $\mu: \mathscr{A} \to G$ is said to be *finitely additive* if

$$\mu(\emptyset)=1, \tag{67}$$

$$\mu(A \cup B)=\mu(A)\mu(B), \qquad \forall A, B \in \mathscr{A}, \qquad A \cap B=\emptyset. \tag{68}$$

The term *countably additive* is defined similarly.

It may seem strange to use the word "additive" in the non-abelian case. However, the range of μ is actually commutative, which justifies the terminology. To see this let A and B belong to $\mathscr{A}$, then $A \cap B$, $A \backslash B$ and $B \backslash A$ all belong to $\mathscr{A}$ and $\mu(A \cap B)$, $\mu(A \backslash B)$ and $\mu(B \backslash A)$ commutes by (69); moreover an easy calculation shows that we have

$$\mu(A)\mu(B)=\mu(A \cup B)\mu(A \cap B)=\mu(B)\mu(A).$$

Thus we have that

$$\left.\begin{array}{l}\text{every finitely additive set function, } \mu: \mathscr{A} \to G \text{ takes values}\\ \text{in a commutative subsemigroup of } G.\end{array}\right\} \tag{69}$$

THEOREM 2.3.9 *Let G be a semigroup with a complete metrizable topology satisfying Axiom* II_0^1, *and let $\{A_n\}$ be a sequence of subsets of G such that*

(a) $1 \in \bar{A}_n, \forall n \geq 1$.

If U is a neighbourhood of 1, *then there exists a sequence $\{x_n\}$ in G, such that $x_n \in A_n$ for all $n \geq 1$, and*

(b) $p=\prod_{j=1}^{\infty} x_j$ *is unconditionally convergent and $p \in U$.*

Proof Let d be a complete metric generating the topology of G, and let $\delta>0$ be chosen so that

$$\{x \in G: d(x, 1) \leq \delta\} \subseteq U.$$

We shall then construct $x_1, x_2, \ldots$ by induction such that $x_n \in A_n$ and

$$d\Big(\prod_{j \in F} x_j, \Big(\prod_{j \in F} x_j\Big) x_{k+1}\Big) \leq \delta 2^{-k-1}, \qquad \forall F \subseteq\{1, \ldots, k\}, \forall k \geq 1. \tag{70}$$

First we choose $x_1 \in A_1$ so that $d(1, x_1) \leq \delta 2^{-1}$, which is possible, since $1 \in \bar{A}_1$. Now suppose that $x_1, \ldots, x_n$ are constructed so that (70) holds for all $k=1, \ldots, n-1$. Let

$$\mu(F)=\prod_{j \in F} x_j$$

for $F \subseteq \{1, \ldots, n\}$, and consider the set

$$V = \{x \in G : d(\mu(F), \mu(F)x) < \delta 2^{-n-1}, \forall F \subseteq \{1, \ldots, n\}\}.$$

Then, by continuity of $x \mapsto \mu(F)x$ at $x = 1$, we have that V is a neighbourhood of 1. Hence by (a) there exists a point x_{n+1} in G such that

$$x_{n+1} \in A_{n+1} \cap V.$$

It is then clear that (70) holds for $k = n$. Thus the induction is completed and we have established the existence of a sequence $\{x_k\}$, satisfying (70) and $x_k \in A_k$ for all $k \geq 1$.

If $T \subseteq \mathbb{N}$ we put $T[n] = T \cap \{1, \ldots, n\}$, and we define

$$\mu_n(T) = \prod_{j \in T[n]} x_j, \qquad \forall n \geq 0.$$

Then, by the very definition of the topology on $2^{\mathbb{N}}$, we have that μ_n is continuous (μ_n is even locally constant). From (70) we deduce that

$$d(\mu_n(T), \mu_{n+1}(T)) \leq \delta 2^{-n-1}, \qquad \forall n \geq 0, \forall T \subseteq \mathbb{N}.$$

Hence we have

$$d(\mu_n(T), \mu_{n+k}(T)) \leq \delta 2^{-n}, \qquad \forall n \geq 0, \forall k \geq 1, \forall T \subseteq \mathbb{N}. \tag{71}$$

This shows that $\{\mu_n(T)\}$ is a Cauchy sequence and so

$$\mu(T) = \lim_{n \to \infty} \mu_n(T) = \prod_{j \in T} x_j$$

exist for all $T \subseteq \mathbb{N}$, and moreover (71) shows that $\{\mu_n\}$ converges uniformly to μ on $2^{\mathbb{N}}$; hence μ is continuous on $2^{\mathbb{N}}$; that is

$$p = \prod_{j=1}^{\infty} x_j$$

is unconditionally convergent.

Finally we note that $\mu_0(\mathbb{N}) = 1$, and so by (71) we have

$$d(1, p) = \lim_{k \to \infty} d(\mu_0(\mathbb{N}), \mu_k(\mathbb{N})) \leq \delta.$$

Then by the choice of δ we have $p \in U$.

Theorem 2.3.9 implies the following simple but often useful criterion for continuity:

COROLLARY 2.3.10 *Let G be a semigroup with a complete metrizable topology satisfying axiom II_0^1. Then a map $f: G \to X$, where X is a Hausdorff space, is continuous at 1 if, and only if,*

(a) $f(x_n) \to f(1)$, *whenever $\prod_{n=1}^{\infty} x_n$ is unconditionally convergent.*

Proof If f is not continuous at 1, then there exists a neighbourhood U of $f(1)$, such that $f(V_k) \not\subseteq U$ for all $k \geq 1$, where V_k is the ball with centre at 1 and radius 2^{-k}. Now let

$$A_k = G \backslash f^{-1}(U), \qquad \forall k \geq 1.$$

Then we have $A_k \cap V_n = V_n \backslash f^{-1}(U) \neq \emptyset$; hence $1 \in \bar{A}_k$ for all $k \geq 1$. So by Theorem 2.3.9 there exists an unconditionally convergent product $\prod_{n=1}^{\infty} x_n$ with $f(x_n) \notin U$. But this contradicts condition (a), and proves the corollary.

Let us finish this section by showing that a finitely additive set function on a σ-algebra is either countably additive or highly "non-measurable". We start with a result due to J. P. R. Christensen (1974), where we only consider $\mathbb{N}$ with the power set $2^{\mathbb{N}}$ as its σ-algebra. But first we need some new notions.

A subset A of a topologized semigroup G is called *totally bounded* (resp. *σ-bounded*) if for every neighbourhood V of 1 there exists a finite (resp. countable) set $T \subseteq G$ such that

$$A \subseteq VT. \tag{72}$$

A map $f: X \to G$ is called *σ-bounded* if $f(X)$ is σ-bounded. And if X is a topological space, then f is called *almost σ-bounded*, if $f(X_0)$ is σ-bounded for some co-meagre set X_0 in X. Note that we have that

in a metrizable group σ-bounded is equivalent to separable; a continuous homomorphic image of a σ-bounded set is σ-bounded. (73)

LEMMA 2.3.11 *Let $(G, +)$ be a topological abelian group, and V a neighbourhood of 0, then there exists a metrizable abelian group H, a continuous homomorphism $\varphi: G \to H$, and a neighbourhood U of 0 in H such that $\varphi^{-1}(U) \subseteq V$.*

Proof Let $\{V_n\}$ be a sequence of open neighbourhoods of 0 in G such that

$$\bar{V}_n - \bar{V}_n \subseteq V_{n-1}, \qquad \forall n \geq 1,$$

where $V_0 = V$. Then $G_0 = \bigcap_{n=1}^{\infty} V_n$ is a closed subgroup of G. Now let $H = G/G_0$, and put

$$W_n = \{x + G_0 : x \in V_n\}, \qquad n = 1, 2, \ldots.$$

Then $\{W_n\}$ generates a metrizable group topology on H such that $\varphi(x) = x + G_0$ is a continuous homomorphism with $\varphi^{-1}(W_1) \subseteq V$.

COROLLARY 2.3.12 *Let X be a topological space and G a topogical abelian group. Then a map $f: X \to G$ is continuous if, and only if, $\varphi \circ f$ is continuous for any continuous homomorphism $\varphi: G \to H$ with values in some metrizable abelian group H.*

We are now ready for the automatic continuity theorem for finitely additive set functions.

THEOREM 2.3.13 *Let $\mu: 2^{\mathbb{N}} \to G$ be a finitely additive set function with values in a topological group G. If the restriction $\mu|2^T$ is $\mathcal{B}_p$-measurable and almost σ-bounded for all $T \subseteq \mathbb{N}$, then μ is continuous and countably additive.*

Proof First we note that by (69) we may assume that G is abelian, and by Corollary 2.3.12 we may assume that G is metrizable and complete, and we shall use additive notation in G. We shall now prove

$$\mu|2^T \text{ has the Baire property}, \qquad \forall T \subseteq \mathbb{N}. \tag{74}$$

Since $\mu|2^T$ is almost σ-bounded, there exists a co-meagre set $\Gamma \subseteq 2^T$ such that $\mu(\Gamma)$ is separable in G (cf. (73)). Moreover, by Proposition 2.2.1 we have that $\mu|\Gamma$ is $\mathcal{B}_p$-measurable. Hence, by Theorem 2.2.10, there exists a co-meagre set Γ_0 in Γ such that $\mu|\Gamma_0$ is continuous, and since Γ is dense in 2^T (2^T is a Baire space!), it follows from Proposition 2.2.1 that Γ_0 is co-meagre in 2^T. Hence (74) is proved.

Our next step is to show that

$$\mu_0 = \mu|2^{(\mathbb{N})} \quad \text{is uniformly continuous.} \tag{75}$$

To see this let R and S belong to $2^{(\mathbb{N})}$. By (74) we can find a co-meagre set $\Sigma \subseteq 2^{\mathbb{N}}$, such that $\mu|\Sigma$ is continuous. The countability of $2^{(\mathbb{N})}$ then shows that there exists $T_0 \in \Sigma$ such that

$$T_0 \triangle S \in \Sigma, \qquad \forall S \in 2^{(\mathbb{N})}.$$

Now let $S, R \in 2^{(\mathbb{N})}$, and put

$$P_0 = R \backslash S \backslash T_0, \qquad P_1 = T_0 \cap R \backslash S,$$
$$Q_0 = S \backslash R \backslash T_0, \qquad Q_1 = T_0 \cap S \backslash R.$$

Then we have

$$\begin{aligned}\mu(R) - \mu(S) &= \mu(R \backslash S) - \mu(S \backslash R) \\ &= \mu(T_0 \triangle P_0) - \mu(T_0 \triangle P_1) - \mu(T_0 \triangle Q_0) + \mu(T_0 \triangle Q_1),\end{aligned}$$

and since P_0, P_1, Q_0 and Q_1 all are contained in $R\triangle S$, we see that the continuity of $\mu|\Sigma$ implies uniform continuity of $\mu|2^{(\mathbb{N})}$.

We can then extend μ_0 uniquely to a continuous map $2^{\mathbb{N}}\to G$, which we shall denote μ_1. The proof will then be concluded if we can show that

$$\mu=\mu_1. \tag{76}$$

(Note that continuity of μ implies countably additivity.)

So let $T\subseteq\mathbb{N}$, and put $\nu=(\mu-\mu_1)|2^T$. We shall then show that $\nu(T)=0$. First we note that μ_1 is finitely additive by continuity of μ_1. So ν is finitely additive. Moreover, we have $\nu(S)=0$ for all $S\in 2^{(T)}$, where $2^{(T)}$ is the set of finite subsets of T. Hence we find

$$\nu(R\triangle S)=\nu(R)+\nu(S)-2\nu(R\cap S)=\nu(R),$$

for all $R\in 2^T$ and all $S\in 2^{(T)}$. Hence from (74) and Theorem 2.3.7 we conclude that ν is constant on some co-meagre set $\Gamma\subseteq 2^T$, say $\nu(S)=g$ for all $S\in\Gamma$. Next we note that the functions

$$(R,S)\mapsto R\backslash S,\qquad (R,S)\mapsto R\cap S,$$

$$(R,S)\mapsto R,\qquad (R,S)\mapsto T\backslash R,$$

are open, continuous and surjective: $2^T\times 2^T\to 2^T$. Hence by Proposition 2.2.1 we have that the set

$$\{(R,S)\in 2^T\times 2^T: R\in\Gamma,\ T\backslash R\in\Gamma,\ R\backslash S\in\Gamma,\ R\cap S\in\Gamma\}$$

is co-meagre in $2^T\times 2^T$. In particular it contains at least one element, say (R_0,S_0). Then we have

$$g=\nu(R_0)=\nu(R_0\backslash S_0)+\nu(S_0\cap R_0)=2g,$$

$$\nu(T)=\nu(T\backslash R_0)+\nu(R_0)=2g.$$

Hence $g=0$ and $\nu(T)=0$, which proves (76) and the theorem.

Note that from Lemma 2.3.11 and (73) we deduce the following useful lemma:

LEMMA 2.3.14 *Let G be a topological abelian group and A a subset of G. Then A is σ-bounded if, and only if, $\varphi(A)$ is separable, whenever φ is a continuous homomorphism from G into a metrizable abelian group H.*

Now since every Borel map from an analytic space into a metric space has separable range, it follows from Lemma 2.3.14 that $\mu|2^T$ is $\mathscr{B}_p$-measurable and almost σ-bounded for all $T\subseteq\mathbb{N}$, whenever μ is Borel-measurable.

Similarly, if $\mu|2^T$ has the Baire property for all $T\subseteq\mathbb{N}$, then $\mu|2^T$ is $\mathscr{B}_p$-measurable by Theorem 2.2.10 and almost σ-bounded by Lemma 2.3.14.

So we conclude that the conclusion of Theorem 2.3.13 holds whenever μ is finitely additive and satisfies one of the following three conditions:

$$\mu \text{ is Borel measurable;} \tag{77}$$

$$\mu|2^T \text{ has the Baire property, } \forall T\subseteq\mathbb{N}\text{; or} \tag{78}$$

$$\mu \text{ is } \sigma\text{-bounded and } \mu|2^T \text{ is } \mathscr{B}_p\text{-measurable, } \forall T\subseteq\mathbb{N}. \tag{79}$$

Note that if μ is continuous and countably additive on $2^{\mathbb{N}}$, then we have

$$\left.\begin{array}{l}\text{if } g_j=\mu(\{j\})\text{, then } \sum_{j=1}^{\infty} g_j \text{ is unconditionally convergent and}\\[1em] \mu(T)=\sum_{j\in T} g_j, \qquad \forall T\subseteq\mathbb{N}.\end{array}\right\} \tag{80}$$

Now let $(X, \mathscr{B})$ be a measurable space, i.e. $\mathscr{B}$ is a σ-algebra on the set X, and let μ be a finitely additive set function on $\mathscr{B}$ with values in a group G. If $\{A_j\}$ are disjoint sets in $\mathscr{B}$ we may consider the finitely additive set function

$$\sigma(T)=\mu\Big(\bigcup_{j\in T} A_j\Big), \qquad \forall T\subseteq\mathbb{N}.$$

Then countably additivity of σ implies

$$\mu\Big(\bigcup_{j=1}^{\infty} A_j\Big)=\sum_{j=1}^{\infty}\mu(A_j).$$

So we obtain the following corollary to Theorem 2.3.13:

COROLLARY 2.3.15 *Let $(X, \mathscr{B})$ be a measurable space and μ a finitely additive set function on $\mathscr{B}$ with values in a group G. Now let Ω be a σ-algebra on $\mathscr{B}$, such that $\mu:(\mathscr{B},\Omega)\to(G,\mathscr{B}(G))$ is measurable. Then μ is countably additive in either of the following two cases:*

(a) *μ is σ-bounded, and $T\mapsto\bigcup_{j\in T}A_j$ is measurable: $(2^{\mathbb{N}},\mathscr{B}_p(2^{\mathbb{N}}))\to(\mathscr{B},\Omega)$ whenever $\{A_j\}$ are disjoint sets from $\mathscr{B}$;*
(b) *$T\mapsto\bigcup_{j\in T}A_j$ is measurable: $(2^{\mathbb{N}},\mathscr{B}(2^{\mathbb{N}}))\to(\mathscr{B},\Omega)$ whenever $\{A_j\}$ are disjoint sets from $\mathscr{B}$.*

Let us remark that Martin's axiom implies the existence of a universally measurable finitely additive set function $\mu: 2^{\mathbb{N}}\to[0, 1]$, which is *not* countably additive.

2.4 Thick sets and automatic continuity

In Section 2.2 we described a method to prove continuity by aid of the Baire property. We shall now describe a second method, which in essence is due to Pettis (1950). The method is based on the following simple observation:

PROPOSITION 2.4.1 *Let G and H be topologized semigroups such that H satisfies Axiom* IV_0^1, *and let $f: G \to H$ be a homomorphism satisfying*

(a) $\forall V \in \mathcal{N}_H(1), \exists A \subseteq G$ *so that* $1 \in \operatorname{int}(AA^{-1})$ *and* $f(A) \subseteq VV^{-1}$.

Then f is continuous at 1.

Proof Let $W \in \mathcal{N}_H(1)$ and choose $V_1, V_2 \in \mathcal{N}_H(1)$, so that $V_2V_2^{-1} \subseteq W$, $V_1V_1^{-1} \subseteq V_2$. Then let $A \subseteq G$ be chosen such that $1 \in \operatorname{int}(AA^{-1}) = U$ and $f(A) \subseteq V_1V_1^{-1}$. Then we have

$$f(U) \subseteq f(AA^{-1}) \subseteq f(A)f(A)^{-1} \subseteq V_2V_2^{-1} \subseteq W.$$

Thus f is continuous at 1.

Despite its extreme simplicity, Proposition 2.4.1 turns out to give very powerful automatic continuity theorems.

Motivated by condition (a) in this proposition, we shall say that a subset A of a topologized semigroup G is *thick* if $1 \in \operatorname{int}(AA^{-1})$. The main idea behind applying Proposition 2.4.1 is then to show that even very small sets may be thick. Our first result in this direction is due to Pettis (1950).

THEOREM 2.4.2 *Let G be a group with a topology satisfying Axiom* II^1, *and let A be a non-meagre subset of X having the Baire property. Then A is thick.*

Proof Let $U = M(A^C)$, then U is open and since $y \mapsto xy$ is a homeomorphism we have

$$\begin{aligned}(xU) \cap U &\subseteq M(xA^C) \cap M(A^C) = M((xA)^C) \cap M(A^C)\\ &\subseteq M((xA)^C \cup A^C) = M(G \backslash (xA \cap A)),\end{aligned}$$

where we have used the obvious inclusion $M(A) \cap M(B) \subseteq M(A \cup B)$. Now A is non-meagre and so G is non-meagre, hence by Proposition 2.2.3 we have that G is a Baire space, and so Proposition 2.2.2 shows that $M(G) = \varnothing$. This leads us to the conclusion

$$UU^{-1} = \{x : xU \cap U \neq \varnothing\} \subseteq \{x : xA \cap A \neq \varnothing\} = AA^{-1}.$$

And since A^C has the Baire property we know (cf. Proposition 2.2.7) that $D(A^C)\backslash A^C = A\backslash U$ is meagre. But $A \subseteq (A\backslash U) \cup U$ is non-meagre so $U \neq \varnothing$ and UU^{-1} is an open neighbourhood of 1, showing that A is thick.

Our next criterion is an extension of the well-known fact that every subset of the real line with positive Lebesgue measure is thick. This was originally proved by Ostrowski (1929) , and later extended to locally compact groups with Haar measure. We shall here extend it further to arbitrary topologized semigroups and arbitrary measures with some suitable modification. But let us first remind the reader about some notation from measure theory.

If (S, Σ, μ) is a positive measure space and $\mathcal{M}$ is a family of positive measures on (S, Σ) we say that $\mathcal{M}$ is *uniformly μ-continuous*, if

$$\forall \varepsilon > 0, \exists \delta > 0 \quad \text{so that} \quad \mu(A) < \delta \Rightarrow \nu(A) < \varepsilon, \forall \nu \in \mathcal{M}.$$

If μ is a measure on (S, Σ), then μ_* denotes *the inner measure generated by μ*, i.e.

$$\mu_*(A) = \sup\{\mu(B): B \in \Sigma, B \subseteq A\}, \qquad \forall A \subseteq S.$$

Thus μ and μ_* coincide on Σ.

A Borel measure μ on a Hausdorff space is called a Radon measure, if μ is *locally finite*, i.e. every point has a neighbourhood of finite measure, and *tight*, i.e.

$$\mu(A) = \sup\{\mu(K): K \text{ compact}, K \subseteq A\}, \qquad \forall A \in \mathcal{B}(X).$$

Our second criterion is stated so that it contains the Haar measure result together with an important result on Gaussian measures on locally convex spaces, which we shall explain in more detail in Example 2.4.2 below.

THEOREM 2.4.3 *Let (G, τ) be a completely regular topologized semigroup and let τ^* denote an auxiliary topology on G. Let m be a Radon measure on (G, τ) and W^* a τ^*-neighbourhood of* 1, *satisfying*

(a) *$y \mapsto xy$ is m-measurable for all $x \in W^*$;*
(b) *$\{m_x: x \in W^*\}$ is uniformly m-continuous;*
(c) *$\psi(x) = \int_K f(xu)m(\mathrm{d}y)$ is τ^*-continuous at* 1, *whenever K is τ-compact and $f: G \to [0, 1]$ is τ-continuous.*

Here $m_x(B) = m(\{y : xy \in B\})$ for $B \in \mathcal{B}(X)$ and $x \in W^*$. Then every set $A \subseteq G$ sdtisfying $m_*(A) > 0$ is thick with respect to (G, τ^*).

Remark Condition (a) implies that m_x is a well-defined Borel measure for all $x \in W^*$, and that the function ψ in condition (c) is defined for all $x \in W^*$.

Proof By the tightness of m there exists a τ-compact set $K \subseteq A$ with $m(K) > 0$. Now let us put

$$\varphi(x) = \int_K 1_K(xu)m(du) \quad \text{for} \quad x \in W^*;$$

we shall then show that φ is τ^*-continuous at 1.

So let $\varepsilon > 0$ be given, then by condition (b) we can choose a $\delta > 0$ so that

$$m(B) \leqslant \delta \Rightarrow m_x(B) \leqslant \tfrac{1}{2}\varepsilon, \qquad \forall x \in W^*.$$

The local finiteness and the tightness of m allows us to pick a τ-open set $V \supseteq K$, so that $m(V \backslash K) \leqslant \delta$. Then by complete regularity of (G, τ) there exists a τ-continuous function $f: G \to [0, 1]$, with

$$1_K \leqslant f \leqslant 1_V.$$

Let ψ be defined as in condition (c), and let $W_0^* \subseteq W^*$ be a τ^*-neighbourhood of 1, such that

$$|\psi(x) - \psi(1)| \leqslant \tfrac{1}{2}\varepsilon, \qquad \forall x \in W_0^*.$$

Now let $x \in W_0^*$; then, noting that $\varphi(1) = \psi(1) = m(K)$, we find

$$\begin{aligned}
|\varphi(x) - \varphi(1)| &\leqslant |\psi(x) - \psi(1)| + |\varphi(x) - \psi(x)| \\
&\leqslant \tfrac{1}{2}\varepsilon + \int_G |f(xu) - 1_K(xu)| m(du) \\
&= \tfrac{1}{2}\varepsilon + \int_G (f(u) - 1_K(u)) m_x(du) \\
&\leqslant \tfrac{1}{2}\varepsilon + m_x(V \backslash K) \\
&\leqslant \varepsilon.
\end{aligned}$$

Thus φ is τ^*-continuous at 1.

We have already noted that $\varphi(1) = m(K) > 0$, thus $U = \{x \in W^* | \varphi(x) > 0\}$ is a τ^*-neighbourhood of 1. And if $x \in U$ then

$$\varphi(x) = \int_G 1_K(u) 1_K(xu) m(du) > 0;$$

hence there exists $u \in K$ with $v = xu \in K$; thus $x \in KK^{-1}$. This shows that $U \subseteq KK^{-1} \subseteq AA^{-1}$, and so A is τ^*-thick.

The condition (c) in Theorem 2.4.3 is a technical condition which is fulfilled in several cases. Note that if $x \mapsto xu$ is sequentially continuous from (G, τ^*) into (G, τ) at $x = 1$, then the Lebesgue dominated convergence theorem we have that ψ is sequentially τ^*-continuous at 1. Thus we have that

$$\left.\begin{array}{l}\text{if } x \mapsto xu: (G, \tau^*) \to (G, \tau) \text{ is sequentially continuous at } x = 1 \\ \text{and if } (G, \tau^*) \text{ is an } s^*\text{-space, then condition (c) holds.}\end{array}\right\} \tag{81}$$

And from (A7) and (A8) in the appendix to Section 2 we find that

$$\left.\begin{array}{l}\text{if } (x, y) \mapsto xy: (G \times G, \tau^* \times \tau) \to (G, \tau) \text{ is separately } k\text{-continuous,} \\ \text{and if } (G, \tau^*) \text{ is a } k^*\text{-space, then condition (c) holds.}\end{array}\right\} \tag{82}$$

Moreover a straightforward argument shows that

$$\left.\begin{array}{l}\text{if } (x, y) \mapsto xy: (G \times G, \tau^* \times \tau) \to (G, \tau) \text{ is continuous at } (x, y) = (1, a) \\ \text{for all } a \in G, \text{ then condition (c) holds.}\end{array}\right\} \tag{83}$$

If (G, τ) is a topological group, then KK^{-1} is compact whenever K is τ-compact. Hence we have that

$$\left.\begin{array}{l}\text{if } (G, \tau) \text{ is a topological group, and } m \neq 0 \text{ a Radon measure sat-} \\ \text{isfying (a)–(c) in Theorem 2.4.3 for some topology } \tau^*, \text{ then } \tau^* \\ \text{has a } \tau\text{-compact neighbourhood of 1.}\end{array}\right\} \tag{84}$$

Let us give two examples indicating the scope of Theorem 2.4.3.

Example 2.4.1 Let (G, τ) be a topological group and let m be a non-zero Radon measure satisfying (a)–(c) in Theorem 2.4.3 for $\tau^* = \tau$. Then by (84) we have that (G, τ) is locally compact; moreover, if G_0 is the subgroup generated by W^*, then G_0 is a clopen subgroup and it is easily checked that we have

$$\mu_x \text{ is } \mu\text{-continuous}, \qquad \forall x \in G_0, \tag{85}$$

where μ is the restriction of m to G_0.

A measure satisfying (85) clearly has the property that μ_x is equivalent to μ_y for $x, y \in G_0$. Now let λ be the Haar measure on G_0 and let A be a Borel subset of G_0, then if K is a compact set in G_0, we have, by (A6) in the appendix to Section 2,

$$\begin{aligned}\int_{K \times K} 1_A(xy)(\lambda \times \mu)(\mathrm{d}x, \mathrm{d}y) &= \int_K \mu((x^{-1}A) \cap K)\lambda(\mathrm{d}x) \\ &= \int_K \lambda((Ay^{-1}) \cap K)\mu(\mathrm{d}y).\end{aligned}$$

But this equality shows that any measure μ satisfying (85) is equivalent to the Haar measure λ.

Conversely if (G, τ) is a locally compact group and λ is the left-invariant Haar measure on some clopen subgroup G_0 of G, then evidently $\tau^* = \tau$, $W^* = G_0$ and $m = \lambda$ satisfies (a)–(c) in Theorem 2.4.3. And since right and left Haar measures are equivalent we find that

$$\left.\begin{array}{l}\text{if } G \text{ is a locally compact group and } A \subseteq G \text{ has} \\ \text{positive Haar measure, then } A \text{ is thick.}\end{array}\right\} \tag{86}$$

So in the case when $\tau = \tau^*$ and (G, τ) is a topological group then Theorem 2.4.3 gives the classical result on the Haar measure and no more. However, our next example shows that if $\tau^* \neq \tau$, then we do get more.

Example 2.4.2 Let (G, τ) be a locally convex linear space and let m be a non-degenerated *Gaussian Radon measure on G with mean* 0, that is

$$\int_G e^{ix'(x)} m(\mathrm{d}x) = \exp\left(-\tfrac{1}{2}R(x', x')\right), \qquad \forall x' \in G', \tag{87}$$

where G' is the dual of G and R is the *covariance function*

$$R(x', y') = \int_G x'(x) y'(x) m(\mathrm{d}x). \tag{88}$$

The assumption that m is *non-degenerated* means

$$R(x', x') > 0, \qquad \forall x' \neq 0; \tag{89}$$

and is equivalent to

$$\operatorname{supp}(m) = G. \tag{90}$$

We can now define *the reproducing kernel* of m, which is a Hilbert space $(H(m), \|\cdot\|)$, defined as follows:

$$\|x\| = \sup\{x'(x)\colon x' \in G', R(x', x') \leq 1\},$$

$$H(m) = \{x \in G\colon \|x\| < \infty\}.$$

It is well known that $\|\cdot\|$ is a quasi-norm (i.e. a norm except that it may take the value $+\infty$), and that $(H(m), \|\cdot\|)$ is a separable Hilbert space, such that $H(m)$ is dense in (G, τ).

Now let τ^* be the topology generated by $\|\cdot\|$, that is the topology generated by the metric

$$\mathrm{d}(x, y) = \arctan\|x - y\| \qquad (\arctan\infty = \tfrac{1}{2}\pi).$$

Then $(G, +, \tau^*)$ is a topological group (but *not* a topological linear space, except if $\dim G < \infty$). Actually (G, τ^*) is a disjoint topological sum of copies of $(H(m), \|\cdot\|)$.

It is well known that m_x is m-continuous if, and only if, $x \in H(m)$, and if so the Radon–Nikodym derivative is given by

$$\frac{dm_x}{dm} = \exp\{\tilde{x} - \tfrac{1}{2}\|x\|^2\},$$

where $\tilde{x}$ is an m-measurable linear function $G \to \mathbb{R}$, satisfying

$$\int_G e^{t\tilde{x}(y)} m(dy) = \exp(\tfrac{1}{2}t^2\|x\|^2), \qquad \forall t \in \mathbb{R}. \tag{91}$$

So by the Cauchy–Schwartz inequality we have

$$\begin{aligned} m_x(B) &= \int_B \exp(\tilde{x} - \tfrac{1}{2}\|x\|^2)\, dm \\ &\leq m(B)^{1/2}\left\{\int_G \exp(2\tilde{x} - \|x\|^2)\, dm\right\}^{1/2} \\ &= m(B)^{1/2} \exp(\tfrac{1}{2}\|x\|^2). \end{aligned}$$

Thus, if we put

$$W(r) = \{x \in G : \|x\| \leq r\},$$

we find that $\{m_x : x \in W(r)\}$ is uniformly m-continuous for all $r < \infty$.

It is easily checked that the τ^*-topology is finer than τ, and so $x \mapsto x + u$ is continuous from (G, τ^*) into (G, τ) for all $u \in G$. So by (81) we have that condition (c) in Theorem 2.4.3 holds. That is (a)–(c) in Theorem 2.4.3 are satisfied for $W^* = W(1)$. Thus by Theorem 2.4.3 and (84) we find that

$$W(r) \text{ is } \tau\text{-compact}, \qquad \forall r < \infty; \tag{92}$$

$$\text{if } m_*(A) > 0, \text{ then } A - A \supseteq W(r), \text{ for some } r > 0. \tag{93}$$

Let us study this a little further. We shall say that a subset A of G is a *Gaussian null set* if A is universally Radon measurable and $\mu(A) = 0$ for some non-degenerate Gaussian Radon measure μ on G with mean zero. Note that in finite-dimensional spaces this notion coincides with the usual notion of a Lebesgue null set. With this notion we find the following:

Let G be a locally convex Polish linear space, and A a universally measurable subset of G. If A is not a Gaussian null set, then A is thick. (94)

To see this let d be a complete metric generating the topology on G, such that

(α) $d(x, y) = \rho(x - y), \qquad \forall x, y \in G$;

(β) $\rho(\lambda x) \leq \rho(x), \qquad \forall x \in G, \forall |\lambda| \leq 1$.

If $B = A - A$ is not a neighbourhood of 0, then an easy argument shows that there exists a sequence $\{x_n\}$ in G, such that

(δ) $\rho(2^n x_n) \leq 2^{-n}, \forall n \geq 1$;
(ε) $x_{2n} \notin B, \forall n \geq 1$;
(ζ) $G = \overline{\text{span}}\,\{x_n : n \geq 1\}$.

Then by (β), (δ) and the completeness of d, we have that

$$T\lambda = \sum_{n=1}^{\infty} \lambda_n 2^n x_n, \qquad \lambda = (\lambda_n) \in l^2,$$

is a continuous linear operator $l^2 \to G$. Now let ν be the Gaussian measure with mean zero on l^2, having covariance function

$$R_0(\lambda, \lambda) = \sum_{n=1}^{\infty} n^{-2} \lambda_n^2, \qquad \forall \lambda \in l^2,$$

and put $\mu = \nu \circ T^{-1}$. Then μ is a Gaussian measure on G with mean zero and covariance function

$$R_1(x', x') = \sum_{n=1}^{\infty} n^{-2} 4^n (x'(x_n))^2, \qquad \forall x' \in G';$$

and μ is non-degenerate by (ζ). Now we note that

$$|x'(x_n)| \leq n 2^{-n} R_1(x', x')^{1/2}, \qquad \forall x' \in G'.$$

So we have

$$\|x_n\| \leq n 2^{-n} \xrightarrow[n \to \infty]{} 0.$$

Thus by (ε) and (93) we find that $\mu(A) = 0$, and so (94) is proved.

Note that our definition of a Gaussian null set differs drastically from the definition of R. R. Phelps (1978), where he requires $\mu(A) = 0$ for *all* non-degenerate Gaussian measures. We think, however, that "universal Gaussian null set" is a more appropriate name for the Gaussian null sets in the sense of Phelps.

Our last criterion is also an extension of the Haar measure result, but in another direction. We shall introduce the notion of a Haar null set in general topologized semigroups satisfying Axiom II.

So let G be a semigroup satisfying Axiom II; we shall call such a semigroup a *separately continuous semigroup*, since it is characterized by separate continuity of the multiplication map $(x, y) \mapsto xy$. Let $\text{Prob}_\tau(G)$ be the set of all *τ-smooth probability measures* on G, and let $\mathcal{B}_{\tau u}(G)$ be the set of all subsets of G which are *measurable with respect to every measure* $\mu \in \text{Prob}_\tau(G)$. Let A be a subset of G, then a τ-smooth probability measure μ is called a *testing measure* for A, if there exist $B \in \mathcal{B}_{\tau u}(G)$, such that $A \subseteq B$ and

$$\mu(x^{-1}By^{-1}) = 0, \qquad \forall x, y \in G.$$

Note that

$$x^{-1}By^{-1} = \{z \in G : xzy \in B\}$$

belongs to $\mathcal{B}_{\tau u}(G)$, since $z \mapsto xzy$ is continuous by Axiom II.

Now let us see what this has to do with Haar null sets in locally compact groups. So let H be a locally compact group with left and right Haar measures λ and ρ. And let G be a Borel-measurable subsemigroup of H. If A is a subset of G with a testing measure μ (with respect to G), then by the extended Fubini theorem (cf. (A6) in the appendix to Section 2), whenever K is a compact subset of G, we have

$$\begin{aligned} 0 &= \int_K \mu((x^{-1}A) \cap G)\lambda(\mathrm{d}x) \\ &= \int_K \int_G 1_A(xy)\mu(\mathrm{d}y)\lambda(\mathrm{d}x) \\ &= \int_G \lambda((Ay^{-1}) \cap K)\mu(\mathrm{d}y). \end{aligned}$$

Now it is well known that $f(y) = \lambda((Ay^{-1}) \cap K)$ is continuous, and since $f(1) = \lambda(A \cap K)$, A is a λ-null set if $1 \in \text{supp}(\mu)$. Conversely, if $A \subseteq G$ is a λ-null set and if $\lambda(G) > 0$, then any λ-continuous probability measure on G is a testing measure for A, and, if

$$\lambda(U) > 0, \qquad \forall U \in \mathcal{N}_G(1), \tag{95}$$

then we can find such a testing measure with $1 \in \text{supp}(\mu)$. Thus we are led to the following definition.

Let G be a separately continuous semigroup, and A a subset of G. Then A is said to be a *Haar null set* if there exist a testing measure $\mu \in \text{Prob}_\tau(G)$ for A such that $1 \in \text{supp}(\mu)$. As we have seen above, this definition extends the usual notion for Borel-measurable subsemigroups of a locally compact group satisfying (95) above.

There does not seem to be much to work with in the general case. In particular the lack of a convolution seems to be an insurmountable obstacle. However, if G is metrizable, then we have shown in the appendix to Section 2 how a convolution may be defined, and it turns out that in this case we can obtain non-trivial results. Let us first collect some elementary results about

Haar null sets. (Recall that δ_a is the Dirac measure at a, and that w denotes the weak topology on $M_\tau^+(X)$; see the appendix to Section 2.)

PROPOSITION 2.4.4 *Let G be a metrizable separately continuous semigroup and A a subset of G. If T is the set of testing measures for A, then T is convex and*

(a) *if* $\mu \in T$ *and* $\nu \in \mathrm{Prob}_\tau(G)$, *then* $\mu * \nu \in T$ *and* $\nu * \mu \in T$;
(b) *if* $\mu \in T$, $\nu \in \mathrm{Prob}(G)$ *and* $\nu \ll \mu$, *then* $\nu \in T$;
(c) $\delta_a \in \mathrm{cl}_w(T) \Leftrightarrow \exists \mu \in T$ *with* $a \in \mathrm{supp}(\mu)$;
(d) *A is a Haar null set* $\Leftrightarrow \delta_1 \in \mathrm{cl}_w(T)$;
(e) *if G is a group and* $T \neq \varnothing$, *then A is a Haar null set and T is weakly dense in* $\mathrm{Prob}_\tau(G)$.

Proof (a) T is evidently convex, and by definition of $\mu * \nu$ we have

$$\mu * \nu(x^{-1}By^{-1}) = \int_G \mu(u^{-1}(x^{-1}B)y^{-1})\nu(du);$$

and, since $u^{-1}(x^{-1}By^{-1}) = (xu)^{-1}By^{-1}$, we find that $\mu * \nu \in T$ if $\mu \in T$. Similarly, $\nu * \mu \in T$.

(b) This is clear.

(c) Suppose that $\delta_a \in \mathrm{cl}_w(T)$, then there exists $\mu_n \in T$ with $\mu_n \xrightarrow{w} \delta_a$, since the weak topology is metrizable. Now let

$$\mu = \sum_{n=1}^{\infty} 2^{-n}\mu_n.$$

Then clearly $\mu \in T$, and since

$$\liminf_{n\to\infty} \mu_n(U) \geqslant \delta_a(U) = 1$$

for every neighbourhood U of a, we have $a \in \mathrm{supp}(\mu)$.

Now suppose that $\mu \in T$ and $a \in \mathrm{supp}(\mu)$. Let $\{U_n\}$ be a neighbourhood base at a, and put

$$\mu_n(B) = \frac{\mu(B \cap U_n)}{\mu(U_n)}, \qquad \forall n \geqslant 1, \forall B \in \mathscr{B}_{\tau u}(G).$$

Then μ_n is a well-defined τ-smooth probability measure ($\mu(U_n) > 0$ since $a \in \mathrm{supp}(\mu)$), and $\mu_n \ll \mu$. Hence $\mu_n \in T$ and $\mu_n \xrightarrow{w} \delta_a$. Thus $\delta_a \in \mathrm{cl}_w(T)$.

(d) Immediate consequence of (c).

(e) Let G be a group and $\mu \in T$. Then μ has a non-empty support so let $a \in \mathrm{supp}(\mu)$, and let b be an arbitrary element in G. If $c = a^{-1}b$ then $\mu_0 = \mu * \delta_c \in T$ and $b = ac \in \mathrm{supp}(\mu_0)$ (cf. (A26) in the appendix to Section

2). Thus $\delta_b \in \text{cl}_w(T)$ for all $b \in G$ and since T is convex we have that T is weakly dense in $\text{Prob}_\tau(G)$.

Note that (a) and (A26) imply that

$$\text{a finite union of Haar null sets is a Haar null set.} \tag{96}$$

If G is complete metrizable we can extend this to

THEOREM 2.4.5 *Let G be a complete metric, separately continuous semigroup. Then every countable union of Haar null sets is a Haar null set.*

Proof Let A_n be Haar null sets for $n \in \mathbb{N}$, and let T_n be the set of testing measures for A_n. Then $\delta_1 \in \text{cl}_w(T_n)$ for all $n \geq 1$ by Proposition 2.4.4.

It will be shown in the appendix, (A23) and (A27), that $(\text{Prob}_\tau(G), *, w)$ is a complete metrizable, separately continuous semigroup. So if W is a given weak neighbourhood of δ_1, then by Theorem 2.3.9 there exist $\mu_n \in T_n$ such that

$$\mu = \mathop{*}_{n=1}^{\infty} \mu_n \in W,$$

and the series converges unconditionally. Now since

$$\mu = \left(\mathop{*}_{j=1}^{n-1} \mu_j\right) * \mu_n * \left(\mathop{*}_{j=n+1}^{\infty} \mu_j\right),$$

we have that $\mu \in T_n$ for all $n \geq 1$ by Proposition 2.4.4. But this clearly implies that μ is a testing measure for $A = \bigcup_{n=1}^{\infty} A_n$, so if T is the set of testing measures for A, then $T \cap W \neq \emptyset$ for any weak neighbourhood W of δ_1. Thus A is a Haar null set by Proposition 2.4.4.

We can now state and prove our third and last criterion for thickness. The criterion was first given by J. P. R. Christensen (1974) for Polish abelian groups, and extended by Hoffmann–Jørgensen to

THEOREM 2.4.6 *Let G be a complete metrizable, separately continuous, abelian semigroup, and let A be a universally Radon measurable subset of G. If A is not a Haar null, then A is thick; more precisely the set*

$$U = \{x \in G: A \cap (x + A) \text{ is not a Haar null set}\}$$

is a neighbourhood of 0 contained in $A - A$ (we use additive notation since G is supposed to be abelian).

Proof Suppose that U is not a neighbourhood of 0, then $0 \in \overline{G \setminus U}$, so by Theorem 2.3.9 and Proposition 2.3.8 there exists a σ-additive G-valued measure f on $(\mathbb{N}, 2^{\mathbb{N}})$, such that

$$a_n = f(\{n\}) \in G \backslash U, \qquad \forall n \geq 1,$$

and $f: 2^{\mathbb{N}} \to G$ is continuous. Let λ be the Haar measure on $2^{\mathbb{N}}$, and put $\mu = \lambda \circ f^{-1}$, then μ is a tight probability on G with supp $(\mu) = f(2^{\mathbb{N}})$; thus $0 \in$ supp (μ). Now let

$$A^* = A \Big\backslash \bigcup_{n=1}^{\infty} (a_n + A),$$

then

$$A = A^* \cup \bigcup_{n=1}^{\infty} A \cap (a_n + A),$$

and $A \cap (a_n + A)$ is a Haar null set for all $n \geq 1$, since $a_n \notin U$. Thus A^* is not a Haar null set (cf. Theorem 2.4.5), and A^* is universally Radon-measurable. Hence there exists $a \in G$, such that $\mu(A^* - a) > 0$. So if

$$B^* = f^{-1}(A^* - a),$$

then B^* is λ-measurable and has positive Haar measure. From Theorem 2.4.3 (see also Example 2.4.1) we then deduce that B^* is thick. Hence for some $k \geq 1$ we have

$$N_{k-1}(0) = \{T \in 2^{\mathbb{N}}: T \cap \{1, \ldots, k-1\} = \emptyset\} \subseteq B^* - B^*.$$

In particular we find that $\{k\} = (T \backslash S) \cup (S \backslash T)$ for some sets $T, S \in B^*$. This implies $k \notin S$ and $T = S \cup \{k\}$ or vice versa; so by additivity of f we find $f(T) = a_k + f(S)$. Hence, if

$$\alpha = f(T) + a, \qquad \beta = f(S) + a,$$

then α and β belong to A^*, since $T, S \in B^* = f^{-1}(A^* - a)$, and $\alpha = a_k + \beta$; thus

$$\alpha \in A^* \cap (A^* + a_k) \subseteq A^* \cap (A + a_k).$$

But the last set is empty by definition of A^* and so we are led to a contradiction. Thus U must be a neighbourhood of 0 and the theorem is proved.

The completeness of G is indispensable, as seen from the following example:

$$G_k = \{(x_n) \in \mathbb{R}^{\mathbb{N}}: x_n = 0, \forall n \geq k\};$$

$$G = \bigcup_{k=1}^{\infty} G_k, \; d(x, y) = \sum_{n=1}^{\infty} |x_n - y_n|;$$

$$A = \bigcup_{k=1}^{\infty} \{x \in G_k: |x_j| \leq 2^{-k}, \forall j = 1, \ldots, k-1\}.$$

Then (G, d) is a separable metric group, G_k is a linear subspace, which is a Haar null set for all k (the normalized Gauss measure on G_{k+1} is a testing measure for G_k), and $G = \bigcup_{j=1}^{\infty} G_j$ is not a Haar null set. Moreover A is not a Haar nullset, and A is not thick.

The same example can be used to show that completeness without metrizability is not enough to make Theorems 2.4.5 and 2.4.6 true. Take the same G and G_k as above, but let the topology be the inductive topology from the injections $G_k \to G$, $k = 1, 2, \ldots$. Then G is a complete nuclear linear space and G is a Lusin space, but again the G_k are Haar null sets, such that their union $G = \bigcup_{k=1}^{\infty} G_k$ is not a Haar null set. And again A is not a Haar null set, and A is not thick.

Having established the three powerful criteria for thickness we may return to the problem of automatic continuity of a homomorphism.

Let G be a topologized semigroup and consider a σ-algebra $\mathscr{B}$ on G satisfying

$$1 \in \bigcup_{n=1}^{\infty} \operatorname{int}(A_n A_n^{-1}), \qquad \forall \{A_n\} \subseteq \mathscr{B} \quad \text{with } G = \bigcup_{n=1}^{\infty} A_n. \tag{97}$$

Theorem 2.4.2 shows that if G is a Baire group satisfying Axiom II^1, then the Baire property σ-algebra, $\mathscr{B}_{\mathrm{p}}(G)$, satisfies (97).

Theorem 2.4.3 shows that the m-measurable sets, $\mathscr{M}(m)$, satisfy (97) (with respect to τ^*) if (G, τ) is a completely regular topologized semigroup and m and τ^* satisfy (a)–(c) in Theorem 2.4.3.

Theorems 2.4.5 and 2.4.6 show that the universally measurable sets, $\mathscr{B}_{\mathrm{u}}(G)$, satisfy (97) if G is a complete metrizable, separately continuous, abelian semigroup.

We shall renounce the opportunity to give σ-algebras satisfying (97) a name and leave the baptism to the reader.

THEOREM 2.4.7 *Let G and H be topologized semigroups such that H satisfies axiom IV_0^1 and admits cancellations from the right (i.e. if $y'z = y''z$ then $y' = y''$). Suppose that $\mathscr{B}$ is a σ-algebra on G satisfying (97) and that $\mathscr{C}$ is a σ-algebra on H satisfying*

(a) *$\exists \mathscr{V}$ a neighbourhood base at 1 in H such that $Vy \in \mathscr{C}$ for all $V \in \mathscr{V}$ and all $y \in H$.*

Let $f: (G, \mathscr{B}) \to (H, \mathscr{C})$ be a measurable homomorphism, such that f is σ-bounded on G, then f is continuous at 1.

Proof Let $V \in \mathscr{N}_H(1)$, then by (a) and axiom IV_0^1 there exist $W \in \mathscr{V}$ such that $WW^{-1} \subseteq V$ and $Wy \in \mathscr{C}$ for all $y \in H$. Now by σ-boundedness of f there exist $y_n \in H$ so that

$$f(G) \subseteq \bigcup_{n=1}^{\infty} Wy_n.$$

Thus if $A_n = f^{-1}(Wy_n)$, then $A_n \in \mathscr{B}$ and $G = \bigcup_{n=1}^{\infty} A_n$. Hence by (97) we have that A_n is thick for some $n \geq 1$. Moreover, right cancellation in H gives

$$(Wy_n)(Wy_n)^{-1} \subseteq WW^{-1} \subseteq V \subseteq VV^{-1}.$$

So taking $A = A_n A_n^{-1}$ it follows from Proposition 2.4.1 that f is continuous at 1.

Specializing to specific σ-algebras $\mathscr{B}$ on G gives various results on automatic continuity, and we shall leave it to the reader to state the corollaries in the following three cases:

(i) $\mathscr{B} = \mathscr{B}_p(G)$, G a Baire group satisfying Axiom II^1;
(ii) $\mathscr{B} = \mathscr{M}(m)$, G, τ, τ^* and m satisfies (a)–(c) in Theorem 2.4.3;
(iii) $\mathscr{B} = \mathscr{B}_u(G)$, G a complete metrizable, separately continuous abelian semigroup.

The results in this section only give continuity at 1, but in many cases this will trivially imply continuity everywhere (see (64), page 358).

The cancellation property in Theorem 2.4.7 is indispensable, as it is seen from the following example.

Example 2.4.3 Let G be the set of probability measures on $\mathbb{R}_+$, with convolution and the weak topology. Then G is a Polish abelian semigroup satisfying Axiom I. Let $H = \mathbb{R} \cup \{\infty\}$ be the one point compactification of $\mathbb{R}$, and extend addition by the usual rules: $x + \infty = \infty$ for all $x \in H$. Then H is a metrizable compact abelian semigroup satisfying Axioms I, II and IV, and Axiom III holds for $x \neq \infty$, but not for $x = \infty$. Now let

$$f(\mu) = \int_0^{\infty} x\mu(dx), \qquad \forall \mu \in G.$$

Note that the integral exists but may be equal to $+\infty$. It is then easily seen that $f\colon G \to H$ is a homomorphism, and that f is lower semicontinuous. But f is not continuous at δ_0 (the neutral element in G is the Dirac measure δ_0). To see this we let

$$\mu_n = \left(1 - \frac{1}{n}\right)\delta_0 + \frac{1}{n}\delta_n,$$

then $\mu_n \xrightarrow{w} \delta_0$ but $f(\mu_n) = 1 \not\to f(\delta_0) = 0$.

Hence f is a lower semicontinuous homomorphism from G into H, such that f is discontinuous at δ_0. G is a Polish abelian jointly continuous semigroup and H is a compact metrizable semigroup, satisfying Axioms I, II and IV, and H is bounded. However, H does not admit cancellations.

σ-boundedness is a weak kind of separability condition and the reader may easily verify that we have (see also Lemma 2.3.14) that

$$\left.\begin{array}{l}\text{every Lindelöf subset and every separable subset of a topological}\\ \text{group is } \sigma\text{-bounded.}\end{array}\right\} \tag{98}$$

Here "separable" means "has a countable dense set". However, the condition is satisfied by even "very large" groups. A fairly straightforward argument shows that we have that

$$\left.\begin{array}{l}\text{every projective limit of } \sigma\text{-bounded topological groups}\\ \text{is itself } \sigma\text{-bounded.}\end{array}\right\} \tag{99}$$

Here H is the *projective limit* of $\{(H_i, p_i): i \in I\}$ if H and H_i are topological groups, $p_i: H \to H_i$ are homomorphisms and

$$\{p_i^{-1}(V_i): V_i \quad \text{open in} \quad H_i, i \in I\} \tag{100}$$

is a subbase for the topology. In particular we have that

$$\text{any product of topological } \sigma\text{-bounded groups is } \sigma\text{-bounded,} \tag{101}$$

and

$$\text{any vector space with a weak topology is } \sigma\text{-bounded.} \tag{102}$$

Here a *weak topology* on a vector space is a topology generated by a family of scalar-valued linear functionals.

Our next result requires a slightly different notion of boundedness. Let H be a topologized semigroup and A a subset of H, then A is called *flatly bounded* (resp. *flatly σ-bounded*), if we have

$$\forall V \in \mathcal{N}_H(1), \exists n \in \mathbb{N} \quad \text{so that} \quad A \subseteq V^n \tag{103'}$$

(resp.

$$\forall V \in \mathcal{N}_H(1) \quad \text{we have} \quad A \subseteq \bigcup_{n=1}^{\infty} V^n), \tag{103''}$$

where $V^n = \{x^n : x \in V\}$, whenever $V \subseteq H$ and $n \in \mathbb{N}$.

We shall say that a map $f: X \to H$ is *flatly bounded* on $A \subseteq X$ if $f(A)$ is flatly bounded, and similarly for *flatly σ-boundedness*.

Note that, if A is a subset of H, then we have

$$\left.\begin{array}{l}A \text{ is flatly } \sigma\text{-bounded if for all } y \in A \text{ there exist } \{y_n\} \subseteq H \\ \text{and } \{\sigma(n)\} \subseteq \mathbb{N}, \text{ such that } y_n \to 1, \text{ and } (y_n)^{\sigma(n)} = y, \forall n \geq 1.\end{array}\right\} \quad (104)$$

In particular we find (put $y_n = y/n$) that

$$\left.\begin{array}{l}\text{every topological linear space, regarded as an additive group,} \\ \text{is flatly } \sigma\text{-bounded.}\end{array}\right\} \quad (105)$$

Note that if H is a locally convex linear space, then we have that our definition of total boundedness coincides with the usual notion of total boundedness, while our definition of flatly bounded coincides with the usual notion of boundedness.

We shall need another group theoretical notion. A semigroup H is called *modular* if $x \mapsto x^n$ is an injective homomorphism for all $n \geq 1$, i.e. if

$$\left.\begin{array}{r}(xy)^n = x^n y^n, \qquad \forall x, y \in H, \forall n \geq 1, \\ x^n = y^n \text{ for some } n \geq 1 \text{ implies } x = y.\end{array}\right\} \quad (106)$$

Putting $n = 2$ we find $x(yx)y = x(xy)y$, so we have that

$$\left.\begin{array}{l}\text{if } H \text{ is a modular semigroup admitting left and right cancellation,} \\ \text{then } H \text{ is abelian.}\end{array}\right\} \quad (107)$$

Clearly we have that

any vector space over the rationals is modular. (108)

The circle group $T = \{z \in \mathscr{C} \mid |z| = 1\}$ is not modular, even though it is *infinitely divisible* (i.e. $x \mapsto x^n$ is surjective for all $n \geq 1$).

On the other hand there do exist modular semigroups which are not infinitely divisible. To see this let G be the set of probability measures μ on $\mathbb{R}$, such that its Fourier transform

$$\hat{\mu}(t) = \int_{-\infty}^{\infty} e^{itx} \mu(dx)$$

never vanishes. Then it is easy to check that $(G, *)$ is modular, but $\mu = p\delta_0 + (1-p)\delta_1 \in G$, if $p \neq \frac{1}{2}$, and μ has no non-trivial factors. If we equip G with its weak topology, then it is fairly straightforward to check that

$$\left.\begin{array}{l}A \subseteq G \text{ is flatly } \sigma\text{-bounded if, and only if,} \\ \mu \text{ is infinitely divisible for all } \mu \in A.\end{array}\right\} \quad (109)$$

Note that the invertible elements of $(G, *)$ are exactly the Dirac measures, and that the infinitely divisible probability measures form a subsemigroup of $(G, *)$.

In analogy to Theorem 2.4.7, we have the following automatic continuity result for modular semigroups and flatly σ-bounded homomorphism:

THEOREM 2.4.8 *Let G be a topologized semigroup such that $x \mapsto x^n$ is continuous at $x = 1$ for all $n \geq 1$, and let H be a topologized modular semigroup satisfying Axiom* IV_0^1. *Suppose that $\mathcal{B}$ is a σ-algebra on G satisfying* (97) *and $\mathcal{C}$ is a σ-algebra satisfying*

(a) *$\exists \mathcal{V}$ a neighbourhood base of 1 in H, such that $V^n \in \mathcal{C}$ for all $V \in \mathcal{V}$ and all $n \geq 1$.*

Let $f: (G, \mathcal{B}) \to (H, \mathcal{C})$ be a measurable homomorphism, such that f is flatly σ-bounded on G, then f is continuous at 1.

Proof Let W be a neighbourhood of 1 in H, then by Axiom IV_0^1 there exist $V \in \mathcal{V}$, such that $VV^{-1} \subseteq W$. Since f is flatly σ-bounded, we have

$$G = \bigcup_{n=1}^{\infty} f^{-1}(V^n),$$

and $f^{-1}(V^n) \in \mathcal{B}$ by (a) and measurability of f. Thus by condition (97) there exists a $k \geq 1$, such that $A = f^{-1}(V^k)$ is thick.

By continuity of $x \mapsto x^k$ at $x = 1$, we can find a neighbourhood U of 1 in G, satisfying

$$U^k \subseteq \operatorname{int}(AA^{-1}).$$

Now suppose that $x \in U$ and let $y = f(x)$. Then

$$y^k = f(x^k) \in f(U^k) \subseteq (V^k)(V^k)^{-1};$$

so there exist $v_1, v_2 \in V$, with $v_1^k = y^k v_2^k$. Hence by modularity of H we have $v_1 = yv_2$, and thus we find $y \in VV^{-1} \subseteq W$ for $y \in U$. But this shows that $f(U) \subseteq W$ and so f is continuous at 1.

Theorems 2.4.7 and 2.4.8 generalize Sierpiński's result on measurable solution to the Cauchy equation. Analysing the proof of Theorem 2.4.8 it is seen that we actually have proved the following generalization of Ostrowski's result:

THEOREM 2.4.9 *Let G be a topologized semigroup, such that $x \mapsto x^n$ is continuous at $x = 1$ for all $n \geq 1$, and let H be a topologized modular semigroup satisfying Axiom* IV_0^1. *If $f: G \to H$ is a homomorphism, which is flatly bounded on some thick set, then f is continuous at* 1.

Note that if H is topological linear vector space, then $(H, +)$ is a modular, flatly σ-bounded topological group and if H is locally convex, then flatly boundedness coincides with the usual concept of boundedness in linear topological linear spaces, thus Theorems 2.4.8 and 2.4.9 are particularly useful, when H is a linear space.

Theorem 2.4.8 has three corollaries, namely the cases

(i) $\mathcal{B} = \mathcal{B}_p(G)$, G a Baire group satisfying Axiom II^1;
(ii) $\mathcal{B} = \mathcal{M}(m)$, G, τ, τ^* and m satisfy (a)–(c) in Theorem 2.4.3;
(iii) $\mathcal{B} = \mathcal{B}_u(G)$, G a completely metrizable, separately continuous abelian semigroup.

We shall leave the statements of these to the reader.

Similarly, Theorems 2.4.2, 2.4.3 and 2.4.5 give natural corollaries to Theorem 2.4.9; again we shall leave the precise statement to the reader. But let us give two applications of Theorems 2.4.8 and 2.4.9.

Example 2.4.4 Let G be a locally compact group with left Haar measure λ, and let H be a topologized semigroup satisfying Axiom IV^1. Suppose that $f: G \to H$ is a homomorphism, then f is continuous in either of the following four cases:

(i) f is σ-bounded and $\mathcal{B}_p$-measurable for some σ-algebra $\mathcal{C}$ satisfying (a) in Theorem 2.4.8;
(ii) f is σ-bounded and λ-measurable for some σ-algebra $\mathcal{C}$ satisfying (a) in Theorem 2.4.8;
(iii) f is flatly bounded on some non-meagre subset of G, and H is modular;
(iv) f is flatly bounded on some set of positive Haar measure, and H is modular.

If H is modular and the σ-algebra $\mathcal{C}$ on H satisfies (a) in Theorem 2.4.8, then

(v) f is continuous if f is flatly σ-bounded and f is either $\mathcal{B}_p$-measurable or λ-measurable with respect to $\mathcal{C}$.

Example 2.4.5 Let G be a locally convex vector space and μ a Gaussian Radon measure on G with supp $(\mu) = G$. If f is a linear map $G \to H$ where H is a topological linear space, then f is continuous with respect to the reproducing kernel norm

$$\|x\| = \sup\left\{|x'(x)|: x' \in G', \int_G |x'|^2 \, d\mu \leq 1\right\}$$

in the following two cases:

(i) f is μ-measurable;
(ii) f is flatly bounded on a set of positive μ-measure.

Moreover, from Example 2.4.2 we find:

(iii) if G_0 is an additive subgroup of G with $\mu_*(G_0)>0$, then $H(\mu)\subseteq G_0$; and
(iv) if f and g are μ-measurable linear maps such that $\mu(f=g)>0$, then $f(x)=g(x)$ for all $x\in H(\mu)$,

where $H(\mu)$ is the reproducing kernel of μ,

$$H(\mu)=\{x\in G: \|x\|<\infty\}.$$

Note that the assumption in (iii) implies that G_0 is μ-measurable and $\mu(G_0)=1$. Similarly, the assumption in (iv) implies that $f=g$, μ-almost everywhere.

2.5 Automatic continuity and integral equations

It is well known that if G is a locally compact abelian group, then $f*g$ is continuous, if $f\in L^p(G)$ and $g\in L^q(G)$ where $1/p+1/q=1$. We shall now show how this extends to an automatic continuity theorem for solutions to the following integral equation:

$$f(x+y)=\int_S g(s,x)h(s,y)\mu(ds), \qquad \forall x, y\in G_0, \tag{110}$$

where G_0 is a subset of the group G. The functions f, g and h in this integral equation will be Banach space-valued functions, so let us recall some of the results from Banach space valued integration.

Let (S, Σ, μ) be a positive measure space and $(E, \|\cdot\|)$ Banach space with dual E'. Then a map $f: S\to E$ is called *weakly measurable* if $x'\circ f$ is μ-measurable for all $x'\in E'$. f is called *strongly measurable* if there exist simple functions

$$f_n=\sum x_{jn}1_{A_{jn}} \qquad (x_{jn}\in E, A_{jn}\in\Sigma),$$

such that $\|f_n-f\|\to 0$, μ-almost everywhere. It is well known that we have

$$\left.\begin{array}{l} f \text{ is strongly measurable if, and only if, } f \text{ is weakly measurable} \\ \text{and } f(S\backslash N) \text{ is a separable subset of } E \text{ for some } \mu\text{-null set } N. \end{array}\right\} \tag{111}$$

The function f is called *Pettis-integrable* with *Pettis integral* $a\in E$ if $x'\circ f$ is μ-integrable for all $x'\in E'$, and if

$$x'(a)=\int_S x'(f(s)\mu(ds), \qquad \forall x'\in E';$$

then we write

$$a = \int_S f(s)\mu(\mathrm{d}s).$$

In the setting of the integral equations (110) we have three Banach spaces $(E_1, \|\cdot\|)$, $(E_2, \|\cdot\|)$ and $(E_3, \|\cdot\|)$ (we use the same symbol for the norm in all three spaces to avoid too many indices). Moreover we have given a bilinear form

$$(x, y) \mapsto xy : E_1 \times E_2 \to E_3,$$

satisfying

$$\|xy\| \leqslant \|x\| \, \|y\|, \qquad \forall x \in E_1, \forall y \in E_2. \tag{112}$$

Then a function $g: (S, \Sigma, \mu) \to E_1$ is called *scalarly measurable*, if $z'(g(\cdot)y)$ is μ-measurable for all $z' \in E_3'$ and all $y \in E_2$. The reader may easily verify that we have that

$$\left.\begin{array}{l}\text{if } G: S \to E_1 \text{ is scalarly measurable and } F: S \to E_2 \text{ is strongly} \\ \text{measurable, then } h(s) = g(s)f(s) \text{ is weakly measurable.}\end{array}\right\} \tag{113}$$

It follows trivially from the definition of the Pettis integral that

$$\left.\begin{array}{l}\text{if } f \text{ is Pettis integrable and } \|f(s)\| \leqslant \varphi(s) \text{ for some} \\ \mu\text{-measurable function } \varphi: S \to \mathbb{R}_+, \text{ then} \\ \qquad \left\| \int_S f(s)\mu(\mathrm{d}s) \right\| \leqslant \int_S \varphi(s)\mu(\mathrm{d}s).\end{array}\right\} \tag{114}$$

Note that $\|f(\cdot)\|$ will *not* in general be μ-measurable for a weakly measurable function $f: S \to E$.

LEMMA 2.5.1 *Let $(G, +)$ be a locally compact abelian group with Haar measure m, and let f be a strongly measurable map from G into a Banach space $(E, \|\cdot\|)$. Let $A \in \mathscr{B}(G)$ and $0 < p < \infty$, if*

(a) $$\int_{A+U} \|f(x)\|^p m(\mathrm{d}x) < \infty,$$

for some open neighbourhood U of 0, then we have

(b) $$\lim_{y \to 0} \int_A \|f(x+y) - f(x)\|^p m(\mathrm{d}x) = 0.$$

Proof The lemma is well known if $A = G$, and the proof in the general case follows the same lines. First, we approximate f by a simple function,

$f_1 = \sum x_j 1_{A_j}$, with $m(A_j) < \infty$; second, 1_{A_j} is approximated by a continuous real function, φ_j, with compact support, and finally we form $f_0 = \sum x_j \varphi_j$, which then approximates f. Formalizing this we can, for any given $\varepsilon > 0$, find a continuous function f_0: $G \to E$ with compact support such that

$$\int_{A+U} \|f(x) - f_0(x)\|^p m(\mathrm{d}x) < \varepsilon. \tag{115}$$

Let C be the support of f_0 and let K be a compact neighbourhood of 0, then we have

$$\int_G \|f_0(x+y) - f_0(x)\|^p m(\mathrm{d}x) = \int_{C-K} \|f_0(x+y) - f_0(x)\|^p m(\mathrm{d}x),$$

for all $y \in K$. Hence f_0 satisfies (b), since f_0 is uniformly continuous. Moreover, if $y \in U$, then

$$\int_A \|f(x+y) - f_0(x+y)\|^p m(\mathrm{d}x) = \int_{A+y} \|f(z) - f_0(z)\|^p m(\mathrm{d}x)$$

$$\leqslant \int_{A+U} \|f(z) - f_0(z)\|^p m(\mathrm{d}x) < \varepsilon,$$

from which (b) follows easily.

We are now ready to prove continuity of solutions to (110). But let us first fix the notation.

Let $(E_j, \|\cdot\|)$, $j = 1, 2, 3$, be *three Banach spaces* with a *bilinear form satisfying* (112). Let (S, Σ, μ) be a *positive measure space* and $(G, +, m)$ a *locally compact abelian group* with its *Haar measure* m. Let g, h, φ and ψ be *functions*:

$$\left.\begin{array}{ll} g: S \times G \to E_1, & \varphi: S \times G \to \mathbb{R}_+, \\ h: S \times G \to E_2, & \psi: S \times G \to \mathbb{R}_+. \end{array}\right\} \tag{116}$$

Finally let G_0 be an *open subset* of G, and put

$$G_1 = G_0 + G_0. \tag{117}$$

Then we have

THEOREM 2.5.2 *Let the situation be as described above and suppose that μ is σ-finite, and that we have*

(a) *$g(\cdot, x)$ is scalarly measurable for all $x \in G$, and $\|g(s, x)\| \leqslant \varphi(s, x)$ for all $(s, x) \in S \times G$;*

(b) *$h(\cdot,\cdot)$ is strongly measurable with respect to $(\mu\times m)$, and $\|h(s,x)\|\leqslant \psi(s,x)$ for all $(s,x)\in S\times G$;*
(c) *$g(\cdot,x)h(\cdot,y)$ is Pettis-integrable for all $x, y\in G$;*
(d) *$\varphi 1_{S\times C}\in L^q(\mu\times m)$ and $\psi 1_{S\times C}\in L^p(\mu\times m)$ for all compact sets $C\subseteq G_0$;*

where $1\leqslant p<\infty$, $1<q\leqslant\infty$ and $1/p+1/q=1$. If f is a function $G_1\to E_3$ satisfying,

$$\text{(e)}\quad f(x+y)=\int_S g(s,x)h(s,y)\mu(\mathrm{d}s),\ \forall x,y\in G_0,$$

then f is sequentially continuous at every point x_0 belonging to $G_0+G_0=G_1$.

Proof Let $x_0\in G_0+G_0$, then $x_0=y_0+z_0$, and there exists a compact neighbourhood K of 0 so that (recall that G_0 is open)

$$y_0+K-K\subseteq G_0,\qquad z_0+K-K\subseteq G_0. \tag{118}$$

Now let $C=y_0+K$, then we have

$$x_0+K-C\subseteq G_0,\qquad C\subseteq G_0. \tag{119}$$

Hence, if $L=x_0+K-C$, and

$$\beta(s)=\begin{cases}\left[\int_C \varphi(s,x)^q m(\mathrm{d}x)\right]^{1/q}, & \text{for } q<\infty,\\ m-\operatorname*{ess\,sup}_{x\in C}\varphi(s,x), & \text{for } q=\infty,\end{cases}$$

$$\gamma(s)=\left[\int_L \psi(s,x)^p m(\mathrm{d}s)\right]^{1/p},$$

then we have (cf. (d) and recall that μ is σ-finite)

$$\beta(s)<\infty,\ \forall s\in S\backslash N,\quad\text{and}\quad \beta\in L^q(\mu), \tag{120}$$

$$\gamma(s)<\infty,\ \forall s\in S\backslash N,\quad\text{and}\quad \gamma\in L^p(\mu), \tag{121}$$

for some μ-null set N. So applying Lemma 2.5.1 with $A=x_0-C$ and $U=\operatorname{int}(K)$, we find

$$\lim_{y\to 0}\int_{x_0-C}\|h(s,y+z)-h(s,z)\|^p m(\mathrm{d}z)=0,\qquad \forall s\in S\backslash N. \tag{122}$$

Now let $y \in K$, and let $\alpha = m(C)$, then we have $x_0 + y - x$, $x_0 - x$ and x belong to G_0 for all $x \in C$. So from (e) we deduce

$$\begin{aligned}
&\alpha\|f(x_0+y)-f(x_0)\| \\
&\quad = \int_C \|f(x_0+y)-f(x_0)\| m(\mathrm{d}x) \\
&\quad = \int_C m(\mathrm{d}x)\left\| \int_S g(s,x)\{h(s,x_0+y-x)-h(s,x_0-x)\}\mu(\mathrm{d}s)\right\| \\
&\quad \leqslant \int_C m(\mathrm{d}x) \int_S \varphi(s,x)\|h(s,x_0+y-x)-h(s,x_0-x)\|\mu(\mathrm{d}s).
\end{aligned}$$

Now φ is $\mu \times m$-measurable on $S \times C$ by assumption (d), and

$$(s,x) \mapsto \|h(s,x_0+y-x)-h(s,x_0-x)\|$$

is $\mu \times m$-measurable by strong measurability of $h(\cdot,\cdot)$. Hence we may interchange the order of integration. So, if

$$\Gamma(s,y) = \int_C \varphi(s,x)\|h(s,x_0+y-x)-h(s,x_0-x)\| m(\mathrm{d}x),$$

then we have (note that $0 < \alpha < \infty$)

$$\|f(x_0+y)-f(x_0)\| \leqslant \alpha^{-1} \int_S \Gamma(s,y)\mu(\mathrm{d}s). \tag{123}$$

We can then estimate $\Gamma(s, y)$ by Hölder's inequality:

$$\begin{aligned}
\Gamma(s,y) &\leqslant \beta(s)\left[\int_C \|h(s,x_0+y-x)-h(s,x_0-x)\|^p m(\mathrm{d}x)\right]^{1/p} \\
&= \beta(s)\left[\int_{x_0-C} \|h(s,z+y)-h(s,z)\|^p m(\mathrm{d}z)\right]^{1/p}.
\end{aligned}$$

Hence by (122) and (120) we have

$$\lim_{y\to 0} \Gamma(s,y) = 0, \qquad \forall s \in S\backslash N. \tag{124}$$

Now let $y \in K$, then continuing the inequality above using Minkowski's inequality, we find

$$\begin{aligned}
\Gamma(s,y) &\leqslant \beta(s)\left\{\left[\int_{x_0-C} \psi(s,z+y)^p m(\mathrm{d}z)\right]^{1/p} + \left[\int_{x_0-C} \psi(s,z)^p(\mathrm{d}z)\right]^{1/p}\right\} \\
&\leqslant \beta(s)\left\{\left[\int_{x_0+y-C} \psi(s,x)^p m(\mathrm{d}x)\right]^{1/p} + \left[\int_{x_0-C} \psi(s,x)^p m(\mathrm{d}x)\right]^{1/p}\right\} \\
&\leqslant 2\beta(s)\gamma(s),
\end{aligned}$$

since $x_0+y-C\subseteq L$ and $x_0-C\subseteq L$ for $y\in K$. In other words, from (120) and (121) we conclude that

$$\Gamma(s,y)\leqslant 2\beta(s)\gamma(s)\,, \tag{125}$$

$$\beta(\cdot)\gamma(\cdot)\in L^1(\mu)\,. \tag{126}$$

Propositions (124)–(126) show that we may use Lebesgue's dominated convergence theorem to deduce

$$\lim_{n\to\infty}\int_S \Gamma(s,y_n)\mu(\mathrm{d}s)=0\,,$$

whenever $y_n\to 0$. So from (123) we find that f is sequentially continuous at x_0.

Note that Lebesgue's theorem on dominated convergence is *not* valid for generalized sequences, so the proof above only gives sequential continuity. However the following version of dominated convergence for generalized sequences,

$$\left.\begin{array}{l} f_\alpha\to f \qquad \text{in } \mu\text{-measure} \\ |f_\alpha|\leqslant g,\ \forall\alpha,\ g\in L^1(\mu) \end{array}\right\}\Rightarrow f_\alpha\to f \quad \text{in} \quad L^1(\mu)\,, \tag{127}$$

shows that, if we have

$$\int_L \|h(s,x+y)-h(s,x)\|^p m(\mathrm{d}x)\xrightarrow[y\to 0]{}\text{in } \mu\text{-measure}\,, \tag{128}$$

for all compact subsets $L\subseteq G_0$, then f is continuous on G_0+G_0.

Example 2.5.1 Let us consider the case where the measure space (S,Σ,μ) is trivial, that is $S=\{0\}$, $\mu=\delta_0$. Then equation (110) becomes

$$f(x+y)=g(x)h(y)\,, \qquad \forall x,y\in G_0\,. \tag{129}$$

Then conditions (a)–(d) of Theorem 2.5.2 are satisfied if g is scalarly m-measurable and h is strongly m-measurable, and if

$$\|g(\cdot)\|\in L^q_{00}(G_0)\,, \qquad \|h(\cdot)\|\in L^p_{00}(G_0)\,, \tag{130}$$

where $L^s_{00}(A)$ is the set of m-measurable functions $\varphi:G\to\mathbb{R}$, such that $\varphi 1_C\in L^s(G)$ for all compact sets $C\subseteq A$. Note also that (128) holds in this case so we get continuity of f, without further hypothesis.

Example 2.5.2 Let $S=\mathbb{N}$ and $\mu=$ the counting measure on $\mathbb{N}$. Then (110) becomes

$$f(x+y)=\sum_{n=1}^{\infty} g_n(x)h_n(y)\,, \qquad \forall x,y\in G_0\,. \tag{131}$$

And (a)–(d) of Theorem 2.5.2 are satisfied if g_n is scalarly m-measurable, h_n is strongly m-measurable, and

$$\left.\begin{aligned} &\sum_{n=1}^{\infty} \|g_n(\cdot)\|^q \in L^1_{00}(G_0)\,, && \sum_{n=1}^{\infty} \|h_n(\cdot)\|^p \in L^1_{00}(G)\,, && \text{if } q<\infty\,, \\ &\sup_n \|g_n(\cdot)\| \in L^{\infty}_{00}(G_0)\,, && \sum_{n=1}^{\infty} \|h_n(\cdot)\| \in L^1_{00}(G)\,, && \text{if } q=\infty\,. \end{aligned}\right\} \tag{132}$$

In this case (128) may fail, so in general we can only conclude sequential continuity of f.

Example 2.5.3 Now let us put $S=G$, $\mu=m$ and consider $g_0 \in L^q(G)$, $h_0 \in L^p(G)$, then $f_0 = g_0 * h_0$ satisfies

$$f_0(x+y) = \int_G g_0(x+z) h_0(y-z) m(\mathrm{d}z)\,, \qquad \forall x, y \in G_0. \tag{133}$$

So if $g(s, x) = g_0(x+s)$ and $h(s, x) = h_0(x-s)$, we are back to equation (110). Moreover we have

$$\begin{aligned} &\int_G \int_L |h(s, x+y) - h(s, x)|^p m(\mathrm{d}x) m(\mathrm{d}s) \\ &\qquad = \int_L m(\mathrm{d}x) \int_G |h_0(x+y-s) - h_0(x-s)|^p m(\mathrm{d}s) \\ &\qquad = m(L) \int_G |h_0(y-z) - h_0(z)|^p m(\mathrm{d}z)\,, \end{aligned}$$

which shows that (128) holds. Hence $g_0 * h_0$ is continuous on G (here we take $G_0 = G$). Note that this extends easily to the case where g_0 and h_0 take values in a pair of Banach spaces (E_1, E_2) with a bilinear form $E_1 \times E_2 \mapsto E_3$.

Appendix : Measures on topological spaces

Complete regularity turns out to be extremely helpful when measures on Hausdorff spaces are studied. So we shall throughout this appendix assume that

> *all topological spaces are completely regular and Hausdorff if not stated explicitly otherwise.*

By a Borel measure on X we shall mean a measure defined on the Borel subsets of X. Baire measures are defined similarly. If μ is a Borel measure on X, then

(a) μ is *τ-smooth* if $\mu(U)=\sup_\alpha \mu(U_\alpha)$ for every family $\{U_\alpha\}$ filtering up to the open set U, such that U_α is open for all α;
(b) μ is *$\mathscr{F}$-regular* if $\mu(B)=\sup \mu(F)$ for all $B\in\mathscr{B}(X)$, where the sup is taken over all closed sets $F\subseteq B$;
(c) μ is *tight* if $\mu(B)=\sup \mu(K)$ for all $B\in\mathscr{B}(X)$, where the sup is taken over all compact sets $K\subseteq B$;
(d) μ is *locally finite*, if every point $x\in X$ has a neighbourhood U with $\mu(U)<\infty$.
(e) μ is a *Radon measure* if μ is tight and locally finite.

The relations between these notions are given by

$$\text{Radon}\Rightarrow\text{tight}\Rightarrow\tau\text{-smooth}\,. \tag{A1}$$

If μ is a Borel measure on X, then its *support*, denoted supp (μ), is defined by

$$\operatorname{supp}(\mu)=\{x\in X:\mu(U)>0,\ \forall U \text{ a neighbourhood of } x\}\,.$$

Then supp (μ) is clearly closed and contained in every closed set F with $\mu(X\backslash F)=0$. Moreover by the definition of τ-smoothness we have

$$\mu(X\backslash\operatorname{supp}(\mu))=0 \quad \text{whenever } \mu \text{ is } \tau\text{-smooth}\,. \tag{A2}$$

Thus the support is really only meaningful when μ is τ-smooth.

Note that a σ-finite measure admits at most countably many disjoint sets of positive measure. This observation gives us the following fact:

$$\left.\begin{array}{l}\text{If } X \text{ is metrizable and } \mu \text{ is } \sigma\text{-finite, then supp}(\mu)\\ \text{is separable. Conversely, if } X \text{ is hereditary Lindelöf}\\ \text{(e.g. second countable), then every Borel measure is } \tau\text{-smooth.}\end{array}\right\} \tag{A3}$$

For tight measures we have the following well-known fact (cf. Part 1 of this volume, Section 3.5):

$$\left.\begin{array}{l}\text{if } X \text{ is } K\text{-analytic and } \mathscr{G}(X)\subseteq S(\mathscr{F}(X)), \text{ then every } \sigma\text{-finite}\\ \text{Borel measure on } X \text{ is tight}\,.\end{array}\right\} \tag{A4}$$

Spaces with this property are sometimes called *Radonian*. If we only require all finite τ-smooth measures to be tight, we say that X is *semi-Radonian.* This is, for instance, the case if X is a Borel subset of a locally Čech complete space.

Let μ (resp. ν) be finite Borel measures on X (resp. Y), then the product measure $\mu\times\nu$ is defined on the product σ-algebra $\mathscr{B}(X)\otimes\mathscr{B}(Y)$. This

σ-algebra is, however, in general a proper sub-σ-algebra of $\mathscr{B}(X \times Y)$. (This will *always* be the case if both X and Y have cardinal strictly greater than the continuum.) To solve this problem note that the function

$$y \mapsto \int_X f(x, y)\mu(\mathrm{d}x)$$

is lower semi-continuous, if μ is a finite τ-smooth measure on X, and f is lower semi-continuous and bounded. This observation leads easily to the following:

if μ and ν are finite τ-smooth measures then $\mu \times \nu$ admits a unique τ-smooth extension $\overline{\mu \times \nu}$ to $\mathscr{B}(X \times Y)$, and we have

$$\iint_{X \times Y} f\,\mathrm{d}(\overline{\mu \times \nu}) = \int_X \mathrm{d}\mu \int_Y f\,\mathrm{d}\nu = \int_Y \mathrm{d}\nu \int_X f\,\mathrm{d}\mu \qquad \text{(A5)}$$

for all bounded Borel functions $f: X \times Y \to \mathbb{R}$.

Similarly,

if μ and ν are Radon measures then $\mu \times \nu$ admits a unique Radon extension $\overline{\mu \times \nu}$ to $\mathscr{B}(X \times Y)$, and if ν is σ-finite on $B \in \mathscr{B}(Y)$ then we have

$$\iint_{X \times B} f\,\mathrm{d}(\overline{\mu \times \nu}) = \int_X \mathrm{d}\mu \int_B f\,\mathrm{d}\nu \qquad \text{(A6)}$$

for all bounded Borel functions $f: X \times Y \to \mathbb{R}$.

Note that if ν is τ-smooth and σ-finite on $B \in \mathscr{B}(Y)$ then

$$x \mapsto \int_B f(x, y)\nu(\mathrm{d}y)$$

is Borel-measurable for all Borel-measurable functions f: $X \times Y \to \mathbb{R}_+$. This and the Fubini theorem in (A6) need not hold if ν is not σ-finite on B. (Take $X = Y = [0, 1]$, μ = Lebesgue measure, ν = the counting measure and f the indicator of the diagonal.)

The result of Glicksberg (1961) and Johnson (1969) allows one to extend the Fubini theorem significantly, when μ and ν are Radon measures. Recall that a function $f: X \to Y$ is called *k-continuous* if its restriction to all compact subsets of X is continuous. Then we have that

a separately k-continuous function $f: X \times Y \to \mathbb{R}$ is λ-measurable for every finite Radon measure λ on $X \times Y$; (A7)

$$\left.\begin{array}{l}\text{if } f: X \times Y \to \mathbb{R} \text{ is bounded and separately } k\text{-continuous} \\ \text{and } \mu \text{ is a finite Radon measure on } X, \text{ then} \\ \qquad y \mapsto \int_X f(x, y)\mu(\mathrm{d}x) \\ \text{is } k\text{-continuous, and if } Y \text{ is compact, then the set} \\ \{f(x, \cdot): x \in \operatorname{supp}(\mu)\} \text{ is a norm-separable subset of } C(Y).\end{array}\right\} \tag{A8}$$

These results were originally stated and proved for locally compact spaces X and Y, but the extension above is immediate and requires only trivial changes in the original proof.

We shall now recall some of the fundamental facts about the weak topology on the sets of measures. Let us introduce three sets of measures:

$$M_\sigma^+(X) = \{\mu : \mu \text{ finite non-negative Baire measure}\},$$

$$M_\tau^+(X) = \{\mu : \mu \text{ finite non-negative } \tau\text{-smooth measure}\},$$

$$M_t^+(X) = \{\mu : \mu \text{ finite non-negative tight measure}\}.$$

Then clearly we have

$$M_t^+(X) \subseteq M_\tau^+(X) \subseteq M_\sigma^+(X), \tag{A9$'$}$$

and we define the weak topology, denoted by w, on $M_\sigma^+(X)$ by

$$\mu_\alpha \overset{w}{\to} \mu \Leftrightarrow \int_X f\,\mathrm{d}\mu_\alpha \to \int_X f\,\mathrm{d}\mu_\alpha, \qquad \forall f \in C(X),$$

where $C(X)$ is the set of bounded continuous real functions on X. Note that a Baire measure is uniquely determined by its integrals of bounded continuous functions, so the weak topology is Hausdorff and completely regular. A simple argument shows that

$$M_t^+(X) \text{ is } w\text{-dense in } M_\sigma^+(X). \tag{A9$''$}$$

Portmanteau's theorem states that $\mu_\alpha \overset{w}{\to} \mu$ if, and only if,

$$\mu_\alpha(X) \to \mu(X) \quad \text{and} \quad \limsup \mu_\alpha(F) \leq \mu(F), \quad \forall F \in \mathcal{F}(X), \tag{A9$'''$}$$

where $\mathcal{F}(X)$ is the set of closed subsets of X. From (A2) and (A3) we obtain that

$$\left.\begin{array}{l}\text{if } X \text{ is metrizable and } \mu \in M_\sigma^+(X), \text{ then } \mu \text{ is } \tau\text{-smooth if, and} \\ \text{only if, } \mu(X) = \mu(\operatorname{supp}(\mu)).\end{array}\right\} \tag{A10}$$

Thus by (A9″) and (A9‴) we have that

$$\left.\begin{array}{l}\text{if } X \text{ is metrizable, then } M_\tau^+(X) \text{ is sequentially } w\text{-closed and}\\ w\text{-dense in } M_\sigma^+(X).\end{array}\right\} \tag{A11}$$

Let f be a Borel map from X into Y, then for every Borel measure μ on X we can define the *image measure* $\mu \circ f^{-1}$ by

$$\mu \circ f^{-1}(B) = \mu(f^{-1}(B)), \qquad \forall B \in \mathscr{B}(Y),$$

and $\mu \circ f^{-1}$ is then a Borel measure on Y. However, $\mu \circ f^{-1}$ may fail to be τ-smooth even when $\mu \in M_\tau^+(X)$, and we shall say that a map $f: X \to Y$ is *τ-measurable* if f is Borel measurable and $\mu \circ f^{-1} \in M_\tau^+(Y)$ for all $\mu \in M_\tau^+(X)$. Clearly we have

$$\left.\begin{array}{l}\text{supp}\,(\mu \circ f^{-1}) \subseteq \overline{f(\text{supp}\,(\mu))} \text{ whenever } \mu \text{ is } \tau\text{-smooth and } f \text{ is}\\ \text{Borel measurable}.\end{array}\right\} \tag{A12}$$

From (A11) it follows easily that

$$\left.\begin{array}{l}\text{if } Y \text{ is metrizable and } f: X \to Y \text{ is a Baire class function, then}\\ f \text{ is } \tau\text{-measurable}.\end{array}\right\} \tag{A13}$$

Here the Baire classes are defined as usual by transfinite induction:

$$B_0(X, Y) = \{f : f \text{ continuous: } X \to Y\},$$

$$B_\alpha(X, Y) = \{f : \exists\{f_n\} \subseteq \bigcup_{\lambda < \alpha} B_\lambda(X, Y) \text{ with } f_n(x) \to f(x), \forall x\},$$

for all ordinals. Note that

$$B_{\omega_1}(X, Y) = \bigcup_{\alpha < \omega_1} B_\alpha(X, Y) = B_\lambda(X, Y), \qquad \forall \lambda \geq \omega_1,$$

and that all functions in $B_\alpha(X, Y)$ are Baire measurable (i.e. measurable with respect to the Baire σ-algebras in X and Y). Hence, if Y is metrizable or, more generally, if Y is perfectly normal, then every Baire class function, $f: X \to Y$, is Borel measurable.

Now suppose that X is metrizable and $f: X \times Y \to Z$ is separately continuous. We can then embed Z into a locally convex linear space Z_0 (e.g. $Z_0 = \mathbb{C}^{C(Z)}$). Now let $\{\varphi_a^n : a \in X\}$ be a locally finite partition of unity in X for each $n \in \mathbb{N}$, such that

$$\varphi_a^n(x) = 0, \quad \text{if} \quad d(x, a) \geq 2^{-n},$$

where d is a metric on X generating the topology. Then

$$f_n(x, y) = \sum_{a \in X} \varphi_a^n(x) f(a, y)$$

is a jointly continuous map $X \times Y \to Z_0$, and

$$f_n(x, y) \xrightarrow[n\to\infty]{} f(x, y), \qquad \forall (x, y) \in X \times Y.$$

Thus

$$\left.\begin{array}{l}\text{if } X \text{ is metrizable and } f: X \times Y \to Z \text{ is separately continuous,}\\ \text{then there exists a completely regular space } Z_0, \text{ with}\\ Z \subseteq Z_0 \text{ and } f \in B_1(X \times Y, Z_0).\end{array}\right\} \tag{A14}$$

If Z is metrizable then Z_0 may be taken to be metrizable (e.g. $Z_0 = (C(Z), \|\cdot\|_\infty)$), and so f becomes Borel-measurable. Thus, from (A13) and (A14),

$$\left.\begin{array}{l}\text{if } X \text{ and } Z \text{ are metrizable and } f: X \times Y \to Z \text{ is separately}\\ \text{continuous, then } f \text{ is } \tau\text{-measurable}.\end{array}\right\} \tag{A15}$$

Let us pursue this a little further. Suppose that X is metrizable and that $f: X \times Y \to Z$ is τ-measurable. Let μ and ν be finite τ-smooth measures on X and Y. Then $\overline{\mu \times \nu}$ is a τ-smooth measure on $X \times Y$, and $\lambda = \overline{(\mu \times \nu)} \circ f^{-1}$ is a τ-smooth measure on Z. Now let $x_0 \in \operatorname{supp}(\mu)$ and $y_0 \in \operatorname{supp}(\nu)$, such that

$$f(x_0, \cdot) \text{ is continuous at } y_0; \tag{A16}$$

$$f(\cdot, y) \text{ is continuous at } x_0 \text{ for all } y \in Y; \tag{A17}$$

and put $z_0 = f(x_0, y_0)$. If V is an open neighbourhood of z_0, then by (A6) we have

$$\lambda(V) = \int_X F \,d\mu, \quad \text{where} \quad F(x) = \int_Y 1_V(f(x, y))\nu(dy).$$

By (A7) and the metrizability of X we have that F is lower semi-continuous at x_0, and since $f_{x_0}^{-1}(V)$ is a neighbourhood of $y_0 \in \operatorname{supp}(\nu)$ by (A16), we have that $F(x_0) > 0$. Hence $\lambda(V) > 0$, since $x_0 \in \operatorname{supp}(\mu)$, and so $z_0 \in \operatorname{supp}(\lambda)$. Thus we have proved the following:

$$\left.\begin{array}{l}\text{Let } X \text{ be metrizable and } f: X \times Y \to Z \text{ be } \tau\text{-measurable, and let}\\ \mu \text{ and } \nu \text{ be } \tau\text{-smooth measures on } X \text{ and } Y. \text{ If } x_0 \in \operatorname{supp}(\mu)\\ \text{and } y_0 \in \operatorname{supp}(\nu) \text{ such that } f(x_0, \cdot) \text{ is continuous at } y_0, \text{ and}\\ f(\cdot, y) \text{ is continuous at } x_0 \text{ for all } y \in Y,\\ \text{then we have}\\ \qquad f(x_0, y_0) \in \operatorname{supp} \overline{((\mu \times \nu) \circ f^{-1})}.\end{array}\right\} \tag{A18}$$

Let X be metrizable, then most likely we have $M_\sigma^+(X) = M_\tau^+(X)$. To see this we note that if $\mu \in M_\sigma^+(X) \backslash M_\tau^+(X)$, then $\mu(X \backslash S) > 0$ where $S = \operatorname{supp}(\mu)$. Then there exists a collection, $\mathcal{U}$, of open subsets of X, such that

$$\mu(U) = 0, \qquad \forall U \in \mathcal{U},$$

$$\bigcup\{U: U \in \mathcal{U}\} = X \backslash S.$$

And we may of course assume that card $\mathcal{U} \leqslant w(X)$, where $w(X)$ is *the weight* of X, i.e.

$$\begin{aligned} w(X) &= \min\{\text{card } \mathcal{V}: \mathcal{V} \text{ is a sub-basis for } X\} \\ &= \min\{\text{card } D: D \text{ is a dense subset of } X\}. \end{aligned}$$

We can then find a locally finite partition of unity $\{\varphi_U: U \in \mathcal{U}\}$ in $X \backslash S$, such that

$$\varphi_U(x) = 0, \qquad \forall x \notin U.$$

If $\mathcal{V}$ is a subset of $\mathcal{U}$ we may define

$$\lambda(\mathcal{V}) = \int_{X \backslash S} \sum_{U \in \mathcal{V}} \varphi_U(x) \mu(\mathrm{d}x),$$

since the function under the integration sign is continuous on $X \backslash S$. Then λ becomes a measure on the paving of all subsets of $\mathcal{U}$, and we have

$$\lambda(\mathcal{U}) = \mu(X \backslash S) > 0,$$

$$\lambda(\{U\}) = \int_{X \backslash S} \varphi_U \,\mathrm{d}\mu = 0, \qquad \forall U \in \mathcal{U}.$$

Thus we have shown the "if" part of the following proposition:

if X is metrizable, then $M_\tau^+(X) = M_\sigma^+(X)$ if, and only if, $w(X)$ is not a real measurable cardinal. (A19)

The "only if" part is easy and we leave it to the reader.

It is well known that the axiom of constructability implies that there exists no real measurable cardinal. Thus under the axiom of constructibility we have that any finite Borel measure on a metric space is τ-smooth.

The propositions above indicate that the weak topology is particularly well behaved when X is metrizable. So let us consider a metrizable space X and let d be a metric on X defining the topology. It is then well known that we may define a metric (*the Prokorov metric*) on $M_\sigma^+(X)$ by

$$d^*(\mu, \nu) = \inf\{\varepsilon > 0: \mu(A) \leqslant \nu(A^\varepsilon) + \varepsilon, \forall A \in \mathcal{B}(X)\},$$

where A^ε is the open ε-neighbourhood of A:

$$A^\varepsilon = \{x \in X : \exists y \in A \text{ with } d(x, y) < \varepsilon\}.$$

Usually the Prokorov metric is only defined for separable metric spaces; however, the definition makes sense for arbitrary metric spaces, and the proof for the fact that d^* is a metric carries over to the general case without any change.

From the definition it follows in a straightforward manner that if $f: X \to \mathbb{R}$ is Borel measurable and $0 \leq f \leq 1$, then

$$\left|\int_X f \, \mathrm{d}(\mu - \nu)\right| \leq t + \int_X \omega_t(f, x)\mu(\mathrm{d}x), \qquad \forall t > d^*(\mu, \nu), \tag{A20}$$

where $\omega_t(f, x)$ is the oscillation function:

$$\omega_t(f, x) = \sup_{d(y,x) < t} |f(x) - f(y)|.$$

Thus we have

$$\left.\begin{array}{l} d^*(\mu_\alpha, \mu) \to 0 \Rightarrow \mu_\alpha \overset{w}{\to} \mu \text{ (i.e. the } d^*\text{-topology is stronger than} \\ \text{the weak topology)}. \end{array}\right\} \tag{A21}$$

When X is separable, it is well known that the d^*-topology equals the weak topology. In the general case,

$$\left.\begin{array}{l} \text{if } \mu_\alpha \overset{w}{\to} \mu \text{ and } \mu \text{ is } \tau\text{-smooth, then } d^*(\mu_\alpha, \mu) \to 0. \text{ Thus the} \\ d^*\text{-topology coincides with the weak topology on } M_\tau^+(X). \end{array}\right\} \tag{A22}$$

The proof of (A22) is based on the use of a suitable partition of unity in very much the same way as we have seen above.

If (X, d) is a complete metric space, then $M_\tau^+(X) = M_t^+(X)$, and in exactly the same way as in the separable case one shows that

$$\text{if } (X, d) \text{ is a complete metric space, then so is } (M_\tau^+(X), d^*). \tag{A23}$$

Note that by (A11) we have that the d^*-topology on $M_\sigma^+(X)$ differs from the weak topology if $M_\sigma^+(X) \neq M_\tau^+(X)$, i.e. if $w(X)$ is a real measurable cardinal.

The results above allow us to define a reasonable convolution of finite τ-smooth measures on certain metrizable semigroups.

Let G be a metrizable semigroup satisfying Axiom II, then by (A15) we have

$$m(x, y) = xy \text{ is } \tau\text{-measurable: } G \times G \to G. \tag{A24}$$

If μ and ν are finite τ-smooth measures on G, we define their *convolution*, $\mu * \nu$, by

$$\mu * \nu = \overline{(\mu \times \nu)} \circ m^{-1} ;$$

that is

$$\int_G \varphi \, \mathrm{d}\mu * \nu = \int_G \mu\,(\mathrm{d}x) \int_G \varphi(xy)\nu(\mathrm{d}y) = \int_G \nu(\mathrm{d}y) \int_G \varphi(xy)\mu(\mathrm{d}x) , \qquad \text{(A25)}$$

for all bounded Borel functions $\varphi : G \to \mathbb{R}$.

Note that $m(x, \cdot)$ is continuous for all $x \in G$, and $m(\cdot, y)$ is continuous for all $y \in G$. So by (A18) and (A12) we have

$$\operatorname{supp}(\mu * \nu) = \overline{\operatorname{supp}(\mu) \,.\, \operatorname{supp}(\nu)} . \qquad \text{(A26)}$$

From (A25) we see that $(M_\tau^+(G), *)$ is a semigroup with unit element equal to δ_1 (the Dirac measure at 1). Moreover, if $f \in C(G)$ and $\mu \in M_\tau^+(G)$, then

$$F(y) = \int_G f(xy)\mu(\mathrm{d}x) ,$$

is continuous by Axiom II and metrizability of G. Hence if $\nu_n \overset{w}{\to} \nu$, then

$$\int_G f \, \mathrm{d}(\mu * \nu_n) = \int_G F \, \mathrm{d}\nu_n \to \int_G F \, \mathrm{d}\nu = \int f \, \mathrm{d}(\mu * \nu) .$$

Thus we have that

$(M_\tau^+(G), *, w)$ is a metrizable semigroup satisfying Axiom II. (A27)

References

Aczél, J.

1969 *On Applications and Theory of Functional Equations*, Birkhäuser Verlag, Basel and Stuttgart.

Alexandrov, P. S.

1924 Sur les ensembles de la première classe et les éspaces abstraits, *C.R. Paris*, **178**, 185.

1927 Über stetige Abbildungen kompakter Räume, *Math. Ann.*, **96**, 555–571.

Alexandrov, P. S. and Urysohn, P. S.

1928 Über nulldimensionale Mengen, *Math. Ann.*, **98**, 89–106.

Baire, R.

1899 Thèse: Sur les fonctions de variable réelle, *Ann. di Math.* (*3*), **3**, 65.

1909 Sur la représentation des fonctions discontinues (2me partie), *Acta Math.*, **32**, 97–176.

Banach, S.
1932 *Théorie des Opérations Linéaires*, reprinted by Chelsea Publishing Co., New York, 1955.
Bourbaki, N.
1966 *General Topology*, Vol. III, Part 2, English edn, Herman, Paris and Addison-Wesley, London.
Brouwer, L. E. J.
1910 On the structure of perfect sets of points, *Proc. Akad. Amsterdam*, **12**, 785–794.
Cantor, G.
1895 Beiträge zur Begründung der transfiniten Mengenlehre, *Math. Ann.*, **46**, 481–512.
Cauchy, A. L.
1821 *Cours d'Analyse de l'École Polytechnique*, Vol. 1, Paris.
Christensen, J. P. R.
1974 *Topology and Borel Structure*, North-Holland, Amsterdam.
Dales, H. G.
1978 Automatic continuity: a survey, *Bull. London Math. Soc.*, **10**, 129–183.
Darboux, G.
1880 Sur le théorème fondamental de la géometrie projective, *Math. Ann.*, **17**, 33–42.
Dudley, R. M.
1961 Continuity of homomorphisms, *Duke Math. J.*, **28**, 587–594.
Ellis, R.
1957 Locally compact transformation groups, *Duke Math. J.*, **24**, 119–125.
Engelking, R.
1977 *General Topology*, PWN, Warsaw.
Fréchet, M.
1910 Les dimensions d'un ensemble abstrait, *Math. Ann.*, **68**, 145–168.
Glicksberg, I.
1961 Weak compactness and separate continuity, *Pacific J. Math.*, **11**, 205–214.
Hamel, G.
1905 Eine Basis aller Zahlen und die unstetigen Lösungen der Functionalgleichung: $f(x+y)=f(x)+f(y)$, *Math. Ann.*, **60**, 459–462.
Hausdorff, F.
1924 Die Mengen $\mathcal{G}_\delta$ in vollständigen Räumen, *Fund. Math.*, **6**, 146.
1927 *Mengenlehre*, Berlin.
Hoffmann-Jørgensen, J.
1970 *The Theory of Analytic Spaces*, Matematisk Institut, Aarhus Universitet, Aarhus.
Hurewicz, W.
1928 Relative perfekte Teile von Punktmengen und Mengen (A), *Fund. Math.*, **12**, 78–109.
Jayne, J. E. and Rogers, C. A.
1980 The potency of Polish spaces: retractions: restrictions, *Proc. London Math. Soc.* (to appear).

Johnson, B. E.
1969 Separate continuity and measurability, *Proc. Amer. Math. Soc.*, **20**, 420–422.

Kerstan, J.
1972 Eine Klasse topologischer Räume die die polnischen Räume und die σ-kompakten lckalkompakten Hausdorffschen Räume verallgemeinert, in *Theory of Sets and Topology*, Berlin, pp. 291–311.

Kuratowski, K.
1923 Sur les fonctions représentables analytiquement et les ensembles de premiere catégorie, *Fund. Math.*, **5**, 75.
1966 *Topology*, Vol. I, Academic Press, New York and London and PWN, Warsaw.

Lebesgue, H.
1905 Sur les fonctions représentables analytiquement, *J. Math. Pures Appl.* (*ser VI*), **1**, 139–216.

Loy, R. J.
1976*a* *Notes on Analytic Spaces*, lecture notes from the University of Edinburgh.
1976*b* Multilinear mappings and Banach algebras, *J. London Math. Soc.* (*2*), **14**, 423–429.

Lusin, N. N.
1917 Sur la classification de M. Baire, *C.R. Paris*, **141**, 91–94.
1927 Sur les ensembles analytiques, *Fund. Math.*, **10**, 1–95.
1930 *Leçons sur les Ensembles Analytiques et leur Applications*, Gauthier-Villars, Paris.

Marczewski, E. (= Szpilrajn-Marczewski, E.)
1929 O mierzalnośei i warunku Baire'a, *C.R. du I. Congrès de Math. des Pays Slaves, Varsovie*, p. 209.
1933 Sur certain invariants de l'opération (A), *Fund. Math.*, **21**, 229–235.

Mazurkiewicz, S.
1917–18 Teorja zbiorów $\mathcal{G}_\delta$, *Wektor*, **6**, 129–185.

McSchane, E. J.
1950 Images of sets satisfying the condition of Baire, *Ann. Math.* (*2*), **51**, 380–386.

Namioka, I.
1974 Separate continuity and joint continuity, *Pacific J. Math.*, **51**, 513–531.

Ostrowski, A.
1929 Matematische Miszellen XIV über die Funktionalgleichung der Exponentialfunktion und verwandte Funktionalgleichungen, *Jbh. Dt. Math. Ver.*, **38**, 54–62.

Oxtoby, J. J.
1961 Cartesian products of Baire spaces, *Fund. Math.*, **49**, 157–166.

Pettis, B. J.
1950 On continuity and openness of homomorphisms in topological groups, *Ann. Math.* (*2*), **51**, 293–308.
1952 A note on everywhere dense subgroups, *Proc. Amer. Math. Soc.*, **3**, 322–326.

Phelps, R. R.
1978 Gaussian null sets and differentiability of Lipschitz maps on Banach spaces, *Pacific J. Math.*, **77**, 523–531.

Saint-Raymond, J.
1978 La structure d'Effros est-elle standard?, *Fund. Math.*, **100**, 201–210.

Shu-Bun Ng and Warner, S.
1972 Continuity of positive linear functionals, *Duke Math. J.*, **39**, 281–284.

Sierpiński, W.
1920*a* Sur une propriété topologique des ensembles dénombrables denses en soi, *Fund. Math.*, **1**, 11–16.
1920*b* Sur l'équation fonctionelle $f(x+y)=f(x)+f(y)$, *Fund. Math.*, **1**, 116–122.
1928*a* Sur les projections des ensembles complementaires aux ensembles (A), *Fund. Math.*, **11**, 117–122.
1928*b* Sur les ensembles complets d'un espace (D), *Fund. Math.*, **11**, 135.

Souslin, M.
1917 Sur une définition des ensembles mesurables B sans nombres transfini, *C.R. Paris*, **141**, 88–91.

Szpilrajn-Marczewski (*see* Marczewski)

Topsøe, F.
1970 *Topology and Measure*, Lecture Notes in Mathematics no. 133, Springer-Verlag, Berlin, Heidelberg and New York.

Wilansky, A.
1970 *Topology for Analysis*, Xerox Publishing Co., New York.

Young, W. H.
1903 Zur Lehre der nicht abgeschlossenen Punktmengen, *Ber. Ges. Wiss. Leipzig*, **55**, 287–293.

Part 4

Infinite games and effective descriptive set theory

D. A. MARTIN AND A. S. KECHRIS

1 Introduction

1.1 Introduction

We aim in this Part of the present volume to show how concepts and techniques of mathematical logic apply to classical descriptive set theory, especially to the theory of analytic sets. We stress two notions: *infinite games* and *effective descriptive set theory*. Section 1–5 (written by Martin) will deal with infinite games. Sections 6–9 (written by Kechris) will deal with the effective theory. Infinite games will play a role in Sections 6–9, but these sections will not depend upon Sections 3–5.

We hope to show the reader how ideas and methods of logic refine and extend the classical theory, make possible simpler proofs of known classical theorems, and yield new theorems of a purely classical nature. We shall also show how new "axioms" of set theory imply results not provable or refutable from the standard axioms.

Descriptive set theory had its origins at the beginning of this century in work of Baire, Borel, and Lebesgue. It was further developed by Souslin, Lusin, Sierpiński, and many others. Beginning in the 1930s, an apparently unrelated theory, based on recursive functions, was developed by logicians, especially Kleene. In the 1950s Mostowski and Addison discovered that the Kleene theory, which we call *effective descriptive* set theory, in fact properly contained the classical theory.

Certain problems in descriptive set theory resisted solution by either classical or effective methods. As an outgrowth of the set-theoretic independence results of Gödel and Cohen, it became clear that these problems were formally undecidable from the Zermelo–Fraenkel axioms. In recent years, new set-theoretic hypotheses have been shown to solve these otherwise unsolvable problems and to yield an elegant and essentially complete theory of projective sets.

1.2 Some definitions

ω denotes the set of all natural numbers, $\{0, 1, \ldots\}$. For any non-empty set X, X^ω is the set of all functions $f: \omega \to X$. We give X^ω the product topology; i.e. we take as basic open sets all sets of the form

$$N_t^X = \{f : f \text{ extends } t\}$$

for $t: \{0, \ldots, n\} \to X$ (i.e. for t a finite sequence of elements of X). By $X^{<\omega}$ we mean the set of all finite sequences of elements of X, which we sometimes think of as functions

$$t: \{0, \ldots, \text{length}(t) - 1\} \to X.$$

An element of X^ω is essentially an infinite sequence of elements of X. N_t^X is thus the collection of infinite sequences of elements of X beginning with the finite sequence t.

We shall develop descriptive set theory mainly in the space ω^ω. We denote elements of this space by

$$\alpha, \beta, \gamma, \alpha', \beta', \text{etc.}$$

We sometimes call them *reals*. (ω^ω is actually homeomorphic to the irrationals.) ω^ω can be made into a complete metric space. Most of our results hold for arbitrary *perfect Polish spaces*, complete separable metric spaces without isolated points (see Moschovakis, 1980). Two theorems useful in showing this are:

Every perfect Polish space is Borel-isomorphic to ω^ω.

Every perfect Polish space is a continuous image of ω^ω.

We shall usually write N_t for N_t^ω.

We sometimes deal with $\{0, 1\}^\omega$ and $\{0, 1\}^{<\omega}$, which we call 2^ω and $2^{<\omega}$. We write N_t^2 for $N_t^{\{0,1\}}$. Note that the topology on 2^ω is the topology it has as a subspace of ω^ω.

One of the many reasons ω^ω is easy to work with is that its dimension is zero, and there is a homeomorphism $h:\omega^\omega \to (\omega^\omega)^2$. To see this, take

$$h(\alpha)=((\alpha)_0, (\alpha)_1),$$

where

$$(\alpha)_0(n)=\alpha(2n);$$

$$(\alpha)_1(n)=\alpha(2n+1).$$

For $A \subseteq X^\omega \times Y^\omega$, we define *the projection of A* pA, by

$$\mathrm{p}A=\{f: (\exists g)((f, g)\in A)\}.$$

A subset A of X^ω is *analytic* if it is the projective of a closed set in $X^\omega \times \omega^\omega$. (Note that, in Part 1 of this volume such a set would be called a Souslin-$\mathcal{F}$ set.)

If s and t are finite or infinite sequences, $s \subseteq t$ means that t *extends* s, i.e. that s is an initial segment of t. A *tree T on X* is a subset of $X^{<\omega}$ such that

$$s \in T \text{ and } t \subseteq s \quad \to \quad t \in T.$$

We regard a tree as a partial ordering by letting *s less than or equal to t* mean that $t \subseteq s$.

A tree on $X \times Y$ is literally a collection of finite sequences of pairs, but we usually regard it as a collection of pairs (s, t) of finite sequences with length (s) = length (t).

If T is a tree on X, $[T]$ is the collection of all $f:\omega \to X$ such that

$$(\forall n)(f \restriction n \in T),$$

where we identify n with the set $\{0, 1, \ldots, n-1\}$. If T is a tree on $X \times Y$ and $f \in X^\omega$,

$$T(f)=\{t: (f \restriction \text{length}(t), t)\in T\}.$$

IMPORTANT FACT 1.2.1 *If T is a tree on $X \times Y$ and $f \in X^\omega$, then $f \in p[T] \leftrightarrow$ $T(f)$ is not well-founded.*

A partial ordering is *well-founded* if every non-empty subset of its field has a minimal element. A weak form of the axiom of choice implies that well-foundedness is equivalent with the non-existence of infinite descending chains. This fact makes 1.2.1 immediate.

If A is a closed subset of $X^\omega \times Y^\omega$, let

$$T=\{(s, t): \text{length}(s)=\text{length}(t) \quad \text{and} \quad (N_s^X \times N_t^Y)\cap A \text{ is non-empty}\}.$$

Then T is a tree on $X \times Y$ and clearly $A=[T]$.

If $A \subseteq X^\omega$ is analytic, we have then that $A = \mathrm{p}[T]$ for T a tree on $X \times \omega$. Thus we have, for all $f \in X^\omega$,

$$f \in A \leftrightarrow T(f) \quad \text{is not well-founded}.$$

Because they can be written in the form

$$\{\alpha : (\exists \beta)(\forall n)(\alpha \restriction n, \beta \restriction n) \in T\},$$

analytic subsets of ω^ω are called $\boldsymbol{\Sigma}^1_1$. (The superscript 1 means we are counting "type 1" quantifiers, quantifiers over ω^ω. The subscript 1 means there is one such quantifier present. The "$\boldsymbol{\Sigma}$" means that the first such quantifier is existential.) Coanalytic subsets of ω^ω are of the form

$$\{\alpha : (\forall \beta)(\exists n)(\alpha \restriction n, \beta \restriction n) \notin T\}$$

and are called $\boldsymbol{\Pi}^1_1$. PCA sets are called $\boldsymbol{\Sigma}^1_2$, since they can be written

$$\{\alpha : (\exists \beta)(\forall \gamma)(\exists n)(\alpha \restriction n, \beta \restriction n, \gamma \restriction n) \notin T\}.$$

CPCA sets are called $\boldsymbol{\Pi}^1_2$, and so on. A subset of ω^ω is *projective* if it is $\boldsymbol{\Sigma}^1_n$ for some n.

Sets which are both analytic and coanalytic are called $\boldsymbol{\Delta}^1_1$. In general, the $\boldsymbol{\Delta}^1_n$ sets are those which are both $\boldsymbol{\Sigma}^1_n$ and $\boldsymbol{\Pi}^1_n$. The *Souslin theorem* says that the $\boldsymbol{\Delta}^1_1$ sets are just the Borel sets.

1.3 Measurable cardinals

A subset A of X^ω is *comeagre* if A contains a countable intersection of dense open sets. By the Baire category theorem, every comeagre set is dense. The complement of a comeagre set is *meagre*. A has the *Baire property* if it differs by a meagre set from an open set. A subset of X^ω is *perfect* if it is non-empty, closed, and has no isolated points. A set of real numbers is $\boldsymbol{\Sigma}^1_1$ if it is the projection of a closed set in the product of the reals with ω^ω. The notion of a $\boldsymbol{\Sigma}^1_n$ set of reals is defined in the obvious way.

Classical descriptive set theory yielded the following results:

(1) Every $\boldsymbol{\Sigma}^1_1$ (analytic) subset of ω^ω has the Baire property (Lusin–Sierpiński).
(2) Every $\boldsymbol{\Sigma}^1_1$ set of reals is Lebesgue measurable (Lusin).
(3) Every uncountable $\boldsymbol{\Sigma}^1_1$ subset of 2^ω has a perfect subset (Souslin).
(4) Every $\boldsymbol{\Sigma}^1_2$ subset of $\omega^\omega \times \omega^\omega$ can be uniformized by a $\boldsymbol{\Sigma}^1_2$ set (Kondô).

Here A *uniformizes* B if $A \subseteq B$, for all α there is at most one β such that $(\alpha, \beta) \in A$, and

$$(\forall a)((\exists \beta)((\alpha, \beta) \in B) \rightarrow (\exists \beta)((\alpha, \beta) \in A)).$$

The following questions were unanswered (and unanswerable):

(1) Does every $\mathbf{\Sigma}_2^1$ set have the Baire property?
(2) Is every $\mathbf{\Sigma}_2^1$ set of reals Lebesgue measurable?
(3) Does every uncountable $\mathbf{\Pi}_1^1$ (CA) subset of 2^ω have a perfect subset?
(4) Can every $\mathbf{\Pi}_2^1$ set be uniformized by a projective set?

R. Solovay proved in about 1965 that, if a measurable cardinal exists (see Section 4 for the definition), then the answers (1)–(3) are positive, even replacing "$\mathbf{\Pi}_1^1$" by "$\mathbf{\Sigma}_2^1$" in (3). Later work of Solovay and Martin showed that, if measurable cardinals exist, then every $\mathbf{\Sigma}_3^1$ set can be uniformized by a $\mathbf{\Delta}_4^1$ set. Unhappily, work of J. Silver showed that measurable cardinals are inadequate for extending these results to higher levels of the projective hierarchy.

1.4 Infinite games

As the use of large cardinals in descriptive set theory came to a temporary halt, a different kind of "axiom" took over.

Let X be a non-empty set and let A be a subset of X^ω. We define a two-person game G of perfect information as follows: Player I chooses an element x_0 of X; then player II chooses $x_1 \in X$; then I chooses $x_2 \in X$, etc.:

I	II
x_0	
	x_1
x_2	
	x_3
. . .	

I *wins* if the resulting infinite sequence of elements of X (the *play*) belongs to A. A strategy for I is a function σ mapping the set of all elements of $X^{<\omega}$ of even length into X. A play or a position is *according to* σ or *consistent with* σ if I has always made the move $\sigma(t)$ at each point t of even length. σ is a *winning strategy* if every play according to σ is a win for I, i.e. belongs to A. The notion of a strategy τ for II and the related concepts are defined similarly. G is *determined* if one of the players has a winning strategy. We call G *open, closed, Borel*, etc., just when A is open, closed, Borel, etc.

The games G were introduced by Gale and Stewart (1953) (although Ulam apparently knew about them earlier). Gale and Stewart also used the axiom of choice to construct a game with $X = \{0, 1\}$ which is not determined. In the positive direction, they proved the following theorem:

THEOREM 1.4.1 *All open games are determined.*

Proof Let A be an open subset of X^{ω}. Let G be the associated game. If $t \in X^{<\omega}$ and t has even length, let $G(t)$ be the game associated with A^t, where

$$f \in A^t \leftrightarrow t^\frown f \in A\,,$$

where $t^\frown f$ is the concatenation of the pair of sequences.

Suppose I has no winning strategy for $G(t)$. Then for every $z_0 \in X$ there is a $z_1 \in X$ such that I has no winning strategy for

$$G(t^\frown (z_0)^\frown (z_1))\,.$$

Otherwise I could win $G(t)$ by playing z_0 and, after II plays some z_1, playing a winning strategy for $G(t^\frown (z_0)^\frown (z_1))$.

If I has no winning strategy for G, inductive use of the preceding paragraph shows that II has a strategy so that, for every position t of even length reached, I has no winning strategy for $G(t)$. If f is a play according to such a strategy, $f \notin A$. This is because $f \in A$ implies that $f \in N_t^X$ for some $N_t^X \subseteq A$; but this is impossible, since I has a trivial winning strategy for such a $G(t)$.

We have the immediate

COROLLARY 1.4.2 *All closed games are determined.*

Proof Let A be a closed subset of X^{ω}, and let G be the associated game. After I has made his first move, say by choosing x^* in X, the subsequent moves can be regarded as those of an open game G_{x^*} in which player II exchanges roles with player I. For each x^* in X, G_{x^*} is determined. If for each x^* in X, player II (who plays first in G_{x^*}) has a winning strategy σ_{x^*} in G_{x^*}, then player II has a winning strategy for G: if I plays x^* at his first move, then play σ_{x^*} in the game G_{x^*} that ensues. If for some x^* in X, player I has a winning strategy τ_{x^*} in G_{x^*}, then player I has a winning strategy in G: first play x^*, then play the strategy τ_{x^*} in the game G_{x^*} that ensues.

The use of hypotheses or "axioms" that games are determined was first introduced by Mycielski and Steinhaus. Work of Banach, Mycielski-Swierczkowski, and Davis showed that, in particular, if all projective games with $X = \omega$ are determined then every projective subset of ω^{ω} has the Baire property, every projective set of reals is Lebesgue measurable, and every uncountable projective set has a perfect subset. See Mycielski (1964) for a survey of early work on infinite games.

In 1967, D. Blackwell discovered a new proof of the separation theorem for analytic sets, a proof based on the determinacy of open games. At this point the application of determinacy hypotheses in descriptive set theory was taken up in earnest by logicians. The picture which emerged is as follows:

(1) Virtually the entire classical theory of projective sets (mostly concerning the first two levels of the hierarchy) can be deduced in a simple manner from open determinacy. We shall illustrate this in Section 2 and also in Sections 7 and 8.
(2) All the important consequences in descriptive set theory of the existence of measurable cardinals follow from the assumption that all analytic games are determined. Some instances of this will be seen in Section 2.
(3) The hypothesis (PD) that all projective games with $X = \omega$ are determined yields essentially a complete structural theory of projective sets.

Examples of (3) will be given later (Section 7), but we outline the general picture here.

If Γ is a class of subsets of ω^ω, *Determinacy* (Γ) means that every game of class Γ is determined. *Reduction* (Γ) means that, for each $A, B \in \Gamma$, there are $A', B' \in \Gamma$ such that $A' \subseteq A$, $B' \subseteq B$, $A' \cup B' = A \cup B$, and $A' \cap B'$ is empty. (This implies first and second separation for the dual of Γ, the collection of complements of elements of Γ.) *Uniformization* (Γ) means that every subset of $\omega^\omega \times \omega^\omega$ belonging to Γ (recall our earlier remark that ω^ω and $\omega^\omega \times \omega^\omega$ are homeomorphic) can be uniformized by a set in Γ.

The classical theory showed that Reduction, Uniformization, and other related properties held for $\mathbf{\Pi}^1_1$ and $\mathbf{\Sigma}^1_2$ but not for $\mathbf{\Sigma}^1_1$ or $\mathbf{\Pi}^1_2$. PD extends these results to the entire projective hierarchy, implying a periodicity of order 2:

THEOREM 1.4.3 (Martin, Moschovakis) *If Determinacy* ($\mathbf{\Delta}^1_{2n}$) *then Reduction* ($\mathbf{\Pi}^1_{2n+1}$) *and Reduction* ($\mathbf{\Sigma}^1_{2n+2}$) *(but neither Reduction* ($\mathbf{\Sigma}^1_{2n+1}$) *nor Reduction* ($\mathbf{\Pi}^1_{2n+2}$)).

THEOREM 1.4.4 (Moschovakis) *If Determinacy* ($\mathbf{\Delta}^1_{2n}$) *then Uniformization* ($\mathbf{\Pi}^1_{2n+1}$) *and Uniformization* ($\mathbf{\Sigma}^1_{2n+2}$) *(but neither Uniformization* ($\mathbf{\Sigma}^1_{2n+1}$) *nor Uniformization* ($\mathbf{\Pi}^1_{2n+2}$)).

For results of a more purely set-theoretic nature, only the first four levels of the hierarchy are understood at present, even with PD. It is a classical result (Sierpiński) that every $\mathbf{\Sigma}^1_2$ set is a union of $\aleph_1$ Borel sets. Assuming a measurable cardinal, Martin showed that every $\mathbf{\Sigma}^1_3$ set is a union of $\aleph_2$ Borel sets. This conclusion also follows from Determinacy ($\mathbf{\Sigma}^1_1$). If PD, then work of Kunen, Moschovakis, and Martin shows that every $\mathbf{\Sigma}^1_4$ set is a union of $\aleph_3$ Borel sets. The problem of $\mathbf{\Sigma}^1_5$ is still open.

See Martin (1977) for a survey of work on determinacy since 1967.

Since determinacy hypotheses are by no means self-evident truths, it is important to see how much determinacy can be proved. In Section 3, we shall sketch the proof of the following theorem of Martin (1975):

THEOREM 1.4.5 *All Borel games are determined.*

A result of Davis implies that Determinacy ($\mathbf{\Sigma}^1_1$) cannot be proved in Zermelo–Fraenkel set theory (ZFC). In Section 4 we shall prove

THEOREM 1.4.6 (Martin) *If a measurable cardinal exists, then Determinacy* ($\mathbf{\Sigma}^1_1$).

A recent result of Martin shows that Determinacy ($\mathbf{\Sigma}^1_2$) follows from a very strong large cardinal axiom.

In Section 5 we shall discuss a theorem of L. Harrington that Determinacy ($\mathbf{\Sigma}^1_1$) is equivalent with the assertion that any two analytic, non-Borel subsets of ω^ω are Borel isomorphic.

In Section 6, we introduce the basic concepts of the effective theory and relate them to the classical theory.

Section 7 is concerned with the prewellordering theorem, a structural property of $\mathbf{\Pi}^1_1$ sets (and of effectively $\mathbf{\Pi}^1_1$, or Π^1_1, sets) which implies many other properties of these sets.

In Section 8, we shall discuss classical uniformization theorems in the light of the effective theory.

In Section 9, we shall use the effective theory to prove a new purely classical theorem:

THEOREM 1.4.7 *A coanalytic equivalence relation in ω^ω has either countably many or $2^{\aleph_0}$ equivalence classes.*

The proof given is due to L. Harrington. The theorem is due to J. Silver. No "classical" proof is known.

2 Open games and their uses

2.1 The perfect subset property

In this section, we use the determinacy of open games (Theorem 1.4.1) to reprove some classical results about analytic and coanalytic sets.

THEOREM 2.1.1 *Every uncountable analytic subset of 2^ω has a perfect subset.*

Proof The proof we give is a combination of an argument of Davis (1964) with a trick, first used by Solovay, called "unfolding". Let $C \subseteq 2^\omega \subseteq \omega^\omega$ be closed and let $A = \mathrm{p}C$. The idea of Davis is for I and II to play a game in which they cooperatively play an element of 2^ω, a game in which I has the advantage in that he is allowed to make any finite number of moves at each turn. I is attempting to make the element α of 2^ω, that is produced, an element of A. The resulting game would be analytic. The unfolding trick is used to replace it by a closed game: we allow I to make extra moves, choosing a $\beta \in \omega^\omega$; I wins if $(\alpha, \beta) \in C$.

More precisely, consider the following game G^*:

I	II
(m_0, s_0)	
	ε_0
(m_1, s_1)	
	ε_1
. . .	

Here, as elsewhere, the notation means that we are construing every move of I as an ordered pair via some appropriate map of X into ordered pairs. We leave the routine problem of defining X and the pairing function to the reader. The choice $X = \omega$ is certainly possible.

We now specify certain *rules*. The players may disobey these rules, but the first player to do so loses. Each m_i must be a natural number. Each ε_i must be 0 or 1. Each s_i must belong to $2^{<\omega}$. In addition, s_{i+1} must extend (not necessarily properly) $s_i{}^\frown(\varepsilon_i)$. Finally, (m_i, s_i) must be chosen so that

$$(\exists\alpha)(\exists\beta)(s_i \subseteq \alpha \;\&\; (m_0, \ldots, m_i) \subseteq \beta \;\&\; (\alpha, \beta) \in C)\,. \tag{1}$$

If all rules are obeyed, I *wins*. If all rules except possibly (1) are obeyed, let $\alpha^* \in 2^\omega$ be the unique sequence extending all the s_i and let $\beta^*(i) = m_i$. Clearly I wins just when $(\alpha^*, \beta^*) \in C$.

G^* is closed. Since open determinacy implies closed determinacy, G^* is determined.

LEMMA 2.1.2 *If I has a winning strategy for G^*, then A has a perfect subset.*

Proof Let σ be a winning strategy for I for G^*. We define a function

$$\varphi : 2^\omega \to 2^\omega$$

as follows. If $\gamma \in 2^\omega$ let $\varphi(\gamma)$ be the α^* which results from the play according to σ in which II plays $\varepsilon_i = \gamma(i)$ for each i. We verify that φ is continuous and one-to-one. The strategy σ being fixed, the values of $\varepsilon_0, \varepsilon_1, \ldots, \varepsilon_i$, by themselves, determine $s_0, s_1, \ldots, s_i, s_{i+1}$. So φ is continuous. Further, if γ and γ' first differ in the ith place, then the corresponding initial segments s_i, ε_i and s'_i, ε'_i of $\varphi(\gamma)$ and $\varphi(\gamma')$ differ. So φ is one-to-one. Since σ is a winning strategy, $\varphi(\gamma) \in A$ for all γ. Now the continuous one-to-one image of 2^ω under φ is a perfect subset of A.

LEMMA 2.1.3 *If* II *has a winning strategy for* G^*, *then* A *is countable.*

Proof Let τ be a winning strategy for II for G^*. Without loss of generality we may assume τ always obeys the rules. Let

$$p = (m_0, s_0, \varepsilon_0, \ldots, m_{i-1}, s_{i-1}, \varepsilon_{i-1}, m_i)$$

be consistent with τ and such that all rules have been obeyed except perhaps (1). (p is not exactly a position: I is halfway through one of his moves.) We say that τ *rejects* an element α of 2^ω *at* p, if $\alpha \supseteq s_{i-1}{}^\frown(\varepsilon_{i-1})$ and there is no s_i extending $s_{i-1}{}^\frown(\varepsilon_{i-1})$, such that, if I plays s_i at p, then τ calls for II to play ε_i such that $\alpha \supseteq s_i{}^\frown\varepsilon_i$. (In other words, whatever initial segment of α I presents II with at p, II obstinately refuses to play the next term of the sequence α.) Since II has at each turn exactly two moves which obey the rules, it is easily seen that, for each p, there is exactly one α such that τ rejects α at p. (Note for future reference that one can *compute* α if one can compute τ.) Thus there are only countably many α such that τ rejects α at *some* p.

Assume $\alpha \in A$. The lemma will be proved if we can show that, for some p, τ rejects α at p. Assume, in order to obtain a contradiction, that this is false. Let β be such that $(\alpha, \beta) \in C$. We define a play of G^* according to τ with $\alpha^* = \alpha$ and $\beta^* = \beta$ as follows. I plays $m_i = \beta(i)$ and he chooses for s_i the shortest $\alpha \restriction n$ such that $n > \text{length}(s_{i-1})$ and such that τ calls for II to reply with $\varepsilon_i = \alpha(n)$. Thus I wins despite II's use of τ. This contradiction completes the proof.

Note that we really used the fact that C is closed only to prove that G^* is determined. Thus we have shown

THEOREM 2.1.4 *If Determinacy* ($\mathbf{\Pi}^1_n$) *then every uncountable* $\mathbf{\Sigma}^1_{n+1}$ *subset of* 2^ω *has a perfect subset.*

2.2 The Baire Property

THEOREM 2.2.1 *Every analytic subset of* ω^ω *has the Baire property.*

Proof We modify an argument of Banach by using the unfolding trick. Let $B \subseteq \omega^\omega$. Let $D = \bigcup\{N_t\colon N_t - B \text{ is meagre}\}$. $D - B$ is a countable union of meagre sets and so is meagre. Assume B is analytic. If we can show that $B - D$ is meagre, then we have shown that B has the Baire property and so have proved the theorem.

LEMMA 2.2.2 *$B - D$ is meagre.*

Proof Let $A = B - D$. A is analytic, since D is open. Let $A = \mathrm{p}C$ with C closed. Consider a game G^{**} played as follows:

I	II
(m_0, s_0)	
	t_0
(m_1, s_1)	
	t_1
...	

m_i must be a natural number. s_i must be an element of $\omega^{<\omega}$, properly extending t_{i-1} if $i > 0$. t_i must be an element of $\omega^{<\omega}$ properly extending s_i. Finally (m_i, s_i) must be chosen so that

$$(\exists\alpha)(\exists\beta)(s_i \subseteq \alpha \;\&\; (m_0, \ldots, m_i) \subseteq \beta \;\&\; (\alpha, \beta) \in C)\,. \tag{2}$$

The first player to disobey these rules *loses*. If all rules are obeyed, I *wins*. If all rules except perhaps (2) are obeyed, let $\alpha^* \in \omega^\omega$ extend the s_i and t_i and let $\beta^*(i) = m_i$. Clearly, I wins just when $(\alpha^*, \beta^*) \in C$.

Suppose τ is a winning strategy for II for G^{**}. We may assume without loss of generality that τ always calls for II to play a $t_i \in \omega^{<\omega}$ properly extending s_i if $s_i \in \omega^{<\omega}$. Let

$$p = (m_0, s_0, t_0, \ldots, m_{i-1}, s_{i-1}, t_{i-1}, m_i)$$

be consistent with τ, with all rules obeyed except perhaps (2). If $\alpha \in \omega^\omega$, τ *rejects* α at p if $\alpha \supseteq t_{i-1}$ and, for every s_i such that $t_{i-1} \subsetneq s_i \subseteq \alpha$, τ calls for II to play a t_i not extended by α.

Let $s \in \omega^{<\omega}$. If neither $s \supseteq t_{i-1}$ nor $t_{i-1} \supseteq s$, then no $\alpha \in N_s$ is rejected by τ at p. Otherwise let s_i extend s and properly extend t_{i-1}. Let t_i be the reply to s_i dictated by τ. If $\alpha \in N_{t_i}$, τ does not reject α at p. (Note for future reference that we can *compute* the map $s \to t_i$ if we can compute τ.) We have thus shown that the set of α not rejected by τ at p has a dense open subset. It follows that the set of α rejected by τ at some p is meagre.

Just as in the proof of Lemma 2.1.3, we can show that every member of A is rejected by τ at some p. It follows that A is meagre.

Suppose now that I has a winning strategy σ for G^{**}. If $p = (m_0, s_0, t_0, \ldots, m_i, s_i)$ is consistent with σ and all rules except perhaps (2) have been obeyed, we define σ *rejects* α at p in the obvious way. As before we argue that the set of α rejected at some p is meagre.

Let s_0 be as dictated by σ. Suppose $\alpha \supseteq s_0$ and σ does not reject α at any p. Then there is a play of G^{**} consistent with σ and with all rules obeyed such that $\alpha^* = \alpha$. $(\alpha^*, \beta^*) \in C$, so $\alpha \in A$. We conclude that $N_{s_0} - A$ is meagre. Since $A \subseteq B$, $N_{s_0} - B$ is meagre. By the definition of D, $N_{s_0} \subseteq D$. By the definition of A, $N_{s_0} \cap A$ is empty. This contradiction completes the proof of the lemma.

Note that the proof of Theorem 2.2.1 establishes the analogue of Theorem 2.1.4:

THEOREM 2.2.3 *If Determinacy* $(\mathbf{\Pi}^1_n)$ *then every* $\mathbf{\Sigma}^1_{n+1}$ *subset of* ω^ω *has the Baire property.*

Games can also be used to establish other regularity properties of analytic sets, e.g. that all analytic sets of real numbers are Lebesgue measurable.

2.3 The reduction property

THEOREM 2.3.1 *Reduction* $(\mathbf{\Pi}^1_1)$.

Proof The following proof is due to Blackwell (1967). Let A and B be $\mathbf{\Pi}^1_1$. Let S and T be trees such that

$$A = \{\alpha : (\forall \beta)(\exists n)(\alpha \restriction n, \beta \restriction n) \notin S\},$$

$$B = \{\alpha : (\forall \gamma)(\exists n)(\alpha \restriction n, \gamma \restriction n) \notin T\}.$$

Let $\alpha \in \omega^\omega$. Consider the following game G_α:

I	II
m_0	
	n_0
m_1	
	n_1
...	

Let $\beta(i) = m_i$ and $\gamma(i) = n_i$. If $(\alpha, \beta) \in [S]$, I wins. Otherwise I wins just when $(\alpha, \gamma) \notin [T]$ and the least j such that $(\alpha \restriction j, \beta \restriction j) \notin S$ is at least as large as the least k such that $(\alpha \restriction k, \gamma \restriction k) \notin T$. Let

$$A' = A \cap \{\alpha : \text{I has no winning strategy for } G_\alpha\};$$

$$B' = B \cap \{\alpha : \text{II has no winning strategy for } G_\alpha\}.$$

Clearly $A' \subseteq A$ and $B' \subseteq B$. If $\alpha \in B - A$, I has a winning strategy for G_α: he just plays a β such that $(\alpha, \beta) \in [S]$. Hence $\alpha \in B - A \to \alpha \in B'$. Similarly $\alpha \in A - B \to \alpha \in A'$. If $\alpha \in A \cap B$, then one of the players does not have a winning strategy, so $\alpha \in A' \cup B'$. Putting these facts together, we conclude that $A' \cup B' = A \cup B$. The games G_α are closed, since II wins if, and only if, for some j, $(\alpha \upharpoonright j, \beta \upharpoonright j) \notin S$ but $(\alpha \upharpoonright j, \gamma \upharpoonright j) \in T$. Thus one of the players has a winning strategy for G_α. This implies that $A' \cap B'$ is empty.

We need only check that A' and B' are $\mathbf{\Pi}^1_1$. Now α belongs to A' just when $\alpha \in A$ and, for all strategies τ for II, there is an $(m_0, n_0, \ldots, m_k, n_k)$ consistent with τ such that $(\alpha \upharpoonright k, (m_i: i < k)) \in S$ but $(\alpha \upharpoonright k + 1, (n_i: i \leqslant k)) \notin T$. Since strategies are essentially number theoretic functions, we can write this last clause in the form $(\forall \gamma')(\exists n)(\alpha \upharpoonright n, \gamma' \upharpoonright n) \notin T'$. This shows that A' is $\mathbf{\Pi}^1_1$. A similar argument shows that B' is $\mathbf{\Pi}^1_1$.

3 Borel games

3.1 Introduction

We now turn our attention to the problem of proving games determined. Wolfe (1955) showed that all $\mathbf{\Sigma}^0_2$ (F_σ) games are determined. Davis (1964) extended this result to $\mathbf{\Sigma}^0_3$ ($G_{\delta\sigma}$) games. Paris (1972) combined the techniques of Davis and those described in Section 4 to prove that all $\mathbf{\Sigma}^0_4$ ($F_{\sigma\delta\sigma}$) games are determined. Friedman (1971) explained this slow progress by proving that proving all Borel games determined was in a certain technical sense difficult: any proof, even for $X = \omega$, would have to use the fact that there are uncountably many cardinal numbers. This was quite surprising, since all known theorems about the continuum could be proved without appeal to anything beyond the continuum. The reader may prefer to read Section 4 before studying this section in detail.

THEOREM 3.1.1 (Martin, 1975) *All Borel games are determined.*

We shall prove only the case of games of finite Borel rank, since the general proof is a bit complicated to state but involves essentially no new ideas. See Martin (1975).

If $A \subseteq X^\omega$, A is $\mathbf{\Sigma}^0_1$ if A is open. If η is a countable ordinal greater than 1, A is $\mathbf{\Sigma}^0_\eta$ if there are sets A_i, $i = 0, 1, \ldots$, such that each A_i is $\mathbf{\Pi}^0_{\eta_i}$ for some $\eta_i < \eta$ and $A = \bigcup_i A_i$. A is $\mathbf{\Pi}^0_\eta$ for $\eta \geqslant 1$ if the complement of A is $\mathbf{\Sigma}^0_\eta$. Thus $\mathbf{\Pi}^0_1$ means closed, $\mathbf{\Sigma}^0_2$ means F_σ, $\mathbf{\Pi}^0_2$ means G_δ, etc. A is *Borel* just when A is $\mathbf{\Sigma}^0_\eta$ for some countable η.

Let $k \geq 1$ be finite. Let us assume that, for every X and every $\mathbf{\Sigma}^0_k$ subset of X^ω, the corresponding game is determined. We shall prove that every $\mathbf{\Sigma}^0_{k+1}$ game is determined. Since we already know that all $\mathbf{\Sigma}^0_1$ games are determined, we shall have shown that, for all finite k, all $\mathbf{\Sigma}^0_k$ games are determined.

3.2 The plan of the proof of Theorem 3.1.1

Let $A \subseteq X^\omega$ be $\mathbf{\Sigma}^0_{k+1}$. Assume for definiteness that k is even. There are open sets $A_{n_1,\ldots,n_k}$ such that

$$A = \bigcup_{n_1} \bigcap_{n_2} \cdots \bigcap_{n_k} A_{n_1,\ldots,n_k}.$$

Let $A^0, A^1, \ldots$ enumerate all the $A_{n_1,\ldots,n_k}$, possibly with repetitions.

Let G be the game corresponding to A. We shall produce a set X^* and a $\mathbf{\Sigma}^0_k$ subset A^* of $X^{*\omega}$ such that, if either player has a winning strategy for the corresponding game G^*, then that same player has a winning strategy for G. By the remarks already made, this will complete our proof.

The idea of G^* is that, in playing G^*, the two players will be playing G and also making certain extra moves. These extra moves will amount to promises about the way the rest of the G-part of the game will be played. In particular, on his ith extra move, player II will promise either that the resulting play of G will belong to A^i or that it will not belong to A^i. The effect (if promises are kept) is to turn the open sets A^i into *clopen* sets.

We now describe the general sort of "promises" which the players will be required to make. If T is a tree on X, a I-*imposed subtree* of T is a tree $S \subseteq T$ such that, if $t \in S$, $x \in X$, $t^\frown(x) \in T$, and t has odd length, then $t^\frown(x) \in S$. To see the meaning of this notion (from Davis, 1964), let us suppose that our players have promised to play in the tree T. If they further promise to play in the I-imposed subtree S of T, then II is making no promises he had not made already. An example of a I-imposed subtree of T is that given by a strategy σ for I, namely the set of $t \in T$ such that t is consistent with σ.
Note, incidentally, that we allow our trees to have *terminal nodes*.

The notion of a II-*imposed subtree of* T is similarly defined.

Before defining G^* in detail, let us indicate how the first two moves of G^* are made.

I begins by choosing a I-imposed subtree S^0 of $X^{<\omega}$ (otherwise he loses).
II now has two options:

(0) II may choose $\varepsilon^0 = 0$. In this case we let T^0 be the II-imposed subtree of S^0 defined by not allowing II to make a move which leaves a position t such that $N_t^X \subseteq A^0$. All subsequent G^* moves which are moves in G must produce positions in T^0. The first player to violate this rule loses. Note that if no player violates this rule, then the resulting play of G will not belong to A^0.

(1) II may choose $\varepsilon^0 = 1$. In this case II must choose a $t^0 \in S^0$ of even length such that $N^X_{t^0} \subseteq A^0$. Let $T^0 = \{t\colon t^0 \subseteq t \text{ or } t \subseteq t^0\}$. Once again, the players must, on pain of instant defeat, play in T^0. Clearly, if the players obey this rule, the resulting play of G will belong to A^0.

Let us suppose for the moment that the first two moves of G^* are as described, but that the rest of G^* consists merely of a play of G. (This will of course not be the case.) Suppose that I wins G^*, provided the rules are obeyed, just when he wins the corresponding play of G.

Suppose I has a winning strategy σ^* for this simplified G^*. Then I can win G as follows. He pretends he is playing G^* according to σ^* and that II chooses $\varepsilon^0 = 0$. If a position t of even length arises such that $t \notin T^0$, then clearly $N_t \subseteq A^0$. I now revises his assumption about G^*. He assumes that II chose $\varepsilon^0 = 1$ and $t^0 = t$. Clearly the moves up to the position t are consistent with σ^*, since they are the only moves consistent with the rules. I now continues to play σ^*. Since all rules are obeyed, he wins the pretended play of G^* and so the actual play of G.

Now suppose II has a winning strategy τ^* for G^*. II wins G as follows. He pretends he is playing G^* according to τ^*. He attributes to I the opening move S, the I-imposed tree defined by not allowing I to make a move, if (a) the empty position t or (b), for some reply by II, the resulting position t satisfies the following: for *some* S^0, τ^* calls for II to play $\varepsilon^0 = 1$ and $t^0 = t$. Note that, for $S^0 = S$, τ^* calls for $\varepsilon^0 = 0$, since otherwise the t^0 given by τ^* must belong to S^0, since τ^* obeys the rules, but $t^0 \notin S^0$ by the definition of S^0.

If a position in G ever arises which is not in the resulting T^0, then II changes his assumption about G^*. For some S^0 he may now assume I played S^0 and II replied $\varepsilon^0 = 1$ and $t^0 = t$, where II can produce t on his next move.

3.3 The proof of Theorem 3.1.1

In the real G^*, we must turn all the A^i, not just A^0, into clopen sets. G^* is played as follows:

I	II
S^0	
	$(\varepsilon^0[, t^0])$
y^1	
	z^1
S^1	
	$(\varepsilon^1[, t^1])$
y^1	
. . .	z^1

The square brackets indicate that t^i is played only if $\varepsilon^i = 1$. T^i is a tree determined by S^i, ε^i, and t^i in a way explained below. Set $T^{-1} = X^{<\omega}$.

S^i must be a I-imposed subtree of T^{i-1} with $(y^0, z^0, \ldots, y^{i-1}, z^{i-1}) \in S^i$.

If $\varepsilon^i = 0$, then T^i is the II-imposed subtree of S^i defined by not allowing II to make a move which produces a position t such that $N_t^X \subseteq A^i$.

If $\varepsilon^i = 1$, then t^i must be a position in S^i of even length such that $N_{t^i}^X \subseteq A^i$. In this case $T^i = \{t \in T^{i-1}: t^i \subseteq t \text{ or } t \subseteq t^i\}$.

y^i must belong to X and $(y^0, z^0, \ldots, y^i)$ must belong to T^i.

z^i must belong to X and $(y^0, z^0, \ldots, y^i, z^i)$ must belong to T^i.

The first player to disobey the rules just given *loses*. Otherwise I *wins* if, and only if,

$$f = (y^0, z^0, y^1, z^1, \ldots) \in A.$$

We leave to the reader the problem of defining a suitable (X^*, A^*) corresponding to G^*.

LEMMA 3.3.1 *G^* is Σ_k^0.*

Proof I wins a play of G^* just when (a) II disobeys the rules first or else (b) both players obey the rules and (c) $f \in A$. Condition (a) is open (and F_σ). Condition (b) is G_δ (and closed). If condition (b) is satisfied, then (c) is satisfied just when

$$(\exists n_1)(\forall n_1) \ldots (\forall n_k)\varepsilon^{h(n_1,\ldots,n_k)} = 1,$$

where $h(n_1, \ldots, n_t)$ is the least i such that $A^i = A_{n_1,\ldots,n_k}$. Since the condition $\varepsilon^i = 1$ is clopen,

$$A^* = \bigcup_{n_1} \bigcap_{n_2} \cdots \bigcap_{n_k} C_{n_1,\ldots,n_k}$$

with $C_{n_1,\ldots,n_k}$ clopen. Hence A^* is Σ_k^0.

Remark We have assumed that k is even. The adaptation for k odd and >1 is straightforward. For $k = 1$, a slight modification of G^* is needed: We do not allow $\varepsilon^i = 0$ and we require that S^i be such that any $t \in S^i$ has a proper extension in S^i.

LEMMA 3.3.2 *If* I *has a winning strategy for G^*, then* I *has a winning strategy for G.*

Proof Let σ^* be a winning strategy for I for G^*. We define a strategy σ for I for G. With each position

$$p = (y^0, z^0, \ldots, y^{i-1}, z^{i-1})$$

in G consistent with σ (and with I to move), we define a position

$$p^* = (S_p^0, \varepsilon_p^0[, t_p^0], y^0, z^0, \ldots, z^{i-1}, S_p^i, \varepsilon_p^i[, t_p^i])$$

in G^* consistent with σ^* and with all rules obeyed. We further arrange that $t_p^j \subseteq p$ for every $j \leq i$ such that $\varepsilon_p^j = 1$. We then set $\sigma(p) = \sigma^*(p^*)$.

Our definition thus proceeds by induction on the length of p. Let $p = p_0{}^\frown(y^{i-1}, z^{i-1})$ if $i > 0$.

Case 1. $p \in T_{p_0}^{i-1}$ (which is trivially true if $i = 0$). Let p^* extend $p_0^*{}^\frown(y^{i-1}, z^{i-1})$ if $i > 0$. Set

$$S_p^i = \sigma^*(p_0^*{}^\frown(y^{i-1}, z^{i-1})),$$

if $i > 0$, and $S_p^i = \sigma^*$ (the empty position) if $i = 0$. If $N_p^X \not\subseteq A^i$, let $\varepsilon_p^i = 0$. This fixes T_p^i. If $N_p^X \subseteq A^i$, let $\varepsilon_p^i = 1$ and let $t_p^i = p$.

Case 2. $p \notin T_{p_0}^{i-1}$. Let j be minimal such that $p \notin T_{p_0}^j$. Let $S_p^{j'} = S_{p_0}^{j'}$, $\varepsilon_p^{j'} = \varepsilon_{p_0}^{j'}$, and (when defined) $t_p^{j'} = t_{p_0}^{j'}$ for $j' < j$. Since p_0 extends $t_{p_0}^j$ if $\varepsilon_{p_0}^j = 1$ and since $p_0 \in T_{p_0}^j$, we must have $\varepsilon_{p_0}^j = 0$. Set (as we must) $S_p^j = S_{p_0}^j$. Let $\varepsilon_p^j = 1$ and $t_p^j = p$. Note that $p \in S_{p_0}^j$ since $p \in T_{p_0}^{j-1}$, σ^* obeys the rules, and $S_{p_0}^j$ is a I-imposed subtree of $T_{p_0}^{j-1}$. Since $p \notin T_{p_0}^j$, by the definition of $T_{p_0}^j$ we must have $N_p^X \subseteq A^j$. This justifies our choice of ε_p^j and t_p^j. We define $(S_p^{i'}, \varepsilon_p^{i'}[, t_p^{i'}])$ for $j < i' \leq i$ by induction on i'. Let $S_p^{i'}$ be as given by σ^*. If there is an s of even length such that $s \subseteq p$, $s \in S_p^{i'}$, and $N_s^X \subseteq A^{i'}$, let $\varepsilon_p^{i'} = 1$ and $t_p^{i'} = s$. Otherwise let $\varepsilon_p^{i'} = 0$.

Suppose that f is a play of G consistent with σ. Let $p_i = f \restriction 2i$. If $(p_{i+1})^*$ does not extend $(p_i)^*$, by our construction, the least j such that $(S_{p_i}^j, \varepsilon_{p_i}^j[, t_{p_i}^j]) \neq (S_{p_{i+1}}^j, \varepsilon_{p_{i+1}}^j[, t_{p_{i+1}}^j])$ satisfies $\varepsilon_{p_i}^j = 0$ and $\varepsilon_{p_{i+1}}^j = 1$. It follows that the sequence $(p_0)^*, (p_1)^*, \ldots$ converges to a play of G^* consistent with σ^* and with all rules obeyed. Since σ^* is a winning strategy, we have $f \in A$.

LEMMA 3.3.3 *If* II *has a winning strategy for* G^*, *then* II *has a winning strategy for* G.

Proof Let τ^* be a winning strategy for II for G^*. We define a strategy τ for II for G. We associate with each position

$$p = (y^0, z^0, \ldots, y^i)$$

(with II to move) in G a position

$$p^* = (S_p^0, \varepsilon_p^0[, t_p^0], y^0, z^0, \ldots, S_p^i, \varepsilon_p^i[, t_p^i], y^i)$$

in G^* consistent with τ^* and with all rules obeyed. We then set $\tau(p) = \tau^*(p^*)$. As before we arrange that if $\varepsilon_p^j = 1$ then $p \supseteq t_p^j$. We also arrange that, if $p = p_0{}^\frown(z^{i-1}, y^i)$, $j < i$, and $p \in T_{p_0}^{j-1}$, then either $p \in T_{p_0}^j$ or else ($\varepsilon_{p_0}^j = 0$

and) there are an S and a z such that, if we replace $S^j_{p_0}$ by S, then τ^* requires that II plays $\varepsilon^j = 1$ and $t^j = p^\frown(z)$.

Let $p = p_0{}^\frown(z^{i-1}, y^i)$ if $i > 0$.

Case 1. $p \in T^{i-1}_{p_0}$. Let p^* extend p_0^*. If there is an S such that, if I plays S at $p_0^{*\frown}(z^{i-1})$, then τ^* calls for II to play $\varepsilon^i = 1$ and either $t^i \subseteq p$ or t^i of the form $p^\frown(z)$, then set $S^i_p = S$. Otherwise let S^i_p be the I-imposed subtree of $T^{i-1}_{p_0}$ defined by not allowing I to make a move so that the resulting position p' satisfies the following: there exist an S and a z such that if I plays S at $p_0{}^\frown(z^{i-1})$ then τ^* calls for II to play $\varepsilon^i = 1$ and $t^i = p'^\frown(z)$. Note that, for this choice of S^i_p, τ^* calls for $\varepsilon^i_p = 0$, since otherwise τ^* calls for $\varepsilon^i_p = 1$ and $t^i_p \in S^i_p$, which is just what the definition of S^i_p rules out.

Case 2. $p \notin T^{i-1}_{p_0}$. Let j be minimal such that $p \notin T^j_{p_0}$. Set $(S^{j'}_p, \varepsilon^{j'}_p[, t^{j'}_p]) = (S^{j'}_{p_0}, \varepsilon^{j'}_{p_0}[, t^{j'}_{p_0}])$ for $j' < j$. By the extra condition stated at the beginning, there are an S and a z such that if we replace $S^j_{p_0}$ by S then τ^* calls for II to play $\varepsilon^j = 1$ and $t^j = p^\frown(z)$. Set $S^j_p = S$. For $j < i' \leq i$, we define $S^{i'}_p$ by induction on i'. The definition is similar to Case 1, so we omit it.

As in the proof of the preceding lemma, any play of G according to τ can be extended to a play of G^* according to τ^* with all rules obeyed. Hence τ is a winning strategy.

3.4 The transfinite case

We now discuss briefly $\mathbf{\Sigma}^0_\eta$ games with η infinite. For simplicity, let G be $\mathbf{\Sigma}^0_\omega$. By iterating the operations $G \to G^*$, we can define a sequence of games $G_0, G_1, \ldots$ with $G_0 = G$ and with G_{n+1}, roughly speaking, equivalent to $(G_n)^*$. In G_1, the open sets which generate the A corresponding to G become clopen. In G_2 the $\mathbf{\Sigma}^0_2$ sets which generate A become clopen, and so on.

By iterating the operations $\sigma^* \to \sigma$ and $\tau^* \to \tau$, we can define functions φ_{ij}, $i \leq j$, and ψ_{ij}, $i \leq j$. If σ_j is a strategy for I for G_j, which obeys the rules, then $\varphi_{ij}(\sigma_j)$ is a strategy for I for G_i, which always obeys the rules, such that any play of G_i according to $\varphi_{ij}(\sigma_j)$, with all rules obeyed, can be extended to a play of G_j according to σ_j, with all rules obeyed. Similarly, ψ_{ij} operates on strategies for II for G_j. The φ_{ij} commute, as do the ψ_{ij}:

$$\varphi_{ij}\varphi_{jk} = \varphi_{ik}, \qquad \psi_{ij}\psi_{ik} = \psi_{ik}.$$

Furthermore the φ_{ij} and ψ_{ij} are continuous with a Lipschitz condition: the action of $\varphi_{ij}(\sigma_j)$ on positions of length $\leq n$ depends only upon the action of σ_j on sequences of length $\leq k_{ij}(n)$. In addition, because of the way in which the extra moves are added in forming G_{j+1} from G_j (not simply every other move, as before), $\lim_j k_{ij}(n)$ always exists, and, for each n, $k_{ij}(n) = n$ for all sufficiently large i. This last fact is because, for each n, there is an i such that every positon of length n in G_i is literally a position in G_j for every $j \geqslant i$.

We define an open game G_ω by letting p be a (legal) position in G_ω just when p is a (legal) position in G_i for all sufficiently large i. We then define $\varphi_{i\omega}$ and $\psi_{i\omega}$, with properties like those of the φ_{ij} and ψ_{ij} as follows. Let σ_ω be a strategy, say for I, for G_ω which obeys the rules. Fix n. Let $k = \lim_j k_{ij}(n)$. We may consider σ_ω restricted to positions of length $\leqslant k$ as a fragment of a strategy for G_j for large enough j. To compute $\varphi_{i\omega}(\sigma_\omega)$ on positions of length $\leqslant n$, simply apply φ_{ij}.

4 Analytic games

4.1 Introduction

It cannot be proved in Zermelo–Fraenkel set theory that all analytic games with $X = \omega$ are determined. By Theorem 2.2.3, Determinacy ($\mathbf{\Pi}_1^1$) implies that every $\mathbf{\Sigma}_2^1$ subset of ω^ω has the Baire property. On the other hand, Gödel's axiom of constructibility implies that there is a $\mathbf{\Delta}_2^1$ well-ordering of ω^ω. A well-ordering of ω^ω cannot have the Baire property.

We shall assume that a measurable cardinal exists and prove that all analytic games with $X = \omega$ are determined. From this result and Theorems 2.1.4 and 2.2.3 follow two of Solovay's results mentioned in Section 1.3. Solovay's original proofs did not use games and were quite different.

4.2 Measurable cardinals

By an *ordinal* we mean a *von Neumann ordinal*, the set of all smaller ordinals. By a cardinal we mean an ordinal which is not equinumerous with any smaller ordinal. For any set X, $P(X)$ is the collection of all subsets of X.

An uncountable cardinal κ is *measurable* if there is a function

$$\mu : P(\kappa) \to \{0, 1\}$$

such that

(1) $\mu(\{\eta\}) = 0$, all $\eta < \kappa$;
(2) $\mu(\kappa) = 1$;
(3) $\mu(A) = 1 - \mu(\kappa - A)$, all $A \subseteq \kappa$;
(4) if each A_i, $i \in I$, satisfies $\mu(A_i) = 1$ and cardinal $(I) < \kappa$, then

$$\mu\Big(\bigcap_{i \in I} A_i\Big) = 1.$$

In other words, μ is a non-trivial zero–one measure defined on all subsets of κ and μ is λ-additive for all $\lambda < \kappa$. If we replace this last condition by the condition that μ is *countably* additive, we get a different notion, but (Ulam) if there is a "measurable cardinal" in this weaker sense then there is one in the stronger sense. Incidentally, the theorem we shall prove about analytic games follows from the still weaker hypothesis that μ is countably additive and *real-valued.*

If μ witnesses that κ is measurable, μ is a *normal measure on* κ if, for every $f: \kappa \to \kappa$, if $\mu\{\eta: f(\eta) < \eta\} = 1$ then, for some $\eta_0 < \kappa$, $\mu\{\eta: f(\eta) = \eta_0\} = 1$.

LEMMA 4.2.1 (D. Scott) *If κ is a measurable cardinal, then there is a normal measure on κ.*

Proof Let μ witness that κ is measurable.

We first show that there is an $f: \kappa \to \kappa$ such that $\mu\{\eta: f(\eta) = \eta_0\} = 0$ for all $\eta < \kappa$ but such that, for all $g: \kappa \to \kappa$ such that $\mu\{\eta: g(\eta) < f(\eta)\} = 1$, there is an $\eta_0 < \kappa$ such that $\mu\{\eta: g(\eta) = \eta_0\} = 1$. If f does not exist, let $f_0(\eta) = \eta$ and let f_{i+1} witness that f_i does not have desired property. Let $A_i = \{\eta: f_{i+1}(\eta) < f_i(\eta)\}$. $\mu(A_i) = 1$ for every i. Let $A = \bigcap_{i\in\omega} A_i$. By countable additivity, $\mu(A) = 1$. If $\eta \in A$ then $f_0(\eta) > f_1(\eta) > \cdots$, which is impossible.

Given such an f, define ν by

$$\nu(A) = \mu(f^{-1}(A)).$$

ν satisfies (1)–(4). Suppose $\nu\{\eta: g(\eta) < \eta\} = 1$. Then $\mu\{\eta: gf(\eta) < f(\eta)\} = 1$. Hence $\mu\{\eta: gf(\eta) = \eta_0\} = 1$ for some η_0. But then $\nu\{\eta: g(\eta) = \eta_0\} = 1$.

LEMMA 4.2.2 (Rowbottom) *Let μ be a normal measure on κ. Let $f: \kappa^n \to \omega$. There is a $k \in \omega$ and an $A \subseteq \kappa$ with $\mu(A) = 1$ such that, if $\eta_1 < \eta_2 < \cdots < \eta_n$ and each $\eta_i \in A$, then $f(\eta_1, \ldots, \eta_n) = k$.*

Proof We proceed by induction on n. The case $n = 1$ follows from the countable additivity of μ. Let $n > 1$ and $f: \kappa^n \to \omega$. By induction we have for each $\eta < \kappa$ a k_η and an A_η such that $\mu(A_\eta) = 1$ and, if $\eta_2 < \cdots < \eta_n$ and each $\eta_i \in A_\eta$, then $f(\eta, \eta_2, \ldots, \eta_n) = k_\eta$. Let $B = \{\rho: (\forall \eta < \rho)(\rho \in A_\eta)\}$. If $\mu(B) = 0$, let $g(\rho)$ be the least $\eta < \rho$ such that $\rho \notin A_\eta$, for each $\rho \notin B$. By normality, there is an η_0 such that $\mu\{\rho: g(\rho) = \eta_0\} = 1$. But then $\mu\{\rho: \rho \notin A_{\eta_0}\} = 1$, which contradicts $\mu(A_{\eta_0}) = 1$. By countable additivity, let k be such that $\mu\{\eta: k_\eta = k\} = 1$. Let

$$A = B \cap \{\eta: k_\eta = k\}.$$

Then $\mu(A) = 1$. If $\eta_1 < \eta_2 < \cdots < \eta_n$ and each $\eta_i \in A$, then each of $\eta_2, \ldots, \eta_n$ belongs to A_{η_1}. Hence $f(\eta_1, \ldots, \eta_n) = k_{\eta_1} = k$.

4.3 $\mathbf{\Pi}_1^1$ sets and well-orderings

Let A be $\mathbf{\Pi}_1^1$, $A \subseteq \omega^\omega$. In Section 1 we showed that there is a tree T on $\omega \times \omega$ such that

$$A = \{\alpha : T(\alpha) \text{ is well founded}\}.$$

The question "$t \in T(\alpha)$?" depends only on $\alpha \restriction \text{length}(t)$. Let us extend the partial ordering of $T(\alpha)$ to a total ordering:

Set $t_1 <_{BK} t_2$ if t_1 properly extends t_2 or if, for some k, $k < \text{length}(t_1)$ and $k < \text{length}(t_2)$ and $t_1(k) \neq t_2(k)$ and, for the least such k, $t_1(k) < t_2(k)$.

LEMMA 4.3.1 *$A = \{\alpha : <_{BK}$ well-orders $T(\alpha)\}$.*

Proof Since $<_{BK}$ is a total ordering and extends the tree partial ordering of $T(\alpha)$, we need only prove that if $<_{BK} \restriction T(\alpha)$ is not well founded then $T(\alpha)$ is not a well-founded tree. Let

$$t_0 >_{BK} t_1 >_{BK} t_2 > \cdots,$$

with all $t_i \in T(\alpha)$. There are a k_0 and an infinite set of i such that $t_i(0) = k_0$. By induction one easily shows that there is for each j a k_j such that the k_j satisfy the following: for all m there are infinitely many i such that $t_i \restriction m = (k_0, \ldots, k_{m-1})$. The $(k_0, \ldots, k_{m-1})$ thus all belong to $T(\alpha)$ and form an infinite descending chain in its partial ordering.

Let $t_0, t_1, \ldots$ enumerate $\omega^{<\omega}$ so that length $(t_i) \leq i$. For each $\alpha \in \omega^\omega$, define a relation $<_\alpha$ on ω by

$$i <_\alpha j \leftrightarrow \begin{cases} t_i \in T(\alpha) \;\&\; t_j \in T(\alpha) \;\&\; t_i <_{BK} t_j, \text{ or} \\ t_i \in T(\alpha) \;\&\; t_j \notin T(\alpha), \text{ or} \\ t_i \notin T(\alpha) \;\&\; t_j \notin T(\alpha) \;\&\; i < j. \end{cases}$$

LEMMA 4.3.2 *$A = \{\alpha : <_\alpha$ well-orders $\omega\}$.*

Proof $<_\alpha$ is just the ordering $<_{BK} \restriction T(\alpha)$ with an ordering of order-type $\leq \omega$ added on at the end.

4.4 Proof of the theorem

THEOREM 4.4.1 (Martin, 1970) *If there is a measurable cardinal, then Determinacy ($\mathbf{\Pi}_1^1$).*

Proof Let μ be a normal measure on κ. Let $A \subseteq \omega^\omega$ be $\mathbf{\Pi}_1^1$. Let G be the game corresponding to A. We define a game G^* as follows:

I	II
(η_0, m_0)	
	n_0
(η_1, m_1)	
	n_1
...	

Each m_i and n_i must be natural numbers. Each η_i must be an ordinal $<\kappa$. For each i and j, we must have

$$i <_\alpha j \leftrightarrow \eta_i < \eta_j,$$

where $\alpha = (m_0, n_0, m_1, n_1, \ldots)$. The first player to disobey these rules *loses*. If all rules are obeyed, I *wins*.

LEMMA 4.4.2 *G^* is determined.*

Proof G^* is closed.

LEMMA 4.4.3 *If* I *has a winning strategy for G^* then* I *has a winning strategy for G.*

Proof The function $i \to \eta_i$ in effect *verifies* that I has won G.

LEMMA 4.4.4 *If* II *has a winning strategy for G^*, then* II *has a winning strategy for G.*

Proof Let τ^* be a winning strategy for II for G^*. We define a strategy τ for II for G. Let $p = (m_0, n_0, \ldots, m_{k-1}, m_k)$ be a position in G of odd length. If $\rho_0 < \rho_1 < \cdots < \rho_k < \kappa$, there is a unique map $\psi: \{0, \ldots, k\} \to \{\rho_0, \ldots, \rho_k\}$ such that, if $\eta_i = \psi(i)$, then I has not already lost at the position

$$p^* = (\eta_0, m_0, n_0, \eta_1, \ldots, \eta_k, m_k).$$

Let $f_p(\rho_0, \ldots, \rho_k) = \tau^*(p^*)$ if $\rho_0 < \cdots < \rho_k$ and let $f(\rho_0, \ldots, \rho_k) = 0$ otherwise. Let $A_p \subseteq \kappa$ be the set whose existence is guaranteed by Lemma 4.2.2. Let $\tau(p)$ be the constant value of $f_p(\rho_0, \ldots, \rho_k)$ for $\rho_0 < \cdots < \rho_k$ and each $\rho_i \in A_p$.

Let $E = \bigcap_p A_p$. Suppose α is a play of G consistent with τ. If $\alpha \in A$, then there is an $F: \omega \to E$ such that $i <_\alpha j \leftrightarrow F(i) < F(j)$. By setting $\eta_i = F(i)$, we can extend α to a play of G^* consistent with τ^*. The resulting play is won by I, and this contradicts the assumption that τ^* is a winning strategy.

Remark Essentially the same proof shows that all analytic games with cardinal $(X) < \kappa$ are determined, where κ is a measurable cardinal. Hence the assumption that there are arbitrarily large measurable cardinals implies that all analytic games, with arbitrary X, are determined.

5 How many analytic, non-Borel sets are there?

5.1 Introduction

In this section we shall discuss work of W. Wadge, J. Steel, and L. Harrington, leading to the following theorem of Harrington:

THEOREM 5.1.1 *Determinacy* $(\mathbf{\Pi}_1^1)$ *is equivalent to the assertion that any two analytic, non-Borel sets are Borel isomorphic.*

Here $A, B \subseteq \omega^\omega$ are *Borel isomorphic* if there is a Borel measurable one–one onto $f: \omega^\omega \to \omega^\omega$ such that $f(A) = B$. Actually the proof constructs an f such that f and f^{-1} are of low Baire class. We shall prove just one direction of the theorem, and from a (*prima facie*) stronger determinacy hypothesis. We shall also deduce from Borel determinacy some related results about levels of the Borel hierarchy.

5.2 Wadge degrees

Suppose $A \subseteq Y^\omega$ and $B \subseteq Z^\omega$. We associate with (A, B) a game $G(A, B)$ played as follows (in effect, over $Y \cup Z$):

I	II
y_0	
	z_0
y_1	
	z_1

. . .

The y_i must belong to Y and the z_i to Z. If this rule is obeyed, II *wins* $G(A, B)$ just when

$$(y_0, y_1, \ldots) \in A \leftrightarrow (z_0, z_1, \ldots) \in B.$$

A is *l-reducible to* B, $A \leq_l B$, just when II has a winning strategy for $G(A, B)$. (This and subsequent definitions depend in general upon Y and Z as well as upon A and B. One can verify that there is no dependence upon Y and Z unless A and B are Boolean combinations of open sets; therefore we usually suppress mention of Y and Z.) $A \sim_l B$ means that $A \leqslant_l B$ *and* $B \leqslant_l A$. The l-degree of A, $[A]_l = \{B: A \sim_l B \text{ or } A \sim_l Z^\omega - B\}$. Partially order the l-degrees:

$$[A]_l < [B]_l \leftrightarrow ([A]_l \neq [B]_l \;\&\; (A \leqslant_l B \text{ or } A \leqslant_l Z^\omega - B)).$$

A is *Wadge-reducible to* B, $A \leqslant_w B$, if there is a continuous $\varphi: Y^\omega \to Z^\omega$ such that $\varphi^{-1}(B) = A$. Since a strategy for II for $G(A, B)$ induces a continuous $\varphi: Y^\omega \to Z^\omega$ with $\varphi^{-1}(B) = A$,

$$A \leqslant_l B \to A \leqslant_w B.$$

(The map induced by a strategy is continuous with a "Lipschitz" condition, and that is the reason for "l".) Define $A \sim_w B$, $[A]_w$, and a partial ordering of the Wadge degrees $[A]_w$ in the obvious way.

LEMMA 5.2.1 (Wadge) *If $G(A, B)$ is determined, then $A \leqslant_l B$ or $(Z^\omega - B) \leqslant_l A$.*

Proof It is enough to show that if I has a winning strategy for $G(A, B)$ then II has a winning strategy for $G(Z^\omega - B, A)$. Let σ be a winning strategy for I for $G(A, B)$. Define a strategy τ for II for $G(Z^\omega - B, A)$ by $\tau(y_0, z_0, \ldots, y_n) = \sigma(z_0, y_0, \ldots, z_{n-1}, y_{n-1})$. (In other words, II pretends he is going first and ignores I's last move.) (Clearly $(z_0, z_1, \ldots) \in A \nleftrightarrow (y_0, y_1, \ldots) \in B$.

COROLLARY 5.2.2 *The l-degrees, and so the Wadge degrees, of Borel sets are linearly ordered. If* PD, *the l-degree of projective subsets of ω^ω or 2^ω are linearly ordered.*

Obviously we have stated only two special cases of a more general theorem. Wadge, assuming Borel determinacy, proved that the l-ordering (and so the Wadge ordering) is in fact a well-ordering for Borel subsets of ω^ω. Wadge proceeded by a detailed analysis of Borel sets. The following theorem was proved by Martin, though the main idea is due to L. Monk.

THEOREM 5.2.3 *Corollary* 5.2.2 *remains true when "linearly ordered" is replaced by "well-ordered".*

Proof It is clearly enough to show that there is no infinite sequence $A_0, A_1, \ldots$ such that, for each i, neither $A_i \leqslant_l A_{i+1}$ nor $A_i \leqslant_l X_{i+1}^\omega - A_{i+1}$, where A_i is given as a subset of X_i^ω. The determinacy hypothesis, together with Theorem 2.2.1, allows us to assume that every $B \subseteq 2^\omega$ which is a continuous pre-image of A_0 has the Baire property.

Assume for a contradiction that the A_i exist. By the determinacy hypothesis, I has winning strategies σ_i^0 for $G(A_i, A_{i+1})$ and σ_i^1 for $G(A_i, X_{i+1}^\omega - A_{i+1})$. Let $\alpha \in 2^\omega$. Consider a sequence of players (0), (1), (2), ... We produce a sequence of elements $f_\alpha^0, f_\alpha^1, \ldots$ of X_i^ω by letting each player (i) play his strategy $\sigma_i^{\alpha(i)}$ against player $(i+1)$. This produces the f_α^i, since the $\sigma_i^{\alpha(i)}$ are strategies for I.

Let $B_0 \subseteq 2^\omega$ be $\{\alpha : f_\alpha^0 \in A_0\}$. B_0 is a continuous pre-image of A_0, and so B_0 has the Baire property. We show that B_0 is meagre. A similar argument shows that $2^\omega - B_0$ is meagre, and this is a contradiction.

If B_0 is not meagre, there is a $t \in 2^{<\omega}$ such that $N_t^2 - B_0$ is meagre. Consider the homeomorphism of 2^ω onto itself defined by sending each α to α', where

$$\alpha'(i) = \alpha(i) \leftrightarrow i \neq \text{length}\,(t).$$

This homeomorphism maps N_t^2 homeomorphically onto itself. If $i > \text{length}\,(t)$, player (i)'s moves are unaffected by replacing α by α', so $f_{\alpha'}^i = f_\alpha^i$. If $i \leq \text{length}\,(t)$, $f_{\alpha'}^i \in A_i \leftrightarrow f_\alpha^i \notin A_i$. Hence our homeomorphism maps B_0 onto $2^\omega - B_0$. This means that $N_t^2 \cap B_0$ is meagre, a contradiction.

The following lemma is the only fact about l-reducibility needed in the next section.

LEMMA 5.2.4 *Let Γ be the class $\mathbf{\Sigma}_\eta^0$ for some η or else the class $\mathbf{\Sigma}_1^1$ (i.e., the class of analytic sets). Let $A \subseteq Y^\omega$ and $B \subseteq Z^\omega$. Assume $A \in \Gamma$ and $B \notin \check{\Gamma}$ where a subset of Z^ω is $\check{\Gamma}$ just when its complement is Γ. Assume further that $G(A, B)$ is determined. We then have $A \leq_l B$.*

Proof Otherwise, by Lemma 5.1.1, $Z^\omega - B \leq_l A$. But then $Z^\omega - B$ is the continuous pre-image of a set in Γ and so belongs to Γ. By definition, $B \in \check{\Gamma}$.

5.3 Borel isomorphism

If E_1 and E_2 are topological spaces and $\varphi: E_1 \to E_2$ is one–one surjection, φ is a *first-level Borel isomorphism* if all images and pre-images of F_σs are F_σ. This notion, and Corollary 5.3.2 (for the case $\Gamma = \mathbf{\Sigma}_1^1$) below, are due to J. E. Jayne and C. A. Rogers (1979).

THEOREM 5.3.1 *Let Γ be $\mathbf{\Sigma}_\eta^0$ for $\eta \geq 3$ or $\mathbf{\Sigma}_1^1$. Assume, in the latter case, determinacy for differences of $\mathbf{\Sigma}_1^1$ sets (for $X = \omega$). Let A and B be subsets of 2^ω*

belonging to $\Gamma - \check{\Gamma}$. *There is a first-level Borel isomorphism* $\varphi: 2^\omega \to 2^\omega$ *such that* $\varphi(A) = B$.

Proof The theorem is due to Harrington. Steel earlier proved the conclusion for $\Gamma = \mathbf{\Sigma}_1^1$ from the hypothesis # stated in the next section.

Let $A^* \subseteq 4^\omega (= \{0, 1, 2, 3,\}^\omega)$ be given by

$$\alpha \in A^* \leftrightarrow \begin{cases} \alpha(n) = 2 & \text{for all sufficiently large } n, \\ \alpha(n) < 2 & \text{for infinitely many } n \text{ and } \alpha^* \in A, \end{cases}$$

where α^* is the sequence resulting from deleting all 2s and 3s from the sequence α. Since $\eta \geqslant 3$ in the case $\Gamma = \mathbf{\Sigma}_\eta^0$, $A^* \in \Gamma$. By Lemma 5.2.4, there is a $\psi: 4^\omega \to 2^\omega$ which is induced by a strategy for II such that $\psi^{-1}(B) = A^*$.

We define a map $\pi: 2^{<\omega} \to 4^{<\omega}$ as follows: π (empty sequence) = empty sequence. Suppose $\pi(s) = t$. Let k be the least number that

$$\psi(N_{t^\frown \underbrace{(2,\ldots,2)}_{k}}) \cap \psi(N_{t^\frown \underbrace{(3,\ldots,3)}_{k}}) \text{ is empty.}$$

k exists, since otherwise $\psi(t^\frown(2, 2, \ldots)) = \psi(t^\frown(3, 3, \ldots))$, and the argument on the left belongs to A^* while the argument on the right belongs to $4^\omega - A^*$. Set

$$\pi(s^\frown(0)) = t^\frown \underbrace{(2, \ldots, 2)}_{k}{}^\frown(0)$$

and

$$\pi(s^\frown(1)) = t^\frown \underbrace{(3, \ldots, 3)}_{k}{}^\frown(1).$$

π induces a map of 2^ω into 4^ω, which we also call π. This π is continuous and satisfies $(\pi(\alpha))^* = \alpha$. By construction, $\psi\pi: 2^\omega \to 2^\omega$ is one–one. Furthermore,

$$\alpha \in A \leftrightarrow \pi(\alpha) \in A^* \leftrightarrow \psi\pi(\alpha) \in B.$$

Combining this construction with a similar one where the roles of A and B are reversed, we get that there are ψ_1, ψ_2: $2^\omega \to 2^\omega$ such that

(1) ψ_1 and ψ_2 are induced by strategies for II, and are continuous;
(2) ψ_1 and ψ_2 are one–one;
(3) $\psi_1^{-1}(B) = A$ and $\psi_2^{-1}(A) = B$.

We now apply a Cantor–Bernstein argument to prove the theorem. Let C be the collection of $\alpha \in 2^\omega$ such that, for some β not in the range of ψ_2 and some k (possibly 0),

$$(\psi_2\psi_1)^k(\beta) = \alpha.$$

Define $\varphi(\alpha) = \psi_1(\alpha)$ for $\alpha \in C$ and $\varphi(\alpha) = \psi_2^{-1}(\alpha)$ for $\alpha \notin C$.

Clearly $\varphi(A)=B$ and φ is one–one and onto. Since 2^ω is compact and ψ_1 and ψ_2 are continuous, C is both F_σ and G_δ. For any D, $\varphi(D)=\psi_1(D\cap C)\cup\psi_2^{-1}(D-C)$ and $\varphi^{-1}(D)=\psi_1^{-1}(D\cap\psi_1(C))\cup(\psi_2(D-\psi_1(C)))$. Since ψ_1 and ψ_2 are continuous and 2^ω is compact, this shows that φ is first-level.

COROLLARY 5.3.2 *Let $A, B\subseteq\omega^\omega$, $A, B\in\Gamma-\check{\Gamma}$, with Γ either $\mathbf{\Sigma}^0_\eta$ for $\eta\geq 3$ or $\mathbf{\Sigma}^1_1$. In the latter case, assume determinacy for differences of analytic sets. There is a first-level Borel isomorphism $\varphi: A\to B$.*

Proof There is a $\Theta:\omega^\omega\to 2^\omega$ which embeds the space ω^ω as a subspace of 2^ω so that $\Theta(A)$ and $\Theta(B)$ belong to $\Gamma-\check{\Gamma}$. By the theorem, let $\psi: 2^\omega\to 2^\omega$ be a first-level Borel isomorphism with $\psi(\Theta(A))=\Theta(B)$. Let $\varphi=\Theta^{-1}\psi\Theta|A$. If $D\subseteq B$ is an F_σ in B, let $D^*\subseteq 2^\omega$ be an F_σ in 2^ω such that $\Theta(D)=D^*\cap\Theta(B)$. $\Theta^{-1}\psi^{-1}(D^*)$ is an F_σ in ω^ω. Thus $\varphi^{-1}(D)=(\Theta^{-1}\psi^{-1}\Theta(D))\cap A$, which is an F_σ. A similar argument shows that φ maps F_σs onto F_σs.

5.4 Harrington's theorem

The proof Theorem 5.1.1 is quite indirect and uses many techniques from mathematical logic. We content ourselves with some general remarks.

If $\alpha\in\omega^\omega$, $L[\alpha]$ is the smallest proper class satisfying all the axioms of ZFC and containing α and all ordinals. A more useful definition is as follows. Let $L_0[\alpha]=\{\alpha\}$. Let

$$L_\lambda[\alpha]=\bigcup_{\eta<\lambda} L_\eta[\lambda]$$

if λ is a limit ordinal. Let $L_{\eta+1}[\alpha]$ be the collection of all subsets y of $L_\eta[\alpha]$ such that, for some formula φ of set theory with only $v_0,\ldots,v_k$ free and some $u_1,\ldots,u_k\in L_\eta[\alpha]$,

$$z\in y\leftrightarrow(z\in L_\eta[\alpha]\ \&\ L_\eta[\alpha]\text{ satisfies }\varphi[z,u_1,\ldots,u_1]).$$

Let

$$L[\alpha]=\bigcup_\eta L_\eta[\alpha].$$

A set Y of ordinals is *a set of indiscernibles for* $L[\alpha]$ if, for every formula of set theory with only $v_0,\ldots,v_k$ free and every $\eta_1<\cdots<\eta_k$ and $\rho_1<\cdots<\rho_k$, all elements of Y,

$$L[\alpha]\text{ satisfies }\varphi[\alpha,\eta_1,\ldots,\eta_k]\leftrightarrow L[\alpha]\text{ satisfies }\varphi[\alpha,\rho_1,\ldots,\rho_k].$$

Let # be the assertion that, for every $\alpha \in \omega^{\omega}$, there is an uncountable set of indiscernibles for $L[\alpha]$. (These notions were introduced by J. Silver and studied by him and R. Solovay.)

A variant of the proof given in Section 4 gives the implication

$$\# \rightarrow \text{Determinacy}(\mathbf{\Pi}_1^1).$$

Friedman (1971*b*) improved this result to get determinacy for a class containing all differences of $\mathbf{\Pi}_1^1$ sets, and Martin improved it to deduce from # the determinacy of a class larger than the Boolean algebra generated by the $\mathbf{\Pi}_1^1$ sets. Combining Friedman's result with Corollary 5.3.2, we get

$$\# \rightarrow \text{BI},$$

where BI is the assertion that any two analytic, non-Borel subsets of ω^{ω} are Borel isomorphic.

Harrington (1980) then proved the following theorems:

$$\text{Determinacy}\ (\mathbf{\Pi}_1^1) \rightarrow \#,$$

$$\text{BI} \rightarrow \#.$$

It follows then that Determinacy ($\mathbf{\Pi}_1^1$), BI, and # are all equivalent. Harrington's proof used a technical lemma originally deduced from a powerful result of Jensen, although Paris later found a direct proof of the technical lemma.

We state Jensen's result (see Devlin and Jensen, 1975) here, because it adds still another equivalence to our list. Let *Cover* be the assertion there is an $\alpha \in \omega^{\omega}$ such that, for every uncountable set X of ordinals, there is a $Y \supset X$ such that $Y \in L[\alpha]$ and cardinal $(Y) = $ cardinal (X). Jensen's theorem is

$$\text{not}\ \# \leftrightarrow \text{Cover}$$

(The difficult direction is $\rightarrow$.) Cover has many interesting consequences. For example, it yields a complete solution to the co-called *singular cardinals problem.* Thus not only is Determinacy ($\mathbf{\Pi}_1^1$) a powerful and interesting assertion but so also is its negation.

6 Introduction to effective descriptive set theory

6.1 Introduction

Effective descriptive set theory has its origins in the pioneering work of Kleene in recursion theory and definability theory, during an almost twenty-year period starting in the mid-1930s. In the 1940s and early 1950s

Mostowski and mainly Addison realized important "analogies" between Kleene's work and that of the classical descriptive set theories. This led to the development of a unified theory, known as effective descriptive set theory, which at the same time extends ordinary recursion theory and also refines the classical theory of analytic and projective sets. By blending the classical methods of descriptive set theory with the new concepts and techniques of recursion theory and logic, effective descriptive set theory not only creates a powerful and elegant machinery for studying problems of classical type and their effective strengthenings, but also provides the stage for the generation of new insights, ideas and questions which are inaccessible or meaningless in the context of the classical theory.

In Sections 6–9 we shall try to give an introduction to effective descriptive set theory, show its connections with the classical subject and some of its applications in proving new theorems as well as effective refinements of classical results by new methods. We shall necessarily have to omit many important results which properly belong to the effective theory but have no immediate counterparts in the classical one. The important role of games will be illustrated in connection with this effective theory.

As with the preceding Sections 2–5, we shall develop the main theory on the Baire space ω^ω, as this involves no essential loss of generality in view of various transfer theorems, and makes the key new ideas more transparent. (For a development of the effective theory in arbitrary "recursively presentable perfect Polish spaces", see Moschovakis' book *Descriptive Set Theory*, 1980.)

6.2 Review of basic recursion theory

(A) Let $\omega = \{0, 1, 2, \ldots\}$ be the set of natural numbers. We shall use letters $e, i, j, k, l, m, n, \ldots$ to vary over ω. The basic notion of computability theory is that of a *computable* function $f: \omega^k \to \omega$, i.e. a function for which there is a *mechanical procedure* or *algorithm* for computing, in finitely many steps, the value $f(n_1, \ldots, n_k)$ for each given argument $\bar{n} = (n_1, \ldots, n_k)$. Via Church's thesis this intuitive notion can be identified with the exact concept of a *recursive* function, which we proceed to define.

DEFINITION The class of *recursive functions* is the smallest class $\mathcal{R}$ of functions $f: \omega^k \to \omega (k = 1, 2, \ldots)$, which contains the *successor function* $n \mapsto n+1$, the *constant* functions $\bar{n} \mapsto c \in \omega$ and the *projection* functions $p_i^k(n_1, \ldots, n_k) = n_i$, $1 \leq i \leq k$, and is closed under the following conditions:

(i) *Composition*, i.e. if $f_1 \ldots f_m$, $g \in \mathcal{R}$, then, if $h(\bar{n}) = g(f_1(\bar{n}) \ldots f_m(\bar{n}))$, $h \in \mathcal{R}$.

(ii) *Primitive recursion*, i.e. if $f_1, f_2 \in \mathcal{R}$ and if $h(0, \bar{n}) - f_1(\bar{n})$, $h(i+1, \bar{n}) = f_2(h(i, \bar{n}), i, \bar{n})$, then $h \in \mathcal{R}$.
(iii) *Minimalization*, i.e. if $f \in \mathcal{R}$ and $\forall \bar{n} \exists m (f(m, \bar{n}) = 0)$, then if $h(\bar{n}) =$ least m such that $f(m, \bar{n}) = 0$, $h \in \mathcal{R}$.

This definition is readily extended to functions $f: \omega^k \to \omega^m$. Such a function is *recursive* if for each $1 \leq i \leq m$, $p_i^m \circ f: \omega^k \to \omega$ is recursive. We call now a set $A \subseteq \omega^m$ *recursively enumerable* (r.e.) if $A = \text{range}(f)$, for some recursive $f: \omega \to \omega^m$ or $A = \emptyset$. Thus speaking intuitively, $A \neq \emptyset$ is r.e. if there is an algorithm for listing, in some order, the elements of A. Finally we call $A \subseteq \omega^m$ *recursive* if its characteristic function $\chi_A: \omega^m \to \omega$ is recursive. Again this means intuitively that there is an algorithm for testing membership in A.

We now have the following simple fact (proofs of this and the other standard results from recursion theory we mention in this lecture can be found in any standard textbook on the subject (e.g. M. Davis, 1958)).

FACT 6.2.1 (i) $A \subseteq \omega^m$ is recursive if, and only if, both A, $\omega^m - A$ are r.e.
(ii) $A \subseteq \omega^m$ is r.e. if, and only if, A can be written in the form

$$\bar{n} \in A \Leftrightarrow \exists k (\bar{n}, k) \in B,$$

with $B \subseteq \omega^{m+1}$ recursive.

(iii) The class of r.e. sets is closed under $\cap$, $\cup$, projection or *existential quantification* (i.e. if $A \subseteq \omega^{m+1}$ is r.e. so is $B = \{\bar{n}: \exists k (\bar{n}, k) \in A\}$) and bounded universal quantification (i.e. if $A \subseteq \omega^{m+1}$ is r.e. so is $B = \{(\bar{n}, l): \forall k \leq l (\bar{n}, k) \in A\}$. The class of recursive sets is closed under $\cap$, $\cup$, bounded quantification of both types and complements.

(iv) The class of r.e. (resp. recursive) sets is closed under recursive pre-images or *recursive substitutions*, i.e. if $A \subseteq \omega^m$ is r.e. (resp. recursive) and $f: \omega^k \to \omega^m$ is recursive then if we define $B = f^{-1}[A]$, i.e.

$$\bar{n} \in B \Leftrightarrow f(\bar{n}) \in A,$$

then B is r.e. (resp. recursive).

A most important structural property of r.e. sets is given by the following basic theorem of Kleene.

THEOREM 6.2.2 (The parametrization theorem for r.e. sets) *For each* $k = 1, 2, \ldots$ *there is an r.e. set* $U^k = U \subseteq \omega^{k+1}$, *such that for* $A \subseteq \omega^k$:

$$A \text{ is r.e.} \Leftrightarrow \text{for some } e,\ A = U_e \ (\stackrel{\text{def}}{=} \{\bar{n}: U(e, \bar{n})\}).$$

Such a set U is called *universal* for the r.e. subsets of ω^k.

COROLLARY 6.2.3 *There is an r.e. non-recursive set.*

Proof By the usual diagonalization argument if $U = U^1$, $\{e: e \in U_e\}$ is r.e. but its complement is not.

(B) All the preceding notions and results can be "relativized" to an arbitrary $\alpha: \omega \to \omega$. Intuitively a function $f: \omega^k \to \omega$ is *computable in* (*or relatively to*) α if there is an algorithm for computing, in finitely many steps, for each $\bar{n} \in \omega^k$ the value $f(\bar{n})$, where the algorithm is allowed to contain instructions requiring the value of α at certain arguments, which (as α itself need not be computable) is understood to be supplied by an "oracle" for α. Note that during the course of the computation of $f(\bar{n})$ from α only finitely many values of α come into play. As an example,

$$f(k) = \begin{cases} 1 + (\alpha(p))^2, & \text{where } p \text{ is the largest prime divisor of } k, \text{ if } k \text{ is odd} \\ \alpha(\alpha(k) + 3(\alpha(k^2))), & \text{otherwise} \end{cases}$$

is computable in α. Again this intuitive notion is identified with the precise notion of a *recursive* in α function, where $f: \omega^k \to \omega$ is recursive in α if f belongs to the smallest class of functions containing α, successor, constants and projections and closed under composition, primitive recursion and minimalization. One can now repeat all of (A) by replacing "recursive" and "r.e." by "recursive in α" and "r.e. in α" respectively, and all the arguments remain valid. In particular one obtains for each fixed α and each k a universal set U^α for the r.e. in α subsets of ω^k, which by the relativization of Fact 6.2.1(ii) will have the form $\exists k(e, \bar{n}, k) \in W^\alpha$ for some W^α recursive in α. There is however a strong "uniform" version of this fact, which we shall state after we have introduced some notation.

Let $f: \omega \times \omega \to \omega$ be any standard recursive bijection of $\omega \times \omega$ with ω, such that $n, m \leqslant f(n, m)$.

For convenience we shall put

$$f(n_0, n_1) = \langle n_0, n_1 \rangle_1$$

and we shall denote by

$$n \mapsto ((n)_0, (n)_1)$$

the inverse function. This can be easily extended to a recursive bijection $\langle n_0 \ldots n_k \rangle_k$ of ω^{k+1} with ω for each $k \geqslant 2$ defined by

$$\langle n_0, n_1, \ldots, n_k \rangle_k = \langle \langle n_0, n_1, \ldots, n_{k-1} \rangle_{k-1}, n_k \rangle_1.$$

Finally consider the following bijection of $\omega^{<\omega}$ and ω:

$$\langle n_0 \ldots n_k \rangle = \langle k, \langle n_0 \ldots n_k \rangle_k \rangle_1 + 1, \qquad \langle \varnothing \rangle = 0,$$

where

$$\langle n \rangle_0 = n.$$

We view $\langle n_0 \ldots n_k \rangle$ as the number coding $(n_0 \ldots n_k) \in \omega^{<\omega}$. At last put for each $\alpha \in \omega^\omega$, $k \in \omega$,

$$\bar{\alpha}(k) = \langle \alpha \restriction k \rangle = \langle \alpha(0) \ldots \alpha(k-1) \rangle.$$

Clearly $k \mapsto \bar{\alpha}(k)$ is recursive in α.

Now we have

THEOREM 6.2.4 (Uniform parametrization theorem for r.e. in α sets) *For each $k = 1, 2, \ldots$ there is recursive (absolutely) $W \subseteq \omega^{k+2}$ such that for any $\alpha \in \omega^\omega$, $A \subseteq \omega^k$:*

$$A \textit{ is r.e. in } \alpha \Leftrightarrow \textit{for some } e \in \omega,$$

$$\bar{n} \in A \Leftrightarrow \exists k(e, \bar{n}, \bar{\alpha}(k)) \in W.$$

(C) As there is a recursive bijection of each ω^k ($k = 1, 2, \ldots$) with ω we shall many times identify ω^k with ω in the sequel without any explicit mentioning. As there is also a very effective bijection between $\omega^{<\omega}$ with ω provided by the coding function $\langle n_0 \ldots n_k \rangle$ we shall also identify a finite sequence $u \in \omega^{<\omega}$ with its code $\langle u \rangle \in \omega$. This allows us automatically to transfer the notions of recursive, r.e. and their relativizations immediately to $\omega^{<\omega}$. For example, if T is a tree on $\omega \times \omega$, we shall call T recursive if $\{(\langle u \rangle, \langle v \rangle): (u, v) \in T\}$ is recursive.

6.3 Effective topology

(A) We shall use letters $\alpha, \beta, \gamma, \delta, \ldots$ for members of ω^ω, i.e. *reals*. Recall that if $s \in \omega^{<\omega}$, $N_s = \{\alpha \in \omega^\omega : s \subseteq \alpha\}$ is the basic neighbourhood determined by s. A typical open set of reals then has the form $G = \bigcup_{n \in \omega} N_{s_n}$, where $n \mapsto s_n$ is some sequence of elements of $\omega^{<\omega}$. By requiring this to be recursive we obtain the notion of an "effective" open set—effective here is used as synonymous to computable or recursive.

DEFINITION A set $A \subseteq \omega^\omega$ is called *effectively open* or $\mathbf{\Sigma}^0_1$ if $A = \bigcup_{n \in \omega} N_{s_n}$, where $n \mapsto s_n$ is recursive (i.e. if A is an effective union of basic neighbourhoods). We are using our identification of $\omega^{<\omega}$ with ω here.

Note that there are only $\aleph_0$ Σ_1^0 sets. This definition of Σ_1^0 sets can be immediately extended in the obvious fashion to an arbitrary product space $(\omega^\omega)^k$, which of course has as neighbourhood basis all products of the form $\prod_{i=1}^k N_{s_i}$, $s_i \in \omega^{<\omega}$ for $1 \leq i \leq k$. But it also makes sense for subsets of ω (or any ω^k), where this is equipped with the discrete topology. In this case the basic neighbourhoods are the singletons $\{i\}$, $i \in \omega$, so an effectively open, i.e. Σ_1^0, set is just an r.e. set. Similarly for each ω^k.

It is convenient at this stage to consider more generally any product space

$$\mathscr{X} = X_1 \times \cdots \times X_k,$$

where $X_i = \omega$ or ω^ω. These spaces will be equipped with the product topology, with ω having the discrete one and ω^ω its standard topology. Elements of these spaces will be called *points* and subsets of these spaces *pointsets*. We often think of them also as properties or relations, of integer and real arguments, and write interchangeably for $x \in \mathscr{X}$, $A \subseteq \mathscr{X}$:

$$x \in A \Leftrightarrow A(x).$$

We call any collection of pointsets a *pointclass*. We can now extend the definition of Σ_1^0 to arbitrary spaces in the obvious way to obtain

DEFINITION The pointclass Σ_1^0 consists of all those pointsets $A \subseteq \mathscr{X}$($\mathscr{X}$ an arbitary space), which are effective unions of basic neighbourhoods in the space $\mathscr{X}$.

For example if $\mathscr{X} = \omega \times \omega^\omega$ then $A \subseteq \mathscr{X}$ is Σ_1^0 if, and only if, $A = \bigcup_n (\{i_n\} \times N_{s_n})$, where $n \mapsto i_n$, $n \mapsto s_n$ are recursive. Also if $\mathscr{X} = \omega^k$, $A \subseteq \mathscr{X}$ is Σ_1^0 if, and only if, A is r.e.

We now have the following representation theorem for Σ_1^0 pointsets.

THEOREM 6.3.1 *Let $\mathscr{X} = X_1 \times \cdots \times X_k$ be a space. A pointset $A \subseteq \mathscr{X}$ is Σ_1^0 if, and only if, there is a recursive $R \subseteq \omega^k$ such that*

$$x \in A \Leftrightarrow \exists n\, R(\bar{x}(n)),$$

where if $x = (x_1, \ldots, x_k)$, $\bar{x}(n) = (\bar{x}_1(n), \ldots, \bar{x}_k(n))$, with $\bar{a}(n) = \langle a, \ldots, a \rangle$, with a repeated n times, if $a \in \omega$.

Proof Take $\mathscr{X} = \omega \times \omega^\omega$ for notational simplicity. If $A \subseteq \mathscr{X}$ is Σ_1^0 then there are recursive $i \mapsto p_i$, $i \mapsto s_i$ such that $A = \bigcup_{i \in \omega} (\{p_i\} \times N_{s_i})$. Then

$$(k, \alpha) \in A \Leftrightarrow \exists i (k = p_i \wedge s_i \subseteq \alpha) \Leftrightarrow \exists i\, \exists j (k = p_i \wedge s_i \subseteq \bar{\alpha}(j))$$

$$\Leftrightarrow \exists n (k = p_{(n)_0} \wedge s_{(n)_0} \subseteq \bar{\alpha}((n)_1)$$

$$\Leftrightarrow \exists n\, R(\bar{k}(n), \bar{\alpha}(n)),$$

with R recursive. We use here the fact $n \geq (n)_0, (n)_1$. The converse is easier.

From this representation we can easily obtain the following closure properties of Σ_1^0.

THEOREM 6.3.2 *The pointclass* Σ_1^0 *is closed under* (i), (ii) *and* (iii) *below.*

(i) *conjunction* $\wedge$, *disjunction* $\vee$ (*i.e. if* R, $S \subseteq \mathscr{X}$ *are* Σ_1^0 *so is* $P(x) \Leftrightarrow^{\text{def}} R(x) \wedge S(x)$ *and* $T(x) \Leftrightarrow^{\text{def}} R(x) \vee S(x)$.)

(Notice our switch to logical notation—if we were to think of pointsets as sets instead of relations we would say that Σ_1^0 is closed under $\cap$, $\cup$).

(ii) *Existential quantification over both* ω *and* ω^ω (*i.e. if* $P \subseteq \mathscr{X} \times \omega^\omega$ *is* Σ_1^0, *so is* $Q(x) \Leftrightarrow^{\text{def}} \exists \alpha P(x, \alpha)$ *and similarly if* $P \subseteq \mathscr{X} \times \omega$ *is* Σ_1^0 *so is* $Q(x) \Leftrightarrow^{\text{def}} \exists n\, P(x, n)$).

(iii) *Bounded universal quantification over* ω.

Proof Let us do as an example (ii) for ω^ω. By Theorem 6.3.1, let $P(x, \alpha) \Leftrightarrow \exists n \; R(\bar{x}(n), \bar{\alpha}(n))$, with R recursive. Then $\exists \alpha \; P(x, \alpha) \Leftrightarrow \exists \alpha \; \exists n \; R(\bar{x}(n), \bar{\alpha}(n)) \Leftrightarrow \exists m \; R'(\bar{x}((m)_0), (m)_1)$ (where we take $(m)_0 = n$, $(m)_1 = \bar{\alpha}(n)) \Leftrightarrow \exists m \, R''(\bar{x}(m))$, for some recursive R', R''. We are using again that $m \geq (m)_0, (m)_1$.

Once we have the notion of an effective open set it is not hard to define the notion of an effectively continuous function.

DEFINITION A function $f: \mathscr{X} \to \omega^\omega$ is *effectively continuous* or *recursive* if the pointset

$$G_f(x, s) \Leftrightarrow f(x) \in N_s$$

is Σ_1^0, i.e. $f^{-1}[N_s]$ is effectively open, uniformly on s. Similarly, $f: \mathscr{X} \to \omega$ is *recursive* if

$$G_f(x, i) \Leftrightarrow f(x) \in \{i\} \Leftrightarrow f(x) = i$$

is Σ_1^0. And in general $f: \mathscr{X} \to \mathscr{Y}$ is *recursive*, where $\mathscr{Y} = Y_1 \times \cdots \times Y_n$ if each $f_i \equiv i$th projection of f, for $1 \leq i \leq n$, is recursive.

Important remark For $\mathscr{X} = \omega^n$, $\mathscr{Y} = \omega^m$ this definition agrees with the one in Section 6.2 as one can easily show that $f: \omega^n \to \omega^m$ is recursive in the sense of Section 6.2 if, and only if, graph (f) is r.e. (i.e. Σ_1^0), i.e. if f is recursive in the present sense.

We now have the following effective analogue of the standard definition of continuity.

FACT 6.3.3 *The pointclass Σ_1^0 is closed under recursive substitution, i.e. if $P \subseteq \mathcal{Y}$ is Σ_1^0, $f: \mathcal{X} \to \mathcal{Y}$ is recursive and $Q(x) \Leftrightarrow P(f(x))$, Q is also Σ_1^0.*

Proof Take $\mathcal{Y} = \omega^\omega$ for simplicity. We have $x \in Q \Leftrightarrow f(x) \in P \Leftrightarrow \exists n(f(x) \in N_{s_n})$, where $n \to s_n$ is recursive, so $x \in Q \Leftrightarrow \exists n\, G_f(x, s_n) \Leftrightarrow \exists m\, \exists n[m = s_n \wedge G_f(x, m)]$ which is Σ_1^0 by Theorem 6.3.2.

(B) One can now relativize all the notions of (A) to an arbitrary $\alpha \in \omega^\omega$, in the obvious way, obtaining the pointclass $\Sigma_1^0(\alpha)$, and the notion of a recursive in α function from one space into another. Clearly all the results in (A) hold relativized to any α, with exactly the same proofs provided one replaces at all the appropriate places "recursive" by "recursive in α", Σ_1^0 by $\Sigma_1^0(\alpha)$, etc.

One can also understand $\Sigma_1^0(\alpha)$ as being obtained from Σ_1^0 by "plugging in" α, in view of the following result.

THEOREM 6.3.4 *A pointset $P \subseteq \mathcal{X}$ is $\Sigma_1^0(\alpha)$ if, and only if, there is $Q \subseteq \omega^\omega \times \mathcal{X}$ in Σ_1^0 such that $P(x) \Leftrightarrow Q(\alpha, x)$.*

Proof From Theorem 6.2.4 and the relativized to α Theorem 6.3.1, we have

$$
\begin{aligned}
P(x) &\Leftrightarrow \exists n\, (R(\bar{x}(n))), \quad R \text{ recursive in } \alpha \\
&\Leftrightarrow \exists n\, \exists m\, U(\bar{x}(n), \bar{\alpha}(m)), \quad U \text{ recursive} \\
&\Leftrightarrow \exists k\, U(\bar{x}((k)_0), \bar{\alpha}((k)_1)) \\
&\Leftrightarrow \exists k\, U'(\bar{x}(k), \bar{\alpha}(k)), \quad U' \text{ recursive} \\
&\Leftrightarrow Q(\alpha, x), \quad Q \in \Sigma_1^0.
\end{aligned}
$$

The converse is simpler, using Fact 6.3.3 relativized and the fact that $f(x) \equiv \alpha$ is recursive in α.

We shall finally clarify the relationship between the "effective" and the "classical" topology.

THEOREM 6.3.5 (i) *$A \subseteq \mathcal{X}$ is open if, and only if, A is $\Sigma_1^0(\alpha)$ for some α.*

(ii) *$f: \mathcal{X} \to \mathcal{Y}$ is continuous if, and only if, f is recursive in some α.*

Proof (⇐) for (i), (ii) are obvious from the definitions. To prove (⇒) of (i), take for simplicity $\mathscr{X}=\omega^\omega$. Then if A is open we have that $A=\bigcup_n N_{s_n}$, for some $n\mapsto s_n$. Put $\alpha(n)=s_n$. Then clearly $n\mapsto s_n$ is recursive in α, so A is $\Sigma_1^0(\alpha)$. For (⇒) of (ii) let $f:\mathscr{X}\to\mathscr{Y}$ be continuous and take $\mathscr{Y}=\omega^\omega$ again. Then $G_f(x,s)\overset{\text{def}}{\Leftrightarrow}f(x)\in N_s$ is obviously an open subset of $\mathscr{X}\times\omega$ (recall that ω has the discrete topology). So by what we just proved, $G_f\in\Sigma_1^0(\alpha)$ for some A, i.e. f is recursive in some α.

(C) Again one can easily verify that if $\mathscr{X}$ is any space containing at least one copy of ω^ω, there is a recursive bijection between $\mathscr{X}$ and ω^ω. This is generated by the following two coding maps:

$$(\alpha,\beta)\mapsto(\alpha(0),\beta(0),\alpha(1),\beta(1),\ldots)\in\omega^\omega,$$

$$(n,\alpha)\mapsto n^\frown\alpha=(n,\alpha(0),\alpha(1),\ldots)\in\omega^\omega.$$

Thus we will now often identify each one of these spaces with ω^ω (and, as we explained in Section 6.2, identify each of the ω^k with ω).

6.4 The Kleene pointclasses

(A) Recall that the classical projective pointclasses A (analytic), CA (coanalytic), PCA, CPCA, ... are defined by starting with the closed pointsets and successively applying projection over ω^ω, i.e. existential quantification over ω^ω and complementation, i.e. negation. Their effective analogues, the *Kleene pointclasses*, are obtained in a similar fashion as follows:

DEFINITION Let Π_1^0 be the pointclass of all negations (complements) of Σ_1^0 pointsets (i.e. for $A\subseteq\mathscr{X}$, $A\in\Pi_1^0\Leftrightarrow\neg A\overset{\text{def}}{=}\mathscr{X}-A\in\Sigma_1^0$). These are clearly the effectively closed pointsets. Let then Σ_1^1 be a pointclass of all pointsets of the form

$$R(x)\Leftrightarrow\exists\alpha\,Q(x,\alpha),$$

with $Q\in\Pi_1^0$. Π_1^1 is the class of all negations of Σ_1^1 pointsets and, inductively for $n\geqslant1$,

$$\Sigma_{n+1}^1=\{P\subseteq\mathscr{X}:\mathscr{X}\text{ any space and there is }Q\subseteq\mathscr{X}\times\omega^\omega\text{ in }\Pi_n^1\text{ such that }P(x)\Leftrightarrow\exists\alpha\,Q(x,\alpha)\},$$

$$\Pi_{n+1}^1=\{\neg P: P\subseteq\mathscr{X},\mathscr{X}\text{ any space}, P\in\Sigma_{n+1}^1\}.$$

Let also

$$\Delta^1_n = \Sigma^1_n \cap \Pi^1_n = \{P \subseteq \mathscr{X} : \mathscr{X} \text{ any space and } P, \neg P \text{ are } \Sigma^1_n\}.$$

The relativized Kleene pointclasses $\Sigma^1_n(\alpha)$, $\Pi^1_n(\alpha)$, $\Delta^1_n(\alpha)$ are defined in the obvious way by starting the above induction from $\Pi^0_1(\alpha)$. Alternatively in view of Theorem 6.3.4 one obtains the $\Sigma^1_n(\alpha)$ pointsets as those of the form

$$P(x) \Leftrightarrow Q(\alpha, x),$$

where Q is Σ^1_n. And similarly for $\Pi^1_n(\alpha)$. One should be careful, however, for $\Delta^1_n(\alpha) = \Sigma^1_n(\alpha) \cap \Pi^1_n(\alpha)$, as it is not the case that every $P \in \Delta^1_n(\alpha)$ is of the form $P(x) \Leftrightarrow A(\alpha, x)$, where $Q \in \Delta^1_n$.

Finally, let

$$\mathbf{\Sigma}^0_1 = \bigcup_\alpha \Sigma^0_1(\alpha), \qquad \mathbf{\Pi}^0_1 = \bigcup_\alpha \Pi^0_1(\alpha),$$

$$\mathbf{\Sigma}^1_n = \bigcup_\alpha \Sigma^1_n(\alpha), \qquad \mathbf{\Pi}^1_n = \bigcup_\alpha \Pi^1_n(\alpha), \qquad \mathbf{\Delta}^1_n = \bigcup_\alpha \Delta^1_n(\alpha).$$

(For distinction when talking about these pointclasses one calls the Σ^i_n, Π^i_n, Δ^i_n "lightface" and the $\mathbf{\Sigma}^i_n$, $\mathbf{\Pi}^i_n$, $\mathbf{\Delta}^i_n$ "boldface").

As expected from Theorem 6.3.5, we have now recaptured the classical projective pointclasses as follows:

THEOREM 6.4.1 (i) $\mathbf{\Sigma}^0_1 =$ *open*, $\mathbf{\Pi}^0_1 =$ *closed*; (ii) $\mathbf{\Sigma}^1_1 = A$, $\mathbf{\Pi}^1_1 = \mathrm{CA}$; (iii) $\mathbf{\Sigma}^1_2 = \mathrm{PCA}$, $\mathbf{\Pi}^1_2 = \mathrm{CPCA}$; etc.

(B) We conclude this section with a useful representation of the Π^0_1 and Σ^1_1 pointsets.

THEOREM 6.4.2 (i) *If $A \subseteq \omega^\omega$, then $A \in \Pi^0_1$ if, and only if, there is a recursive tree T on ω such that*

$$\alpha \in A \Leftrightarrow \alpha \in [T] (\overset{\text{def}}{\Leftrightarrow} \forall n (\alpha \restriction n \in T)).$$

Similarly for $A \subseteq (\omega^\omega)^k$. For example, if $A \subseteq \omega^\omega \times \omega^\omega$, A is Π^0_1 if, and only if,

$$A = [T] = \{(\alpha, \beta) : \forall n (\alpha \restriction n, \beta \restriction n) \in T\},$$

where T is a recursive tree on $\omega \times \omega$.

(ii) *If $A \subseteq \omega^\omega$, then A is Σ^1_1 if, and only if, there is a recursive tree T on $\omega \times \omega$ such that $A = \mathrm{p}[T] = \{\alpha : \exists \beta (\alpha, \beta) \in [T]\}$.*

Proof (ii) Clearly follows from (i). To prove (i) let $\alpha \notin A \Leftrightarrow \exists n\, R(\bar{\alpha}(n))$, R recursive by Theorem 6.3.1. Then $\alpha \in A \Leftrightarrow \forall n\ \neg R(\bar{\alpha}(n)) \Leftrightarrow \forall n (\forall m \leqslant n\ \neg R(\bar{\alpha}(m))$. Put for $s = (a_0 \cdots a_{l-1}) \in \omega^{<\omega}$,

$$s \in T \Leftrightarrow \forall m \leqslant l \neg R(\langle a_0 \cdots a_{m-1} \rangle) .$$

Then T is recursive and $A = [T]$, as $\alpha \in A \Leftrightarrow \forall n (\alpha \upharpoonright n \in T)$.

6.5 Basic properties of the Kleene pointclasses

We shall first mention the closure properties of the pointclasses Σ^1_n, Π^1_n, Δ^1_n. (Those of Π^0_1 can be easily deduced from Theorem 6.3.2.)

THEOREM 6.5.1 (i) *The pointclasses Σ^1_n are closed under* $\wedge$, $\vee$, *existential and universal quantification over* ω, *existential quantification over* ω^ω *and recursive substitution.*

(ii) *The pointclasses Π^1_n are closed under* $\wedge$, $\vee$, *existential and universal quantification over* ω, *universal quantification over* ω^ω *and recursive substitution.*

(iii) *The pointclasses Δ^1_n are closed under* $\wedge$, $\vee$, *existential and universal quantification over* ω, *recursive substitution and negation.*

Proof By induction on n. Let us take, as an illustration, a couple of cases for $n = 1$. Let $P(x, n)$ be Σ^1_1 say,

$$P(x, n) \Leftrightarrow \exists \alpha\, Q(x, n, \alpha) ,$$

with $Q \in \Pi^0_1$. Then $\forall n\, P(x, n) \Leftrightarrow \forall n\ \exists \alpha\, Q(x, n, \alpha) \Leftrightarrow \exists \alpha\ \forall n\, Q(x, n, (\alpha)_n)$, where $(\alpha)_n(m) = \alpha(2^n \times 3^m)$. As $\mathrm{p}(\alpha, n) = (\alpha)_n$ is recursive and Π^0_1 is closed under recursive substitution and universal quantification over ω the result follows.

Let now $P(x, \alpha)$ be Σ^1_1 say,

$$P(x, \alpha) \Leftrightarrow \exists \beta\, Q(x, \alpha, \beta) .$$

Then $\exists \alpha\ P(x, \alpha) \Leftrightarrow \exists \alpha\ \exists \beta\ Q(x, \alpha, \beta) \Leftrightarrow \exists \gamma\ Q(x, (\gamma)_0, (\gamma)_1)$ and so $\exists \alpha\ P(x, \alpha)$ also Σ^1_1.

This closure theorem allows us to compute efficiently using logical notation with relations, in order to show that a certain pointset we are interested in belongs to a desired pointclass—this kind of computation generally goes by the name of Kuratowski–Tarski algorithm. For example repeated applications of Theorem 6.5.1 show that, if f is recursive and P, Q, R are Π^1_1, Π^1_1, and Σ_1^1 respectively, then T defined by

$$T(\alpha, \beta, m) \Leftrightarrow \forall\gamma\,[R(f(\gamma, \alpha), m) \Rightarrow \exists l\,(P(l) \vee \forall t\, Q(t, \alpha, \beta))],$$

is Π_1^1. Recall here that "$\varphi \Rightarrow \psi$" is logically equivalent to "$\neg\varphi \vee \psi$".

Another basic fact about the Kleene pointclasses is

THEOREM 6.5.2 (Parametrization theorem for the Kleene pointclasses) (i) *For each space $\mathscr{X}$ there is $U^{\mathscr{X}} = U \subseteq \omega \times \mathscr{X}$ which is universal for the Σ_1^0 subsets of $\mathscr{X}$, i.e. $U \in \Sigma_1^0$ and for any $A \subseteq \mathscr{X}$*

$$A \in \Sigma_1^0 \Leftrightarrow \exists e\,(A = U_e =^{\text{def}} \{x : U(e, x)\}).$$

Similarly for Π_1^0.

(ii) *Fix $n \geqslant 1$. For each $\mathscr{X}$ there is $U^{\mathscr{X}} = U \subseteq \omega \times \mathscr{X}$ in Σ_n^1 which is universal for the Σ_n^1 subsets of $\mathscr{X}$. Similarly for Π_n^1.*

Proof (i) follows immediately from the representation theorem (Theorem 6.3.1) and the parametrization theorem for r.e. ($=\Sigma_1^0$) subsets of ω^k (Theorem 6.2.2).

(ii) follows by an easy induction on n, noticing that if U is universal for the Γ subsets of $\mathscr{X}$, Γ any pointclass, then $\neg U$ (= the negation of U) is universal for the *dual pointclass* $\neg\Gamma$ of Γ. This dual pointclass is the class of all negations of pointsets in Γ. Further, if $U \subseteq \omega \times \mathscr{X} \times \omega^\omega$ is universal for the Γ subsets of $\mathscr{X} \times \omega^\omega$, then $U'(e, x) \Leftrightarrow \exists\alpha\, U(e, x, \alpha)$ is universal for the subsets of $\mathscr{X}$ in the pointclass consisting of pointsets of the form $\exists\alpha\, P(x, \alpha)$, $P \in \Gamma$.

COROLLARY 6.5.3 *The pointclasses Σ_1^0, Π_1^0, Σ_n^1, Π_n^1 are not closed under negation.*

So we have the following picture:

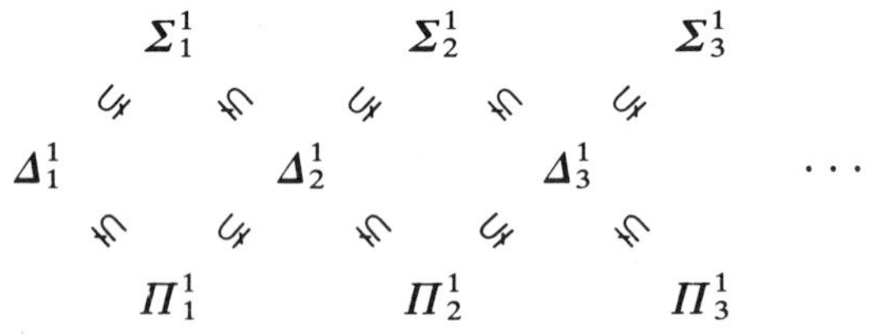

Of course all the preceding results hold when relativized to an arbitrary α.

It is a classical result that all the boldface pointclasses $\mathbf{\Sigma}_n^1$, $\mathbf{\Pi}_n^1$ admit universal sets in the sense that, fixing $n \geqslant 1$, for each $\mathscr{X}$ there is a $W^{\mathscr{X}} = W \subseteq \omega^\omega \times \mathscr{X}$ in $\mathbf{\Sigma}_n^1$ such that, for $A \subseteq \mathscr{X}$

$$A \in \mathbf{\Sigma}_n^1 \Leftrightarrow \text{for some } \varepsilon \in \omega^\omega, \qquad A = W_\varepsilon,$$

and similarly for $\mathbf{\Pi}^1_n$. We can now actually state the following stronger version.

THEOREM 6.5.4 *Fix $n \geq 1$. For each space $\mathscr{X}$ there is $W \subseteq \omega^\omega \times \mathscr{X}$ in ("lightface") Σ^1_n such that for $A \subseteq \mathscr{X}$,*

$$A \in \mathbf{\Sigma}^1_n \Leftrightarrow \textit{for some } \varepsilon , \qquad A = W_\varepsilon .$$

Similarly for $\mathbf{\Pi}^1_n$ and $\mathbf{\Sigma}^0_1$, $\mathbf{\Pi}^0_1$.

Proof It is enough again to prove this for $\mathbf{\Sigma}^0_1$. Take $\mathscr{X} = \omega^\omega$; for simplicity, let $N^*_0 = \varnothing$, $N^*_{t+1} = N_t$ (recall our identification of $\omega^{<\omega}$ with ω here), and put

$$\begin{aligned} W(\varepsilon, \alpha) &\Leftrightarrow \alpha \in \bigcup_n N^*_{\varepsilon(n)} \\ &\Leftrightarrow \exists n \, (\alpha \in N^*_{\varepsilon(n)}) . \end{aligned}$$

Then clearly W is Σ^0_1 and, for $A \subseteq \omega^\omega$,

$$A \text{ is } \mathbf{\Sigma}^0_1 \,(= \text{open}) \Leftrightarrow \text{for some } \varepsilon \in \omega^\omega, \qquad A = W_\varepsilon .$$

6.6 The Souslin–Kleene theorem

A fundamental classical result of Souslin identifies the analytic-coanalytic sets with the Borel sets; in our notation it says that

$$\text{Borel} = \mathbf{\Delta}^1_1 .$$

What is the effective version of the theorem? Clearly Δ^1_1 is the effective version of $\mathbf{\Delta}^1_1$: but what is the effective version of Borel? Recall that Borel ramifies in the hierarchy $\{\mathbf{\Sigma}^0_\xi\}_{\xi<\omega_1}$, where ω_1 is the first uncountable ordinal, defined as follows:

$$\mathbf{\Sigma}^0_1 = \text{open} ;$$

$$\mathbf{\Sigma}^0_{\xi+1} = \text{all countable unions of pointsets in } \mathbf{\Pi}^0_\xi, \text{ where } \mathbf{\Pi}^0_\xi = \neg \mathbf{\Sigma}^0_\xi ;$$

$$\mathbf{\Sigma}^0_\lambda = \text{countable unions of pointsets in } \bigcup_{\xi<\lambda} \mathbf{\Sigma}^0_\xi, \text{ if } \lambda \text{ is limit} .$$

The effective version of this hierarchy is defined by (1) formulating a notion of "effective countable union", which involves ideas which we will not try to explain here and (2) replacing the notion of arbitrary countable ordinal by its effective analogue, i.e. the notion of a *recursive ordinal,* where a recursive ordinal is just the order type of a recursive well-ordering on ω. Put $\omega_1^{\text{ck}} = \text{Church–Kleene } \omega_1 \stackrel{\text{def}}{=} \text{first non-recursive ordinal}$. Clearly $\omega_1^{\text{ck}} < \omega_1$

and ω_1^{ck} is the effective version of ω_1. One now defines the effective hierarchy $\{\Sigma_\xi^0\}_{\xi<\omega_1^{ck}}$ by

$$\Sigma_1^0 = \text{what we have called before } \Sigma_1^0\ ;$$

$$\begin{aligned}\Sigma_{\xi+1}^0 &= \text{all "effective countable unions" of } \Pi_\xi^0 \text{ sets, where } \Pi_\xi^0 = \neg\Sigma_\xi^0\\ &= \text{all pointsets of the form } \exists n\, P(x,n), \text{ with } P\in\Pi_\xi^0\ ;\end{aligned}$$

$$\Sigma_\lambda^0 = \text{all "effective countable unions" of pointsets in } \bigcup_{\xi<\lambda}\Sigma_\xi^0, \text{ if } \lambda \text{ is limit}.$$

The pointclass

$$\mathrm{HYP} = \bigcup_{\xi<\omega_1^{ck}} \Sigma_\xi^0$$

is the pointclass of *hyperarithmetic sets* (the pointclass $\bigcup_{n<\omega}\Sigma_n^0$ being the class of *arithmetic* sets, which are the effective analogues of the finite level Borel sets).

We have now

THEOREM 6.6.1 (The Souslin–Kleene theorem) (Kleene) *For any* $A \subseteq \mathscr{X}$,

$$A \textit{ is } \mathrm{HYP} \Leftrightarrow A \textit{ is } \Delta_1^1\ .$$

This result of course relativises to an arbitrary α. (Note: in the definitions of HYP (α) one must also replace ω_1^{ck} by $\omega_1^{ck}(\alpha)$; clearly $\xi<\omega_1 \Leftrightarrow$for some α, $\xi<\omega_1^{ck}(\alpha)$.) Thus, as it is easy to see that

$$A \text{ is Borel} \Leftrightarrow \text{for some } \alpha,\ A \text{ is } \mathrm{HYP}(\alpha)\,,$$

we obtain as a corollary the classical Souslin theorem. The proof of the Souslin–Kleene theorem can be found in J. R. Shoenfield (1967) or in Moschovakis' book.

6.7 A general convention

In the remaining sections we shall usually state the various effective theorems in their "absolute" form, i.e. the one concerning the "lightface" pointclasses Σ_n^1, Π_n^1, Δ_n^1. *It will be always understood, however, that the same result and its proof hold when relativized to an arbitrary* α *by making the obvious notational changes of inserting everywhere* α *in the appropriate places.* In view of the fact that $\boldsymbol{\Sigma}_n^1 = \bigcup_\alpha \Sigma_n^1(\alpha)$. etc. this will also imply immediately a corresponding result about the classical projective pointsets, which many times we will not state explicitly.

7 Basic prewellordering theory

7.1 Introduction

A very important role in the classical theory of $\mathbf{\Pi}_1^1$ and $\mathbf{\Sigma}_1^1$ sets is played by the theory of the Lusin–Sierpiński index. This is the study and use of the properties of a certain ordinal assignment to the members of an arbitrary $\mathbf{\Pi}_1^1$ set. In an informal sense this can be understood as a kind of representation theory, whereby a complicated object like a typical $\mathbf{\Pi}_1^1$ set of reals is mapped in a nice fashion into a set of ordinals, which from our point of view has a very simple structure, and important information is carried back via this map to our $\mathbf{\Pi}_1^1$ set. This vague remark will become more clear as we proceed and we give various illustrations. Similarly ordinal assignments to points in a $\mathbf{\Sigma}_2^1$ set play a key role in P. S. Novikov's basic work on the theory of $\mathbf{\Sigma}_2^1$ and $\mathbf{\Pi}_2^1$ sets.

In the modern theory one is looking at these ordinal assignments from an abstract point of view, which is summarized in the following notion of "Γ-norm". This was first formulated explicitly in its current form (or actually a somewhat different but equivalent one) by Moschovakis.

7.2 The prewellordering theorem

Let $A \subseteq \mathscr{X}$ be a pointset. A *norm* on A is a map $\varphi : A \to$ ordinals. To each such norm φ we associate the prewellordering on A given by

$$x \leqslant y \Leftrightarrow \varphi(x) \leqslant \varphi(y)\,.$$

Here by a *prewellordering* $\leqslant$ on A we understand a reflexive, transitive, connected relation on A, with the property that

If $X \subseteq A$ is non-empty, then there is $x \in X$ such that $\forall y \in X (x \leqslant y)$; (*)

$\leqslant$ does not however have to be antisymmetric. Thus a prewellordering is essentially a wellordering of the equivalence classes of the equivalence relation $x \simeq y \Leftrightarrow x \leqslant y \wedge y \leqslant x$. Note also that (*) is easily equivalent to the following:

there is no infinite descending $\leqslant$ chain, i.e. there is no sequence $x_0 > x_1 > \cdots$, where $x < y \Leftrightarrow x \leqslant y \wedge y \not\leqslant x$. (**)

Conversely to each prewellordering $\leqslant$ on A corresponds a norm $\varphi : A \to$ ordinals such that

$$x \leqslant y \Leftrightarrow \varphi(x) \leqslant \varphi(y)\,.$$

If we insist that φ maps A *onto an initial segment* of ordinals, such a φ is unique.

One more bit of notation: if Γ is a pointclass, $\neg\Gamma$ is its *dual*, i.e. $\neg\Gamma = \{A \subseteq \mathscr{X}: \neg A = \mathscr{X} - A \in \Gamma\}$, $\Delta = \Gamma \cap \neg\Gamma$ its *ambiguous* part and for any point $y \in \mathscr{Y}$, $\Gamma(y) = \{P \subseteq \mathscr{X}$: for some $Q \subseteq \mathscr{Y} \times \mathscr{X}$, $Q \in \Gamma$, $P(x) \Leftrightarrow Q(y, x)\}$ and $\Delta(y) = \Gamma(y) \cap \neg\Gamma(y)$. Finally

$$\mathbf{\Gamma} = \bigcup_{\alpha \in \omega^\omega} \Gamma(\alpha) \quad \text{and} \quad \mathbf{\Delta} = \mathbf{\Gamma} \cap \neg\mathbf{\Gamma}.$$

DEFINITION Let Γ be a pointclass, φ a norm on a pointset A. Then we say that φ is a *Γ-norm* if for each $y \in A$ the set of φ-predecessors of y, i.e.

$$\{x \in A: \varphi(x) \leqslant \varphi(y)\}$$

is in $\Delta(y)$ "uniformly on y", i.e. there are pointsets P, Q in Γ, $\neg\Gamma$ respectively such that, when $y \in A$,

$$x \in A \wedge \varphi(x) \leqslant \varphi(y) \Leftrightarrow P(x, y) \Leftrightarrow Q(x, y).$$

Note that this is weaker than requiring that the prewellordering $\leqslant$ associated with φ is in Δ. (That would imply most of the time that A itself is in Δ.)

Finally let us say that a pointclass Γ is *normed* or has the *prewellordering property* if every pointset in Γ admits a Γ-norm.

One can now verify easily that the classical Lusin–Sierpiński index (see for instance Part 2 in this volume) provides for each $\mathbf{\Pi}^1_1$ set a $\mathbf{\Pi}^1_1$-norm, and one can also check that essentially the same proof effectivizes and shows that every Π^1_1 set carries a Π^1_1-norm, so we have

THEOREM 7.2.1 (The prewellordering theorem) *The pointclass Π^1_1 is normed.*

This is actually an effective strengthening of the corresponding fact about $\mathbf{\Pi}^1_1$ as it is trivial to check that if Γ is normed so is $\mathbf{\Gamma}$.

P. S. Novikov's work on the theory of $\mathbf{\Sigma}^1_2$ sets provides an ordinal assignment to the elements of every $\mathbf{\Sigma}^1_2$ pointset, which can be easily verified to be a $\mathbf{\Sigma}^1_2$-norm. This proof can be immediately effectivized to show that Σ^1_2 *is normed*, a result first proved by Moschovakis when he first formulated the prewellordering property. The argument is as follows. If $P(x)$ is Σ^1_2 then there is $Q(x, \alpha)$ in Π^1_1 so that $P(x) \Leftrightarrow \exists\alpha\, Q(x, \alpha)$. Let φ be a Π^1_1-norm on Q. Then define the following norm ψ on P: for $x \in P$, $\psi(x) = \min\{\varphi(x, \alpha): (x, \alpha) \in Q\}$. It is easy to verify that ψ is a Σ^1_2-norm.

In the later part of Section 7 we shall give a different proof of the prewellordering theorem, which is based on the determinacy of open games and provides a far reaching extension of Blackwell's result discussed in the Section 2. This new proof is due to Moschovakis.

We shall concentrate now on presenting some elementary uses and applications of the prewellordering theorem to the theory of Π^1_1 (and Σ^1_1) sets, which will illustrate some of its elegance and power, especially in the context of the effective theory.

7.3 Applications

Let us first reformulate slightly the crucial property of a Γ-norm given in Section 7.2 (provided Γ has some simple closure properties).

A norm $\varphi : A \to$ ordinals, where $A \in \Gamma$ is a Γ-norm if, and only if, the two relations below are in Γ:

$$x \leqslant^*_\varphi y \Leftrightarrow x \in A \wedge [y \notin A \vee \varphi(x) \leqslant \varphi(y)],$$

$$x <^*_\varphi y \Leftrightarrow x \in A \wedge [y \notin A \vee \varphi(x) < \varphi(y)].$$

To see this let P, Q be as in the definition of Γ-norm and note that

$$x \leqslant^*_\varphi y \Leftrightarrow x \in A \wedge (P(x, y) \vee \neg Q(y, x)),$$

$$x <^*_\varphi y \Leftrightarrow x \in A \wedge \neg Q(y, x).$$

Conversely if $\leqslant^*_\varphi$, $<^*_\varphi$ as above are in Γ put

$$P(x, y) \Leftrightarrow x \leqslant^*_\varphi y,$$

$$Q(x, y) \Leftrightarrow \neg(y <^*_\varphi x).$$

As a first immediate application we have

THEOREM 7.3.1 *The pointclass Π^1_1 has the reduction property, i.e. if $A, B \subseteq \mathscr{X}$ are in Π^1_1 then we can find $A^*, B^* \subseteq \mathscr{X}$ in Π^1_1, reducing A, B, i.e. $A^* \subseteq A$, $B^* \subseteq B$, $A^* \cap B^* = \varnothing$, $A^* \cup B^* = A \cup B$.*

Proof Consider $C = (A \times \{0\}) \cup (B \times \{1\}) \subseteq \mathscr{X} \times \omega$. Let φ be a Π^1_1-norm on C and let

$$x \in A^* \Leftrightarrow (x, 0) \leqslant^*_\varphi (x, 1),$$

$$x \in B^* \Leftrightarrow (x, 1) <^*_\varphi (x, 0).$$

COROLLARY 7.3.2 *The pointclass Σ^1_1 has the* separation *property, i.e. if A, B are disjoint Σ^1_1 sets there is $C \in \Delta^1_1$ such that $A \subseteq C$, $B \cap C = \varnothing$.*

Proof To separate two disjoint sets, reduce their complements.

COROLLARY 7.3.3 *Σ^1_1 does not have the reduction property so it does not have the prewellordering property; Π^1_1 does not have the separation property.*

Proof We shall show that if a pointclass Γ satisfies the parametrization theorem and has some simple closure properties (that Π_1^1, Σ_1^1 have), then if reduction holds for Γ, separation fails for Γ. This and the previous two corollaries (or their proofs) imply all the desired conclusions.

Let $\mathscr{P}(n, \alpha)$ be in Γ and universal for the Γ subsets of ω^ω. Let $\mathscr{P}^0(n, \alpha) \Leftrightarrow \mathscr{P}((n)_0, \alpha)$, $\mathscr{P}^1(n, \alpha) \Leftrightarrow \mathscr{P}((n)_1, \alpha)$ here $n \to ((n)_0, (n)_1)$ is a recursive bijection of ω with $\omega \times \omega$. Then $\mathscr{P}^0, \mathscr{P}^1 \in \Gamma$ and they are *doubly universal*, i.e. if $A, B \subseteq \omega^\omega$ are in Γ there is $n \in \omega$ with $A = \mathscr{P}^0_n$, $B = \mathscr{P}^1_n$. Since Γ has the reduction property let $\bar{\mathscr{P}}^0, \bar{\mathscr{P}}^1$ reduce $\mathscr{P}^0, \mathscr{P}^1$ respectively. If now also Γ has the separation property, find D in Δ such that $\bar{\mathscr{P}}^0 \subseteq D$, $D \cap \bar{\mathscr{P}}^1 = \varnothing$. Then $D \subseteq \omega \times \omega^\omega$ is universal for the Δ subsets of ω^ω, a contradiction as Δ is closed under negation.

As the argument before is fairly general we also have that Π_2^1 does not have the prewellordering property, since Σ_2^1 does. So we have the following picture:

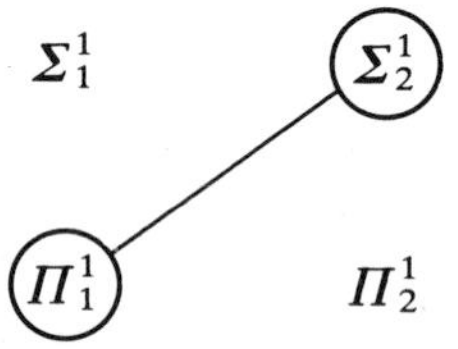

where the circled pointclasses have the prewellordering (and reduction) property but the others do not (but have the separation property).

In the classical theory one also has a simultaneous reduction theorem for a sequence of $\mathbf{\Pi}_1^1$ sets. This says that if $A_0, A_1, A_2, \ldots$ is a sequence of $\mathbf{\Pi}_1^1$ sets there is a sequence $A_0^*, A_1^*, A_2^*, \ldots$ of $\mathbf{\Pi}_1^1$ sets such that $A_i^* \subseteq A_i$, $A_i^* \cap A_j^* = \varnothing$ if $i \neq j$ and

$$\bigcup_{i=0}^{\infty} A_i = \bigcup_{i=0}^{\infty} A_i^* .$$

The following is the effective strengthening of this theorem

THEOREM 7.3.4 (The number uniformization theorem) (Kreisel) *Let $P \subseteq \mathscr{X} \times \omega$ be Π_1^1. Then there is a Π_1^1 pointset $P^* \subseteq \mathscr{X} \times \omega$, $P^* \subseteq P$ (see Figure 1) such that*

$$\exists n\, P(x, n) \Leftrightarrow \exists! n\, P(x, n) .$$

In particular if $\forall x\, \exists n\, P(x, n)$ there is a function $F : \mathscr{X} \to \omega$ with Δ_1^1 graph such that $\forall x\, P(x, F(x))$.

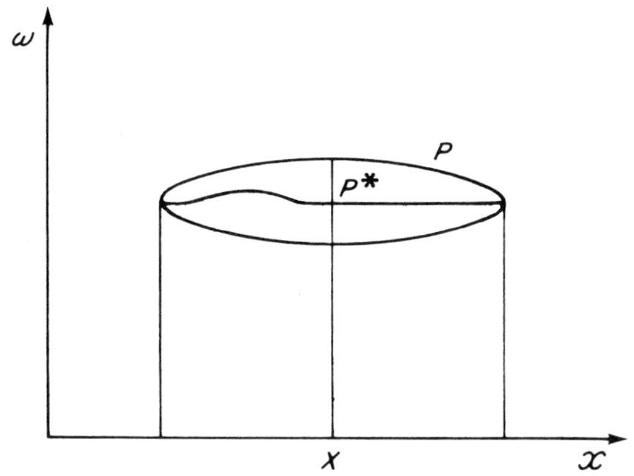

Figure 1 (Here $\exists! \equiv \exists$ unique).

Proof Let φ be a Π^1_1-norm on P and put

$$P^*(x, n) \Leftrightarrow P(x, n) \wedge \forall m\, [(x, n) \leqslant^*_\varphi (x, m)]$$
$$\wedge\, \forall m < n[(x, n) <^*_\varphi (x, m)],$$

i.e. $P^*(x, n) \Leftrightarrow P(x, n) \wedge \varphi(x, n)$ is least among all $\varphi(x, m)$, with $P(x, m)$ and n is least among all m with $P(x, m)$ and $\varphi(x, m) = \varphi(x, n)$.

For the second conclusion let $F(x) = n \Leftrightarrow P^*(x, n)$. As dom $(F) = \mathscr{X}$,

$$F(x) \neq n \Leftrightarrow \exists m\ (F(x) = m \wedge m \neq n)\,.$$

To see that this is indeed an effective strengthening of the classical reduction theorem mentioned before, notice that if $A_0, A_1, \ldots$ are $\mathbf{\Pi}^1_1$, then

$$P(x, n) \Leftrightarrow x \in A_n$$

is in $\mathbf{\Pi}^1_1$, since $P = \bigcup_n (A_n \times \{n\})$. Say $P \in \Pi^1_1(\alpha)$. By applying (the relativization of) Theorem 7.3.4 we obtain $P^*(x, n)$ in $\Pi^1_1(\alpha)$ as in the statement of that theorem. Put then $x \in A^*_n \Leftrightarrow P^*(x, n)$.

Definition A real $\beta \in \omega^\omega$ is called Δ^1_1 in a given real $\alpha \in \omega^\omega$ if $\{(k, m)\colon \beta(k) = m\} \in \Delta^1_1(\alpha)$. Write then $\beta \in \Delta^1_1(\alpha)$. Note now the following

Lemma 7.3.5 *Let $\alpha \in \omega^\omega$. Then the following are equivalent*:

(i) $\beta \in \Delta^1_1(\alpha)$,
(ii) $\{\beta\} \in \Delta^1_1(\alpha)$,
(iii) $\{\beta\} \in \Sigma^1_1(\alpha)$.

Proof (i)⇒(ii). $\gamma \in \{\beta\} \Leftrightarrow \forall k\ \forall m\ (\gamma(k) = m \Leftrightarrow \beta(k) = m)$.
(ii)⇒(iii). Obvious.
(iii)⇒(i). $\beta(k) = m \Leftrightarrow \forall \gamma\ [\gamma \in \{\beta\} \Rightarrow \gamma(k) = m] \Leftrightarrow \exists \gamma\ [\gamma \in \{\beta\} \wedge \gamma(k) = m]$.

The number uniformization theorem (Theorem 7.3.4) allows us to code $\Delta_1^1(\alpha)$ reals by numbers (uniformly on α), as follows:

THEOREM 7.3.6 *There is a Π_1^1 relation $C(n, \alpha)$ (read: n is an α-code) and a map $(n, \alpha) \mapsto \delta_n^\alpha \in \omega^\omega$ defined for $(n, \alpha) \in C$ such that $\delta \in \Delta_1^1(\alpha) \Leftrightarrow \exists n\ [(n, \alpha) \in C \wedge \delta_n^\alpha = \beta]$ and the relation "$\beta = \delta_n^\alpha$" is Δ_1^1, uniformly on $(n, \alpha) \in C$, i.e. there are Q, R in Π_1^1, Σ_1^1 respectively, such that, for $(n, \alpha) \in C$,*

$$\beta = \delta_n^\alpha \Leftrightarrow Q(n, \alpha, \beta) \Leftrightarrow R(n, \alpha, \beta)\,.$$

Proof Let $\mathscr{P}(n, m, k, \alpha)$ be Π_1^1 and universal for the Π_1^1 subsets of $\omega \times \omega \times \omega^\omega$. By Theorem 7.3.4 let $\mathscr{P}^* \subseteq \mathscr{P}$ be Π_1^1 such that

$$\exists k\, \mathscr{P}(n, m, k, \alpha) \Leftrightarrow \exists! k\, \mathscr{P}^*(n, m, k, \alpha)\,.$$

Put $(n, \alpha) \in C \Leftrightarrow \forall m\ \exists k\ \mathscr{P}^*(n, m, k, \alpha)$ and put for $(n, \alpha) \in C$: $\delta_n^\alpha(m) =$ the unique k such that $\mathscr{P}^*(n, m, k, \alpha)$. Then let

$$Q(n, \alpha, \beta) \Leftrightarrow \forall m\ \forall k\ [\beta(m) = k \Rightarrow \mathscr{P}^*(n, m, k, \alpha)]\,,$$

$$R(n, \alpha, \beta) \Leftrightarrow \forall m\ \forall k\ [\mathscr{P}^*(n, m, k, \alpha) \Rightarrow \beta(m) = k]\,.$$

Finally note that if β is $\Delta_1^1(\alpha)$, then $\beta(m) = k \Leftrightarrow S(m, k)$, where S is $\Pi_1^1(\alpha)$, and therefore $\beta(m) = k \Leftrightarrow S'(m, k, \alpha)$, where $S' \in \Pi_1^1$. Therefore for some n, $\beta(m) = k \Leftrightarrow \mathscr{P}(n, m, k, \alpha)$, and therefore $\beta(m) = k \Leftrightarrow \mathscr{P}^*(n, m, k, \alpha)$. Hence $\beta = \delta_n^\alpha$.

To see how this coding is used in the effective theory let us prove the following result which is the effective strengthening of the well-known classical theorem that if $R \subseteq \omega^\omega \times \omega^\omega$ is Borel ($= \mathbf{\Delta}_1^1$) then $\{\alpha : \exists! \beta\ (R(\alpha, \beta)\}$ is $\mathbf{\Pi}_1^1$, which easily implies that the one–one Borel image of a Borel set is also Borel.

THEOREM 7.3.7 *Let $R \subseteq \omega^\omega \times \omega^\omega$ be Δ_1^1. If $\alpha \in Q \Leftrightarrow \exists! \beta\, R(\alpha, \beta)$, then $Q \in \Pi_1^1$. In particular if $\exists \beta\, R(\alpha, \beta) \Leftrightarrow \exists! \beta\, R(\alpha, \beta)$,* p$[R] =$ *$\{\alpha : \exists \beta\ (\alpha, \beta) \in R\}$ is Δ_1^1; i.e. the one–one projection of a Δ_1^1 set is Δ_1^1.*

Proof Note that $\alpha \in Q \Leftrightarrow \exists! \beta\, R(\alpha, \beta) \Leftrightarrow \exists \beta\, [\beta$ is $\Delta^1_1(\alpha) \wedge R(\alpha, \beta) \wedge \forall \gamma\, (R(\alpha, \gamma) \Rightarrow \gamma = \beta)] \Leftrightarrow \exists \beta \in \Delta^1_1(\alpha) P(\alpha, \beta)$, where P is Π^1_1. So it is enough to prove the following "bounded quantification lemma" (due to Kleene):

LEMMA 7.3.7 *If $P(\alpha, \beta)$ is Π^1_1, so is $\exists \beta \in \Delta^1_1(\alpha) P(\alpha, \beta)$.*

Proof $\exists \beta \in \Delta^1_1(\alpha) P(\alpha, \beta) \Leftrightarrow \exists n\ ((n, \alpha) \in C \wedge P(\alpha, \delta^\alpha_n)) \Leftrightarrow \exists n\ [(n, \alpha) \in C \wedge \forall \beta\ (\beta = \delta^\alpha_n \Rightarrow P(\alpha, \beta))]$.

7.4 Proof of the prewellordering theorem

We start from the following observation:

$$\textit{the pointclass } \Sigma^0_1 \textit{ is normed}.$$

Indeed if $P(\alpha) \Leftrightarrow \exists n\, Q(\bar{\alpha}(n))$, where Q is recursive and for $\alpha \in P$ we let

$$\varphi(\alpha) = \text{least } n \text{ such that } Q(\bar{\alpha}(n)),$$

then it is trivial to verify that φ is a Σ^0_1-norm on P.

For convenience call a pointclass Γ *adequate* if $\Gamma \supseteq \Sigma^0_1$ and Γ is closed under $\wedge$, $\vee$, bounded quantification over ω (of both kinds) and recursive substitution. Thus Σ^0_1 itself is adequate. As Σ^0_1 is also closed under existential quantification and all $\mathbf{\Sigma}^0_1$ = open games are determined, it is clearly sufficient to establish the following general result of Martin and Moschovakis.

THEOREM 7.4.1 (The first periodicity theorem) (Martin, 1968; Moschovakis (see Addison and Moschovakis, 1968)) *Assume Γ is an adequate pointclass closed under existential quantification over ω^ω. Then*

$$\Gamma \textit{ is normed} \Rightarrow \Gamma' \textit{ is normed},$$

where

$\Gamma' = \{Q \subseteq X$: for some $P \subseteq X \times \omega^\omega$ in Γ, $Q(x) \Leftrightarrow \forall \alpha P(x, \alpha)\}$, provided all the games in $\mathbf{\Delta} = \Gamma \cap \neg \mathbf{\Gamma}$ are determined.

Proof (Moschovakis) Fix $P(x, \alpha)$ in Γ with $Q(x) \Leftrightarrow \forall \alpha\, P(x, \alpha)$. Let φ be a Γ-norm on P. We shall define, using a game, a relation $\leqslant$ on Q, prove that it is a prewellordering and then show that its associated norm ψ *is a* Γ'-norm.

To define $\leqslant$ fix $x, y \in Q$ and consider the following game, $G(x, y)$:

I	II	II wins if
α	β	$\varphi(x, \alpha) \leqslant \varphi(x, \beta)$

(Note that since $x, y \in Q$, for all α, β, $(x, \alpha) \in P$, $(y, \beta) \in P$.) Then put

$$x \leqslant y \Leftrightarrow \text{II has a winning strategy in } G(x, y)$$
$$\Leftrightarrow \exists \tau \, \forall \alpha \, (\varphi(x, \alpha) \leqslant \varphi(y, [\alpha] * \tau)) ,$$

where $[\alpha] * \tau$ is the real that II plays following τ, when I plays α. Thus $x \leqslant y$ means that $\sup \{\varphi(x, \alpha): \alpha \in \omega^\omega\} \leqslant \sup\{\varphi(y, \beta): \beta \in \omega^\omega\}$, in the strong sense that there is a uniform method, provided by a strategy τ for II, to produce for each $\alpha \in \omega^\omega$ a $\beta = [\alpha] * \tau$ with $\varphi(x, \alpha) \leqslant \varphi(y, [\alpha] * \tau)$. This should be compared with the dual comparison of "minima" of Π_1^1-norms used in the proof that Σ_2^1 is normed. We now verify that this is indeed a prewellordering on Q:

① $\forall x \in Q (x \leqslant x)$.

Obvious as II can win $G(x, x)$ by just copying I's moves.

② $\forall x, y, z \in Q (x \leqslant y \wedge y \leqslant z \Rightarrow x \leqslant z)$.

A winning strategy for II is illustrated in the following diagram:

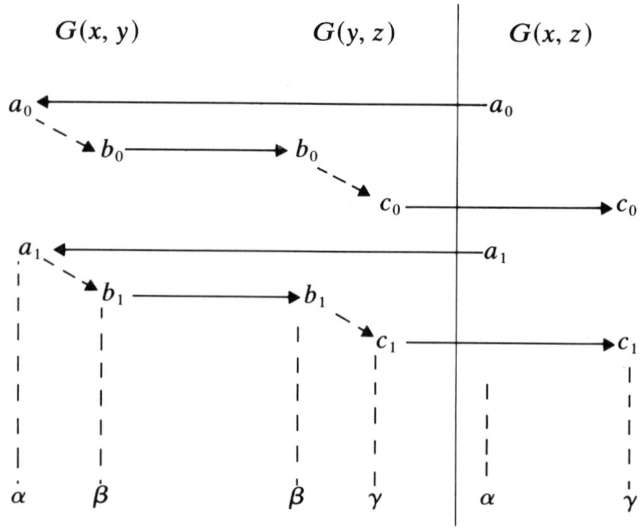

$\varphi(x, \alpha) \leqslant \varphi(y, \beta)$ $\qquad$ $\varphi(y, \beta) \leqslant \varphi(z, \gamma) \therefore \varphi(x, \alpha) \leqslant \varphi(z, \gamma)$.

Straight lines indicate copied moves and dotted lines moves played by II following II's winning strategies in the appropriate games.

③ $\forall x, y \in Q(x \leqslant y \vee y \leqslant x)$.

As $x, y \in Q$ it is clear that the game $G(x, y)$ is in $\mathbf{\Delta} = \mathbf{\Gamma} \cap \neg \mathbf{\Gamma}$, so it is determined. Thus if $x \leqslant y$ fails, I has a winning strategy in $G(x, y)$. The diagram below then shows how to obtain a winning strategy for II in $G(y, x)$:

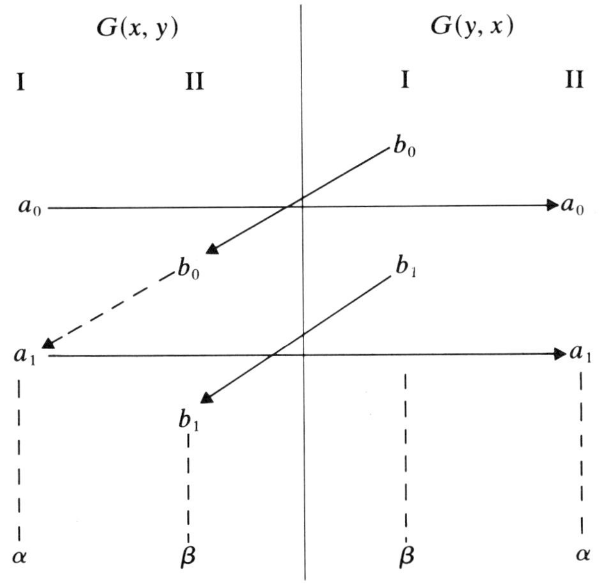

$$\varphi(x, \alpha) > \varphi(y, \beta) \Rightarrow \varphi(y, \beta) \leqslant \varphi(x, \alpha)$$

④ There is no $x_0 > x_1 > x_2 > \cdots$ in Q, where $x > y \Leftrightarrow x \geqslant y \wedge x \not\leqslant y$.

In order to prove this we first characterize $x > y$ for $x, y \in Q$ by

$$x > y \Leftrightarrow \text{I has a winning strategy in } G(x, y)\,. \tag{*}$$

Proof of (*) If $x > y$ then $x \not\leqslant y$, so II does not have a winning strategy in $G(x, y)$, so I has a winning strategy in $G(x, y)$. Conversely, if I has a winning strategy in $G(x, y)$ then as above $x \geqslant y$ and clearly also II has no winning strategy in $G(x, y)$, thus $x \not\leqslant y$, i.e. $x > y$.

Assume now $x_0 > x_1 > \cdots$ towards a contradiction. Consider then the following diagram:

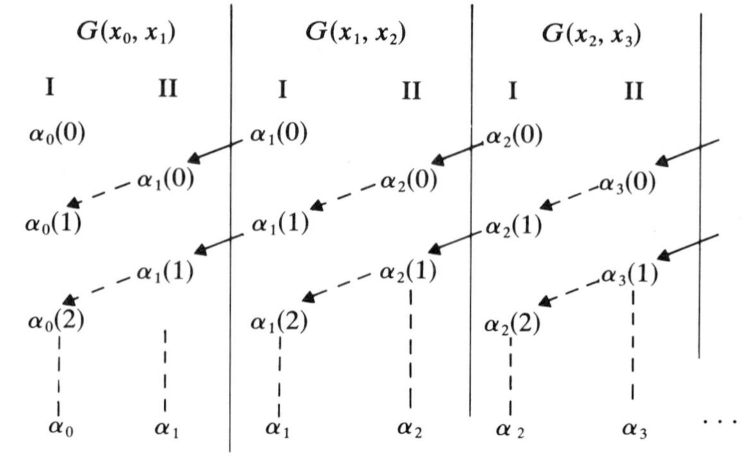

$$\varphi(x_0, \alpha_0) > \varphi(x_1, \alpha_1) \quad > \quad \varphi(x_2, \alpha_2) \quad > \quad \varphi(x_3, \alpha_3) \ldots,$$

a contradiction.

Finally we have to compute that ψ, the norm corresponding to $\leqslant$, is a Γ'-norm. For that extend the above definition of $G(x, y)$ to apply to arbitrary $x, y \in \mathscr{X}$ by replacing the payoff "$\varphi(x, \alpha) \leqslant \varphi(y, \beta)$" by "$\neg P(y, \beta) \vee [(x, \alpha) \in P \wedge (y, \beta) \in P \wedge \varphi(x, \alpha) \leqslant \varphi(y, \beta)]$". Then it is easy to check, by considering cases as $y \in Q$ or not, that this game, denoted also by $G(x, y)$, is determined. Now we can easily check that

$$\begin{aligned} x \leqslant^*_\psi y &\Leftrightarrow Q(x) \text{ and II has a winning strategy in } G(x, y) \\ &\Leftrightarrow Q(x) \text{ and I does not have a winning strategy in } G(x, y) \\ &\Leftrightarrow Q(x) \text{ and } \forall\sigma\, \exists\beta\, [(x, \sigma * [\beta]) \leqslant^*_\varphi (y, \beta)], \end{aligned}$$

where $\sigma * [\beta]$ is what I plays following a strategy σ, while II plays β, and also

$$\begin{aligned} x <^*_\psi y &\Leftrightarrow Q(x) \text{ and I has a winning strategy in } G(y, x) \\ &\Leftrightarrow Q(x) \text{ and II does not have a winning strategy in } G(y, x) \\ &\Leftrightarrow Q(x) \wedge \forall\tau\, \exists\alpha\, ((y, \alpha) <^*_\varphi (x, [\alpha] * \tau)). \end{aligned}$$

7.5 The periodicity picture

By combining the first periodicity theorem and the argument used to show that $\boldsymbol{\Sigma}^1_2$ is normed we immediately see that if we assume that every $\boldsymbol{\Delta}^1_{2n}$ game is determined then $\boldsymbol{\Pi}^1_{2n+1}$, $\boldsymbol{\Sigma}^1_{2n+2}$ are normed (while $\boldsymbol{\Sigma}^1_{2n+1}$, $\boldsymbol{\Pi}^1_{2n+2}$ are not).

So assuming the hypothesis of projective determinacy, i.e. that every projective set is determined, we have the following picture:

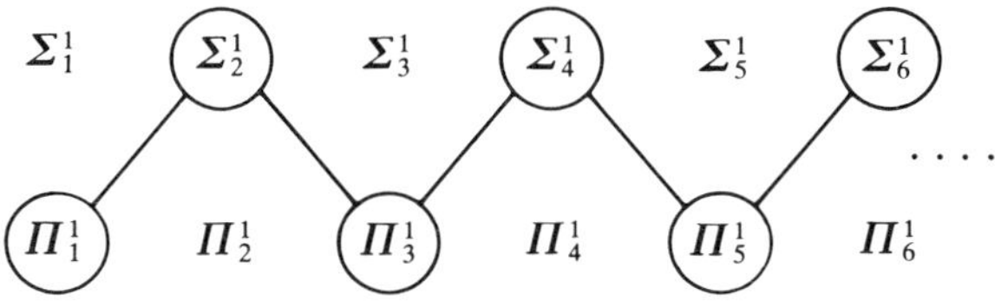

where those, and only those, circled pointclasses have the prewellordering (and reduction) property.

8 Uniformization and basis theorems

8.1 Introduction

Let $A \subseteq \omega^\omega \times \omega^\omega$. Let $B = \mathrm{p}[A] = \{\alpha : \exists \beta\ (\alpha, \beta) \in A\}$ be the projection of A (Figure 2). The set A^* *uniformizes* A if $A^* \subseteq A$ and $\forall \alpha \in B\ \exists! \beta\ (\alpha, \beta) \in A^*$,

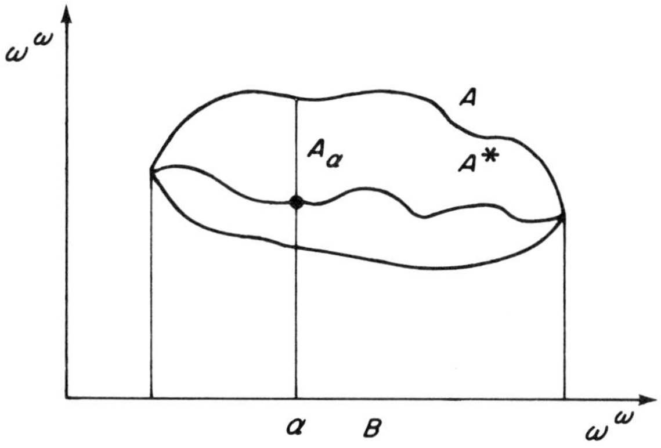

Figure 2

i.e. A^* is the graph of a function with domain B which picks for each $\alpha \in B$ a unique element of the section $A_\alpha = \{\beta : (\alpha, \beta) \in A\}$. Of course by the axiom of choice there are sets A^* uniformizing A, for which however one does not have the slightest information. One of the central problems of both classical and effective descriptive set theory is then the following:

Uniformization problem Does every projective set admit a projective uniformization?

In other words given projective A as above can we find a uniformizing A^* which is also projective?

We shall restrict ourselves in this work to various aspects of this problem only as far as they are related to the theory of Π_1^1, Σ_1^1 and Δ_1^1 sets.

8.2 Uniformization of Π_1^1 sets

One of the deepest classical results in descriptive set theory solves completely the uniformization problem for Π_1^1 sets. This is the well-known theorem of Novikov and Kondô that asserts that every $\mathbf{\Pi}_1^1$ set can be uniformized by a $\mathbf{\Pi}_1^1$ set. This was later effectivized by Addison, who showed the following:

THEOREM 8.2.1 (Novikov–Kondô–Addison) *Every Π_1^1 set can be uniformized by a Π_1^1 set.*

The proof of this result requires concepts and techniques going deeper and beyond the "prewellordering" methods sketched in Section 7. What is needed is a strengthening of the prewellordering property which we shall now explain.

DEFINITION Let P be a pointset. A *scale* on P is a sequence $\{\varphi_n\}_{n\in\omega}$ of norms on P with the following *limit property*:

If $x_0, x_1, \ldots$ is a sequence of points in P and $x_i \mapsto x$ and for each n and all large enough i (depending on n) $\varphi_n(x_i) = \text{constant} \equiv \lambda_n$, *then* $x \in P$, and $\varphi_n(x) \leqslant \lambda_n$, for all n.

If Γ is a pointclass and $\{\varphi_n\}_{n\in\omega}$ a scale on P, we shall say that $\{\varphi_n\}_{n\in\omega}$ is a Γ-*scale* if the norms φ_n are Γ-norms, uniformly on n, i.e. there are S, T in Γ so that

$$x \leqslant^*_{\varphi_n} y \Leftrightarrow S(n, x, y). \qquad x <^*_{\varphi_n} y \Leftrightarrow T(n, x, y).$$

Finally we say that a pointclass Γ satisfies the *scale property* if every $P \in \Gamma$ admits a Γ-scale.

The notion of scale was formulated by Moschovakis as an abstraction from the key properties of sequences of ordinal assignments used in the classical proof of the uniformization theorem for $\mathbf{\Pi}_1^1$ sets and its effective version. In fact one views now the first half of this classical proof as showing the following:

THEOREM 8.2.2 *The pointclass Π^1_1 has the scale property.*

The second half of this proof then amounts to a fairly general argument that shows that for sufficiently closed pointclasses Γ (Π^1_1 surely being one of them), if Γ has the scale property then Γ has the *uniformization property*, i.e. every set in Γ can be uniformized by a set in Γ. A brief sketch of this argument is as follows.

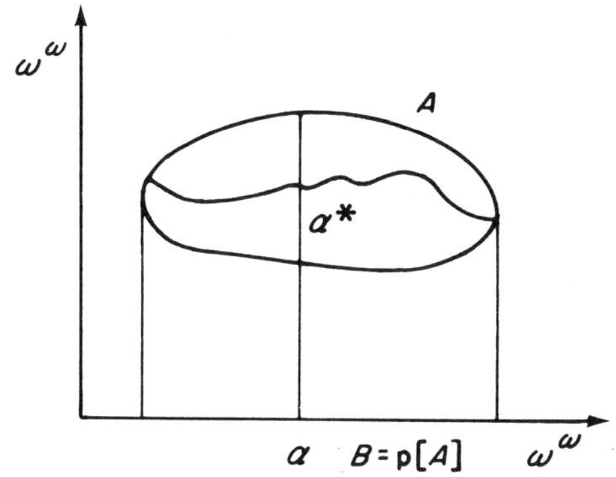

Figure 3

Let $\{\varphi_n\}_{n\in\omega}$ be a Γ-scale on $A \in \Gamma$. Then given $\alpha \in B = \mathrm{p}[A]$, the limit property of the scale $\{\varphi_n\}$ allows us to choose an $\alpha^* \in A_\alpha$ so that the sequence $\langle \varphi_0(\alpha^*), \alpha^*(0), \varphi_1(\alpha^*), \alpha^*(1), \ldots \rangle$ is lexicographically least among all corresponding sequences $\langle \varphi_0(\beta), \beta(0), \varphi_1(\beta), \beta(1), \ldots \rangle$ for all $\beta \in A_\alpha$ (see Figure 3). This is done by successively minimizing $\varphi_0(\alpha^*), \alpha^*(0), \varphi_1(\alpha^*), \alpha^*(1), \ldots$, etc. Now put

$$A^*(\alpha, \beta) \Leftrightarrow \beta = \alpha^* .$$

Then the fact that $\{\varphi_n\}_{n\in\omega}$ is a Γ-scale allows us to show that A^* is also in Γ.

A different proof of Theorem 8.2.2 can be given based again on open games. In fact one has the exact analogue of Theorem 7.4.2 for scales which is called the *second periodicity theorem* and is due to Moschovakis (1971) (for an exposition see Kechris and Moschovakis (1978)).

Remark The idea of picking an element of a set by lexicographically minimizing certain sequences of ordinals associated with it is also used in the proof of the standard Jankov–von Neumann uniformization theorem for $\mathbf{\Sigma}^1_1$ sets. As an exception we shall state it below in its classical form as it is more convenient to do so here.

THEOREM 8.2.3 *Every $\mathbf{\Sigma}_1^1$ set can be uniformized by a set in the smallest σ-algebra S containing the $\mathbf{\Sigma}_1^1$ sets. In fact the corresponding uniformizing function is S-measurable.*

Sketch of the proof Let $A \subseteq \omega^\omega \times \omega^\omega$ be $\mathbf{\Sigma}_1^1$ and let B be closed, $B \subseteq \omega^\omega \times \omega^\omega \times \omega^\omega$ such that

$$(\alpha, \beta) \in A \Leftrightarrow \exists \gamma\, (\alpha, \beta, \gamma) \in B\,.$$

For each $\alpha \in \mathrm{p}[A] = \{\alpha : \exists \beta\, \exists \gamma\, (\alpha, \beta, \gamma) \in B\}$ we can pick a unique pair (β^*, γ^*) in $B_\alpha = \{(\beta, \gamma) : (\alpha, \beta, \gamma) \in B\}$ such that the sequence $(\beta^*(0), \gamma^*(0), \beta^*(1), \gamma^*(1), \ldots)$ is lexicographically least among all $(\beta(0), \gamma(0), \beta(1), \gamma(1), \ldots)$ in B_α. But then if we define

$$(\alpha, \beta) \in A^* \Leftrightarrow \beta = \beta^*$$

we can verify by a straightforward calculation that the function $\alpha \mapsto \beta^*$ (with domain $\mathrm{p}[A]$) is S-measurable, i.e. the inverse image of every open set is in S.

8.3 Uniformization of Δ_1^1 sets

Although as a special case of the results in Section 8.2 every Δ_1^1 set can be uniformized by a Π_1^1 set, it is not true that every Δ_1^1 set can be uniformized by a Δ_1^1 (or even a Σ_1^1) set. There are however many particularly important special cases under which such a uniformization is possible. We mention two examples:

THEOREM 8.3.1 (Harrison) *If $A \subseteq \omega^\omega \times \omega^\omega$ is Δ_1^1 and for each $\alpha \in B = \mathrm{p}[A]$ the section A_α is countable, then A can be uniformized by a Δ_1^1 set.*

This is of course the effective strengthening of a classical result of Lusin and Novikov to the effect that every $\mathbf{\Delta}_1^1$ (= Borel) set with countable sections can be uniformized by a $\mathbf{\Delta}_1^1$ (= Borel) set.

THEOREM 8.3.2 (Hinman, Thomason) *If $A \subseteq \omega^\omega \times \omega^\omega$ is Δ_1^1 and for each $\alpha \in B = \mathrm{p}[A]$ the section A_α is not meagre, then A can be uniformized by a Δ_1^1 set.*

Other examples we could mention here include Δ_1^1 sets with K_σ sections (the effective version of the Kunugui–Arsenin theorem) or Δ_1^1 sets with positive measure non-empty sections, etc., but we shall only concentrate on the proofs of these two in order to illustrate some important techniques of the effective theory. The proofs of the others generally follow the same pattern, with extra ingredients mixed up in each particular case.

8.4 The uniformization criterion

The standard way in which one handles uniformization problems in effective descriptive set theory is by reducing them to "localized" basis problems. We shall explain this method, as well as the concept of "basis" involved, by showing how it applies in the case of uniformization of Δ_1^1 sets.

Let us recall a definition. A real $\beta \in \omega^\omega$ is $\Delta_1^1(\alpha)$ if its graph, i.e. $\{(n, m): \beta(n) = m\}$, is $\Delta_1^1(\alpha)$.

Recall also the following simple

LEMMA 8.4.1 *Let $\alpha \in \omega^\omega$. Then the following are equivalent*:

(1) *β is $\Delta_1^1(\alpha)$*;
(2) *$\{\beta\}$ is $\Delta_1^1(\alpha)$*;
(3) *$\{\beta\}$ is $\Sigma_1^1(\alpha)$*.

Using this and some of the prewellordering theory developed in Section 7 we can now prove

THE UNIFORMIZATION CRITERION FOR Δ_1^1 SETS *Let $A \subseteq \omega^\omega \times \omega^\omega$ be Δ_1^1. Then the following are equivalent*:

(1) *A can be uniformized by a Δ_1^1 set A^**;
(2) *For every $\alpha \in \mathrm{p}[A]$, the section A_α contains a β which is $\Delta_1^1(\alpha)$*.

Proof (1)$\Rightarrow$(2). Clearly if β is the unique element of A_α^*, $\{\beta\}$ is a singleton in $\Delta_1^1(\alpha)$ (as A^* is Δ_1^1) so β is $\Delta_1^1(\alpha)$.

(2)$\Rightarrow$(1). Recall from Theorem 7.3.6 the coding of $\Delta_1^1(\alpha)$ reals by numbers, uniformly on α. This will reduce our problem to the number uniformization theorem (Theorem 7.3.4).

Assume the hypothesis of (2). Put, using the notation of Theorem 7.3.6:

$$B(\alpha, n) \Leftrightarrow C(n, \alpha) \wedge A(\alpha, \delta_n^\alpha)$$
$$\Leftrightarrow C(n, \alpha) \wedge \forall \beta\, [\beta = \delta_n^\alpha \Rightarrow A(\alpha, \beta)]\,.$$

Then clearly $B \in \Pi_1^1$.

Use now the number uniformization theorem (Theorem 7.3.4) to find $B^*(\alpha, n)$ in Π_1^1 such that

$$B^* \subseteq B;\ \exists n\, B(\alpha, n) \Leftrightarrow \exists! n\, B^*(\alpha, n) (\Leftrightarrow \alpha \in \mathrm{p}[A])\,.$$

Put

$$A^*(\alpha, \beta) \Leftrightarrow \exists n\, [B^*(\alpha, n) \wedge \beta = \delta_n^\alpha]$$
$$\Leftrightarrow A(\alpha, \beta) \wedge \forall n\, [B^*(\alpha, n) \Rightarrow \beta = \delta_n^\alpha]\,.$$

Then $A^* \in \Delta_1^1$ and uniformizes A.

In view of this criterion a uniformization statement of the form

"every Δ_1^1 subset of $\omega^\omega \times \omega^\omega$, whose non-empty sections have some particular property $\mathbb{P}$, can be uniformized by a Δ_1^1 set"

is reduced to the "local" statement:

"for all α, every non-empty $\Delta_1^1(\alpha)$ subset of ω^ω which has property $\mathbb{P}$ contains a $\Delta_1^1(\alpha)$ member". (*)

This is a typical "basis result". In general given a collection of pointsets $\mathscr{C}$ and a collection of points B, we say that B is a *basis* for $\mathscr{C}$ if every non-empty set in $\mathscr{C}$ contains an element of B.

So statement (*) above says that the $\Delta_1^1(\alpha)$ reals form a basis for the collection of $\Delta_1^1(\alpha)$ subsets of ω^ω which have property $\mathbb{P}$.

In view of all this, Theorems 8.3.1 and 8.3.2 are now reduced to

THEOREM 8.4.1 *If $\varnothing \neq A \subseteq \omega^\omega$ is Δ_1^1 and countable, then A contains a Δ_1^1 member.*

THEOREM 8.4.2 *If $A \subseteq \omega^\omega$ is Δ_1^1 and non-meagre, then A contains a Δ_1^1 member.*

(As usual we do not explicitly mention the relativized form of these theorems, which however is also implicitly understood to hold as well, according to our conventions from Section 6.)

We shall actually show the following stronger statements, of which the first also serves the role of providing the effective version of the classical theorem of Souslin on the cardinality of $\mathbf{\Sigma}_1^1$ sets.

THEOREM 8.4.3 (Harrison) *Every Σ_1^1 set which contains no perfect subset contains only Δ_1^1 members, in fact there is $F:\omega \to \omega^\omega$ with Δ_1^1 graph such that $A \subseteq \{F(n): n \in \omega\}$.*

THEOREM 8.4.4 *If a Σ_1^1 set A is meagre then $\neg A \supseteq \bigcap_n G_n$, where for each G_n is dense open and for some $F:\omega \to \omega^\omega$ with Δ_1^1 graph, if $F(n) = \alpha_n$, $G_n = \bigcup_n N_{\alpha_n(k)}$.*

That Theorem 8.4.3$\Rightarrow$Theorem 8.4.1 is obvious. To see that Theorem 8.4.4$\Rightarrow$Theorem 8.4.2 notice that by Theorem 8.4.4 every Π_1^1 non-meagre set B contains a Δ_1^1 member. Indeed assume without loss of generality that B is comeagre (since it must be comeagre on some basic open set anyway). Let $A = \omega^\omega - B$. Then by Theorem 8.4.4 $B \supseteq \bigcap_n G_n$ where each G_n is dense and for some $F:\omega \to \omega^\omega$ with Δ_1^1 graph if $F(n) = \alpha_n$ we have $G_n = \bigcup_n N_{\alpha_n(k)}$.

Then define inductively $s_0, s_1, \ldots, s_n \in \omega^{<\omega}$ so that s_n is the least proper extension of s_{n-1} such that for some k, $\alpha_n(k) \subseteq s_n$. Then if $\alpha = \lim_{n\to\omega} s_n$, $\alpha \in \Delta_1^1$ and $\alpha \in \bigcap_n G_n \subseteq B$.

8.5 Computation of strategies for Σ_1^0 games

Let $A \subseteq \omega^\omega$ be Σ_1^1 and let T be a recursive tree on $\omega \times \omega$ such that $A = \mathrm{p}[T]$, i.e.

$$\alpha \in A \leftrightarrow \exists\beta\ (\alpha, \beta) \in [T].$$

As in the second section, let us associate with A the following two games G^*, G^{**} (these are exactly the two games given there, except that the roles of the two players I, II have been switched).

GAME G^* We assume here that $A \subseteq 2^\omega$.

I	II	
		II plays $m_0 \in \omega$, $s_0 \in 2^{<\omega}$;
	(m_0, s_0)	I plays $\varepsilon_0 \in \{0, 1\} = 2$;
ε_0		I plays $m_1 \in \omega$, $s_1 \in 2^{<\omega}$ extending $s_0{}^\frown\varepsilon_0$,
	(m_1, s_1)	II plays $\varepsilon_1 \in 2$, etc.
ε_1		Let $\alpha \supseteq s_i$ for all i, $\beta = (m_0, m_1, \ldots)$.
...		Then II wins if, and only if, $(\alpha, \beta) \in [T]$.

GAME G^{**}

I	II	
		II plays $m_0 \in \omega$, $s_0 \in \omega^{<\omega}$; I plays
	(m_0, s_0)	$t_0 \in \omega^{<\omega}$ properly extending s_0;
t_0		II $m_1 \in \omega$, $s_1 \in \omega^{<\omega}$ properly extending t_1, etc.
	(m_1, s_1)	Then II wins if, and only if, $(\alpha, \beta) \in [T]$.
t_1		
...		

Now in Section 2 it was shown that if player II has a winning strategy in the game G^*, then A contains a perfect subset, while if he has a winning strategy in game G^{**}, A is non-meagre. Thus we have

(i) If $A \subseteq 2^\omega$ is countable then I has a winning strategy in G^*. Moreover, as it follows from Section 2, if σ is a winning strategy for I in G^* every member of A is recursive in σ. (Here we view σ as a member of ω^ω, which we can easily do by identifying $\omega^{<\omega}$ with ω.)

(ii) If $A \subseteq \omega^\omega$ is meagre, then I has a winning strategy in G^{**}. Moreover, as it follows from Section 2, if σ is a winning strategy for I in G^{**}, then $\neg A \supseteq \bigcap_n G_n$, where $G_n = \bigcup_k N_{\alpha_n(k)}$ is dense open and if $F(n) = \alpha_n$, F has $\Delta_1^1(\sigma)$ graph.

Note The restriction that $A \subseteq 2^\omega$ in (i) is of no consequence, as it is enough to prove Theorem 8.4.3 for Σ_1^1 subsets of 2^ω. This is because there is Δ_1^1 embedding of ω^ω into 2^ω (i.e. an $f: \omega^\omega \rightarrowtail\!\!\!\!\twoheadrightarrow 2^\omega$ such that graph (f) is Δ_1^1).

Now since T is a recursive tree it is easy to verify that the games G^*, G^{**} are not just open but actually Σ_1^0. Thus it is evident that to establish Theorems 8.4.3 and 8.4.4 it would be enough to show in (i) and (ii) that σ can be taken to be itself Δ_1^1 (as it is easy to verify that if A is a pointset and $\alpha \in \omega^\omega$ then

$$A \text{ is } \Delta_1^1(\alpha) \quad \text{and} \quad \alpha \text{ is } \Delta_1^1 \Rightarrow A \text{ is } \Delta_1^1).$$

We can now state our final result

THEOREM 8.5.1 *Let $G \subseteq \omega^\omega$ be a Σ_1^0 set. If player* I *has a winning strategy in the game determined by G then he actually has a Δ_1^1 winning strategy.*

Proof We shall try to imitate the proof of open game determinacy as explained in the second section. Recall that if $G \subseteq \omega^\omega$ is any set, or ambiguously the corresponding game, and $u \in \omega^{<\omega}$, we let $G(u) = \{\alpha : u^\frown \alpha \in G\}$. If G is Σ_1^0, so is each $G(u)$. Now the key fact behind the proof of open game determinacy was:

$$\left.\begin{array}{l}\textbf{if } G \text{ is open, } u \in \omega^{<\omega} \text{ has even length and player I has no winning} \\ \text{strategy of } G(u) \text{ then } \forall n\ \exists m \text{ (I has no winning strategy in} \\ G(u^\frown n^\frown m)).\end{array}\right\} \quad (*)$$

This provides an obvious strategy for player II in the game G, provided that I has no winning strategy in G, which since G is open is also winning. It is thus clear that by the same argument it will suffice, if we can show the following effective version of $(*)$:

$$\left.\begin{array}{l}\text{If } G \in \Sigma_1^0,\ u \in \omega^{<\omega} \text{ has even length, and player I has no } \Delta_1^1 \textit{ winning} \\ \textit{strategy in } G(u), \text{ then } \forall n\ \exists m \text{ (I has no } \Delta_1^1 \text{ winning strategy in} \\ G(u^\frown n^\frown m)).\end{array}\right\} \quad (**)$$

Let us now prove $(**)$. Assume its hypothesis and also (towards a contradiction) that

$$\exists n\ \forall m \text{ (I has a } \Delta_1^1 \text{ winning strategy in } G(u^\frown n^\frown m)).$$

Fix such an n, say n_0; then we have

$$\forall m\ \exists \sigma\ [\sigma \text{ is } \Delta_1^1 \wedge \forall \beta\ (u^\frown n_0{}^\frown m^\frown (\sigma * \beta) \in G)] \Leftrightarrow \forall m\ \exists \sigma \in \Delta_1^1 P(u, n_0, m, \sigma).$$

Where $\sigma * \beta$ is the result of the run of the game in which II plays β and I follows σ. Clearly P is Π_1^1. If we could now find a map $F: \omega \to \omega^\omega$ with Δ_1^1

graph such that $\forall m\, P(u, n_0, m, F(m))$, then the result would follow, because we would have produced the following Δ_1^1 winning strategy for I in $G(u)$: I starts with n_0 in $G(u)$. If II then plays m, I continues in the rest of the game by following $F(m)$.

So it is enough, in order to complete our proof, to show the following result, a basic fact of the effective theory which should be understood as an effective version of the countable axiom of choice:

$$\forall i\, \exists x\, P(i, x) \Rightarrow \exists f\, \forall i\, P(i, f(i)) .$$

THEOREM 8.5.2 *Assume $P(i, \alpha)$ is Π_1^1 and $\forall i\, \exists \alpha \in \Delta_1^1 P(i, \alpha)$. Then there is $F: \omega \to \omega^\omega$, with graph $(F) \in \Delta_1^1$ such that $\forall i\, P(i, F(i))$.*

Proof Recall from Section 7 that we have a Π_1^1 set $C \subseteq \omega$ and for each $n \in C$ a real δ_n such that

$$\alpha \text{ is } \Delta_1^1 \Leftrightarrow \exists n\, (n \in C \wedge \alpha = \delta_n)$$

and there are Π_1^1, Σ_1^1 relations Q, R respectively such that if $n \in C$:

$$\delta_n(k) = m \Leftrightarrow Q(n, k, m) \Leftrightarrow R(n, k, m) .$$

Now our hypothesis implies that

$$\forall i\, \exists n\, [n \in C \wedge \forall \alpha\, [\alpha = \delta_n \Rightarrow P(i, \alpha)] .$$

As the statement following "$\exists n$" is Π_1^1, the number uniformization theorem (Theorem 7.3.4) implies that there is $g: \omega \to \omega$, with graph (g) in Δ_1^1 (i.e. g is Δ_1^1 viewed as a real) such that $\forall i\, [g(i) \in C \wedge P(i, \delta_{g(i)})]$. Put $F(i) = \delta_{g(i)}$.

9 $\mathbf{\Pi_1^1}$ equivalence relations

9.1 Introduction

The purpose of this is to prove a fairly recent (1974) result of Silver which states the following:

THEOREM 9.1.1 (Silver, 1980) *Let E be a $\mathbf{\Pi}_1^1$ equivalence relation on ω^ω. Then E has either $\leqslant \aleph_0$ or $2^{\aleph_0}$ equivalence classes.*

One can view this result also as an extension of the classical theorem of Souslin on the cardinality of $\mathbf{\Sigma}_1^1$ sets, which asserts that every $\mathbf{\Sigma}_1^1$ subset of ω^ω

has either $\leqslant \aleph_0$ or $2^{\aleph_0}$ elements. Indeed, let $A \subseteq \omega^\omega$ be $\mathbf{\Sigma}_1^1$ and define the following equivalence relation on ω^ω:

$$\alpha \mathrel{E} \beta \Leftrightarrow (\alpha \notin A \text{ and } \beta \notin A) \vee \alpha = \beta.$$

The equivalence classes of E are the singletons $\{\alpha\}$ for $\alpha \in A$ and the set $\omega^\omega - A$. As E is clearly $\mathbf{\Pi}_1^1$ it has, by Silver's theorem, either $\leqslant \aleph_0$ or $2^{\aleph_0}$ equivalence classes, so that A has $\leqslant \aleph_0$ or $2^{\aleph_0}$ elements.

Surprisingly enough Silver's original proof of the above theorem was very complicated both in the technical and in the metamathematical sense, as it required the existence of $\aleph_1$ iterations of the power set operation, a fact that was suggestive of a phenomenon analogous to that of Borel Determinacy. As it turned out however that was not the case. A few years later, Harrington (1976) found a remarkably simple proof of Silver's theorem, which moreover was carried through in the usual context of analysis and did not require the existence of objects of higher type than the reals. Harrington's proof is an excellent example of the power and elegance of effective descriptive set theory and will be given in detail here as a final illustration of how this theory can be used to obtain results of undoubtedly classical character and content.

Our presentation below owes much to some seminar notes by A. Louveau, a seminar talk by R. Van Wesep and private conversations with Y. N. Moschovakis.

9.2 Some motivation

Let E be an equivalence relation on ω^ω. There is a special case in which one can easily show that E has $2^{\aleph_0}$ equivalence classes, namely when E, as a subset of the plane $\omega^\omega \times \omega^\omega$, is meagre or more generally if it is meagre on some square $V \times V$, where V is a non-empty open subset of ω^ω. What in fact is true is that, if $R \subseteq \omega^\omega \times \omega^\omega$ is comeagre on some $V \times V$, where $V \neq \varnothing$ is open, then there is a perfect subset $P \subseteq \omega^\omega$, such that, if $\alpha, \beta \in P$ and $\alpha \neq \beta$, then $(\alpha, \beta) \in R$. By applying this to $R = \neg E$, i.e. the negation of the equivalence relation, we have that if E is meagre on $V \times V$, there is a perfect set P such that any two distinct members of P are E-inequivalent—we summarize this by saying that E has *perfectly many equivalence classes.* In this case clearly E has $2^{\aleph_0}$ equivalence classes.

Before we proceed let us prove the above mentioned fact. So assume V is a non-empty open set in ω^ω, $R \subseteq \omega^\omega \times \omega^\omega$ and R is comeagre on $V \times V$. Let $\{G_n\}_{n=1}^\infty$ be a decreasing sequence of open sets, dense on $V \times V$, such that

$$\bigcap_{n=1}^{\infty} G_n \subseteq (V \times V) \cap R.$$

Assume that $V = N_s$ is a basic neighbourhood, without any loss of generality. We shall inductively define for each $\sigma \in 2^{<\omega}$ (= set of all binary finite sequences) a finite sequence $f(\sigma) \in \omega^{<\omega}$ such that

(1) $\sigma \subseteq \tau \Rightarrow f(\sigma) \subseteq f(\tau)$.
(2) $\mathrm{lh} f(\sigma) \geq \mathrm{lh}\sigma$ ($\mathrm{lh}\sigma$ = length of σ).
(3) If σ, τ are incompatible of the same length n then $f(\sigma)$, $f(\tau)$ are incompatible and $N_{f(\sigma)} \times N_{f(\tau)} \subseteq G_n$.

If then we let for $\alpha \in 2^\omega$, $f^*(\alpha) = \lim_{\sigma \subseteq \alpha} f(\sigma)$ = unique common element of all $N_{f(\alpha \upharpoonright n)}$, then $P = f^*[2^\omega]$ is perfect and if γ, δ are distinct members of P, say $\gamma = f^*(\alpha)$, $\delta = f^*(\beta)$ and if $\alpha \upharpoonright m$, $\beta \upharpoonright m$ are incompatible for all $m \geq n$, then $(\gamma, \delta) \in N_{f(\alpha \upharpoonright m)} \times N_{f(\beta \upharpoonright m)} \subseteq G_m$ for all $m \geq n$, therefore $(\gamma, \delta) \in R$ and the result follows.

We shall illustrate the construction of $f(\sigma)$ for σ of length 0, 1, 2, from which the general construction by induction on $lh\sigma$ will become evident.

Let $f(\emptyset) = s$. Let s_0, s_1 be two incompatible extensions of s. As $N_{s_0} \times N_{s_1} \subseteq V \times V$ and G_1 is dense open in $V \times V$ we can find two further extensions s^0, s^1 of s_0, s_1 such that $N_{s^0} \times N_{s^1} \subseteq G_1$. Put $f((0)) = s^0$, $f((1)) = s^1$. Now let s_{00}, s_{01} be two incompatible extensions of s^0 and s_{10}, s_{11} be two incompatible extensions of s^1. As $N_{s_{00}} \times N_{s_{01}} \subseteq V \times V$ we can find extensions s'_{00}, s'_{01} of s_{00}, s_{01} respectively such that $N_{s'_{00}} \times N_{s'_{01}} \subseteq G_2$. As $N_{s'_{00}} \times N_{s_{10}} \subseteq V \times V$, we can find extensions s''_0, s'_{10} of s'_{00}, s_{10} respectively such that $N_{s''_{00}} \times N_{s'_{10}} \subseteq G_2$ and similarly we can find extensions s'''_{00}, s'_{11} of s''_{00}, s_{11} respectively so that $N_{s'''_{00}} \times N_{s'_{11}} \subseteq G_2$. Put $s^{00} = s'''_{00}$. Now repeat this with s'_{01}. Find extensions s''_{01}, s''_{10} of s'_{01}, s'_{10} respectively so that $N_{s''_{01}} \times N_{s''_{01}} \subseteq G_2$ and then extensions s'''_{01}, s''_{11} of s''_{01}, s'_{11} respectively so that $N_{s'''_{01}} \times N_{s''_{11}} \subseteq G_2$. Put $s^{01} = s'''_{01}$. Finally, let s'''_{10}, s'''_{11} be extensions of s''_{10}, s''_{11} respectively so that $N_{s'''_{10}} \times N_{s'''_{11}} \subseteq G_2$ and put $s^{10} = s'''_{10}$, $s^{11} = s'''_{11}$. Put then $f(\sigma) = s^{i,j}$, where $\sigma = (i, j)$, with $0 \leq i, j \leq 1$.

All this of course is fine, but unfortunately, given a $\mathbf{\Pi}_1^1$ equivalence relation E with more than $\aleph_0$ equivalence classes, it is not in general true that E must be meagre in some square $V \times V$. Harrington's idea amounts then to the following, in the light of the above comments:

Given E, a $\mathbf{\Pi}_1^1$ equivalence relation on ω^ω, find first an $\alpha \in \omega^\omega$ such that $E \in \Pi_1^1(\alpha)$. Then consider a new topology on ω^ω whose basic open sets are now the $\Sigma_1^1(\alpha)$ subsets of ω^ω. Show that this is a well-behaved topology, especially that it satisfies the Baire category theorem, and then proceed to show, utilizing some effective theory, that either E has at most countably many equivalence classes or else it is meagre on some $V \times V$, where V is non-empty open *in this new topology*, and "meagre" means in the product of this new topology. In the latter case repeat, *mutatis mutandis*, the preceding argument to obtain perfectly many equivalence classes of E, thus completing the proof.

9.3 The topology of Σ_1^1 sets

We shall now turn the preceding outline into a complete proof. We assume that E is a Π_1^1 equivalence relation on ω^ω and we will show that either E has $\leqslant\aleph_0$ or perfectly many equivalence classes. By relativizing the proof below to an arbitrary $\alpha\in\omega^\omega$ (i.e. replacing everywhere below $\Pi_1^1, \Sigma_1^1, \Delta_1^1$ by $\Pi_1^1(\alpha)$, $\Sigma_1^1(\alpha)$, $\Delta_1^1(\alpha)$ resp.) we obtain the result for any $\mathbf{\Pi}_1^1$ equivalence relation.

As we explained in Section 9.2 the first step in the proof will be to define a new topology with basic open sets being now all the Σ_1^1 subsets of ω^ω. Clearly this is a larger topology than the standard one as every basic open set in the latter is obviously Σ_1^1 (in fact Σ_1^0). Call the new topology the Σ_1^1-*topology* for convenience. It is clearly separable. We now have

THEOREM 9.3.1 *The Σ_1^1-topology satisfies the Baire category theorem, i.e. the intersection of countably many dense open sets in this topology is still dense.*

Proof Let $\{D_i\}_{i=1}^\infty$ be a sequence of dense open sets in the Σ_1^1-topology. Thus $D_i=\bigcup_{m=1}^\infty A_{im}$, where each A_{im} is Σ_1^1 and if $A\in\Sigma_1^1$ and is non-empty then for some m, $A\cap A_{im}\neq\varnothing$. Let $B\neq\varnothing$ be an arbitrary Σ_1^1 set. We have to show $B\cap\bigcap_{i=1}^\infty D_i\neq\varnothing$. For that fix T^0, a recursive tree on $\omega\times\omega$, such that $B=\mathrm{p}[T^0]=\{\alpha: \exists\beta(\alpha,\beta)\in[T^0]\}$. The plan is to define inductively for each n a recursive tree T^n on $\omega\times\omega$, a finite sequence $s_n\in\omega^{<\omega}$ and finite sequences $t_n^i\in\omega^{<\omega}$ for $0\leqslant i\leqslant n$ such that for each n:

(1) $\varnothing=s_0\subseteq s_1\subseteq\cdots\subseteq s_n$, $\mathrm{lh}s_n=n$.
(2) $(s_n, t_n^i)\in T^i$, $0\leqslant i\leqslant n$ (so also $\mathrm{lh}t_n^i=n$) and

$$\bigcap_{i=0}^{n}\{\alpha:\alpha\supseteq s_n\wedge\exists\beta\supseteq t_n^i(\alpha,\beta)\in[T^i]\}\neq\varnothing.$$

(3) $\{\alpha:\alpha\supseteq s_n\wedge\exists\beta\supseteq t_n^i(\alpha,\beta)\in[T^i]\}\subseteq D_i,\qquad 1\leqslant i\leqslant n.$
(4) $t_i^i\subseteq t_{i+1}^i\subseteq\cdots\subseteq t_n^i$, $0\leqslant i\leqslant n$.

Then if we let $\alpha=\lim_{n\to\infty}s_n$, $\beta^i=\lim_{n\to\infty}t_n^i$, we have $(\alpha,\beta^i)\in[T^i]$; therefore for $n\geqslant i\geqslant 1$, $\alpha\supseteq s_n\wedge\beta^i\supseteq t_n^i$; therefore $\alpha\in D_i$ by (3), and also $(\alpha,\beta^0)\in[t^0]$; therefore $\alpha\in B$, so $\alpha\in B\cap\bigcap_{i=1}^\infty D^i$ and the proof will be complete. Clearly for $n=0$ we can take $s_0=\varnothing$, $t_0^0=\varnothing$. Assume now that we have defined T^m for $m\leqslant n$, s_m for $m\leqslant n$ and t_m^i for $0\leqslant i\leqslant m\leqslant n$, satisfying (1)–(4) above. Then

$$C=\bigcap_{i=0}^{n}\{\alpha:\alpha\supseteq s_n\wedge\exists\beta\supseteq t_n^i(\alpha,\beta)\in[T^i]\}\neq\varnothing,$$

and $C \in \Sigma_1^1$, so $C \cap D_{n+1} \neq \varnothing$, therefore we can find m so that $C \cap A_{n+1,m} \neq \varnothing$. Let T^{n+1} be a recursive tree such that $p[T^{n+1}] = A_{n+1,m}$. Fix $\alpha \in C \cap A_{n+1,m}$. Find then $\beta_0, \ldots, \beta_n, \beta_{n+1}$ such that $(\alpha, \beta_i) \in [T^i]$, $\alpha \supseteq s_n$, $\beta_i \supseteq t_n^i$ for $0 \leqslant i \leqslant n$ and $(\alpha, \beta_{n+1}) \in [T^{n+1}]$. Finally write

$$s_{n+1} = \alpha \restriction (n+1)(\supseteq s_n),$$
$$t_{n+1}^i = \beta_i \restriction (n+1)(\supseteq t_n^i),$$
$$t_{n+1}^{n+1} = \beta_{n+1} \restriction (n+1).$$

Clearly conditions (1)–(4) above are also satisfied at stage $n+1$.

There is another basic fact about the Σ_1^1-topology that we shall use later on.

LEMMA 9.3.2 (Louveau) *Let $D \subseteq \omega^\omega$ be comeagre in the Σ_1^1-topology. Then if $A \subseteq \omega^\omega \times \omega^\omega$ is Σ_1^1 and non-empty, $A \cap (D \times D) \neq \varnothing$.*

Similarly, if D is comeagre in some set V open in the Σ_1^1-topology and $A \subseteq V \times V$.

Proof The result just says that $D \times D$ is dense on the Σ_1^1-topology of $\omega^\omega \times \omega^\omega$, i.e. the topology on $\omega^\omega \times \omega^\omega$ whose basic open sets are the Σ_1^1 subsets of $\omega^\omega \times \omega^\omega$. As there is a recursive bijection between ω^ω and $\omega^\omega \times \omega^\omega$ this topology is homeomorphic to the Σ_1^1-topology of ω^ω. So the Baire category theorem holds for it. (Notice that the product of the Σ_1^1-topology on $\omega^\omega \times \omega^\omega$ is smaller than the Σ_1^1-topology of $\omega^\omega \times \omega^\omega$.)

Now let $\{D_n\}_{n=1}^\infty$ be a sequence of dense open sets of the Σ_1^1-topology of ω^ω such that $\bigcap_{n=1}^\infty D_n \subseteq D$. If $\varnothing \neq A \subseteq \omega^\omega \times \omega^\omega$, $A \in \Sigma_1^1$ then $p[A] = \{\alpha : \exists \beta (\alpha, \beta) \in A\}$ is a non-empty Σ_1^1 subset of ω^ω, so $p[A] \cap D_n \neq \varnothing$, $\forall n \geqslant 1$. Thus $(D_n \times \omega^\omega) \cap A \neq \varnothing$, $\forall n \geqslant 1$; i.e. $D_n \times \omega^\omega$ is dense in the Σ_1^1-topology of $\omega^\omega \times \omega^\omega$, for all $n \geqslant 1$, and similarly for $\omega^\omega \times D_n$, $\forall n \geqslant 1$. So

$$\bigcap_{n=1}^\infty (D_n \times \omega^\omega) \cap \bigcap_{n=1}^\infty (\omega^\omega \times D_n) \subseteq D \times D$$

is dense in this topology, by the Baire category theorem, and the result follows.

9.4 Completion of the proof

After these preliminaries on the Σ_1^1-topology, let us complete Harrington's proof. We have that E is a Π_1^1 equivalence relation on ω^ω, that E has more than $\aleph_0$ equivalence classes and we want to show that E has perfectly many equivalence classes. For that we shall find a non-empty set V in the Σ_1^1-topology of ω^ω such that E is meagre on $V \times V$, in the product of the Σ_1^1-topology on $\omega^\omega \times \omega^\omega$. Once such a V has been found, one can produce a perfect set of pairwise E-inequivalent members of ω^ω, by mixing the same kind of construction as in Section 9.2, and the proof of the Baire category theorem in Section 9.3; the details of this can be safely left to the reader, so we shall concentrate on finding the appropriate V.

The set $E \subseteq \omega^\omega \times \omega^\omega$ is Π_1^1, so in particular $\mathbf{\Pi}_1^1$, thus its complement is the result of applying the Souslin operation to closed sets of the standard topology of $\omega^\omega \times \omega^\omega$, thus also closed in the product of the Σ_1^1-topology on $\omega^\omega \times \omega^\omega$. So E has the property of Baire in this new product topology on $\omega^\omega \times \omega^\omega$. So by the Kuratowski–Ulam theorem, in order to check that $E \cap (V \times V)$ is meagre, for a prospective V, it is sufficient to check that for each $\alpha \in \omega^\omega$, $V \cap E_\alpha = \{\beta \in V: \alpha E \beta\}$ is meagre in the Σ_1^1-topology of ω^ω, where E_α is the equivalence class of α. To check this again, it is sufficient to show that $V \cap E_\alpha$ is not comeagre in any $U \subseteq V$, $U \neq \emptyset$, $U \in \Sigma_1^1$, as E_α being $\mathbf{\Pi}_1^1$ again has the property of Baire in the Σ_1^1-topology. By Lemma 9.3.2, if $U \subseteq V$, $U \neq \emptyset$, $U \in \Sigma_1^1$ and $V \cap E_\alpha$ is comeagre on U, then

$$(U \cap E_\alpha) \times (U \cap E_\alpha)$$

intersects $(U \times U) \cap (\omega^\omega \times \omega^\omega - E)$ if this latter is non-empty, i.e. if $U \times U \not\subseteq E$, a clear contradiction as this implies that there are $\beta, \gamma \in E_\alpha$ such that $\neg \beta E \gamma$. So it will suffice if we can choose $V \neq \emptyset$, V open in the Σ_1^1-topology such that for all $\emptyset \neq U$, $U \subseteq V$, $U \in \Sigma_1^1$, U meets more than one equivalence class of E. So let us put

$$A = \bigcup \{U: U \text{ is } \Sigma_1^1 \text{ and } U \text{ is contained in some equivalence class of } E\}.$$

As there are only countably many Σ_1^1 sets and E has uncountably many equivalence classes, $A \not\subseteq \omega^\omega$. Let $V = \neg A$. If we can show that V is Σ_1^1, then we have clearly found the required V.

This is indeed shown easily by using some effective theory. First note that if $\emptyset \neq U$ is Σ_1^1 and $U \subseteq E_\alpha$, then actually $E_\alpha \in \Pi_1^1$ (lightface!), since

$$\beta \in E_\alpha \Leftrightarrow \forall \gamma [\gamma \in U \Rightarrow \gamma E \beta].$$

But then by the Separation Theorem we can find $\bar{U} \in \Delta_1^1$ such that $U \subseteq \bar{U} \subseteq E_\alpha$. Thus

$$A = \bigcup \{U: U \text{ is } \Delta_1^1 \text{ and } U \text{ is contained in some equivalence class}\}.$$

Using this we shall calculate now that A is Π_1^1 as follows:

First we shall code Δ_1^1 subsets of ω^ω by numbers as we did in Section 7 for reals. Let $P(n, \alpha)$ be Π_1^1 and universal for the Π_1^1 subsets of ω^ω. Put

$$P^0(n, \alpha) \Leftrightarrow P((n)_0, \alpha),$$

$$P^1(n, \alpha) \Leftrightarrow P((n)_1, \alpha),$$

where $n \mapsto ((n)_0, (n)_1)$ is a recursive bijection between ω and $\omega \times \omega$. Then P^0, P^1 are doubly universal, i.e. given $A, B \subseteq \omega^\omega$ both Π_1^1, there is $n \in \omega$ such that $P_n^0 = A$, $P_n^1 = B$, where $P_n = \{\alpha : (n, \alpha) \in P\}$. Now say "$n$ is a code" ($\Leftrightarrow^{\text{def}} D(n)$) if, and only if, $P_n^0 \cup P_n^1 = \omega^\omega$. Clearly this is a Π_1^1 subset of ω. By the reduction theorem for Π_1^1 let $\bar{P}^0$, $\bar{P}^1$, both Π_1^1, reduce P^0, P^1 respectively. For $n \in D$, let

$$D_n = \bar{P}_n^0 (= \neg \bar{P}_n^1).$$

Clearly for $A \subseteq \omega^\omega$

$$A \in \Delta_1^1 \Leftrightarrow \exists n \ (n \in D \text{ and } A = D_n).$$

We now compute

$$\alpha \in \bigcup\{U : U \text{ is } \Delta_1^1 \wedge U \text{ is contained in some } E\text{-equivalence class}\}$$

$$\Leftrightarrow \exists n \ [n \in D \wedge \alpha \in \bar{P}_n^0 \wedge \forall \beta \ \forall \gamma (\beta, \gamma \notin \bar{P}_n^1 \Rightarrow \beta E \gamma)],$$

which is clearly Π_1^1.

This finishes Harrington's proof.

As a conclusion it should be mentioned that no classical-type proof of Silver's theorem is known. Silver's original proof uses, directly or indirectly, metamathematical forcing techniques.

For $\mathbf{\Sigma}_1^1$ equivalence relations, Burgess (1978) showed, using Silver's theorem, that if E is a $\mathbf{\Sigma}_1^1$ equivalent relation on ω^ω then E has $\leqslant \aleph_1$ or perfectly many (therefore $2^{\aleph_0}$) equivalent classes. This result is best possible as there is a $\mathbf{\Sigma}_1^1$ equivalence relation on ω^ω with exactly $\aleph_1$ equivalence classes.

References

Addison, J. W. Jr and Moschovakis, Y. N.

1968 Some consequences of the axiom of definable determinateness, *Proc. Nat. Acad. Sci. USA*, **59**, 708–712.

Blackwell, D.

1967 Infinite games and analytic sets, *Proc. Nat. Acad. Sci. USA*, **58**, 1836–1837.

Burgess, J.
1978 Equivalences generated by families of Borel sets, *Proc. Amer. Math. Soc.*, **69**, 323–326.
Davis, Martin.
1958 *Computability and Unsolvability*, McGraw-Hill, New York.
Davis, Morton.
1964 Infinite games of perfect information, *Ann. Math. Studies*, **52**, 85–101.
Devlin, K. I. and Jensen, R. B.
1975 Marginalia to a theorem of Silver, in *Logic Conference, Kiel 1974*, Springer-Verlag, Berlin, pp. 115–142.
Friedman, H.
1971*a* Higher set theory and mathematical practice, *Ann. Math. Logic*, **2**, 326–357.
1971*b* Determinateness in the low projective hierarchy, *Fund. Math.*, **72**, 79–95.
Gale, D. and Stewart, F. M.
1953 Infinite games with perfect information, *Ann. Math. Studies*, **28**, 245–266.
Harrington, L.
1978 Analytic determinacy and $0^{\#}$, *J. Symbolic Logic*, **43**, 685–693.
Jayne, J. E. and Rogers, C. A.
1979 Borel isomorphisms at the first level, *Mathematika*, **26**, 125–179; and a correction to appear in *Mathematika*, **27**.
Kechris, A. S. and Moschovakis, Y. N.
1978 Notes on the theory of scales, in *Cabal Seminar 76–77*, Lecture Notes in Mathematics no. 689, Springer-Verlag, Berlin, Heidleberg, New York.
Martin, D. A.
1968 The axiom of determinateness and reduction principles in the analytical hierarchy, *Bull. Amer. Math. Soc.*, **74**, 687–689.
1970 Measurable cardinals and analytic games, *Fund. Math.*, **66**, 287–291.
1975 Borel determinacy, *Ann. Math.*, **102**, 363–371.
1977 Descriptive set theory: projective sets, in *Handbook of Mathematical Logic*, North-Holland, Amsterdam, pp. 783–815.
Moschovakis, Y. N.
1971 Uniformization in a playful universe, *Bull. Amer. Math. Soc.*, **77**, 731–736.
1980 *Descriptive Set Theory*, North-Holland, Amsterdam.
Mycielski, J.
1964 On the axiom of determinateness, *Fund. Math.*, **53**, 204–224.
Paris, J. B.
1972 $\mathrm{ZF} \vdash \Sigma_4^0$ determinateness, *J. Symbolic Logic*, **37**, 661–667.
Shoenfield, J. R.
1967 *Mathematical Logic*, Addison-Wesley, Reading, Mass.
Silver, J. H.
1980 Counting the number of equivalence classes of Borel and coanalytic equivalence relations, *Ann. Math. Logic*, **18**, 1–28.
Wolfe, P.
1955 The strict determinateness of certain infinite games, *Pacific J. Math.*, **5**, 841–847.

Part 5

Analytic sets in non-separable metric spaces

A. H. STONE

Although the K-analytic sets of Choquet, Sion and Frolík give in many ways a good generalization of the "classical" theory (of analytic sets in Polish spaces), they are—from the point of view of general metrizable spaces—rather too good. In fact, they are Lindelöf, hence separable if metrizable. But there is a respectable theory of non-separable metric Borel sets, and one naturally wishes to have a theory of non-separable metric analytic sets to go with it. Such a theory, apart from its intrinsic interest, might lead to simplifications and improvements of the Borel theory. This indeed actually happens; examples will be mentioned later. However, the following discussion is mostly concerned with analytic sets for their own sakes. But since the analytic theory is an outgrowth from the Borel theory, we begin with a summary of some of the principal non-separable Borel results. For simplicity, it is assumed throughout that *all spaces considered are metrizable*, apart from a few clearly stated exceptions, although many of the results hold in greater generality.

The definitions of "Borel set" and "Borel set of additive or multiplicative class α" ($\alpha < \omega_1$) are the same as in the "classical" separable case. We have, by the usual argument (based on Lavrentiev's theorem), that if a space Y is Borel (of additive class $\alpha \geq 1$, or multiplicative class ≥ 2) in one complete space in which it is (topologically) embedded, then it is so (of the same class) in *every* space in which it is embedded. Such spaces Y are called "absolutely Borel". Most of the basic properties of Borel sets, familiar in the separable case (as in K. Kuratowski, 1966), still go through, though often different proofs are required (see A. H. Stone, 1962). The roles of the Cantor set and the space $\mathbb{N}^{\mathbb{N}}$ of irrationals, in the classical case, are taken by the "Baire spaces" $B(k)$, where k is an infinite cardinal; here $B(k) = T(k)^{\aleph_0}$, where $T(k)$ is a discrete set of cardinal k, and $B(k)$ is given the product topology. (Thus $B(\aleph_0) = \mathbb{N}^{\mathbb{N}}$.) It is also convenient to let $B(1)$ denote a 1-point space.

Two spaces X, Y are said to be "Borel isomorphic" if there is a bijection $f: X \twoheadrightarrow Y$ such that both f and f^{-1} take Borel subsets to Borel subsets. A "generalized homeomorphism" $f: X \twoheadrightarrow Y$, of class (α, β), is a bijection such that, for all open subsets U of X and V of Y, $f^{-1}(V)$ and $f(U)$ are Borel of additive classes α, β respectively. It follows easily that such an f is a Borel isomorphism. (The converse is also true when X and Y are "nice", and Fleissner (1979) has given a model of set-theory in which the converse is always true. We return to this point later.) In any case, generalized homeomorphism and Borel isomorphism give the same classification of the absolute Borel sets. To illustrate the general result, which is a little complicated: it is a familiar "classical" fact that every absolute Borel space of weight $\aleph_0$ is Borel isomorphic to just one of the two spaces $B(\aleph_0)$, $\aleph_0 B(1)$ (the latter meaning the discrete union of $\aleph_0$ copies of the singleton $B(1)$). Analogously, every absolute Borel space of weight $\aleph_1$ is Borel isomorphic to just one of the four spaces $B(\aleph_1)$, $\aleph_1 B(\aleph_0)$, $\aleph_1 B(1)$ and $B(\aleph_0)+\aleph_1 B(1)$, where the "+" refers to discrete union. For absolute Borel sets of weight $\aleph_\alpha$ there are exactly $2^{\alpha+1}$ corresponding types if α is finite, $\|\alpha\|$ otherwise (see Stone, 1972).

The proof that an absolute Borel set X is generalized homeomorphic to at least one of the appropriate types goes by first showing that X (if of weight k) is Borel isomorphic to a closed subset of $B(k)$; and this is done by a transfinite induction over the Borel class of X. One then shows that this closed subset either contains a copy of $B(k)$ or is "σ-locally-of-weight-less-than-k" (σLW($<k$)); that is, is the union of a countable family of sets, each locally of weight less than k. A further transfinite induction over k completes the proof. (See Stone, 1962, for details.) The proof that no two of the types are Borel isomorphic requires the finding of properties invariant under Borel isomorphism. One such property is having a given weight. The local weight of a space is *not* an invariant; but it turns out that, for each infinite cardinal h, the σLW($<h$) property is a Borel invariant, and this is enough to discriminate between the Borel types. The first proof of the Borel invariance of σLW($<h$) was an elaborate transfinite induction; a much simpler one comes from the theory of analytic sets. The sets needed here are simply the Souslin-$\mathscr{F}$ sets, but it will save time later to give some more general definitions at this point.

Let k be an infinite cardinal, and $\mathscr{H}$ an arbitrary family of subsets of a space X. Let $\mathbf{S}_k(\mathscr{H})$ denote the family of all sets expressible as

$$\bigcup_{t \in B(k)} \bigcap_{n=1}^{\infty} H(t|n),$$

where (as usual) $t=(t_1, t_2, \ldots)$ with each $t_i \in T(k)$, and $t|n=(t_1, \ldots, t_n)$, and where each $H(t|n) \in \mathscr{H}$. The classical arguments are easily modified to show that $\mathbf{S}_k(\mathbf{S}_k(\mathscr{H}))=\mathbf{S}_k(\mathscr{H})$, and that $\mathbf{S}_k(\mathscr{H})$ is closed under countable intersections and unions of $\leqslant k$ sets. It follows easily that $\mathbf{S}_k(\mathscr{F})=\mathbf{S}_k(\mathscr{G})$ (where

$\mathscr{F}$, $\mathscr{G}$ denote the families of closed and of open subsets of X); we call these sets the k-*Souslin* subsets of X. If a set E is k-Souslin in one complete metric space X, it is so in all (metrizable) spaces that contain E; that is, E is "absolutely k-Souslin". For brevity, I shall call such spaces "k-analytic".

(This terminology is intended to agree with current majority usage, in which an "analytic" set is a "nice" Souslin one. Unfortunately it differs from that used in Freiwald (1970, 1972), Hansell (1971*b*, 1972, 1973*a*, *b*, 1974*a*, *b*) and Stone (1962, 1972), where "k-analytic" is used to mean the present "k-Souslin", and "absolutely k-analytic" is used to mean the present "k-analytic".)

It is easily seen that, if $\aleph_0 \leq h < k$, every Borel set is h-Souslin, and every h-Souslin set is k-Souslin; and similarly for "analytic". Also "$\aleph_0$-Souslin" is just "Souslin" in the usual sense. The $\aleph_0$-analytic sets are often called "analytic" or "A-sets" (El'kin, 1967).

The invariance of the σLW($<h$) property under Borel isomorphism follows from the following more general result.

THEOREM *If X and Y are $\aleph_0$-analytic spaces that are Borel isomorphic, and if X is σLW($<h$), then so is Y.*

Outline of proof Suppose $F: X \twoheadrightarrow Y$ is a Borel isomorphism. Because X is σLW($<h$), one can arrange to have

$$X = \bigcup_{n=1}^{\infty} X_n,$$

where each X_n is *closed* and locally of weight less than h. Take a σ-discrete refinement of a (relatively) open cover of X_n by sets of weights less than h, say

$$\{U_{n\lambda m}: \lambda \in \Lambda_{mn}, m = 1, 2, \ldots\},$$

where for fixed m and n these sets form a discrete system; thus each $U_{n\lambda m}$ is Borel in X and of weight $<h$. Now every union of $U_{n\lambda m}$'s (still keeping m and n fixed) is open in X_n; hence every union of $f(U_{n\lambda m})$'s ($\lambda \in \Lambda_{nm}$) is Borel (and therefore $\aleph_0$-analytic) in Y; and these sets are pairwise disjoint. Now an important theorem of Hansell (1971*b*) asserts that a system of pairwise disjoint sets, every union of which is $\aleph_0$-analytic, is "σ-discretely decomposable" (σ-d.d.); in the present case this says that we can write

$$f(U_{n\lambda m}) = \bigcup\{V_{n\lambda mp}: p = 1, 2, \ldots\}$$

where (for fixed m, n, p) the system $\{V_{n\lambda mp}: \lambda \in \Lambda_{mn}\}$ is discrete. Put

$$Y_{nmp} = \bigcup\{V_{n\lambda mp}: \lambda \in \Lambda_{mn}\};$$

this is locally of weight less than h (because each $f(U_{n\lambda m})$ has weight less than h), and

$$Y = \bigcup\{Y_{nmp}: n, m, p \in \mathbb{N}\},$$

a countable union.

Remark In the above theorem, "Borel isomorphism" can be weakened to "$\aleph_0$-analytic isomorphism". Moreover, Hansell's theorem has been extended to point-finite (rather than pairwise disjoint) completely-additive analytic collections (J. Kaniewski and R. Pol, 1975).

A similar argument, combining Hansell's theorem with a theorem of Preiss (1975), shows that every Borel-measurable map between $\aleph_0$-analytic spaces is of bounded class, and thus that every Borel isomorphism between $\aleph_0$-analytic spaces is a generalized homeomorphism.

In what follows we give a sampling of results about k-Souslin and k-analytic sets, and mention a few unsolved problems.

The elementary properties, familiar when $k = \aleph_0$, extend to the general case without difficulty. For instance: if $Y \subset X$, the k-Souslin subsets of Y are the intersections of Y with the k-Souslin subsets of X. A k-Souslin subset of a k-Souslin set is k-Souslin; and similarly for k-analytic. A locally k-Souslin set is k-Souslin; and similarly for k-analytic. Further (see Stone, 1962), we have

THEOREM *A non-empty (metric) space is a continuous image of $B(k)$ if, and only if, it is k-analytic and of weight $\leqslant k$.*

COROLLARY *The class of k-analytic spaces of weight $\leqslant k$ is closed under continuous maps.*

Note, however, that this theorem has a weakness. We have seen that the k-Souslin subsets of a given space are closed under unions of k sets. Hence *every* set of $\leqslant k$ points is k-analytic. Now a space of weight k has $\leqslant k^{\aleph_0}$ points, and for "most" k we have $k^{\aleph_0} = k$. Thus for "most" k (on GCH, for all but sequential k) the above theorem merely says that every set of $\leqslant k$ points is a continuous image of $B(k)$, which is a triviality. There are arbitrarily large cardinals k for which $k^{\aleph_0} > k$, and for which the above theorem thus has content; but even then, the k-analytic sets will include (when $k > \aleph_0$) some quite pathological sets. Nevertheless, they have been studied in some depth by Čoban (1974) (even in non-metrizable spaces) and Freiwald. Some specimen results (from Freiwald, 1970, 1972) are

THEOREM *X is k-analytic if (and only if) X is the intersection of a $\mathcal{G}_\delta$-set and a k-Souslin set in some compactification Y* (not *necessarily metrizable*) *of X; further, one can always take Y to be βX here.*

The analogous properties are known to hold for Borel sets (of given class) in stronger form (Hansell, 1971*a*; Willard, 1969); to complete the analogy one would like to know: if Z is an arbitrary compactification (not necessarily metrizable) of the k-analytic metric space X, must X be the intersection of a k-Souslin set and a $\mathcal{G}_\delta$-set in Z? When $k = \aleph_0$, the answer is "yes" (Hansell, 1971*b*); but in general the question seems to be open.

Now suppose a surjective map $f: X \to Y$ is given (where, as usual, X and Y are metrizable). We have that

if f is perfect and Y is k-analytic, then so is X.

More generally, it suffices that f be continuous and closed and have k-analytic point-inverses $f^{-1}(y)$, $y \in Y$.

In the opposite direction, suppose X is k-analytic and f is Borel-measurable of bounded class. Then if f is closed, or open and with complete point-inverses (in a fixed metric on X), Y is k-analytic. If the point-inverses are k-analytic, the graph of f is k-analytic.

Freiwald also proved

THEOREM *If X is k-analytic, then either X is "k-discrete" or X contains a closed subset of cardinal $k^{\aleph_0}$, homeomorphic to some $B(p)$ or to a Cantor set. (Both alternatives can happen at once, in general.)*

Here X is said to be "k-discrete" when it is the union of k discrete subsets. The case $k = \aleph_0$ (when the alternatives are mutually exclusive) is a theorem of El'kin (1967), generalizing the classical theorem of Alexandrov and Hausdorff (that an uncountable Borel set in a Polish space contains a Cantor set).

It is of interest to compare the "k-discrete" and "σLW($<h$)" properties. Using the fact that metrizable spaces have σ-discrete bases, it is not hard to show that a (metrizable) space X is k-discrete if, and only if, it is σ-locally-of-cardinal-k-or-less (that is, $X = \bigcup\{X_n: n = 1, 2, \ldots\}$, where each point of X_n has a neighbourhood in X_n with at most k points). Thus, on writing k^+ to denote the successor of the infinite cardinal k, we have

$$k\text{-discrete} \Rightarrow \sigma\mathrm{LW}(<k^+) \Rightarrow {}^{\aleph_0}\text{-discrete}.$$

Since $k^{\aleph_0} = k$ for "most" cardinals, we have that "usually" k-discrete $\Leftrightarrow$ σLW($<k^+$).

R. Pol (1978) has observed that Freiwald's generalization of El'kin's theorem can be improved to a "sharp" result for many values of k; but the general problem remains open. When $k = \aleph_0$ there is a much better generalization of El'kin's theorem (Stone, 1972):

THEOREM *If X is $\aleph_0$-analytic, and h is an infinite cardinal, then either X is σLW$(<h)$ or X contains a subset homeomorphic to $B(h)$; but not both.*

But this does not extend to k-analytic X when $k > \aleph_0$.

From the preceding considerations it follows that, on the generalized continuum hypothesis, X has the property that every subset of X is k-analytic if, and only if, X is k-discrete. Examples show that the GCH is needed here (more precisely, some hypothesis beyond ZFC is needed).

Further mapping results seem to depend on the notions of "σ-discrete map" and "co-σ-discrete map". A map $F: X \to Y$ is said to be "σ-discrete" (Hansell, 1971*b*) when there is a σ-discrete collection $\mathscr{B}$ of subsets of X such that, for each open V in Y, $f^{-1}(V)$ is a union of members of $\mathscr{B}$; we say $\mathscr{B}$ is a "σ-discrete base" for f. This condition holds automatically if f is continuous, or if Y is separable, or if X is σ-discrete. Hansell (1971*b*) shows that, if X is $\aleph_0$-analytic, then every map $f: X \to Y$ that is Borel (or even analytically) measurable is σ-discrete (having a σ-discrete base consisting of $\aleph_0$-analytic sets), and that properties of maps of $\aleph_0$-analytic spaces can be generalized to σ-discrete maps of general (metric) spaces. For instance, if $f: X \to Y$ and $g: X \to Z$ are maps, and f is σ-discrete, then $(f, g): X \to Y \times Z$ is Borel (of class α) if, and only if, f and g are Borel (of class α); this is a typical "bread-and-butter result", used in proving others. Baire's theorem on "pointwise discontinuous" functions takes the following form:

THEOREM *The set of points of discontinuity of a σ-discrete Borel-measurable map of class 1 is meagre.*

The Lebesgue–Hausdorff theorem on "analytically representable" functions can be extended (roughly) as follows (for details see Hansell, 1974*b*):

THEOREM *The Borel-measurable functions are those obtained from the continuous ones by repeatedly taking limits of convergent sequences of functions, provided that we restrict attention to σ-discrete functions.*

This proviso is automatically fulfilled when X is $\aleph_0$-analytic; it is an open question whether it is always superfluous. (Presumably it is superfluous in Fleissner's model.) This is closely related to another open question; is the limit of a (pointwise) convergent sequence of σ-discrete maps necessarily σ-discrete?

A map $f: X \to Y$ is said to be "co-σ-discrete" provided, for each discrete system $\{A_\lambda : \lambda \in \Lambda\}$ of sets in X, the image system $\{f(A_\lambda): \lambda \in \Lambda\}$ has a "σ-discrete base" in Y (that is, a σ-discrete system $\mathscr{B}$ of subsets of Y such that each $f(A_\lambda)$ is a union of members of $\mathscr{B}$). Although this is dual, in a sense, to the "σ-discrete map" defined above, there is a significant difference; the hypothesis of co-σ-discreteness is definitely needed for certain results, and cannot be eliminated by a set-theoretic axiom. However, every $f: X \to Y$ where X is separable will be automatically co-σ-discrete (for Λ will be countable). It is also not hard to see that f is co-σ-discrete whenever f is a closed surjection, or an open map with separable point-inverses. But it is not known whether every $f: X \to Y$ on an $\aleph_0$-analytic space X, taking closed sets to $\aleph_0$-analytic ones, is necessarily co-σ-discrete. (If f is injective, co-σ-discreteness does follow.) Hansell (1974*a*) has used this notion to characterize the $\aleph_0$-analytic sets as follows:

THEOREM *A non-empty (metrizable) space X is $\aleph_0$-analytic and of weight $\leq k$ if, and only if, it is the image of $B(k)$ under a continuous and co-σ-discrete map.*

Note that "co-σ-discrete" is essential in this theorem; without it we merely obtain the k-analytic sets.

COROLLARY *If X is $\aleph_0$-analytic and $g: X \to Y$ is a closed continuous co-σ-discrete surjection, then Y is $\aleph_0$-analytic.*

In this corollary, the continuity requirement on g can be weakened to "analytic measurability" by an argument, of a standard type, involving the graph of the map.

One of the main challenges in the non-separable theory was to generalize the first separation theorem, and in particular Souslin's theorem that a set that is both analytic and co-analytic must be Borel. Here the fact that all sets of cardinal $\leq k$ are k-analytic presents a serious obstacle; it is easy for a set and its complement both to be k-analytic, and hard for them to be Borel. Hansell (1972) overcame this difficulty by enlarging the family of Borel sets, and restricting the k-analytic sets. Define the family of *extended Borel sets* of a space X to be the smallest family of subsets of X containing all open (or closed) sets, closed under the operations of countable intersection and *σ-discrete union*. (Thus, for instance, if X is $\aleph_1 B(\aleph_0)$ and we take a Borel set E_α of exact class α in the αth copy of $B(\aleph_0)$, $\alpha < \omega_1$, $\bigcup_\alpha E_\alpha$ will be extended

Borel, though clearly not Borel.) Define the "*restricted k-Souslin sets*" in X to be those having k-Souslin representations

$$A = \bigcup_{t \in B(k)} \bigcap_{n=1}^{\infty} F(t|n) \qquad \text{(each } F(t|n) \text{ being closed)}$$

for which, for each fixed n, the sets $F(t|n)$, $t \in B(k)$, form a σ-discrete family. (In Hansell's terminology, A is "k-Souslin".) Again most of the elementary properties of Borel and Souslin sets have the expected analogues. In particular, the extended Borel sets can be assigned "Borel classes" in much the usual way, with discrete unions occurring just at the limit ordinals. One defines k-Borel sets to be extended Borel sets in which no more than k sets occur in the σ-discrete unions, and easily sees that, in a space of weight $\leqslant k$, they coincide with the extended Borel sets. The notion of "absolute restricted k-Souslin set" is defined in the obvious way, and has the usual properties; let us call such sets (provisionally) "restricted k-analytic". Hansell showed that the first separation theorem holds in the following form:

THEOREM *In a space of weight $\leqslant k$, two disjoint restricted k-analytic sets can be separated by disjoint k-Borel sets.*

He also showed that $B(k)$, for instance, contains extended Borel sets of all possible exact classes; this is a difficult theorem, and the proof combines generalizations of both of the standard methods for constructing sets of arbitrary exact class in (say) the Hilbert cube—the "universal set" method and the method of Engelking *et al.* (1966). On the other hand, it surprisingly turns out (Hansell, 1973*b*) that (because there is so much discreteness present in their definition) the restricted k-Souslin sets are precisely the $\aleph_0$-Souslin (= Souslin-$\mathscr{F}$) sets. Thus the "restricted k-analytic sets" are just the $\aleph_0$-analytic sets; and the Souslin theorem takes the form

THEOREM *In an arbitrary metric space X, if both A and $X \backslash A$ are $\aleph_0$-analytic, then A is extended Borel.*

The notion of "Lusin set" has a reasonable analogue, giving rise to the following theorem:

THEOREM *The absolute extended Borel sets of weight $\leqslant k$ are the images of closed subsets of $B(k)$ under injective, continuous, co-σ-discrete maps.*

(See Hansell, 1974*a*).) Here "continuous" can be weakened to "extended-Borel-measurable".

Finally, one sees by the usual construction that the $\aleph_0$-Souslin subsets of a space X are the projections of Borel (or $\aleph_0$-Souslin) subsets of $X \times \mathbb{N}^{\mathbb{N}}$. Analogously, the k-Souslin subsets of X are the projections of the k-Borel (or k-Souslin) subsets of $X \times B(k)$.

In summary, while k-analytic sets (in non-separable metric spaces) seem to deserve attention for all k, they are certainly at their best when $k = \aleph_0$. Perhaps other values of k for which $k^{\aleph_0} > k$ (for instance $k = \aleph_\omega$) may also be of special interest; but that remains to be seen.

References

Čoban, M. M.

1974 Continuous images of complete spaces, *Trudy Moskov. Mat. Obšč.*, **30**, 23–59 (= *Trans. Moscow Math. Soc.*, **30**, 25–64).

El'kin, A. G.

1967 *A*-sets in complete metric spaces, *Dokl. Akad. Nauk SSSR*, **175**, 517–520 (= *Soviet Math. Dokl.* **8**, 847–877).

Engelking, R., Holsztyński, W. and Sikorski, R.

1966 Some examples of Borel sets, *Colloq. Math.*, **15**, 271–274.

Fleissner, W.

1979 An axiom for nonseparable Borel theory, *Trans. Amer. Math. Soc.*, **251**, 309–328.

Freiwald, R. C.

1970 Images of Borel sets and k-analytic sets, Ph.D. thesis, University of Rochester.

1972 Images of Borel sets and k-analytic sets, *Fund. Math.*, **75**, 35–46.

Hansell, R. W.

1971*a* General embedding properties of absolute Borel and Souslin spaces, *Proc. Amer. Math. Soc.*, **27**, 343–352.

1971*b* Borel measurable mappings for nonseparable metric spaces, *Trans. Amer. Math. Soc.*, **161**, 145–169.

1972 On the nonseparable theory of Borel and Souslin sets, *Bull. Amer. Math. Soc.*, **78**, 236–241.

1973*a* On the nonseparable theory of k-Borel and k-Souslin sets, *Gen. Topology Appl.*, **3**, 161–195.

1973*b* On the representation of nonseparable analytic sets, *Proc. Amer. Math. Soc.*, **39**, 402–408.

1974*a* On characterizing non-separable analytic and extended Borel sets as types of continuous images, *Proc. London Math. Soc.* (*3*), **28**, 683–699.

1974*b* On Borel mappings and Baire functions, *Trans. Amer. Math. Soc.*, **194**, 195–211.

Kaniewski, J. and Pol, R.

1975 Borel measurable selections for compact-valued mappings in the non-separable case, *Bull. Acad. Polon. Sci. Sér. Sci. Math. Phys.*, **23**, 1043–1050.

Kuratowski, K.
1966 *Topology, Vol. I,* Academic Press, New York and London and PWN, Warsaw.

Pol, R.
1978 Note on decompositions of metrizable spaces II, *Fund. Math.*, **100**, 129–143.

Preiss, D.
1975 Completely additive disjoint system of Baire sets is of bounded class, *Comm. Math. Univ. Carolinae*, **15**, 341–344.

Stone, A. H.
1962 Non-separable Borel sets, *Rozprawy Mat. = Dissertationes Math.*, **18**.
1972 Non-separable Borel sets II, *Gen. Topology Appl.*, **2**, 249–270.

Willard, S.
1969 Embedding metric absolute Borel sets in completely regular spaces, *Colloq. Math.*, **20**, 83–88.

Problems

Asterisks indicate problems to which further comments are added on pages 491–492

1 Given a sequence $f_n\colon \mathbb{R}^{\mathbb{N}} \to [0, 1]$ of continuous functions, is there a subsequence (f_{n_i}) of (f_n), and are there perfect non-empty subsets P_m of $\mathbb{R}$ such that (f_{n_i}) converges pointwise on $\prod_m P_m$? [This is a translation (and possibly a weakening) of a problem of Halpern.] (L. Harrington)

2 Does there exist a non-trivial analytic (Hausdorff) space X such that $\varnothing$ and X are the only open $\mathscr{F}_\sigma$-sets? (J. Hoffmann-Jørgensen)

3* Does every analytic space admit a stronger metrizable analytic topology? This is true if the space is regular. (J. Hoffmann-Jørgensen)

4 Let $f\colon X \to Y$ be a continuous function of a complete metric space X, into a complete metric space Y, such that f takes σ-discrete sets to σ-discrete sets. Is $f(X)$ in $S(\mathscr{F}(Y))$? This is true if X has a base of cardinality less than or equal to $\aleph_1$. (R. Pol)

5* Let $F\colon X \to 2^Y$, where X and Y are complete metric spaces and 2^Y is the space of closed non-empty subsets of Y, be such that $\{x\colon F(x) \cap U \neq \varnothing\}$ is Borel class α whenever U is open and each $F(x)$ is separable. Is there a Borel selection for F? There is, if X has a base of cardinality less than or equal to $\aleph_1$, or if each $F(x)$ is compact; even without the assumption that X is complete, Fleissner has given a model of set theory in which there is a Borel selection. (R. Pol)

6 Let $f\colon X \to Y$, where X is a complete metric space and Y is a metric space, be such that $f^{-1}(U)$ has the Baire property whenever U is open in Y. Is f necessarily continuous apart from a meagre set? This is the case if the cardinality of Y is not Ulam-measurable and if $f^{-1}(U)$ has the Baire property in the restrictive sense; or if X has a base of cardinality less than or equal to $2^{\aleph_0}$. (R. Pol)

7 Let X be a separable metric space which contains a set $A \in S(\mathscr{F}(X))$ which is not a Borel subset of X. Does X contain Borel sets of arbitrarily high class? (R. D. Mauldin)

8 Let X and Y be Polish spaces. Suppose Φ_N is a δs-operation with base N Borel. Let B be a Borel set in $X \times Y$ such that for each $x \in X$ there exist closed sets $E_n(x)$ in Y such that

$$B^x \equiv \{y \in Y: (x, y) \in B\} = \Phi_N(\{E_n(x)\}).$$

Are there Borel sets B_n in $X \times Y$ with $B_n^x \equiv \{y \in Y: (x, y) \in B_n\}$ closed in Y for each $x \in X$ and $n = 1, 2, 3, \ldots$, such that $B = \Phi_n(\{B_n\})$? (A. Maitra)

9 Is there a Borel subset of $\mathbb{R}^2$ which meets every straight line in exactly two points? This is an old problem. (D. H. Fremlin)

10* Let $\mathscr{A}$ be a point-finite family of subsets of a metrizable space X, satisfying the following condition: if $\mathscr{A}' \subset \mathscr{A}$, then $\bigcup\{A: A \in \mathscr{A}'\}$ is a Borel set in X. Is it true that the members of A are of bounded Borel class? Preiss has shown this to be the case if A is a disjoint family. It is trivially not the case for a point-countable family. (J. Kaniewski)

11* Let X be an uncountable Polish group. For what countable ordinals α does X have subgroups of exactly additive, multiplicative or ambiguous class α? (R. D. Mauldin)

12 By beginning with the family of open sets of a topological space and iterating the operations of adjoining countable unions and complements we obtain all Borel sets after exactly α steps for some $\alpha \leqslant \omega_1$. What possible values can α have for compact Hausdorff spaces? (R. D. Mauldin)

DEFINITION Proposition P is the assertion that whenever a family of subsets $\mathscr{A} = \{A_i: i \in I\}$ of a metrizable space X has properties (a) and (b) below, then A has property (c).

(a) For every $x \in \bigcup_{i \in I} A_i$, $\{i: x \in A_i\}$ is finite.
(b) For every $J \subset I$, $\bigcup_{i \in J} A_i \in S(\mathscr{F}(X))$.
(c) There is a family $\mathscr{A}^* = \{A_{i,n}: i \in I,\ n = 1, 2, \ldots\}$ such that, for each n, $\{A_{i,n}: i \in I\}$ is discrete, and, for each i, $\bigcup_{n=1}^{\infty} A_{i,n} = A_i$.

It is known that in a certain model of set theory Proposition P is true, and that Proposition P implies that every Borel isomorphism between metric spaces is a generalized homeomorphism.

13 Does $V = L$ imply Proposition P? (W. G. Fleissner)

14* In ZFC is every Borel isomorphism between metric spaces a generalized homeomorphism? (A. H. Stone)

15* Let X be any set and $\mathscr{C}$ a family of subsets of X. Let $\boldsymbol{\Sigma}^0_\alpha$ and $\boldsymbol{\Pi}^0_\alpha$ be the levels of the σ-algebra generated from $\mathscr{C}$. Suppose $\mathscr{A} = \{A_i: i \in I\}$ is disjoint and for every $J \subset I$,

$$\bigcup_{i \in J} A_i \in \bigcup_{\alpha < \omega_1} \boldsymbol{\Sigma}^0_\alpha.$$

D. Preiss showed that there is an $\alpha < \omega_1$ such that $\mathscr{A} \subset \boldsymbol{\Sigma}^0_\alpha$.

Does there exist $\alpha < \omega_1$ such that for all $J \subset I$,

$$\bigcup_{i \in J} A_i \in \boldsymbol{\Sigma}^0_\alpha?$$

(W. G. Fleissner)

16* Is every analytic space homeomorphic to a subset of a Lusin space?
(J. Hoffmann-Jørgensen)

17* Let X and Y be compact metric spaces and $A \subset X \times Y$ an analytic set such that for each $y \in Y$ the section $A^y = \{x \in X: (x, y) \in A\}$ is a K_σ-set.

(a) Is A the countable union of analytic sets with compact sections?
(b) Does there exist an analytic subset B of $X \times Y$ such that, for all $y \in Y$, B^y is a non-empty $\mathscr{K}_\sigma$-set, and, if $A^y \neq \emptyset$, then $A^y = B^y$?

(C. Dellacherie)

18* Suppose $A \subset \mathbb{N}^\mathbb{N}$ is such that $A \cap K$ is analytic for every compact $K \subset \mathbb{N}^\mathbb{N}$. Does it follow that A is analytic? (Certainly not subject to CH; but suppose we have MA + not CH?) (D. H. Fremlin)

19* Let X be a K-analytic space. Then the following are equivalent: (i) there is a coarser second countable T_1 topology on X, (ii) the diagonal Δ in $X \times X$ is in $S(\mathscr{G}(X \times X))$ (i.e.

$$\Delta = \bigcup_{\sigma \in \mathbb{N}^\mathbb{N}} \bigcap_{n=1}^{\infty} G(\sigma|n)$$

with each $G(\sigma|n)$ open in $X \times X$). Does it follow that X is an analytic space (i.e. the continuous image of $\mathbb{N}^\mathbb{N}$)? (If MA + not CH, then the answer is yes. What if $V = L$?) [See D. H. Fremlin (1977) K-analytic sets with metrizable compacta, *Mathematika*, **24**, 257–261.] (D. H. Fremlin)

20* If X is a compact Hausdorff space such that every uncountable Borel set contains a copy of $2^\mathbb{N}$, then is X necessarily metrizable? (D. H. Fremlin)

21* If X is a compact Hausdorff space which is Borel isomorphic to the unit interval, then is X necessarily metrizable? (If CH, then the answer is no; but otherwise?) Juhasz and Kunen have given an example, subject to CH, of a non-metrizable compact Hausdorff topology on $[0, 1] \cup \{\infty\}$ with the same Borel sets as usual. (D. H. Fremlin)

22 Let H be a separable Hilbert space and $B(H)$ the space of continuous linear operators of H into itself. Let $A = \{T \in B(H): \exists x \in H$ such that span $\{x, Tx, T^2x, \ldots\}$ is dense in $H\}$. Is A an analytic non-Borel set? (R. D. Mauldin)

23. Let $A \subset \omega^\omega \times \omega^\omega$ be $\mathbf{\Sigma}^1_1$. If there is a set $Q \subset \omega^\omega$ of cardinality greater than $\aleph_1$ such that for all $\alpha, \beta \in Q$, $\alpha \neq \beta$, we have $(\alpha, \beta) \in A$, does it follow that there is a perfect set $P \subset \omega^\omega$ such that for all $\alpha, \beta \in P$, $\alpha \neq \beta$, we have $(\alpha, \beta) \in A$? It is known that card $Q > \aleph_0$ does not suffice. This is a problem of Silver. (A. S. Kechris)

24* Suppose X is a Polish space and E is an equivalence relation on X such that each equivalence class is a $\mathcal{G}_\delta$-set and the saturation of each basic open set is an ambiguous class α Borel set. Does there exist a Borel class α measurable selector for E? The answer is "yes" if the equivalence classes are closed, or if $\alpha = 1$. (D. E. Miller)

25 Suppose a Polish topological group G acts continuously (or Borel measurably) on a metric space. Let $H \subset G$ be a Borel subgroup of G. Are the H-orbits Borel sets? "Yes", if H is a closed subgroup. (D. E. Miller)

26* (Vaught's conjecture) Suppose a Polish group G acts continuously (or Borel measurably) on a Polish space X. Let $B \subset X$ be an invariant Borel set containing uncountably many orbits.

CONJECTURE *There is a perfect set of G-inequivalent elements of B.*

This is known to be so if G is locally compact. The following special case is of particular interest: $G = \mathbb{N}!$ (the permutations of $\mathbb{N}$), $X = 2^{\mathbb{N} \times \mathbb{N}}$, with the canonical action $g\alpha(n, m) = \alpha(gn, gm)$ for $\alpha \in 2^{\mathbb{N} \times \mathbb{N}}$. (D. E. Miller)

27 Suppose a Polish group G acts Borel measurably on a countably separated Borel space X. Is there a metric (or even T_1) topology on X which generates the Borel structure and makes the action continuous? This is known to be so if G is locally compact. (D. E. Miller)

28 Let $G = \mathbb{N}!$, $X = 2^{\mathbb{N} \times \mathbb{N}}$ as in Problem 26, and let $\mathcal{L}$ be the smallest field of sets which contains the clopen sets in X and is closed under saturations by open subgroups H of G. ($\{g\alpha: g \in H, \alpha \in B\}$ is the saturation of B by H.) Suppose F is an autohomeomorphism of X which is weakly invariant (i.e. $(\forall g, \alpha)(\exists h)(F(g, \alpha) = h(F\alpha))$. Is $F(B) \in \mathscr{L}$ for each $B \in \mathscr{L}$? (D. E. Miller)

29 Suppose P_n, $n \geq 1$, are $\mathbf{\Sigma}_2^1$ subsets of ω^ω such that $\lim\sup P_n = \varnothing$. Are there $\mathbf{\Delta}_2^1$ subsets Q_n of ω^ω such that, for all n, $P_n \subset Q_n$ and $\lim\sup Q_n = \varnothing$? (A. Maitra)

30 Let B be a weakly sequentially complete Banach space, B^* its topological dual and $w(B^*, B)$ the weak-star topology on B^*. Is it true that every $w(B^*, B)$-Borel (or Baire) measurable linear form on B^* is $w(B^*, B)$-continuous? The answer is "yes", if B is a Banach lattice. [See Wickstead (1976) *Ann. Inst. Fourier* (*Grenoble*), **26** 25–28.] (J. E. Jayne)

31–36 These six problems are the main remaining unsolved problems in J. P. R. Christensen's monograph *Topology and Borel Structure* (North-Holland, Amsterdam, 1974). The original numbers are given in parentheses.

31*(1) Let C be a coanalytic subset of $\mathbb{N}^\mathbb{N}$ with the property that every injective Borel measurable map of C onto a subset of $\mathbb{N}^\mathbb{N}$ is a Borel isomorphism. Is C necessarily analytic? (J. P. R. Christensen)

32 (3) If a completely regular Hausdorff space is the continuous image of a separable metric space, then is it homeomorphic to a subset of an analytic space? (J. P. R. Christensen)

33 (7) Consider $C(\mathbb{N}^\mathbb{N})$ with the Borel σ-algebra generated by the topology of compact convergence. Does there exist an injective Borel measurable map of $C(\mathbb{N}^\mathbb{N})$ onto $\mathbb{N}^\mathbb{N}$ or onto an analytic subset of $\mathbb{N}^\mathbb{N}$? (J. P. R. Christensen)

34*(10) Let (X, d) be a metric space and $(\hat{X}, d^*)$ its space of non-empty closed sets with the metric d^* defined by

$$d^*(A, B) = \sup\{d(a, B), d(A, b): a \in A, b \in B\}.$$

We say that a net $\{F_a: a \in D\}$ in X converges to $F \in X$, if each $x \in X \backslash F$ has a neighbourhood which is eventually disjoint from (F_α), and every neighbourhood of each $x \in F$ eventually intersects (F_α). The convergence topology on $\hat{X}$ has as closed sets those which are closed under this notion of convergence. When does the convergence topology coincide with the infimum of all d^*-topologies over all precompact metrics d on X compatible with the topology? (J. P. R. Christensen)

35 (15) The Effros Borel structure on the space of non-empty closed subsets $\hat{X}$ of a topological space X is the smallest σ-algebra containing $\{F \in \hat{X} : F \subset A\}$ for each $A \in \hat{X}$. To what extent can the results known for the Effros Borel structure of metrizable analytic spaces be generalized to non-metrizable analytic spaces? (J. P. R. Christensen)

36 (17) Let H be a non-closed hyperplane in a Polish real topological vector space X. Is H necessarily second category in X? Can H be universally measurable? (J. P. R. Christensen)

37 Let X be a completely regular Hausdorff space and $C_p(X)$ the space of continuous real-valued functions on X with the topology of pointwise convergence. If X is metrizable, then $C_p(X)$ is analytic, if, and only if, X is σ-compact. There exists a σ-compact non-metrizable analytic space such that $C_p(X)$ is not analytic. If $C_p(X)$ is analytic, is X necessarily a σ-compact analytic space? It is enough to show that X is analytic.

(J. P. R. Christensen)

38 Let I be a maximal dense ideal in the space $H(\mathbb{C})$ of entire functions on the complex plane $\mathbb{C}$. Can I be analytic or universally measurable?

(J. P. R. Christensen)

39 Suppose k is a cardinal number and $B(k)$ the product of a countable number of discrete sets of cardinality k. If A is a k-Souslin set in some complete metric space X (i.e.

$$A = \bigcup_{\sigma \in B(k)} \bigcap_{n=1}^{\infty} H(\sigma|n),$$

where each $H(\sigma|n)$ is closed in X) and Z is an arbitrary compactification of A, then is A the intersection of a k-Souslin set and a $\mathscr{G}_\delta$-set in Z? When $k = \aleph_0$, the answer is "yes". (A. H. Stone)

40 A map $f: X \to Y$ between metric spaces X and Y is called σ-discrete if there is a σ-discrete collection $\mathscr{H}$ of X such that, for each open V in Y, $f^{-1}(V)$ is a union of members of $\mathscr{H}$. Is the limit of a pointwise convergent sequence of σ-discrete maps necessarily σ-discrete? (A. H. Stone)

41 Give an example of a Banach space which is K-analytic in its weak topology, but not a $\mathscr{K}_{\sigma\delta}$-subset of its second dual with its weak-star topology.

(M. Talagrand)

42 Let X_w denote a Banach space X with its weak topology. If there exists a separable metric space S and an upper semi-continuous map $K: S \to \mathscr{K}(X_w)$ such that $X_w = K(S)$, is X_w necessarily K-analytic? (M. Talagrand)

43 Let X and Y be Polish spaces, $B \subset X \times Y$ a Borel set and $M(B)$ the union of σ-compact sections of B over X. Do there exist Borel sets $B_1, B_2, \ldots$ in $X \times Y$ such that

(a) for each $x \in \pi_X B$ and each $n \in N$

$$B_n^x = \{y: (x, y) \in B_n\}$$

is compact, if, and only if, $x \in \pi_X(M(B))$;

(b) $\bigcup_n B_n = B$?

(C. A. Rogers)

44* Suppose B is a Borel set of exactly multiplicative class α in a compact metric space X. Does there exist a zero-dimensional compact subset K of X such that $K \cap B$ is exactly multiplicative class α? (J. E. Jayne)

45 Suppose $\mathscr{H}$ is a family of subsets of a set X such that $\mathscr{H}_c \subset \mathscr{H}_{(\sigma\delta)}$ (or stronger $\mathscr{H}_c \subset \mathscr{H}_\sigma$). Does it follow that $\mathscr{H}_{(\sigma\delta)} = \mathscr{H}_{(+\delta)}$? [See Theorem 5.8.3(b) of Part 1 of this volume, and J. E. Jayne (1977) Generation of σ-algebras, Baire sets, and descriptive Borel sets, *Mathematika*, **24**, 241–256.]

(J. E. Jayne)

46 Suppose that a completely regular space is a $\mathscr{K}_{\sigma\delta}$-subset of some compactification. Is it then a $\mathscr{K}_{\sigma\delta}$-subset of every compactification? This is true if it is a $\mathscr{Z}_{\sigma\delta}$-subset of some compactification. ($\mathscr{K}$ denotes the class of compact sets and $\mathscr{Z}$ the class of zero sets of continuous real-valued functions.) [See theorem 5.8.7 and 5.9.14 in Part 1 of this volume.]

(J. E. Jayne)

47 Is $\mathbb{N}^{\mathbb{N}}$ second-level Borel isomorphic to $[0, 1]^{\mathbb{N}}$; i.e. is there a bijection which takes $\mathscr{F}_{\sigma\delta}$-sets to $\mathscr{F}_{\sigma\delta}$-sets in both directions? [See Theorem 6.3.5 of Part 1 of this volume.] (J. E. Jayne)

48* If X and Y are first-level Borel isomorphic Polish spaces, do we necessarily have $\dim X = \dim Y$? This is true if X is σ-compact. [See Theorem 6.3.3 of Part 1 of this volume.] (J. E. Jayne)

49* Suppose a compact metric space X is first-level Borel isomorphic to the Hilbert cube $[0, 1]^{\mathbb{N}}$. Does X necessarily contain a homeomorphic copy of

$[0, 1]^{\mathbb{N}}$? The converse is true; namely, if X contains $[0, 1]^{\mathbb{N}}$ then X is first-level isomorphic to $[0, 1]^{\mathbb{N}}$. (J. E. Jayne)

50 Consider a Banach space X canonically embedded in its second dual X^{**}. Let $X_1 \subset X^{**}$ be the space of limits of weak-star convergent sequences in X. For each countable ordinal α let $X_\alpha \subset X^{**}$ be the space of limits of weak-star convergent sequences in $\bigcup_{\beta<\alpha} X_\beta$. For each $\alpha < \omega_1$ does there exist a separable Banach space X such that $X_\alpha \neq \bigcup_{\beta<\alpha} X_\beta$ and $X_\alpha = X_{\alpha+1}$? [See J. M. Anderson and J. E. Jayne (1973) The sequential stability index of a function space, *Mathematika*, **20**, 210–213.] (J. E. Jayne)

51 Does every non-reflexive Banach space contain a closed bounded convex set whose set of extreme points is not a Borel set? [See J. E. Jayne and C. A. Rogers (1977) The extremal structure of convex sets, *J. Functional Anal.*, **26**, 251–288.] (J. E. Jayne)

52 Does every non-reflexive Banach space contain two closed bounded convex sets whose convex hull is not a Borel set? [See the paper by Jayne and Rogers mentioned in Problem 51.] (J. E. Jayne)

53 Is the set of exposed points of a compact convex subset of a Banach space necessarily a Borel set? It is necessarily analytic, and if the Banach space is finite dimensional it is necessarily Borel. [See p. 255 of the paper by Jayne and Rogers mentioned in Problem 51.] (J. E. Jayne)

54 Let B be a Baire subset of a compact Hausdorff space X. Is B Baire isomorphic to some compact Hausdorff space? This is true if B is either metrizable or locally compact. It would suffice to consider only those B which are the countable intersection of cozero sets in their Stone-Čech compactifications. [See theorem 5.1.7 in Part 1 of this volume.]
(J. E. Jayne)

55 Let X be a compact Hausdorff space and Y a completely regular space such that $Y \in \mathscr{Z}_{(+,\delta)}(\beta Y)$. If X and Y are first-level Baire isomorphic, then is Y necessarily σ-compact? [See Section 6 in Part 1 of this volume.]
(J. E. Jayne)

56 Let $B_\alpha^*[0, 1]$ be the space of bounded real-valued class α Baire functions on the unit interval with the supremum norm. If $1 < \alpha < \beta < \omega_1$, then is $B_\alpha^*[0, 1]$ linearly homeomorphic to $B_\beta^*[0, 1]$? [See Section 6 in Part 1 of this volume and Dashiell and Lindenstrauss (1973) Some examples concerning strictly convex norms on $C(K)$ spaces, *Israel J. Math.*, **16**, 329–342.] (J. E. Jayne)

57 Let $\mathscr{C}$ be the class of Choquet's capacities on [0, 1] which are subadditive (i.e. those capacities I for which $I(A \cup B) \leqslant I(A) + I(B)$). Is the statement "Every subset of [0, 1] is capacitable for every capacity in $\mathscr{C}$" consistent with ZF? (A. Louveau)

59 Let X be a Polish space, **C** the family of calibrations on X. To each calibration C we denote by Φ_C the family of sets of calibration 0. We let $\mathscr{R}$ be a family of subsets of X and let

$$\Phi_C^{\mathscr{R}} = \{A : \exists B \in \Phi_C \cap \mathscr{R}, A \subset B\}.$$

Now consider the following statement on $\mathscr{R}$: For each Polish space Y and each calibration C on X, if B is a Borel subset of $X \times Y$ with sections in $(\Phi_C^{\mathscr{R}})_\sigma$, then B is the countable union of a sequence of Borel sets with sections $\Phi_C^{\mathscr{R}}$.

(a) Is this statement true when $\mathscr{R}$ is $\mathbf{\Sigma}_1^1$?
(b) Is this statement consistent with ZF when $\mathscr{R}$ is the class of all subsets of X?

((a) and (b) are true if the class $\{\Phi_C : C \in \mathbf{C}\}$ is replaced by the more restrictive class of $\mathbf{\Pi}_1^1$-monotone families of Cenzer and Mauldin.) [See Part 2 of this volume.] (A. Louveau)

59 For a set A of ultrafilters, let Σ_A be the σ-algebra of subsets of $\{0, 1\}^{\mathbb{N}}$ generated by the elements of A. Does there exist a $u \in \beta\mathbb{N}$ such that $u \in \Sigma_{\beta\mathbb{N} \setminus \{u\}}$? (M. Talagrand)

60 Does there exist a sequence (u_n) in $\beta\mathbb{N}$ of distinct ultrafilters and $u \in \beta\mathbb{N}$ such that the filter $\mathscr{F}$ on $\mathbb{N}$ consisting of the sets $\{n : u_n \in V\}$, where V is a neighbourhood of u is contained in a Borel filter? (M. Talagrand)

61 Let I be an uncountable set. Is the space of bounded linear maps $\mathscr{L}(l_2(I), C([0, 1]))$ with the pointwise-weak convergence topology, a Lindelöf space? (It is not K-analytic.) (M. Talagrand)

62 Let X be a compact Hausdorff space, $F \subset X$ a closed set, and $U \subset X$ an open set. Suppose $F \cup U$ is K-analytic. Is it then a $\mathscr{K}_{\sigma\delta}$-set? (M. Talagrand)

63 (Due to G. Choquet) Let K be a compact convex set whose set $\mathscr{E}(K)$ of extreme points is K-analytic. Is $\mathscr{E}(K)$ then a $\mathscr{K}_{\sigma\delta}$-set? (M. Talagrand has shown that

$$\mathscr{E}(K)=\bigcap_{n=1}^{\infty}(F_n\cup U_n),$$

where F_n is closed, U_n is open, and $F_n\cap U_n=\varnothing$; see Theorem 5.10.5 in Part 1 of this volume.) (M. Talagrand)

64 Let X and Y be completely regular Hausdorff spaces such that $X\in\mathscr{Z}_{(+,\delta)}(\beta X)$ and $Y\in\mathbf{S}(\mathscr{Z}(\beta Y))$. If X and Y are Baire isomorphic, is $Y\in\mathscr{Z}_{(+,\delta)}(\beta Y)$? (J. E. Jayne)

65* Let X be an analytic space (i.e. a Hausdorff space which is the continuous image of a Polish space). Is the diagonal in $X\times X$ always a $\mathscr{G}_\delta$-set? The diagonal is always in $\mathbf{S}(\mathscr{G}(X\times X))$, and it is a $\mathscr{G}_\delta$-set when X is completely regular. (J. E. Jayne)

66 Let A be an analytic subset of $[0,1]\times[0,1]$, and, for each $x\in[0,1]$, let $A^x=\{y\colon (x,y)\in A\}$. Is there a bijection $g\colon [0,1]\times[0,1]\to A$ such that:

(a) $g(x,\cdot)$ maps $[0,1]$ onto A^x; and
(b) both g and g^{-1} are measurable with respect to $\mathscr{B}(\mathscr{A})$ (the σ-algebra generated by the analytic subsets of $[0,1]\times[0,1]$)?

The answer is "yes" if $\mathscr{B}(\mathscr{A})$ is enlarged to the smallest σ-algebra $\mathscr{S}$ containing $\mathscr{B}(\mathscr{A})$ and closed under operation A; i.e. $\mathbf{S}(\mathscr{S})=\mathscr{S}$. [See D. Cenzer and R. D. Mauldin (1978) *Trans. Amer. Math. Soc.*, **245**, 399–408.] (R. D. Mauldin)

67 Let B be a Borel subset of $[0,1]\times[0,1]$ such that $B^x=\{y\colon (x,y)\in B\}$ is an uncountable compact set for each $x\in[0,1]$. Are there $2^{\aleph_0}$ pairwise disjoint Borel uniformizations of B? Is this true if each B^x is an uncountable σ-compact set? (It is known that there are $\aleph_1$ pairwise disjoint Borel uniformizations. [See Larman (1973) *Mathematika*, **20**, 233–246.] Under special conditions it is known that B is filled up by disjoint Borel uniformizations. [See R. D. Mauldin (1979) *Trans. Amer. Math. Soc.*, **250**, 223–234.] The second question cannot be reduced to the first in the following sense: there is A Borel subset B of $[0,1]\times[0,1]$ such that for each $x\in[0,1]$, B^x is an uncountable $\mathscr{F}_\sigma$-set and B contains no Borel set C such that each C^x is an uncountable compact set.) (R. D. Mauldin)

Further comments

3 D. H. Fremlin has found a counterexample.

5 R. W. Hansell has shown that the answer is yes if one further assumes that f is a quotient map with separable point inverses.

10 R. W. Hansell has given a positive answer.

11 The answer is "for all countable ordinals", if X is connected, abelian and has an element of infinite order. See R. D. Mauldin (1980) On the Borel subspaces of algebraic structures, *Indiana J. Math.*, **29**, 261–265. The problem remains unsolved otherwise.

14 R. W. Hansell has shown that $MA + \neg CH$ implies a positive answer.

15 R. W. Hansell has shown that the answer is yes in the case when either the σ-algebra is countable generated or $MA + \neg CH$ holds.

16 D. H. Fremlin's counterexample to problem 3 also gives a negative answer to this question.

17 The answer to both questions is no. See J. Steel, A note on analytic sets, to appear.

18 K. Kunen has shown that $MA + 2^\omega = \omega_2 + \exists x \in 2^\omega \omega_1 = \omega_1^{L[x]}$ implies $\exists X \subset \omega^\omega \forall K^{\text{compact}}\ X \cap K$ is analytic but X is not analytic. A. W. Miller has shown that Con(ZF) implies Con($ZFC + \forall X \subset \omega^\omega$ if $\forall K^{\text{compact}}\ X \cap K$ is PCA ($\mathbf{\Sigma}^1_2$), then X is PCA); and that Con($ZFC + \exists$ a very large cardinal) implies Con($ZFC + \forall X \subset \omega^\omega$ if $\forall K^{\text{compact}}\ X \cap K$ is analytic, then X is analytic) (the large cardinal needed is the one D. A. Martin used to show $\mathbf{\Sigma}^1_2$-determinacy); and conjectures that this large cardinal assumption is unnecessary.

19 S. Graversen has given a positive answer.

20 D. H. Fremlin and K. Kunen have independently given counterexamples.

21 D. H. Fremlin has shown that $MA + \neg CH$ implies the answer is yes.

24 There exists a Borel class α^* measurable selector for E, where $\alpha^* = \sup\{\alpha+\beta : \beta < \alpha\}$. See D. E. Miller, Borel selectors for separated quotients, to appear.

26 R. Sami has shown the conjecture to be true when G is abelian.

31 D. H. Fremlin has observed that the answer is no if we suppose $MA + \neg CH + \exists$ coanalytic set of cardinality $\aleph_1$. J. E. Jayne has observed that the answer is yes if we suppose the axiom of analytic determinacy.

34 D. H. Fremlin has observed that this is true precisely when X is compact.

44 K. Kunen has given a positive answer.

48 The answer is yes. See Jayne and Rogers, First level Borel functions and isomorphisms, to appear.

49 The answer is yes. See Jayne and Rogers, First level Borel functions and isomorphisms, to appear.

65 D. H. Fremlin has constructed an analytic space X such that the diagonal in $X \times X$ is not a $\mathcal{G}_\delta$-set.

Author Index

References P1, P2, . . . , P67 *are to the problems on pages* 481–490.

Subject Index